THE JAMES A. DE ROTHSCHILD COLLECTION
AT WADDESDON MANOR
GENERAL EDITOR: SIR ANTHONY BLUNT

Glass and Enamels

THE JAMES A. DE ROTHSCHILD COLLECTION
AT WADDESDON MANOR

Glass and Stained Glass

R. J. Charleston
Michael Archer

Limoges and other painted Enamels

Madeleine Marcheix
R. J. Charleston

PUBLISHED FOR THE NATIONAL TRUST BY
OFFICE DU LIVRE
1977

Designed by George Rainbird Ltd, London
Printed in Great Britain by
Lund Humphries, Bradford and London
Text printed in Monotype Baskerville
on Royal Sword Mills paper

ISBN 0 7078 0066 8

Contents

Foreword

The objects forming the material of the present catalogue are perhaps more heterogeneous in character than those included in any previous volume in the Waddesdon Catalogue series. They are nevertheless bound together by their common character as products of the 'arts of fire', the material of enamels being no more than glass fused to a backing of metal. Within the four individual categories of objects covered by this volume, there is a certain coherence but no pretence at completeness. The glass is strong in certain categories of Venetian, Venetian-style and German-Bohemian glass-making, but while including some Roman glasses, contains not a piece of English, Spanish or Dutch glass. The stained-glass collection, apart from two English roundels, is confined to German and Swiss armorial panels of the sixteenth and seventeenth centuries. The enamels likewise, apart from a fairly homogeneous but not wholly representative series of sixteenth-century Limoges pieces, are divided between the French and Genevan schools of the seventeenth century, and the English 'commercial' enamels of the eighteenth.

These objects were collected mainly by Baron Ferdinand de Rothschild (1839–1898), who built Waddesdon Manor, and by his sister Miss Alice de Rothschild (1847–1922), who inherited Waddesdon on her brother's death. Certain of the works of art, however, passed by inheritance from Baron Edmond de Rothschild of Paris (1845–1934) to his son Mr James A. de Rothschild, to whom Miss Alice had bequeathed Waddesdon in 1922. Many of them bear labels with the series of numbers prefixed by 'R' allocated to sequestered de Rothschild property by the Nazis during the 1939–1945 war. On his death in 1957 Mr James de Rothschild left the house and the greater part of its collection to the National Trust. The glass and nearly all the enamels which are the subject of this catalogue were not included in the bequest and most have been added by Mrs James de Rothschild and accepted by the Treasury in part satisfaction of Estate Duty; they are distinguished here by an asterisk against their inventory number. Others have been given by Mrs de Rothschild to the National Trust, and these are indicated by two asterisks.[1]

Some notes on provenance have been taken from written information originally furnished by Miss Alice de Rothschild in 1910, or by Baron Edmond de Rothschild, and these can be distinguished by the formula 'Miss A.' or 'Baron E.' which accompanies them. The Eythrope Inventory which is occasionally quoted, refers to the contents of Miss Alice's pavilion at Eythrope.

The author of the first and last parts of this Catalogue would like to express his cordial thanks to Dr Donald B. Harden, doyen of English studies on ancient glass, for reading through the section on Roman glass and making most useful comments on it; to his daughter, Mrs J. Stringer, for drawing the Roman glasses (*1–12*); to his colleagues Frances Cooper, Charles Gibbs-Smith and John Harthan for help with questions of heraldry; to Géza Fehérvári, of the

[1] For a full account of Waddesdon Manor and the formation of its collections, see Mrs de Rothschild's preface – 'Waddesdon and the Rothschild Family' – in the first volume of the Waddesdon Catalogue series, *Paintings*, by Ellis Waterhouse, Office du Livre, 1967, pp.11–22.

School of Oriental and African Studies of London University, for transcribing and translating the inscription on the Islamic beaker (*14*); and to various friends and colleagues for answering questions, providing information on glasses in their care, or generally sharing their knowledge – notably, Dr Fritz Biemann, of Zurich; Dr Guido Donatone, of Naples; Dr Olga Drahotová, of the Museum of Applied Arts, Prague; Dr Astone Gasparetto, of Venice; Mrs Madeleine Ginsburg, of the Victoria and Albert Museum, London; Dr J. Horschik, of Dresden; Mr Dwight Lanmon, of the Corning Museum of Glass, New York; Dr Tasnádi Marik, of the National Museum, Budapest; Mr A. V. B. Norman, of the Wallace Collection, London; Dr H. Günter Rau, of Berlin; Dr Rainer Rückert, of the Bayerisches Nationalmuseum, Munich; Dr Axel von Saldern, of the Museum für Kunst und Gewerbe, Hamburg; Dr Jiřina Vydrová, of the Museum of Applied Arts, Prague; Dr. G. Wegener, of the Historisches Archiv, Cologne; and last only by reason of his initial, Ingegnere Luigi Zecchin, of Murano. To those who have helped but who may inadvertently have been omitted from this list, thanks equally sincere are extended.

As far as the Miscellaneous Enamels section of the book is concerned, the writer's colleague Ronald Lightbown kindly read the text of the Introduction and made suggestions for its improvement in the light of his great knowledge of the subject. Professor Sir Anthony Blunt not only gave the text the benefit of his editorial wisdom, but greatly helped with the identification of some of the subjects on the seventeenth-century pieces, as did Miss Jennifer Montagu, of the Warburg Institute. It should not be forgotten that the original text describing the watch cases was written by M. Serge Grandjean of the Musée du Louvre, Paris, while Dr F. A. B. Ward, lately of the Science Museum, London, wrote the horological descriptions: much of the merit is theirs, any faults entirely the responsibility of the author.

To those who helped with both these sections of the catalogue a double measure of thanks is due – Dr F.-A. Dreier, of the Kunstgewerbemuseum, Berlin, who shed light from these two facets of his many-sided knowledge; and Claude Blair and Charles Truman, both of the Metalwork Department of the Victoria and Albert Museum, who gave invaluable advice on goldsmiths' marks.

The author of the catalogue entries on the English stained glass in the collection is much indebted to Mr David King, of the University of East Anglia; Mr Dennis King, of Norwich, and Dr Peter Newton, of York University. Valuable assistance with the dating and identification of the German armorial roundels was received from Dr Rüdiger Becksmann and Dr Fritz Herz, both of the German Corpus Vitrearum Medii Aevi, from Dr Rainer Kahsnitz, of the Germanisches Nationalmusum, Nuremberg, and Professor Dr Hans Wentzel, of Stuttgart University. The great knowledge of Dr Jenny Schneider of the Schweizerisches Landesmuseum, Zurich, was invaluable for the entries relating to the Swiss panels. Finally, thanks are due to colleagues in the Victoria and Albert Museum: Michael Holmes of the Library for help with heraldic descriptions; Dr Michael Kauffmann of the Prints and Drawings Department for advice on the translation of inscriptions and on iconography; and Robert Charleston, of the Ceramics Department, for continuing advice and encouragement.

The author of the catalogue of the Limoges enamels would like to express her gratitude to Sir Anthony Blunt for his valuable advice and his patience; to Miss Elizabeth Hasloch of the Courtauld Institute of Art, for translating the text into English; and to Mr Roger Pinkham of the Victoria and Albert Museum, particularly for suggesting additions to the Bibliography. M. Philippe Verdier, author of the Catalogue of Limoges Enamels in the Walters Art Gallery,

Baltimore, has given kind support, and the staff of the Bibliothèque Municipale at Limoges have provided every assistance in making available the many documents in their care. The author's especial gratitude goes to M. Martial Lachenaud, son of that learned scholar the late Emile Lachenaud, who has made his father's invaluable notes, records, photographs and collection of engravings available to the Musée de Limoges.

Madame Raymond Chambon has kindly permitted the reproduction of two drawings for glasses in her late husband's collection (pp.106, 114), and the Terry-Engell Gallery, London, has allowed a painting by Juan van der Hamen y Leon once in their possession to be reproduced, as have the Newhouse Galleries Inc., New York, in the case of the Recco flower-piece, on pp.22–3. The following museums have also allowed works of art in their collections to be illustrated in this catalogue: the Gabinetto Disegni e Stampe degli Uffizi, Florence; the Muzeum skla a bižutérie, Jablonec nad Nisou, Czechoslovakia; the British Museum, London; the Museo Poldi-Pezzoli, Milan; the Bibliothèque Nationale, Paris; the Kunsthistorisches Museum, Vienna; and the Henry Francis du Pont Winterthur Museum.

All who have worked at Waddesdon will know what a debt of gratitude is owed to Miss Mary Brassey for putting at their disposal her unequalled knowledge of the de Rothschild family and their collections at Waddesdon; and to Colonel A. R. Waller and his devoted staff, particularly Miss Rosamund Griffin and Mr Frank Grenet, for innumerable acts of kindness and hospitality.

The photographs, both in monochrome and colour, which illustrate this book – as all the other volumes in the series – were taken by Mr S. A. R. Eost and Mr P. F. Macdonald, whose helpfulness and patience over a considerable span of years is here gratefully acknowledged. The details of the French enamel box illustrated on p.441 and following pages were kindly provided by Mr Francis Carver, who also made the detailed photographs necessary for research.

Special thanks are also due to Mrs J. Easson, who kindly undertook the task of compiling the index, and, with Mrs I. R. Taylor, typed the manuscript of the Stained Glass and Limoges Enamels sections.

R. J. C.
D. M. A.
M. M.

FAMILY RELATIONSHIPS OF THE ROTHSCHILDS CONCERNED WITH WADDESDON

FRANKFORT

- Mayer Amschel Rothschild 1744–1812
  - FRANKFORT: Amschel 1773–1855
    - No issue
  - VIENNA: Salomon 1774–1855
    - Anselm 1803–1874 m. Charlotte (London)
      - Mathilde 1832–1924 m. Wilhelm (Naples & Frankfort)
      - FERDINAND 1839–1898 m. Evelina (London)
      - ALICE 1847–1922
    - Betty 1805–1886 m. James (Paris)
  - LONDON: Nathan Mayer 1777–1836
    - Lionel 1808–1879 m. Charlotte (Naples & Frankfort)
      - Evelina 1839–1866 m. Ferdinand (Vienna)
    - Charlotte 1807–1859 m. Anselm (Vienna)
  - NAPLES & FRANKFORT: Carl Mayer 1788–1855
    - Charlotte 1819–1884 m. Lionel (London)
    - Wilhelm 1828–1901 m. Mathilde (Vienna)
      - Adelaide 1853–1935 m. Edmond (Paris)
  - PARIS: James 1792–1868 m. Betty (Vienna)
    - EDMOND 1845–1934 m. Adelaide (Naples & Frankfort)
      - JAMES A. 1878–1957 m. Dorothy Pinto

GLASS

R. J. Charleston

Introduction

The glass collection at Waddesdon Manor is of a very mixed and varied character. In the main, it was formed by Miss Alice de Rothschild after the death of her brother, Baron Ferdinand, in 1898. He too had collected glass, but his finest pieces were included in his bequest to the British Museum. A little of his glass, however, remains at Waddesdon, where the collection now also includes a number of Near Eastern glasses of Roman and medieval date, formerly owned by the late Baron Edmond de Rothschild of Paris (*1–12*, *14*). Except for this inheritance from Baron Edmond, the glass at Waddesdon is exclusively of the Renaissance and later periods.

The glasses from Antiquity form a small collection apart (*1–12*), put up in a case lined with chamois-leather to take the individual pieces. They were apparently presented to Baron Edmond de Rothschild, father of Mr James A. de Rothschild, and are characterized as 'verres irisés trouvés en Palestine aportés en 1897'.[1] It is known that Baron Edmond systematically supported excavations in the Near East, and specifically those undertaken by Raymond Weill in Palestine, and by Eustache de Lorey in Syria,[2] but there is no evidence to connect the glasses in question with Raymond Weill's dig. On the contrary, the presence in this group of a palpably nineteenth century production (*13*) suggests that part at least was acquired by purchase, and throws considerable doubt on the homogeneity of the group, a doubt which is reinforced by the heterogeneous nature of the glasses themselves. Unbroken glasses of this sort normally come from burials, and although groups of tombs in the Near East can contain glasses from a number of different centuries, here not only is the range – from the first to the fourth century A.D. – unusually extended, but there is even some question as to whether all the glasses in the group originated in the Syro-Palestinian area.

The character and importance of the Roman glasses at Waddesdon do not warrant an extended survey of the history of glass-making in the Roman Empire. It should suffice to say that it was an Oriental art which probably originated in Mesopotamia about the middle of the third millennium B.C., and thereafter spread to Egypt. In the course of the first millennium B.C. centres on the Syrian-Phoenician coast, and probably also in other parts of the Eastern Mediterranean area, took up the making of small perfume-bottles formed usually from dark opaque-blue glass modelled round a removable core, and decorated with threads of yellow, white and sometimes red opaque glass.[3] It is the glasses of this character that were imitated by the small *alabastron* at Waddesdon (*13*) made in modern times.

It was the invention of glass-blowing, probably shortly after the middle of the first century B.C., that gave to Roman glass its readily recognizable character and caused an enormous

1 In the Waddesdon inventory under the number 'E. de R.2051 ter' they are described as 'Objets gréco-romains en verre irisé du II-ième siècle avant Jésus Christ trouvés en Palestine (aportés en 1897)'.

2 André Blum in *Bulletin de l'Académie des Beaux-Arts*, *20*, July–December, 1934, a source kindly communicated by Miss M. Brassey.

3 The most up-to-date, concise and authoritative account of this phase of glass-history is given by D. B. Harden, 'Ancient Glass, I: Pre-Roman', *The Archaeological Journal*, CXXV, 1969, pp.46–72.

expansion in the production of glass in the Roman Empire. This crucial discovery was almost certainly made in the Syro-Palestinian area, the technique spreading thence first to Italy and subsequently to Gaul and other western parts of the Empire, in the course of the first century A.D.[4] This East-West flow of techniques and ideas was characteristic of Roman glass-making until at least the middle of the fourth century A.D., and was then inhibited, perhaps partly as a consequence of the political division of the Empire begun by Diocletian at the end of the third century. The constant interchange guaranteed to Roman glass its remarkable international style and character, but nevertheless within this unity there were subtle local nuances which have only recently begun to be recognized and formulated.[5] The glasses at Waddesdon are mainly of types which can be identified as of 'Syrian' origin, but the group also contains glasses which seem to be commoner in Cyprus than in the Syro-Palestinian area. It is clear that Cyprus had a glass-industry of its own,[6] but it is also evident that the island, as its geography dictates, formed part of the general sphere of influence of 'Syrian' glass-making, and the difficulty of making distinctions between them is obviously great; it is not lessened by the paucity of publications dealing with finds in the Syro-Palestinian area. Nevertheless, the strong likelihood that some of the Waddesdon glasses are of specifically Cypriot origin throws further doubt on the homogeneous nature of the group of which they form a part. The arguments as to the attributions of these glasses are given in the individual entries.

Doubt as to the origins of these glasses, however, does nothing to diminish their attractiveness. Although the group contains none of the luxury glasses decorated by cutting, engraving, or enamelling, it well exemplifies the virtues of the simpler, workaday glass of the period. The bowls, dishes and flasks of which it is composed mainly have a pleasant light green colour caused by the natural iron impurities in the sands used, and rely for their aesthetic appeal on their harmonious and sometimes noble forms, and on the occasional use of trailed threading, an economy of means which admirably achieves its end.

Although the making of glass in no way came to an end with the break-up of the Roman Empire,[7] the standards – at least in the northern and western provinces – declined disastrously, and only rough vessel-glass, usually of an imperfect greenish 'metal' continued to be made. In the eastern provinces, however, there was a continuity of production which was only momentarily broken by the Arab invasions of the seventh century A.D. (Syria was overrun between 635 and 638, and Egypt conquered in 641.) Thereafter, the skills of the local craftsmen were bent on satisfying the wants of their conquerors, and a different range of shapes evolved, and decorative techniques were developed which, although not entirely new, were now transformed. Among these was wheel-engraving, both in *intaglio* and relief, in the practice of which a notable school arose in Iran, and probably also in Iraq, in the ninth-tenth centuries.[8]

[4] The best conspectus of Roman glass-making is to be found in the continuation of the article referred to in n.3 above – D. B. Harden, 'Ancient Glass, II: Roman', *The Archaeological Journal*, CXXVI, 1970, pp.44–77.

[5] See, e.g., Glady D. Weinberg, 'Evidence for Glass Manufacture in Ancient Thessaly', *American Journal of Archaeology, 66*, 1962, pp.129–33. An unguentarium of markedly idiosyncratic shape frequently occurs in this area, and was presumably made locally.

[6] Olof Vessberg, 'Roman Glass in Cyprus', *Opuscula Archaeologica*, VII, 1952, pp.154–61.

[7] D. B. Harden, 'Ancient Glass, III: Post-Roman', *The Archaeological Journal*, CXXVIII, 1972, pp.78 ff.

[8] See, e.g., R. J. Charleston, 'A Group of Near Eastern Glasses', *The Burlington Magazine*, LXXXI, No.474, September, 1942, pp.212–18; Corning Museum of Glass, *Glass from the Ancient World*, Corning, 1957, Nos.530 ff.; Prudence Oliver, 'Islamic Relief Cut Glass: a suggested Chronology', *Journal of Glass Studies*, III, 1961, pp.9–28 etc.

In Egypt, a technique of painting on the surface with a variety of metallic pigments was carried over from the pre-Islamic period and elaborated in a variety of ways.[9] It was perhaps an offshoot from this tradition, transplanted by way of craftsmen who left Egypt at the fall of the Fatimid dynasty in 1171, which grew into the technique of enamelling and gilding glass which arose in Syria towards the end of the twelfth century. The Waddesdon Collection contains no example of the Mesopotamian-Iranian wheel-engraved glasses, or of the Egyptian 'lustre' glasses, but it does have one unusually well-preserved gilt and enamelled specimen (*14*).

The Syrian enamelled and gilt glasses of the late twelfth-early fourteenth centuries have been divided into groups which have been tentatively attributed to the cities of Raqqa (actually in north-western Mesopotamia, but formerly a peripheral part of the Syrian sphere), Aleppo and Damascus.[10] It has not been proved that glass was actually made in Raqqa, although this is highly likely – it was certainly a pottery-town: but Aleppo and Damascus are well established as centres of glass manufacture. The geographer Yāqūt, who died in 1229, writing of a mountain in the desert, says it produces a white sand used for glass-making in Aleppo, and the traveller al-Qaswīnī (d. 1283) wrote: 'Among the most notable things in this city is the glass-bazaar. He who enters there will be reluctant to leave, on account of the astonishing multitude of notable and exquisite objects which he sees there, and which are exported into every other country . . .'[11]. Similar texts substantiate the renown of Damascus in the glass-industry. The Waddesdon beaker belongs to the group of enamelled and gilt glass which has been tentatively ascribed to Aleppo.

Although there is evidence that enamelled glass was being made in Venice as early as the thirteenth century, there is little to indicate the extent to which this art may have been influenced, either technically or stylistically, by the enamelled and gilt glasses of the Near East.[12] The strong trading ties of Venice with the Near East at this time are a matter of common knowledge, and it is established that the Venetian glass-makers obtained some of their materials from that source.[13] Nothing is certainly known of Venetian glass with enamelled decoration in the fourteenth or the first half of the fifteenth century, and in this period Near Eastern glass-making was in decline – so much so that when in the middle of the fifteenth century the art of enamelling and gilding seems to have experienced a dramatic revival in Venice, the

[9] See, e.g., C. J. Lamm, *Oriental Glass of Medieval Date found in Sweden, and the early history of Lustre Painting*, Stockholm, 1941; R. H. Pinder Wilson and George T. Scanlon, 'Glass Finds from Fustat, 1964–71', *Journal of Glass Studies*, xv, 1973, pp.28–9.

[10] The fundamental study of this subject is C. J. Lamm, *Mittelalterliche Gläser und Steinschnittarbeiten aus dem Nahen Osten*, I, Berlin, 1930, pp.241 ff; II, 1929, Pls.89–205. This has been supplemented, but not superseded, by various archaeological reports – e.g., Arthur Lane, 'Medieval Finds at Al Mina in North Syria', *Archaeologia*, LXXXVII, 1938, pp.19–78 (glass on pp.61–74); P. J. Riis and Vagn Poulsen, *Hama . . . Les Verreries et Poteries Médiévales*, Copenhagen, 1957.

[11] Lamm, *Mittelalterliche Gläser*, I, p.492, excerpts 49 and 51.

[12] L. Zecchin, 'Vetrarie muranesi dal 1276 al 1300', *Rivista della Stazione sperimentale del Vetro*, I, No.4, July–August, 1971, p.3 of the offprint. For some discussion of this question, see R. J. Charleston's report on the glass in Stephen Moorhouse, 'Finds from Excavations in the Refectory at the Dominican Friary, Boston', *Lincolnshire History and Archaeology*, I, No.7, 1972, pp.45–8.

[13] In the much-quoted treaty of 1277 between the Doge Jacopo Contarini and Bohemund VII, Prince of Antioch, reference is made to 'verre brizé' (i.e. cullet) brought from Syria to Venice (R. Schmidt, *Das Glas*, Berlin and Leipzig, 2nd ed., 1922, p.63).

countries of the Near East were happy to obtain even lamps for their mosques from Murano.[14] In the first stages of this revival, glasses of coloured 'metal' (blue, green or manganese-purple) were painted with elaborate figural scenes of triumphs and processions, often with mildly erotic connotations, and sometimes combined with portraits of a man and a woman in medallions, suggesting that they were made specifically for betrothals or marriages (*coppe nuziali*).[15] Somewhat later, probably in the third quarter of the fifteenth century, enamelling began to be used on the clear colourless glass which, although certainly known in Italy in the thirteenth century and probably made at that time in Venice, appears (like the enamelling technique) to have experienced a revival from about the middle of the fifteenth century.[16] On these *cristallo* glasses, the palette of enamels was extended, particularly by the use of a turquoise-green, and the choice of motifs turned rather to the 'grotesques' inspired ultimately by Roman wall-painting, which had recently been discovered for the first time. Centaurs, mermen, winged *putti*, and sea-monsters disport themselves amid scrolling late-Gothic foliage, and subsequently Renaissance trophies, swags of flowers and leaves, candelabrum-motifs, and classical urns appear. The Waddesdon Collection contains one example of this rare kind, and one of the best (*16*).

The enamelled glasses of the quarter century on either side of 1500 are almost invariably embellished with borders formed of dots of different colours, all laid on a gold-leaf ground. The goblet already alluded to has such a decoration at the rim, and a second Waddesdon glass (*15*) combines with an imbricated border (formed by etching the gold-leaf with a point) Renaissance ornaments at the sides and a coat-of-arms at front and back. Glasses enriched with coats-of-arms are among the most commonly found types of the early sixteenth century, but it is hard to isolate examples earlier than 1490. A goblet closely similar in shape to *15* here, bearing the arms of Hungary on one side and of Bohemia on the other, may have been made for King Matthias Corvinus of Hungary, who died in 1490; but might also have been made for his successor, King Wladislaw II Jagello (1490–1516).[17] Matthias Corvinus, however, was a particular patron of Italian art and artists, and it would have been entirely in character for him to have commissioned such a glass from Venice.[18] A pair of pilgrim-flasks in the Museo Civico, Bologna, bear the arms of Bentivoglio and Sforza, and were probably made for the marriage of

[14] R. Rückert, 'Venezianische Moscheeampeln in Istanbul', *Festchrift für Harald Keller*, Darmstadt, 1963, pp.223–34; R. J. Charleston, 'The Import of Venetian Glass into the Near-East; 15th–16th century', *Annales du 3e. Congrès des 'Journées Internationales du Verre'*, Liège, n.d. (1965), pp.158–68.

[15] R. Schmidt, 'Die Venezianischen Emailgläser des XV. und XVI. Jahrhunderts', *Jahrbuch der Königlich Preuszischen Kunstsammlungen*, XXXII, 1911, pp.249–86, is still the best account of the different styles in Venetian enamel-painted glass; see also L. Zecchin, 'Cristallini dorati e smaltati', *Vetro e silicati*, XII, No.68, 1968, pp.22–5.

[16] Astone Gasparetto, *Il Vetro di Murano dalle origini ad oggi*, Venice, 1958, pp.83–4; see also L. Zecchin, 'Il Vetro 'Cristallino' nelle Carte del Quattrocento', *Vetro e silicati*, VII, No.38, 1963, pp.21–4, where the word *cristallini* is documented as early as 1409, becoming frequent about the middle of the century; also id., 'Nascita del Cristallo Veneziano', ibid., XI, No.66, 1967, pp.20–3.

[17] Schmidt, *Das Glas*, p.96; Gasparetto, op. cit., p.86 and Fig.30; Erik Zahle, 'Venetiansk Glaspokal' in *C. L. Davids Samling: Tredie Del*, Copenhagen, 1958, p.108, pointed out that this combination of arms could indicate Wladislaw II Jagello as easily as Matthias Corvinus himself.

[18] A number of enamelled and gilt Venetian glasses were found on the site of the palace in Budapest, including a fragmentary goblet with the arms of Beatrix of Aragon (1457–1508), wife of Matthias (L. Gerevich, 'Les Fouilles de Badavár', *Acta Historiae Artium Academiae Scientiarum Hungaricae*, I, fasc.1–2, 1953, Pl. X,4). It is recorded that among gifts sent to the Queen from Ferrara in 1486 were included Venetian crystal glass (B. Borsos, *Glass-making in old Hungary*, Budapest, 1963, p.21).

Alessandro Bentivoglio with Ippolita Sforza in 1492.[19] A number of pieces with the conjoined arms of Louis XII of France and Anne of Brittany may well have been made at the time of their marriage in 1499, and certainly before Anne's death in 1514.[20] Examples in the early sixteenth century are too numerous to cite. Into this general range of dates the Waddesdon goblet fits well enough.

Bearing little decoration other than the dots of enamel on a gold ground are two pilgrim-flasks in the Waddesdon Collection (*17–18*). The first of them has the slightly sagging shape and the cordon foot-motif of the Bentivoglio-Sforza flasks already alluded to, but the lighter decoration of the body probably puts it in the early years of the sixteenth century. The second (*18*) exemplifies this somewhat impoverished style at its simplest. It is uncertain how early it was introduced, but most of the glasses of this kind are probably to be dated in the first twenty-five years of the sixteenth century.[21]

Little is known of the artists who decorated these glasses, and little has been done to divide the known glasses into stylistic groups. Not a single glass can be attributed with any certainty to a known artist, the only ascription being that of the well-known blue *coppa nuziale* in the Museo Vetrario at Murano to the famous glass-maker Angelo Barovier (d. 1460).[22] The credit for perfecting enamelling on glass was given to Angelo Barovier on the authority of a Latin translation commissioned by King Matthias Corvinus about 1489 of Filarete's *Trattato di Architettura.* This included the phrase 'depicto intus vitro' ('glass painted on the inside'). The passage in question, however, in the original Italian of the book, refers to 'flat glasses, on which may be seen figures, animals and other things sculpted inside' ('vetri . . . piani, e dentro vi si vedrà scolpite figure et animali e varie cose . . .'), and, in the context of Filarete's description of the Utopian city of 'Sforzinda', may be taken as a flight of fancy.[23] Famous though Angelo Barovier unquestionably was, as 'the best Venetian maker of cristalline vases' ('Venetum optimum artificem crystallinorum vasorum')[24] as well as much else, there is no positive

19 Gasparetto, op. cit., p.83 and Fig.21. The suggestion that the arms might refer to the marriage of Giovanni II Bentivoglio and Ginevra Sforza in 1464 may safely be discounted.

20 Corning Museum, *Three Great Centuries of Venetian Glass*, Corning, 1958, No.28; Victoria and Albert Museum, *Glass: a Handbook . . . and a Guide to the Museum Collection*, by W. B. Honey, London, 1946, p.67, suggests the possibility that these glasses might be of French origin, but this seems unlikely.

21 A goblet with no more than a scale-border of this type, found in a tomb at Somogyvár in southern Hungary, is said to date from about the middle of the fifteenth century – Béla Borsos, *Glass-making in Old Hungary*, Budapest 1963, pp.18–19, Fig.6. The archaeological evidence on which this dating rests, however, seems very questionable. I am indebted to Madame Klara Tasnádi Marik for this information, based on the report in *Acta Historiae Artium Academiae Scientiarum Hungaricae*, 1959, p.344.

22 Gasparetto, op. cit., pp.77–9, Fig.21, I–V. L. Zecchin, in 'Una Fornace Muranese all' Insegna dell' Angelo', *Vetro e silicati*, XI, No.62, 1967, p.23, corrects the previously recorded date of death of 1461.

23 Schmidt, 'Die Venezianischen Emailgläser . . .', p.250; Gasparetto, op. cit., p.78; L. Zecchin, 'Angelo Barovier . . .', (see next note), pp.25–6. It should be noted that Venetian enamelled glasses are painted on the *outside.* That Filarete's observations should perhaps be taken seriously is suggested by recent research – see Anthony Radcliffe and Charles Avery, 'The "Chellini Madonna" by Donatello', *Burlington Magazine*, CXVIII, June, 1976, pp.377–87.

24 From the Epigram on Angelo Barovier composed by Ludovico Carbone, quoted by Gasparetto, op. cit., p.79; on the Barovier family, see also W. E. S. Turner, 'The Oldest Glass-making Family', *News and Reviews of the Society of Glass Technology*, *34*, 1950, pp.47–51; L. Zecchin, 'Una Fornace Muranese all' Insegna dell' Angelo', *Vetro e silicati*, XI, No.62, 1967, pp.23–6; id., 'Angelo Barovier, Vetraio del Rinascimento', ibid., X, No.58, 1966, pp.24–6; id., 'I Barovier e il Vetro Cristallino', ibid., XII, No.67, 1968, pp.23–6.

evidence linking him with the art of enamelling on glass. The only apparently signed glass bears the initials 'P.A.', which cannot be connected with any known artist; the dish in question, moreover, resembles stylistically no other glass of this general category.[25]

One member of the Barovier family (Maria or Mariet(t)a), however, certainly obtained permission in 1487 to build a 'little furnace' ('parvula fornace'), which may have been a furnace for the firing of enamelled glass.[26] Another woman glassmaker, one Elena of the Lando family, is recorded in 1447 among the earliest decorators of glass, and was probably an enamel-painter.[27] The Venetian Pietro di Giorgio *cortiner* (that is, a painter of curtains), also recorded as 'a perfect master in the gilding and enamelling of glasses' ('perfetto maistro da dorar e smaltar lavorieri de vero'), was authorized in 1470 and 1471 to open a furnace at Murano:[28] and Giovanni Maria Obizzo specialized in the enamel-decoration not only of transparent glass but also of the opaque-white *lattimo*, which he produced for Bernardino Ferro.[29]

The Venetian technique of enamelling glass is nowhere described in its entirety, and can only be reconstructed by inference.[30] The first point to be observed is that the enamelling always overlies the gilding, which is produced by the application of gold-leaf, as the overlapping of the rectangular leaves, easily observable on some glasses, demonstrates beyond a doubt. A Swedish priest named Peder Månsson, who travelled to Rome in 1508, and in 1524 returned to his native land and compiled an account of what he had seen, is the main witness. Having described the making of a glass, he continues: 'When the glass is thus made ready, cut some gold leaf and lay it round the edge. Hold the glass in the furnace so that the gold is at once welded to the glass. Then lay the finished glass up in the other chamber to cool off, so that it does not get cold too quickly and break . . .' The 'other chamber' referred to is the annealing-chamber or *lehr*, which formed the vaulted uppermost storey of the typical three-tier Venetian furnace, receiving from below, through a central hole in its floor, such heat as had not already been absorbed in the reverberatory founding-storey of the furnace below. There are Venetian glasses which are decorated with gilding alone, but they are rare, and it is reasonable to assume that some at least were intended for subsequent enamelling. Månsson's mention of a zone of gilding round the rim of a glass, where enamelled borders are nearly always found, also suggests this likelihood.

25 Schmidt, 'Die Venezianischen Emailgläser. . .', pp. 261–2, Abb. 6.

26 Gasparetto, op. cit., pp.79–80; Corning, *Three Centuries . . .*, p.15, states categorically that this furnace was 'for firing painted glasses'. L. Zecchin, however, in 'Maria Barovier e le "Rosette" ', *Journal of Glass Studies*, X, 1968, pp.105–9, argues very persuasively that Maria Barovier's furnace was for the making of decorative canes. See also n.24.

27 Gasparetto, op. cit., pp.80–1, giving the name as De Laudo; see also L. Zecchin, 'Cristallini Dorati e Smaltati', p.22, which also lists dates between 1443 and 1445.

28 Gasparetto, op. cit., p.81. This practice had been forbidden in 1469 – see L. Zecchin, 'Cristallini Dorati e Smaltati', p.22.

29 Gasparetto, op. cit., p.81; see also L. Zecchin, 'Le Avventure del Lattimo', *Vetro e silicati*, VII, No.40, 1963, p.22, and 'Cristallini Dorati e Smaltati', pp.24–5. Another 'depentor' was Giovanni Antonio da Lodi (ibid., p.24). Yet another enameller working in the last decade of the fifteenth century and the first of the sixteenth may have been a certain Filippo (Serena) di Catani 'depentor de vedri' – see L. Zecchin, 'Famiglie Vetrarie Famose: I Serena', *Vetro e silicati*, VIII, No.44, 1964, p.17. See also T. H. Clarke, 'Lattimo – a Group of Venetian Glass enameled on an opaque-white Ground', Journal of Glass Studies, XVI, 1974, pp.22–36.

30 The whole question is dealt with by R. J. Charleston, 'Enamelling and Gilding on Glass', *The Glass Circle*, I, Newcastle upon Tyne, 1972, pp.18 ff.

The enamelling process itself is described in a fifteenth century MS. preserved in the Library of the Canonici Regolari in the Convent of S. Salvatore in Bologna. It runs: 'To paint glass, that is to say, cups or any other works in glass with smalti [enamels] of any colour you please. – Take the smalti you wish to use, and let them be soft and fusible, and pound them upon marble or porphyry in the same way that the goldsmiths do. Then wash the powder and apply it upon your glass as you please and let the colour dry thoroughly; then put the glass upon the rim of the chamber in which glasses are cooled, on the side from which the glasses are taken out cold, and gradually introduce it into the chamber towards the fire which comes out of the furnace, and take care you do not push too fast lest the heat should split it, and when you see that it is thoroughly heated, take it up with the "pontello" [pontil] and fix it to the "pontello" and put it in the mouth of the furnace, heating it and introducing it gradually. When you see that the smalti shine and that they have flowed well, take the glass out and put it in the chamber to cool, and it is done . . .' The general accuracy of this picture is confirmed for us by the first printed book to contain a coherent account of the art of glass-making – Vannoccio Biringuccio's *Pirotechnia* (*Art of Fire*), published in Venice in 1540. Biringuccio wrote: 'They finish them [i.e. glass vessels] by affixing feet or handles or other features of various kinds of glass, or truly gilding them with fine gold, and also if they wish they ornament them with paintings and with beautiful and fine enamels . . .' The division between gilding and enamelling here is to be noted. Biringuccio condenses his account of the whole process, which would in fact have gone as follows: first, the glass would have been gilt as the last process of its making, and would then have been placed in the annealing-chamber to cool, being gradually moved out to the point where the lehr-man could transfer it, now sufficiently cooled, to the outside atmosphere in what was later known as the 'Sarrosel' room: secondly, the glass would go to the decorator's studio, whether inside or outside the glasshouse,[31] there to be painted (over the gilding in the appropriate passages) with a mixture of the powdered enamel and a medium such as oil of lavender: thirdly, the glass would be cautiously pushed into the annealing-chamber once again, and when sufficiently heated, taken up on a pontil and introduced into the 'glory hole' of the main working-furnace, being rotated until the enamels were seen to melt and shine, when the glass would be knocked off the pontil and replaced in the annealing-chamber as before; after cooling, the operation was complete.

This process is described as having been performed in the full-scale Venetian furnace where the glasses themselves were made, but if the decorator's shop wished to be independent of a main furnace, a smaller furnace consisting of a heating-chamber and an annealing-chamber would suffice. Possibly Maria Barovier's 'parvula fornace' of 1487, mentioned above, was of this kind.

Enamelled and gilt glass seems to have begun to go out of favour in Venice in the course of the first half of the sixteenth century, except for her clientele in Central and Northern Europe, where finally the technique was taken over by native craftsmen and continued without break right through to the eighteenth century and even – but by now degenerated into a peasant craft – into the nineteenth century (see pp.26–30 below). Its place as the main decorative

[31] These decorators may have worked as salaried employees in the glasshouses, or, more likely, in outside studios. The refiring, however, was probably done in the glasshouses which made the glasses. An edict of 1469 forbade the use of outside furnaces for this purpose – see L. Zecchin, 'Cristallini Dorati e Smaltati', p.22 (see also n.28).

technique of glass-making in the Venetian manner was taken by that of using opaque-white *lattimo* canes or flattened tubes in striped embellishments of greater or less complexity.[32] Opaque-white threads had been used as superficial decoration, trailed on the surface or round the rim of a glass, at a far earlier date, but the new technique normally incorporated the threads in the colourless matrix of the glass. This was probably done by lining the inside of a mould with lengths of cane – alternately opaque-white and colourless – placed vertically, and then picking them up on a gather of glass, and 'marvering' them into the surface on the slab of marble or metal (the 'marver') on which the first gather was normally rolled and smoothed before further working. The 'paraison', or gather with its first bubble of inflation, could then be worked like any other, the longitudinal *lattimo* threads having first been pinched together into one point of junction at the end of the paraison, and the surplus cut away, thus leaving a neat termination. The white 'threads' could be either solid canes of circular section; or tubes which collapsed to give flat broad stripes of solid whiteness; or circular canes themselves made up of a spiral of finer canes.[33] These were produced by standing canes in a mould, as described, picking them up on a gather, marvering, and finally pulling the resultant cylinder out and twisting at the same time. Cables made in this way could alternate with plain canes, and occasionally coloured canes were used. A further complication was introduced when a paraison made in the way described was twisted in one direction and then blown inside a second paraison with canes twisted in the opposite direction, giving a mesh of intersecting canes with a bubble of air trapped in each lozenge of the trellis.[34] This 'lace glass' or *vetro di trina*[35] was often spun into flat shapes, usually dishes and often of considerable diameter (*21*). Once the canes were incorporated in the paraison, this could be worked like any other, the only limitation being set by the worker's skill and taste.

It is difficult, if not impossible, to determine just when this technique was introduced into Venetian glass-making. It is first referred to by Vannoccio Biringuccio in his *De la Pirotechnia* (1540). In this he alludes to 'the drinking vessels in which one actually sees twisted designs of thorn branches and other criss-cross inlays . . . Look, too, at the large things, as well as the small, that they make of white or coloured glass and that seem to be woven of osier twigs equally spaced with the greatest uniformity and exactness of termination'. Here Biringuccio is evidently describing the more elaborate manifestations of this type of decoration, and it is known that the decoration *a retortoli* had already evolved by 1527, when the brothers Filippo and Bernardo Serena applied to the Serenissima for a patent of monopoly, stating that they had discovered a new method of working within their craft which called for 'stripes with twists of

[32] The normal Venetian word for opaque-white glass was *lattimo*, and the technique described here has usually been referred to as *latticinio*. This, however, is a nineteenth-century coinage, and as such has been avoided here – see L. Zecchin, 'Fortuna d'una Parola sbagliata', *Journal of Glass Studies*, X, 1968, pp.110–13; see also article 'Fadenglas' in *Reallexikon zur Deutschen Kunstgeschichte*, Stuttgart, 69. Lieferung, 1972, Cols. 1048–55.

[33] Glasses decorated in this way appear to have been called in the old Venetian parlance *a retortoli* or *a retorti*, as opposed to *a fili*, where plain threads were used – see Gasparetto, op. cit., pp.94–5; L. Zecchin, 'Fortuna d'una Parola sbagliata', pp.112–13: see also n.35.

[34] These processes are described in detail with diagrams, by, e.g., G. Bontemps, *Guide du Verrier*, Paris, 1868, pp.602–13; this is partially resumed in E. Garnier, *Histoire de la Verrerie et de l'Emaillerie*, Tours, 1886, pp.98–104, and more shortly in the article 'Glass' in the *Encyclopaedia Britannica*, 1947.

[35] The old terms appear to have been *a redexolo* (*redesello*=*reticello*), or *a redexin* both implying a mesh-pattern. The expression *a filigrana* could probably be used of both glasses *a retortoli* and glasses *a reticello*; see works cited n.33.

cane' ('facete(=fascette) con retortoli a fil').[36] If it is reasonable to suppose that the simpler forms preceded the more complicated, one might conclude that the evolution of the technique lay mainly in the first quarter of the sixteenth century, were it not for the fact that one glass exists, decorated already *a retortoli*, which has claims to antedate 1490. This is a large goblet, originally of biconical form, the foot of which was at one time broken and was replaced by a silver foot bearing an inscription. The inscription records that 'This was King Matthias' goblet', received from Venetians. It was alleged to have passed to King Louis II, who made a present of it to Ferenc Batthyány (d.1566), his Grand Cup-bearer, in 1524.[37] If the goblet did indeed belong to Matthias Corvinus, it is the most precocious example recorded of a glass decorated with *lattimo* bands (both *a fili* and *a retortoli*). On the whole it seems rather more likely that the tradition was mistaken; and the gift of 1524 may have been at the root of the error. Be that as it may, a picture dated 1537, by Peter Gaertner, shows a covered beaker of apparently fairly complicated technique[38]; while Henry VIII's inventory of the glass in the 'Glasse Housse' at Westminster, taken in 1542, lists 'bolles of glasse with oone cover to them all wrought with diaper worke white',[39] a description which suggests decoration *a filigrana*. Probably none of the glasses at Waddesdon, with the possible exception of *22B* dates from as early as this. Most derive probably from the late sixteenth or early seventeenth century. One of them (*23*) probably betrays, in its impure 'metal' and uncouth shape, manufacture in one of the many glasshouses working outside Italy in the *façon de Venise*.[40]

Another of these glasses (*20*) shows, in addition to its white-striped decoration, a mould-blown embellishment of the base of the body, a technique quite commonly seen in goblets with plain everted rims[41] in place of the complicated foliations of the Waddesdon piece. Moulds for this purpose were probably made of metal, in from two to four sections, probably hinged together, and the ornament they conferred was as little as possible modified by subsequent working.[42] Moulds of essentially this type were used for the vessel in the form of a *Lion* (*19*), as well as for the vase (*28*) in blue glass with added gilt embellishments.

Of a different character were the moulds used for the somewhat later and heavier coloured glasses with gilt base-metal mounts (*29–35*). These moulds were vertically fluted, and imparted vertical relief-ribbing to the glasses blown in them. In subsequent working, this ribbed decoration followed the contours of the vessel, swelling and thinning, sometimes almost disappearing.

This group is of problematic origin. The colours are normally emerald-green, deep blue, or

[36] Their petition runs: '. . . supplichemo che havendo trovato certo modo et nova invention di lavorar del mestier nostro, il qual modo si domanderà a facete con retortoli a fil, non mai più fatto . . .', cit. L. Zecchin, 'Famiglie Vetrarie Famose: i Serena', p.20, referring to Archivio di Stato, 'Consiglio dei X', *Parti Comuni*, fila 6, n.84; see also id., 'Fortuna d'una Parola sbagliata', pp.111 and 113.

[37] Borsos, op. cit., p.20 and Pl.5.

[38] Corning Museum of Glass, *Three Great Centuries of Venetian Glass*, Corning, 1958, p.20, Fig.2.

[39] Albert Hartshorne, *Old English Glasses*, London, 1897, p.464.

[40] On the spread of Venetian glass-making in the fifteenth–seventeenth centuries, see Axel von Saldern, article, 'Façon de Venise' in *Reallexikon zur Deutschen Kunstgeschichte*, Stuttgart, 69. Lieferung, 1972, Cols.1024–48.

[41] e.g., Kunstgewerbemuseum der Stadt Köln, *Glas*, by Brigitte Klesse, Cologne, 1963, No.209; Museum für Kunsthandwerk, Frankfurt-am-Main, *Europäisches und Aussereuropäisches Glas*, by Annaliese Ohm, Frankfurt-am-Main, 1973, No.150. These glasses, however, normally have a depressed knop between mereses above the pedestal foot.

[42] On glasses of this type blown in a fixed mould, see R. Chambon, 'Verres de la Renaissance soufflés dans un moule fermé', *Glastechnische Berichte*, Bd.32 K, Heft VIII, 1959, pp.49–56. See also *28*.

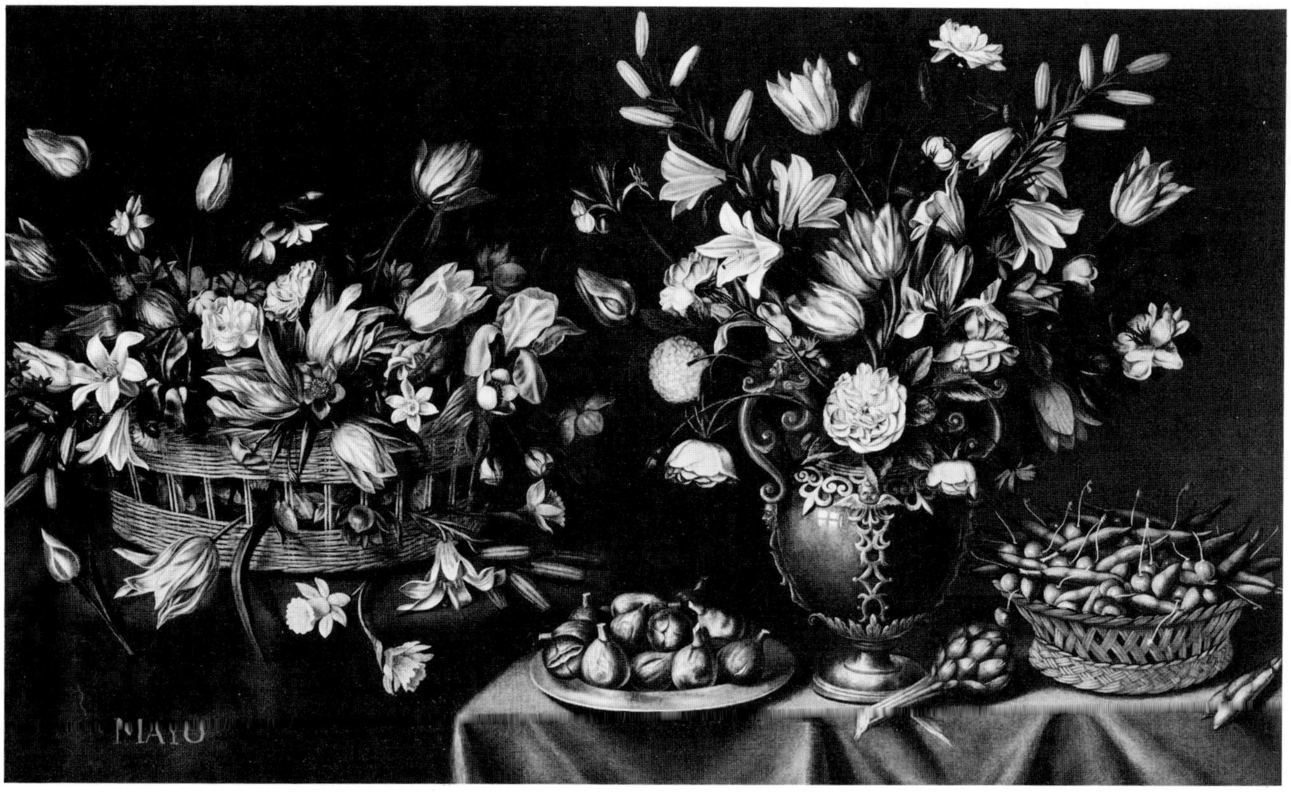

blue-green. The gilt metal mounts run true to type, being characterized by horizontal bands with lappets of various kinds, often in the form of a *fleur-de-lis*, and vertical bands of strapwork often incorporating *putto*-heads; the handles are usually ear-shaped, with or without external beading, but are also often of a fanciful character, incorporating more elaborate scrollwork, or term-figures; the feet are frequently lathe-turned.

These vessels have traditionally been attributed to Venice, but the glass-metal is uncharacteristically thick and heavy, and the vessels concerned are simple and quite free of the fantasy and delight in manipulation characteristic of the Venetian glass of this era.[43] The quality of solidity observable in these glasses has tempted the notion that they might be of German origin, but none of the certainly German coloured glasses of the seventeenth century have mounts of this character – virtually all the ruby glasses, for example, having silver-gilt mounts if they have mounts at all. There are also other contra-indications. Glasses of this character are represented in at least two paintings of the seventeenth century. The first (*above*) is a still-life of flowers and fruits emblematic of *May*, by the Spanish-Flemish painter Juan van der

[43] A glass of colourless material in the Murano Museum, however, has a mount which appears to bring it into the group under discussion – see A. Gasparetto, *Il Vetro di Murano*, Venice, 1958, Fig.73. It has evidently been broken and mounted in order to give it a foot, whereas all the coloured glasses appear to have been made for mounting. This piece, which shows every sign of being a Venetian product, was perhaps mounted in the place where it was broken. Another colourless piece is in the Museo Civico, Turin (see L. Mallé, *Vetri-Vetrate-Giadi* . . ., Turin, 1971, pp.77–8, Pl. 68).

(Right) Flower painting by Giuseppe Recco (1634–95)
By courtesy of Newhouse Galleries Inc., New York

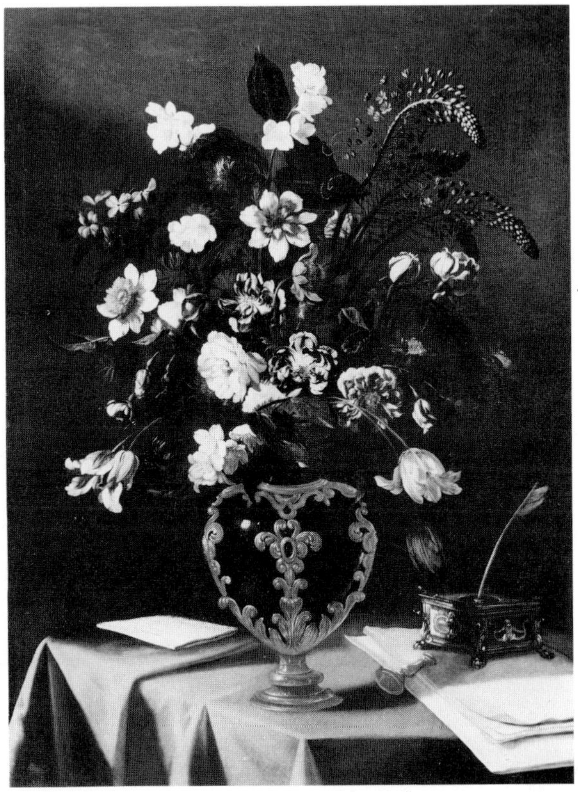

(Opposite) Juan van der Hamen y Leon (1596–before 1632), May Flowers
By courtesy of the Terry-Engell Gallery, London

Hamen y Leon (born in Madrid 1596, died before 1632).[44] The second (*above*) is a flower-piece by the Neapolitan painter Giuseppe Recco (1634–95).[45] Here the style of the mounts is less clearly affiliated to that of the mounts on the glass under discussion, but the kinship is unmistakable. The likelihood of German glasses finding their way at this period to both Spain and Italy is slim; but the possibility of Italian glasses making their way to Spain is considerable, at a time when the glass-industry of Barcelona was past its peak, and the remaining glass-centres of Spain produced an essentially peasant type of glass.

[44] He was the son of Jan van der Hamen, of Brussels. One of the earliest Spanish writers on art (Palomino, *Parnaso Español Pintor*, ed. 1797, III, p.473 f) surmised that Juan van der Hamen y Leon died in 1660, which would not only have given him a longer working life, but would fit more easily with the presumed date of these glasses. A second picture by the same artist, signed and dated 1629, shows a plate of figs identical in every detail with that shown above, together with a small ribbed double-gourd bottle, evidently of glass, now showing black in colour but once probably dark-blue or green, which would have consorted well with the group of ribbed glasses now under discussion (see *Apollo*, April 1970, p. cxiv, Fig.14, picture in the possession of H. Schickman Gallery, New York). It should be added that a still-life of flowers by Abraham Mignon, in the Ashmolean Museum, Oxford, shows a vase apparently mounted in the same way (*Catalogue of . . . Dutch and Flemish Still-Life Pictures bequeathed by Daisy Linda Ward*, Oxford, 1950, pp.130–1, No. 58). Mignon (1640–79) was a pupil of Jan Davidsz. de Heem, and although of German origin, worked mainly in the Netherlands (P. Mitchell, *European Flower Painters*, London, 1973, pp.171–2).

[45] Born in Naples, he was summoned to Spain by Charles II, at the suggestion of Luca Giordano, and died in Alicante shortly after his arrival (29 May 1695). He worked in Milan until his twentieth year (Thieme-Becker, *Allgemeines Lexikon der bildenden Künstler*, Leipzig, Bd. XXVIII, 1934, s.v.).

The notion of an Italian origin for these glasses is given additional strength by the fact that one of the characteristic forms within this group – a bowl pinched-in to give a boat-shape (*29*) is shown in one of the drawings attributed to Giacomo Ligozzi (*c.*1547–1626)[46] in the Gabinetto Disegni e Stampe degli Uffizi, Florence (see page 128). This not only gives the shape, but shows a mount in character essentially like those actually found on these glasses, except that the foot (obviously intended to be of *repoussé* metal)[47] is a true stem instead of a low pedestal. Another shape found within this family of glasses is also echoed in the Ligozzi drawings.[48]

The association with Florence, and the presence in the collections there of very numerous drawings for glass – by Stefano della Bella and Jacques Callot, as well as by Ligozzi – raises the question whether these glasses may have been made there. It is known that in 1567 Cosimo I had tried to get a Venetian master to come and work in Florence.[49] The first attempt failed, but by 1579 a furnace was working under the Master Bortolo of the Venetian glasshouse which went under the sign of the Three Moors.[50] This Florentine furnace was located in the Medici gardens between S. Marco and Via S. Gallo, and remained there during the reigns of Francesco I and Ferdinando I.[51] In 1618 Cosimo II had a furnace built in the gardens of the Pitti Palace, and in August of that year another Venetian master, Jacomo Della Luna, arrived in Florence, soon to be joined by his brother Alvise, and a year later by their uncle, Ludovico of the same name.[52] There is also a hint that the original furnace near the Via S. Gallo was resuscitated at this date (1620), and in 1629 a Venetian, Pietro Giardini, opened a furnace in Florence, to be followed in 1657 by Tommaso di Filippo Romolini and Vincenzo di Giovanni Bastardello, who sought to make glass in accordance with the basic Venetian formula.[53]

There is very little evidence to show what was made in these furnaces. In the documents the emphasis is on *cristallo*, although other kinds of glass are also referred to,[54] and it is evident that the official furnaces existed mainly to produce unusual glasses for the delectation of the Grand Duke,[55] as one or two glasses surviving in the collections of the Museo di Storia della Scienza,[56]

46 Ligozzi was born in Verona, and was summoned to Florence in 1578 in the time of the Grand Duke Francesco I (1549–87) – see L. Zecchin, 'Disegni Vetrari del Seicento', *Sulla Storia dell' Arte Vetraria* (Camera di Commercio Industria e Agricoltura di Venezia), Venice, 1961, pp.57 ff; id., 'La Toscana e il vetro muranese' (ibid., 1954), p.25; Thieme-Becker, op. cit., Bd. XXIII, 1929, s.v.

47 *Repoussé* feet in place of the more normal turned feet are by no means uncommon on these glasses – see, e.g., a boat-shaped green bowl in the Victoria and Albert Museum (Mus. No.19–1867).

48 Gabinetto Disegni e Stampe degli Uffizi, 97313, *recto*, No.129 (but with side-handles and foot, which in the mounted pieces are replaced in gilt metal); compare with this a green flask in the Victoria and Albert Museum (Mus. No.7445–1861 – see G. Savage, *Glass*, London, 1965, Fig.125).

49 Guido Taddei, *L'Arte del Vetro in Firenze e nel suo Dominio*, Florence, 1954, p.48.

50 Ibid., p.49; see also L. Zecchin, 'La Toscana e il vetro muranese,' p.17; id., 'Bortolo d'Alvise ai Tre Mori', *Vetro e silicati*, Anno XIII, No.76, July–August, 1969. Bortolo in Murano appears to have made enamelled and gilt glasses, white-striped glasses (*a reticelli*), 'ice-glass' and glasses engraved with the diamond-point (ibid., pp.3–4 of the offprint); there is no notice of coloured glasses.

51 G. Taddei, op. cit., p.50. This glasshouse *al Casino* was also frequented by Antonio Neri, author of the first handbook of glass technology (see n.58 below).

52 Ibid., pp.52–3; L. Zecchin, 'La Toscana e il vetro muranese', pp.21–5; id., 'La Vetreria Medicea di Palazzo Pitti', *Giornale Economico*, Venice, No.3, March, 1961, pp.289–95; id., 'Vetrai muranesi a Firenze nel 1619', ibid., No.6, June, 1961, pp.662–9; id., 'La Nascita di una vetreria a Firenze nel 1618', *Technica Vetraria*, Anno VII, No.1, January-February, 1962, pp.14–16; id., 'I Muranesi Luna nella vetreria di Cosimo II', ibid., Anno VII, No.2, March-April, 1962, pp.1–4.

53 G. Taddei, op. cit., p.54.

54 Ibid., pp.50–1.

55 Ibid., pp.53–4.

56 e.g. Giovanni Mariacher, *Italian Blown Glass*, London, 1960, Fig.XXXVI.

and the Ligozzi drawings, indicate. The public collections in Florence, moreover, appear to contain none of the mounted coloured glasses. For these reasons it seems somewhat less likely that the coloured ribbed glasses were made here, although they might have issued from the private furnaces set up in 1629 and 1657.[57] It may well transpire that they were made at some other Italian centre[58] and their presence in the paintings of Hamen y Leon and Recco suggests that Naples may have been the place, the Kingdom of Naples being at this time united under the Spanish crown. The career of Recco, a Neapolitan artist who was summoned to Spain at the end of his career, is indicative of the close artistic ties between the two kingdoms; and it would not be unnatural to find a Neapolitan glass in a Spanish painting such as that of Hamen y Leon. There are stated to have been glass-makers working in the *façon de Venise* at Naples.[59]

Further manifestations of the *façon de Venise* at Waddesdon are the 'goblet-vase' (*24*) with hollow mould-blown stem and applied parallel notched cordons punctuated by applied moulded masks; and the lamp (*25*) with mould-blown body and hollow stem, and fanciful embellishments in blue-green glass, applied and snipped with the shears into the forms of wattles and wings. Of more pronounced provincial flavour are the two goblets (*26* and *27*), the stems of which have been pincered from a single cable-cane into an inverted heart-shape enclosing a symmetrical coil-design; the profile is enlivened with pincered 'wings', and the two terminals have been worked into dragons' heads with pincered crests. These elaborate *verres de parade* with their stems worked in a single plane – probably composed on the marver before the addition of bowl or foot – contrast with other goblets of the same period and general type having rather more three-dimensional and lively stem-formations. These last are perhaps to be credited to the glasshouses in Venice itself,[60] while the goblets with flat pastry-cook stems are usually attributed to glass-factories in the Netherlands.

57 See above, p.24. It is worth bearing in mind that the fleur-de-lis, which appears almost invariably on these mounts, is the emblem of Florence.

58 It is worth bearing in mind that Cosimo I also founded a glasshouse at Pisa (G. Taddei, op. cit., pp.50, 52), and that it was here that the priest Antonio Neri conducted many of his experiments on coloured glasses. In his work *L'Arte Vetraria*, published in Florence in 1612, Book I contains a number of recipes for 'aquamarine' (Caps. 22–3, 26, 29–31) and 'emerald' (Caps. 32–5); and Book III (Caps. 49–50) for blue (*Zaffiro*). Of the three aquamarine recipes he says that he tried one out in the Casino in Florence in 1602, another at Pisa in the same year, and the third in Pisa; of two of the recipes for emerald-green he says that he had proved them in Pisa; the same is true of one of the recipes for blue. See also L. Zecchin, 'Lettere a Prete Neri . . .', *Vetro e silicati*, Anno VIII, No.43, January–February, 1964, pp.17–20.

59 R. Schmidt, *Das Glas*, p.113. Schmidt unfortunately does not quote his sources, and it is uncertain whether his dating of '16th and 17th centuries' refers to all of the centres named. A Muranese document of 1554 refers to 'zottole napolitane con do maneghi col pè d'oro' ('Neapolitan bowls with two handles and gilt stem (or foot)') – L. Zecchin, 'Una Fornace muranese all'insegna della Sirena', *Rivista della Stazione Sperimentale del Vetro*, *1*, March–April, 1971, p.4 of the offprint. Glass was certainly made at Naples in the 16th century – see G. Donatone, *La Maiolica Napolitana dell'Età Barocca*, Naples, n.d. (?1974), pp.7, 15; L. Zecchin, 'Vetrerie italiane nel Quattro e Cinquecento (a Napoli e altrove)', *Tecnica Vetraria*, Anno VII, No.3, May–June, 1962, pp.1–4. There was a glasshouse in Naples in the eighteenth century – see G. Donatone, *La Farmacia degli Incurabili e la maiolica Napoletana del Settecento*, Naples, n.d. (1974?), pp.23–4.

60 See Axel von Saldern, 'Spiral-Schlangenpokale à la Façon de Venise', *Jahrbuch der Hamburger Kunstsammlungen*, Bd. 17, 1972, pp.67–80. Dr von Saldern comes down on the side of a Netherlands origin for these glasses, as forerunners of the flat-stemmed goblets under discussion. He perhaps underestimates the extent to which Venetian style was adapted to the needs of the northern markets. A Venetian origin for the three-dimensional stemmed glasses is proposed by Anna-Elisabeth Liederwald, *Niederländische Glasformen des 17. Jahrhunderts* (Dissertation), Freiburg-im-Breisgau, 1964, pp.96 ff. Doubt has recently been cast on the authenticity of these glasses.

German enamelled glass

Ever since the Middle Ages there has been a lively traffic, both commercial and artistic, between northern Italy and the Germanic lands. This intercourse is reflected in the glass-making sphere too. German packmen carrying glasses on their backs are among the earliest means of transporting Venetian glass mentioned in the medieval records,[61] and the first glasshouses working outside Venice in the *façon de Venise* are heard of in Austria (Vienna in 1486,[62] then Hall in the Tyrol in 1534[63]) in Southern Bohemia (1577),[64] and thereafter in a number of localities in Germany and Bohemia.[65]

Among the most sought-after Venetian glasses north of the Alps were those decorated by enamelling and gilding (see pp.16–19 above). Often these were embellished with coats-of-arms and other forms of symbolism specific to northern European customers and were made to their commission. In due course, with the establishment of glasshouses working in those countries in the *façon de Venise*, the techniques of enamelling and gilding were taken up too, at a time when this mode of decoration was falling out of favour in Venice itself; and it becomes extremely difficult to distinguish between a Venetian glass made to the order of a northern customer and one made locally in the Venetian style for a local customer. This difficulty is mainly to be experienced in the middle years of the sixteenth century.[66] In the third quarter of that century the balance appears to have tipped decisively in favour of the local glasshouses. Johann Mathesius, who was a pastor in the Joachimsthal, in north-western Bohemia, published in 1562 a book entitled *Sarepta, oder Bergpostill*, the fifteenth chapter (or sermon) of which was devoted to glass-making ('Predigt vom Glasmachen'). This account, both learned and topical, is one of the liveliest descriptions of contemporary glass-making available to us. Mathesius writes: 'Thereafter imagination has invented one new thing after another: some have caused all sorts of pictures and inscriptions to be fired in colours on the colourless glass in the annealing-furnace, as the portraits and coats-of-arms of important gentlemen had been painted on window panes.'[67] This account suggests that glass-enamelling by then was established, but perhaps not long established, in his area of Bohemia. Archaeological evidence has recently been adduced to demonstrate that glass-enamelling was a Bohemian speciality by the third quarter of the six-

[61] A. Gasparetto, *Il Vetro di Murano*, p.53.

[62] Still surviving in 1563. A previous glasshouse run by an Italian had been founded there as early as 1428 – see R. Schmidt, *Das Glas*, Berlin and Leipzig, 2nd ed., 1922, p.120; also H. Zedinek, 'Wiener Glashütten des 15. und 16. Jahrhunderts', *Altes Kunsthandwerk*, I, Nr.6, 1927, pp.236–56. Venetian-style glasshouses were established in a number of places in Austria, but little is known of their history (except for Hall – see n.63 below).

[63] H. Zedinek, 'Die Glashütte zu Hall in Tirol', *Altes Kunsthandwerk*, I, Nr.3, 1927, pp.98–117; E. Egg, *Die Glashütten zu Hall und Innsbruck im 16. Jahrhundert*, Innsbruck, 1962, pp.18 ff.

[64] On the Rožmberk estate in southern Bohemia – see K. Hetteš, 'Venetian Trends in Bohemian Glassmaking in the 16th and 17th Centuries', *Journal of Glass Studies*, V, 1963, p.41, n.7, citing F. Mareš, *Ceské sklo* (*Bohemian Glass*), Prague, 1893, pp.23, 168. Other Venetian-style glasshouses were founded in Prague.

[65] See A. von Saldern, *German Enameled Glass*, Corning, 1965, pp.39–40; id., 'Façon de Venise', *Reallexikon zur Deutschen Kunstgeschichte*, Stuttgart, 69. Lieferung, 1972, Cols.1034–6.

[66] A. von Saldern, *German Enameled Glass*, pp.37 ff.

[67] 'Nachmals hat fürwitz immer ein newes uber das ander erdacht/etliche haben an die weisse gleser farben allerley bildwerck unnd Sprüche im külofen brennen lassen/wie man auch grosser Herrn contrafactur unnd wappen auff scheiben gemalet . . .', Johann Mathesius, *Sarepta oder Bergpostill*, Nuremberg, 1562, p. CCLXXIIII.

teenth century, and perhaps earlier.[68] Most enamelled glasses of Germanic character with dates in the second half of the century, therefore, are *ipso facto* likely to be of Bohemian origin, although some may have been made on the other side of the mountains in the Bayrischer Wald (cf. *36* here).[69] The evidence for the spread of the technique to other centres which later became famous for it does not much antedate 1600, and a very persuasive case would have to be made out for attributing a late sixteenth century glass of this type to any area other than Bohemia. When the tradition did spread, it fanned out from the Bohemian to the German side of the great mountain barrier formed by the Böhmer Wald, the Fichtelgebirge and the Erzgebirge;[70] or was deliberately transplanted, as when in 1602 the Elector of Brandenburg set up a glasshouse at Grimnitz, to the north-east of Berlin, specifically enlisting for the purpose a team of Bohemian glassmen under a certain Martin Friedrich, who in all likelihood came from Kreibitz, in the same north-western area of Bohemia as the Joachimsthal.[71] There is usually no danger of identifying as Venetian a glass enamelled in Germany or Bohemia in the decades on either side of 1600. To distinguish between the glasses made in different centres in those areas, however, is very much more difficult, and often impossible with the criteria at present established.[72] Types of glass apparently peculiar to Bohemia which did not survive long into the seventeenth century include a restricted range of shapes in blue 'metal',[73] most frequently decorated with a hunting-frieze, as is the case with the Waddesdon example of the commonest shape within this family (*40*). The Collection also includes two of the relatively slender cylindrical glasses called *Stange*,[74] decorated with the coats-of-arms or emblems of two men, both Germans and both living in areas (Bavaria and Saxony) which in the seventeenth century produced enamelled glass of their own; both glasses, however, are more likely to be of Bohemian than local origin (*36* and *41*). Virtually certainly Bohemian also is the extremely rare jug in the form of an owl (*39*).

The technique used in the production of these glasses probably differed little from that already described in the case of the Venetian industry (see pp.18 and 19 above). Indeed, pastor Mathesius specifically says that the glasses were re-fired in the annealing-chambers of

68 K. Hetteš, loc. cit., p.43, where it is suggested that enamelling began about the middle of the sixteenth century; pp.45 ff, where it is demonstrated that the glasses found in Prague were of potash-lime type and therefore essentially of Bohemian and not Venetian character, although the decoration is technically inspired by Venetian prototypes.

69 R. Schmidt, *Das Glas*, pp.175–6. This argument is based on the fact that a number of the armorial *Stangen* bear the coats of Bavarian nobility; and that three vessel glasshouses are known to have been active in this area in 1566 (see also on this A. von Saldern, *German Enameled Glass*, p.47). There is, however, no direct evidence that enamelled glasses were made here, and A. von Saldern has preferred to substitute the concept of 'South German', covering 'Southern Germany and the central western part of the Habsburg Empire' (*German Enameled Glass*, p.39). Nowhere else, however, does there appear to be such unambiguous evidence for the practice of the technique as there is in Bohemia. The possibility of a Bavarian origin for a glass such as *36* here has, however, to be borne in mind. The authenticated enamelled glasses made at Hall in the Tyrol have a distinctive character which marks them out as apart from the mainstream (A. von Saldern, *German Enameled Glass*, pp.47–8; E. Egg, op. cit., pp.67–71, Figs.56–7, 59; cf., however, ibid., p.38 for earlier (1558) documentary evidence of enamelling at Hall).

70 For the regional distribution of German/Bohemian enamelled and gilt glasses, see A. von Saldern, *German Enameled Glass*, pp.39–50 and 159–229; R. Schmidt, *Das Glas*, pp.175–215. The attributions proposed in the present *Catalogue* are argued in the individual entries.

71 A. von Saldern, op. cit., pp.163–4 and 222–3; R. Schmidt, op. cit., pp.138, 177, 208–11.

72 A. von Saldern, op. cit., pp.159–60.

73 See A. von Saldern, op. cit., pp.165–9, 246–51, 254–5, Figs.296–8, 302, 304–26.

74 See A. von Saldern, op. cit., pp.26–7.

the furnace: 'some have caused all sorts of pictures and inscriptions in colours to be fired on the colourless glass in the annealing-furnace'.[75] It seems likely that at least at first the enamellers worked in the glasshouses themselves rather than in outside *ateliers*, as was probably more frequently the case in Venice. Since the glasshouses tended to be the economic centres of isolated communities in the woodlands, this was natural enough. Later in the seventeenth century the glass-makers' guild-ordinances prescribed the ability to paint a *Reichsadlerhumpen* in one-and-a-half days as one of the skills necessary for the attainment of a Master's status (see below and p.29); which also suggests an intimate connexion between enameller and glasshouse.[76] On the technical side it may be said that the Bohemian and German enamellers in the seventeenth century used a limited range of bright colours, composed mainly of the primaries with black and white, a palette conducive to bold and simple effects. Indeed, the naivety and the chromatic gaiety of these glasses constitute their main charm. As time went on, the gold used for borders tended to be replaced by yellow enamel.

By far and away the commonest form of enamelled glass in the seventeenth century was the large cylindrical *Humpen*, often a truly 'two-handed' glass (see p.30 below). As the shape was stereotyped, so often was the range of decoration. Apart from the armorial glasses already referred to, which continue throughout the period during which enamelled glasses were in favour,[77] a certain limited range of themes is displayed on these glasses, and most of them are to be found on the glasses at Waddesdon. Some are self-explanatory, such as the ubiquitous hunting-scenes (*40*), or a glass decorated with a cavalier and a sententious inscription (*42*). Some, however, fall within a somewhat more complex frame of reference. Among these may be mentioned the glasses decorated with representations of the Twelve Apostles, Scenes from the Passion, Four Cardinal Virtues,[78] or the Ages of Man, all probably copied from prints. Of this last theme there are two representatives in the Waddesdon Collection (*37* and *38*). They illustrate human development from the age of ten to a hundred,[79] and by the rotation of the glass emphasize the cycle of growth and decay.

Even more elaborate are the glasses which celebrate the institutions of the Holy Roman Empire – the *Reichsadlerhumpen* ('Imperial Eagle beakers' – *43* and *44*), and the *Kurfürstenhumpen* ('Elector beakers' – *45*). The former, commemorating an ideal order which never really existed, display the crowned and haloed double-eagle of the Imperial insignia, with the orb (earlier the crucifix) on its breast, and coats-of-arms forming the feathers of its wings. Along the tops of the wings run the arms of the Electors of the Holy Roman Empire – on the spectator's left the ecclesiastical Electors (Trier, Cologne, Mainz) with the Holy See (*Potestat zu Rom*) for the sake of symmetry; while on the right are the secular Electors (Bohemia, the Palatinate, Saxony and Brandenburg). Below these the groups of shields run vertically in fours, each group representing a so-called 'Quaternion', or constituent part of the Empire: thus, there are four *Seyl* (that is, *Säulen* or 'pillars', the Dukes) of Brunswick, Bavaria, Swabia, and Lotharingia; the four Markgraves of Moravia, Brandenburg, Meissen and Baden; the

75 See n.67.

76 A. von Saldern, op. cit., p.67; R. Schmidt, op. cit., pp.177–8. The latter authority gives some reason to think, however, that at least some of the painters may have worked in *ateliers* in towns away from the glasshouses (op. cit., p.184).

77 A. von Saldern, op. cit., pp.125, ff.

78 Ibid., pp.86–7; p.91, Fig.110; p.101, Fig.143, etc.

79 Ibid., pp.96–100, Figs.129–34, 138.

four Burggraves of Magdeburg, Nuremberg, Rieneck and Stromberg; the Freiherrn, the cities, the peasants, vicars, landgraves, counts, knights, villages and burgs. This codification was first published about 1460 in a work entitled *De Imperio Romano* by one Peter von Andlau. The graphic sources for the double-eagle and crucifixion design, with the 'Quaternions', go back as early as the late fifteenth century,[80] and a version very close to that on the *Humpen* was produced by Hans Burgkmair in 1510. The prototype most probably used by the glass-painters, however, was a version engraved, probably about 1570, by Nickel Nerlich and published in Leipzig.[81] The double eagle with orb instead of crucifix occurs on an engraving of 1587 by Jan Bussemaker, but the rendering is not very close to that used on the glasses, and since a glass with this theme is known which is dated 1585, the source must be looked for elsewhere.[82]

The Waddesdon Collection contains two of these beakers, one a notable example standing 11⅜ in. high, dated 1661 and probably of Bohemian origin (*44*); the other very much smaller, dated 1645 and certainly Franconian (*43*).

The second type of glass commemorating the political solidarity of the Holy Roman Empire was the so-called *Kurfürstenhumpen* ('Elector beaker'), showing the Emperor accompanied by his four temporal and three ecclesiastical Electors.[83] The earliest commonly show the Emperor enthroned and the Electors standing to either side of him. Later, however, from about 1600 right through the seventeenth century, the Emperor and Electors are shown on horseback in two registers, on the upper and lower halves of the *Humpen*, sometimes merely separated from each other by sprays of flowers and leaves (*45*), sometimes in architecturally defined panels. This general scheme is varied in numerous ways, the King of Bohemia often being replaced by the imperial eagle (as on *45*), and other insignia and inscriptions being variously introduced.[84] The close connection of the *Kurfürstenhumpen* with the *Reichsadlerhumpen* is demonstrated by a composite beaker once in the Museum für Kunsthandwerk in Dresden, where the cover of the *Reichsadlerhumpen* is surmounted by a smaller *Kurfürstenhumpen*.[85]

Of all the areas in Germany which became notable for their production of enamelled glass, Franconia was probably the most prolific. Although its earliest productions cannot be identified with any certainty, there is one family of glasses which is peculiar to it and inalienable – those celebrating the riches and beauty of the Fichtelgebirge, and in particular the area of the 'Ochsenkopf' mountain (*47*). The glasses of this group date from the 1650s onwards and provide, together with the analogous Kreussen stonewares of the early years of the seventeenth century, some idea of the Franconian enamelled glasses of the first half of that century. Production continued well on into the eighteenth century (*47*) and it is probable that a further Waddesdon piece was made in this area (*46*).[86]

A word as to the uses of these glasses may not be out of place. It may be taken as axiomatic that glasses so elaborately decorated were not intended for everyday use, and the dedications or family and guild inscriptions on many of them indicate that they were made to special

[80] Ibid., pp.56–7, Figs.40–1.
[81] Ibid., p.57 and Figs.43–5.
[82] Ibid., pp.53, 57–8 and Fig.46.
[83] Ibid., pp.69–79.
[84] Ibid.
[85] Ibid., p.69 and Fig.57.
[86] Ibid., pp.182–196; R. Schmidt, op.cit., pp.185–194.

commission as presentation-pieces. They were certainly intended for the service of drinks on special and ceremonial occasions, and were probably kept for that purpose on the cornice of the panelling of the room. The French traveller F. M. Misson, writing in 1688 from Nuremberg, commented: 'You know the *Germans* are strange Drinkers; there are no People in the World more obliging, civil, and officious; but they have terrible Customs as to the Point of Drinking . . . Every Draught must be a Health, and as soon as you have emptied your Glass, you must present it full to him whose Health you drank. You must never refuse the Glass, which is presented, but drink it off to the last Drop. Do but reflect a little on these Customs, and see how it is impossible to leave off drinking: and indeed, they never make an End, but carouse, in a perpetual Round: to drink in *Germany* is to drink eternally. You must further know, that the Glasses are as much respected in this Country as the Wine is beloved. They place them all *en Parade*. The greatest Part of the Chambers are wainscotted, to two third of the Walls, and the Glasses are ranged all about upon the Cornish of the Wainscot, like Pipes of Organs. They begin with the little, and end with the great Ones; and these great Ones are always used, and must be emptied at a Draught, when there is any Health of Importance . . .'[87] *Humpen*, sometimes in the sixteenth century called *Luntz*, were frequently in the seventeenth century referred to as *Willkommen*,[88] which sufficiently indicates their function of holding a drink to welcome the stranger, a gesture which formed a part of European courtesy from at latest the Middle Ages onwards.[89] As the passage from Misson indicates, however, no such excuse was really required to instigate a drinking-party. In 1616 an author with the bacchic pseudonym of Blasius Multibibus advised his fellow-topers: 'one should summon to oneself merry fellows and good friends, wipe the dust off the "Römische Reich" and other drinking-vessels, and thus inspire and set on foot a jolly toping and tippling-party . . .' – especially desirable in hot weather.[90] The *Römische Reich* (that is, the *Reichsadlerhumpen* or *Kurfürstenhumpen*) was an obvious glass for the drinking of loyal toasts, and it was fitting that its capacity should be above the average. As the same Blasius Multibibus records: 'Then there is the great monstrous Humpe which is called the "Römische Reich", the power and force of which are so great and mighty that it might well trip up the most powerful Hercules or Tipple-knight and throw him down on God's earth.'[91] Good manners demanded that the glass should be taken in both hands, and small wonder – some of these *Humpen* held as much as five quarts.[92]

One glass in the Waddesdon Collection (*48*) was designed for a special mode of drinking favoured in Germany. This was drinking from the *Passglas*, where the cylindrical body of the vessel is divided by horizontal bands, whether in the form of applied threads or painted lines. The toper had to drink down to the first *Pass*, but if he failed to hit it he had to proceed on to the next, with obvious possibilities of deep drinking.[93]

[87] R. J. Charleston, 'Three Glasses of Welcome', *Annales du 5e. Congrès de l'Association Internationale pour l'Histoire du Verre*, Liège, 1972, p.222.
[88] A. von Saldern, op. cit., pp.27–9. See also *47* here.
[89] R. J. Charleston, loc. cit., p.218.
[90] Ibid., p.223
[91] A von Saldern, op. cit., pp.28–9.
[92] Ibid., p.29.
[93] Ibid., pp.145–7.

Coloured and opaque-white glasses

The first of all systematic glass-making text-books, Antonio Neri's epoch-making *L'Arte Vetraria*, first published in Florence in 1612, was translated into German, with copious annotations, by one Johann Kunckel, whose book, *Ars Vitraria Experimentalis* came out in a first edition in 1679, and in a second enlarged version in 1689. Neri had devoted much space to the colouring of glasses, although most of the colourants were already known from experience to the majority of practising glass-makers – cobalt for blue (*40* and *50* here), copper for green (*49* here), manganese for purple, etc. Far more difficult and arcane was the production of a ruby-red by means of a precipitate of gold. The credit for this is usually given to Kunckel, although, as will be seen, others before him had had sporadic success in this field.

Johann Kunckel (?1630–1703), a member of an old German glass-making family,[94] was appointed Private Groom of the Chamber (*Geheimer Kammerdiener*) to the Elector of Brandenburg in 1678. His function was not that of a normal Court official but of a chemical research-worker, and his investigations were specifically directed towards the ancestral field of glass-making, in which he already had considerable experience.

In 1679 Kunckel took a lease of the glasshouse at Drewitz (south-east of Potsdam), where he had been at work since August, 1678; and a few years later occupied a glasshouse built specially for him in the neighbourhood of the latter town.[95] He had probably already been making ruby-glass during the Drewitz period, for he mentions it in his *Ars Vitraria Experimentalis*, published in 1679.[96] In 1685 the Elector made over to Kunckel an experimental establishment on the Pfaueninsel, on the River Havel to the north-east of Potsdam, and it is likely that he worked here in secret on the perfection of his ruby-glass, for which he received a special patent from his Electoral master.[97]

Although it is likely that Kunckel was the first glass-maker to succeed in making ruby-glass consistently and in quantity, there had been in fact a long previous history of sporadic successes in colouring glass red with gold.[98] It has to be remembered that in the Middle Ages, and thereafter

94 W. Mylius, 'Johann Kunckel', *Keramische Rundschau*, *35*, 1927, p.38; H. Maurach, 'Johann Kunckel (1630–1703)', Deutsches Museum, *Abhandlungen und Berichte*, 5. Jahrg., Heft 2, 1933, pp.33–4; Karl Hucke, 'Oswald Kunckel – Glasmeister auf Stendorf', *Jahrbuch für Heimatkunde im Kreis Oldenburg/Holstein*, 1962, pp.101–5; id., 'Glasmacherei', in (ed. E. Schlee) Catalogue of the Exhibition *Die Gottorfer Kultur im Jahrhundert der Universitätsgründung*, Flensburg, 1965, pp.428–34; this authority suggests that the date of Kunckel's birth must lie between 1637 and 1642, and not about 1630, as traditionally given (loc. cit., p.432).

95 R. Schmidt, *Brandenburgische Gläser*, Berlin, 1914, pp.27 ff.

96 Johann Kunckel, *Ars Vitraria Experimentalis oder vollkommene Glas-macher-Kunst*, Amsterdam and Danzig, also Leipzig and elsewhere, 1679, p.101. A second edition appeared in 1689 (Frankfurt and Leipzig), a third in 1743 (Nuremberg), and a fourth in 1756 (*Johann Kunckels . . . vollständige Glasmacher Kunst . . . Nürnberg . . . bei Christoph Riegels seel. Wittib.*). A French translation (by the Baron Holbach) appeared in Paris in 1752 as *L'Art de la Verrerie de Neri, Merret et Kunckel.*

97 R. Schmidt, *Das Glas*, pp.308–9. Recent excavations on the island have brought to light fragments of many types of glass, including some of ruby. These do not appear to correspond to known types of ruby vessel-glass. Rib-moulded glasses of other colours, however, were found (kindly communicated to me by Dr H. Günter Rau – see also his article 'Das Glaslaboratorium des Johann Kunckel auf der Pfaueninsel in Berlin', *Ausgrabungen in Berlin*, *3*, 1972, pp.148–71; this, however, does not include mention of the ruby fragments, which were found in a subsequent season's digging).

98 Ludwig F. Fuchs, 'Johann Kunckels Erfindung des Goldrubins', *Der Kunstwanderer*, December, 1928, pp.153–8; Herbert Kühnert, 'Das Gold-Rubin-Glas im Lichte der "Alchemia" des Andreas Libavius', *Jahrbuch der Coburger Landesstiftung*, 1968, pp.197–220.

into the seventeenth century and even later, gold was a metal regarded as having special magical properties, and the same was true of the ruby, which brought its possessor strength and health, and ensured him against witchcraft.[99] It was believed that a glass coloured red with gold was in effect a ruby.[100] Transformations of this kind were part of the alchemical theory of the Middle Ages and Renaissance, and it is no surprise that Kunckel, before he turned seriously to glass-making, was in fact an alchemist.[101] The essential feature in Kunckel's formula was the use of gold precipitated by means of tin, a process perfected earlier in the century by Andreas Cassius of Hamburg. In a later book (*Laboratorium Chymicum*, 2nd edition 1722) Kunckel expressly gives Cassius the credit for this, but suggests that he was unable successfully to exploit his method because he did not fully appreciate the necessity of re-heating the glass to get the gold to 'strike'.[102]

In this same passage, Kunckel, who died in 1703, refers to the fact that the making of ruby-glass had become common-place, and it is now known that a number of centres did in fact produce it. As early as 1699 Hans Christoph Fidler, who in 1677 had started a glasshouse in Munich for the Elector of Bavaria, reported that he was in a position to make ruby-glass, and later in the year he presented his princely patron with two glasses as a proof of his ability.[103] A glasshouse at nearby Freising was stated in a contemporary source (1684) to have produced gold ruby-glass,[104] and the Nuremberg faience- and glass-painter Abraham Helmhack is reported to have revived the process in 1717: it is unlikely, however, that much was produced by Helmhack's formula.[105] The precious substance was certainly also produced in Saxony, probably in Dresden;[106] and possibly also in a glasshouse at 'Schlackewerdt' (Schlackenwerth, between Joachimsthal and Karlsbad, in Bohemia), owned by the Duke of Sachsen-Lauenburg, who had suborned a member of Kunckel's staff.[107] Another of Kunckel's men may have divulged the secret at Bayreuth.[108] Ruby-glass seems also to have been made in various other centres in Bohemia – at Reichstadt (Zákupy), also a glasshouse belonging to the Duke of Sachsen-Lauenburg (before 1688); on the Eisenstein (Železná Ruda) estate, belonging to the

[99] Fuchs, loc. cit., p.153.
[100] Ibid., p.154.
[101] R. Schmidt, *Brandenburgische Gläser*, pp.28–9.
[102] Kunckel's formula for making ruby-glass is probably reflected in the later prescription preserved in the Zechlin glasshouse – see R. Schmidt, *Brandenburgische Gläser*, p.63. This question is briefly discussed by R. W. Douglas and S. Frank, *A History of Glassmaking*, Henley-on-Thames, 1972, pp.18–19. The first edition of the *Laboratorium Chymicum* appeared in 1716.
[103] R. Berliner, 'Eine Münchner Glashütte im letzten Viertel des 17. Jahrhunderts', *Münchner Jahrbuch der bildenden Kunst*, N.F., Bd. I, 1924, pp.112–13. Fidler may have picked up this knowledge in the Juliusthal glasshouse at Reichstadt, where he was from December 1687 until early in 1688 (kindly communicated by Dr Olga Drahotová – see also n.108 below).
[104] F.-A. Dreier, 'Fünf Silbermontierte Prunkvasen aus Goldrubinglas', Berliner Museen, *Berichte aus den Staatlichen Museen des Preussischen Kulturbesitzes*, N.F. XIX, Heft 2, 1969, pp.81–3: R. Berliner, loc. cit., pp.118–19, n.21.
[105] Fuchs, loc. cit., p.155, quoting J. G. Doppelmayer, *Historische Nachricht von den Nürnbergischen Mathematicis und Künstlern . . .*, Nuremberg, 1730.
[106] W. Holzhausen, 'Sächsisches Rubinglas und Steingefässe von J. F. Böttger', *Belvedere*, XII. Jahrgang, 1934/37, pp.16–22. The typical Saxon vessels seem to have been in ruby cased with crystal, or vice versa.
[107] Fuchs, loc. cit., p.157; F.-A. Dreier, 'Ein Paar Schraubflaschen', *Europäisches Kunsthandwerk, Kataloge des Kunstgewerbemuseums Berlin*, IV, Berlin, 1970, No.121.
[108] Ibid.: Olga Drahotová, in 'Bohemian Ruby Glass at the turn of the 17th and 18th Centuries', *Glass Review*, 4, 1973, pp.8–11, suggests that Kunckel's words 'bayreuther Hof' ('Court of Bayreuth') may have been corrupted from 'Bayrischer Hof' ('Bavarian Court') referring to the Munich glasshouse.

Count Heinrich Nothaft, where Hans Fidler was employed from 1691 until 1695; at Schreiberhau, in Silesia, in the glasshouse of Johann Christoph Preussler, where the secret appears to have become known about 1700; and at other centres later in the eighteenth century.[109]

With so many centres potential producers of ruby-glass, it is evident that attribution must be very difficult. Originally, almost all ruby-glass was credited to Kunckel at Potsdam, but this portmanteau ascription soon became untenable.[110] It was observed that many of the silver-gilt mounts on ruby-glass objects bore the hallmarks of South German silversmiths and cities – Nuremberg, and especially Augsburg. Many of these glasses were decorated with vertical mould-blown ribbing, whereas the certainly Potsdam pieces were wheel-engraved.[111] It has also been observed that the records of the Munich glasshouse frequently refer to 'ribbed' glasses, while there is mention of a ribbed mould amongst the equipment there.[112] It therefore seems reasonable to assume, at least as a working-hypothesis, that such pieces were made in South Germany and sent for mounting to the famous metal-working cities of Nuremberg and Augsburg.

A number of ruby-glasses at Waddesdon bear these silver-gilt mounts, among them pieces by the Augsburg masters using the mark 'TB' in an oval (*53, 54*) and 'MB' in an oblong label (*56*): both silversmiths made a speciality of this kind of work, for their marks are the most frequently-recurring on the mounted ruby-glasses. It is clear that the ruby-glasses themselves in many instances were made expressly to be mounted, since their shapes are incomplete in themselves (see *51*, *52* and *56*).

In one other field Kunckel again appears to have played an important part as a standardizer and popularizer, if not as inventor. This was the making of opaque-white glass.

The opacifiers used in glass have varied throughout history in accordance with time and place. In dynastic Egypt, and later in the early years of the Roman Empire, an antimony compound was used to produce a milk-white glass.[113] At some point about A.D. 400 its place was taken by tin-oxide,[114] and this was the standard opacifier of the Venetian glass-industry, prescribed by Neri.[115] In the eighteenth century a lead-arsenic-oxide compound supervened both in Venice and in England.[116] The method prescribed by Kunckel, however, differed from all of these. In an appendix to his commentary on Book I of Neri, he adds a section on 'The preparation of beautiful Porcelain-glass', the secret of which he freely admits was communicated to him by a certain Daniel Krafft, who was the inventor, and who had worked for the Elector of Mainz. The recipe consisted of compounding 10 lb of burned bones or stag's horn with 60 lb of white flint or fine sand, and 40 lb of purified potash. This mixture would come clear and colourless from the furnace, and needed to be re-heated, when it turned opalescent

109 Drahotová, loc. cit. in previous note.

110 cf. R. Schmidt, *Die Gläser der Sammlung Mühsam*, Neue Folge, Berlin, n.d. (1926), p.47.

111 See e.g. R. Schmidt, *Brandenburgische Gläser*, Pls. 16, 4 and 5; 17, 2 and 4, etc.

112 Berliner, loc. cit., p.120: Dreier, loc. cit. in note 104, p.82.

113 W. E. S. Turner and H. P. Rooksby, 'A Study of the Opalising Agents in Ancient Glasses . . .', *Glastechnische Berichte*, VIII, *Sonderband, V. Internationaler Glaskongress*, 32K, 1959, pp. VIII, 17–28.

114 H. P. Rooksby, 'Opacifiers in Opal Glasses through the Ages', *G.E.C. Journal of Science and Technology*, *29*, No. 1, 1962, pp.23–4.

115 *L'Arte Vetraria*, Book III, Chaps. 54–5.

116 W. E. S. Turner and H. P. Rooksby, loc. cit., pp.21, 23–5, 27; H. P. Rooksby loc. cit., pp.25–6.

or opaque-white in accordance with the amount of bone used and the duration of the reheating.[117]

Ever since the early years of the sixteenth century opaque-white glass, usually with enamelled decoration, had been made to simulate Oriental porcelain, mainly in Venice.[118] By the seventeenth century the porcelain-mania had spread to northern Europe, and Kunckel's characterization of his opaque-white glass as *Parcelleinglass* is revealing. In the records of the Munich glasshouse between 1677 and 1681 the substance is referred to as *Porzelän*, and very numerous objects of it were made there.[119] In a book published in 1692 a writer on the Fichtelgebirge District of Franconia refers to the glasshouses in Bischofsgrün where 'all sorts of common, but also beautiful porcelain- and crystal-glass' were manufactured ('allerley gemeines, auch schönes Porcellan- und crystall-glas gemachet . . .').[120] That opaque-white glass was made in a number of centres in Bohemia and Germany well before this, however, seems demonstrated by the existence of several earlier dated objects, even if their provenance is sometimes uncertain. Particularly precocious is a goblet of distinctive Bohemian form enamelled with a figure of St Peter and dated 1636.[121] A jug dated 1651 in the British Museum seems both in form and decoration to be of 'Central European' character,[122] and a jug dated 1685 in the Helfried Krug Collection is painted with black enamel (*Schwarzlot*) in a style which suggests a Bohemian origin.[123] An opal beaker dated 1681 in the British Museum is enamelled in a style indicating Saxon manufacture.[124] Similarly, one or two pieces of opaque-white glass with unmistakable Nuremberg enamelling of the third quarter of the seventeenth century suggest that the material was current during this period in South Germany.[125] With so many potential centres of production, attribution of a particular glass can only be based on its own intrinsic characteristics of shape and decoration (*60* and *61*).

117 J. Kunckel, op. cit., 2nd ed., 1689, pp.57–8. In the first edition (1679) Kunckel had specified merely 'The ashes of burned houses and barns' (p.68). An opaque-white glass used as an enamel on an Islamic glass of the fourteenth century has been shown to be a calcium fluophosphate, and therefore essentially of the type of Kunckel's glass (W. E. S. Turner and H. P. Rooksby, loc. cit., pp.21 (No.6a), 24 and 25): further examples dating between the fifth and seventeenth centuries are cited without details by H. P. Rooksby, loc. cit., pp.24–6.

118 T. H. Clarke, 'Lattimo – a Group of Venetian Glass enameled on an opaque-white Ground', *Journal of Glass Studies*, XVI, 1974, pp.22–36.

119 R. Berliner, loc. cit., pp.117 ff.

120 J. Willen, *Das Teutsche Paradeiss in dem vortrefflichen Fichtelberg*, cit. A. von Saldern, *German Enameled Glass*, pp.184–5.

121 Karel Hetteš, 'Venetian Trends in Bohemian Glassmaking in the 16th and 17th centuries', *Journal of Glass Studies*, V, 1963, p.52, Fig.26; also in colour, Fig.1 in A. Fricky, 'Remains of Ancient Glassworks in the Vicinity of Bardejov, Slovakia', *Glass Review*, *3*, 1970, p.86.

122 British Museum, *Masterpieces of Glass*, p.186, No.261.

123 B. Klesse, *Glassammlung Helfried Krug*, Munich, 1965, p.166, No.140.

124 British Museum, *Masterpieces of Glass*, p.186, No.262. Its Saxon origin seems assured by two parallel pieces, one also of opal glass, with Saxon Electoral arms (A. von Saldern, *German Enameled Glass*, pp.204–5, Figs.363 and 365). A second beaker of the same type and date (1681) is in the Museum at Frankfurt-am-Main (*Europäisches und Aussereuropäisches Glas*, No.456).

125 e.g., a little cup attributed to Johann Schaper (d.1670) in the Victoria and Albert Museum (Honey, *Glass*, p.80); a rectangular bottle with screw-top, painted in enamels probably by Georg Strauch (d.1675) – see I. Schlosser, *Das Alte Glas*, Pl. IV).

Bohemian wrought and engraved glass of the late seventeenth century

Much of the glass in the *façon de Venise* had been made in a material basically similar to that used in Venice itself – a soda-lime glass in which the fluxing agent was an alkali derived from the burning of marine vegetation (soda). In Bohemia, however, a quite good-quality colourless glass was produced before the end of the sixteenth century by the traditional indigenous method of using the ashes of local plants (potash).[126] This development coincided fortunately (perhaps it was no coincidence) with the first European essays in wheel-engraving on glass since Roman times. Wheel-engraving was executed by means of a treadle-driven spindle into which were fitted metal (usually copper) wheels of varying sizes and profiles, the actual cutting being effected by an abrasive powder suspended in a greasy medium which was applied to the wheel. By this means cuts of varying shapes could be made into the surface of the glass which gave an illusion of depth and modelling when held to the light. After the initial cuts, the rough surfaces were polished as required by the use of progressively softer wheels (of wood, cork, etc.) and finer abrasives.[127]

The techniques of wheel-engraving were taken over from the workshops of the engravers of hardstones, and it is natural enough that the earliest glass-engravers should also have been hardstone-engravers.[128] The best-known, although it is by no means certain that he was the first, engraver to transfer his skill from hardstones to glass, was a certain Caspar Lehmann, a native of Ülzen, in the territory of Lüneburg, who left home in 1586, to turn up two years later in Prague. In 1601, he was there given the title of 'Imperial Gem-engraver'. After a sojourn at the Court of Saxony in 1606–08, he was back in Prague in 1608 as 'Imperial gem-engraver and glass-engraver', a significant addition to his title. In 1606 Lehmann is recorded in the Saxon archives as having engraved four'Christellin Tafflen' ('crystalline panels') with portraits of members of the ruling house, and since such panels survive and are exclusively of glass, it is reasonable to ascribe them to Lehmann.[129] They give us a good idea of his style, which is confirmed by a signed glass beaker, dated 1605, in the Museum of Applied Art in Prague.[130] Lehmann was already far advanced in the development of his technique, building up his effects by the juxtaposition of cuts of different depths, thus achieving the three-dimensional effect of modelling, rather than the linear effect of a purely graphic method, such as diamond-point engraving.

Lehmann had a number of contemporaries who may well have practised the art of wheel-engraving on glass, but we know nothing of their work. In 1609 Lehmann obtained an Imperial Privilege empowering him alone to practice the art of glass-engraving within the Imperial dominions, since he had 'discovered the art and business of glass-engraving, and put it into practice . . .' Whether this monopoly was effective we do not know, but in fact most of the glass

126 K. Hetteš, loc. cit. in n.121, pp.39–42.
127 R. J. Charleston, 'Wheel-engraving and -cutting: some early Equipment, I. Engraving,' *Journal of Glass Studies*, VI, 1964, pp.83–100.
128 Ibid., pp.88–91.
129 Ibid., p.91; Z. Pesatová, *Bohemian Engraved Glass*, Feltham, 1968, pp.10–16; E. Meyer-Heisig, 'Caspar Lehmann: ein Beitrag zur Frühgeschichte des deutschen Glasschnittes', *Anzeiger des Germanischen Nationalmuseums*, 1963, pp.116–31.
130 Ill. Pesatová, op. cit., Figs.3–6; K. Hetteš, *The Bohemian Glass*, (?) Jablonec, n.d., Fig.9; O. Drahotová, *Bohemian Glass*, Prague, 1970, Fig. on p.34.

engravers of the next generation seem to have come from Lehmann's orbit. Most important of these was Georg Schwanhardt, who inherited his master's privilege on the latter's death in 1622, and moved back to his native city of Nuremberg, there to found a school of glass-engravers which dominated the German scene until the early years of the eighteenth century.[131] Another pupil may have been Georg Schindler, who was in Prague in 1610, but who later became, like two other members of his family, a citizen of Dresden.[132] A further pupil may have been Johann Hess, who is alleged to have fled from Bohemia to Frankfurt-am-Main during the Thirty Years' War, and who founded a dynasty of glass-engravers there which lasted into the eighteenth century.[133] Ironically, less is known of the vicissitudes of the art in Bohemia itself. The political and religious upheavals leading to the Battle of the White Mountain in 1620, and the subsequent ravages of the Thirty Years' War left Bohemia depopulated and in virtual economic ruin. Little is known of the glass-industry and the art of engraving in this period, but after the Treaty of Westphalia (1648) strenuous efforts were made by the great landowners to improve the economy of the country, and in this endeavour the revival of the glass-industry, which made profitable use of the woodlands, played a predominant part. During the second half of the seventeenth century the industry increased to such a point that the home-market could not absorb all its products, and the foundations of a flourishing export-market were laid.[134]

Little is known of the means by which glass-engraving was kept alive or revived during the troubled years of the mid-century. Apart from Caspar Lehmann's pupils, however, who appear to have scattered to various parts of Germany, there was one other possible source of the necessary knowledge and skill. This was in the workshop of the Miseroni family of rock-crystal and hardstone engravers in Prague. At least three members of the Miseroni family moved to Prague at the end of the sixteenth or in the early years of the seventeenth century,[135] the most important of whom was Ottavio. More significant in the present context was his son Dionisio, who had worked for the Emperor Rudolph II and continued to work for the Emperor Ferdinand III. Ottavio had been created Keeper of the Treasury (*Schatzmeister*) and Dionisio in turn was made Warden of the Treasury and an Inspector of Prague castle. It may well have been these Imperial functions in Prague which held the Miseroni family there at a time when others were leaving. Dionisio died in 1661 and was succeeded by his son Ferdinand Eusebio, until the latter's death in 1684 brought an end to the dynasty.[136]

The Miseroni workshop was at Bubeneč, on the outskirts of Prague, and this fact may be taken to lend some substance to the suggestion that Martin Friedrich, who owned a glassworks at Kreibitz in northern Bohemia, and was alleged to have learned the processes of glass-making at Bubeneč in 1605, might also have become familiar with the technique of glass-engraving

[131] E. Meyer-Heisig, 'War Georg Schwanhardt wirklich der erste Glasschneider Nürnbergs?', *Bewahren und Gestalten. Festschrift für Günther Grundmann*, Hamburg, 1962, pp.103 ff.
[132] Pesatová, op. cit., p.12; W. Holzhausen, 'Dresden-Prager Glas- und Steinschnitt um 1600', *Neues Archiv für Sächsische Geschichte und Altertumskunde*, Dresden, 1934, pp.86–118. (See p. 37).
[133] Schmidt, *Das Glas*, pp.254–7.
[134] Ada Polak, *Glass: its Makers and its Public*, London, 1975, pp.106–11.
[135] Giovanni Ambrogio, appointed Imperial gem-cutter in 1593, and certainly in Prague by 1600; Alessandro, who worked in Prague 1605–12; and Ottavio, brother of Giovanni Ambrogio, employed in Prague from 1588 until his death in 1624. He was ennobled in 1608 with the name 'de Lisone' – see Jana Hofmeisterová, 'A Family of Italian Rock Crystal Carvers in Prague', *Glass Review*, *4*, 1970, pp.118–22.
[136] Ibid.

before moving to northern Bohemia.[137] Friedrich died in 1612, however, and a wide gap yawns between this date and the middle years of the century. A further thread of continuity has been discerned in the career of Georg Schindler, who was no doubt a member of the prolific Dresden family of glass-engravers of that name. This Georg Schindler is stated to have been a pupil of Lehmann's and an important glass-engraver at the Grünewald glasshouse, near Gablonz, in northern Bohemia, before finally settling in Dresden. Since this, however, appears to have taken place in 1628, it is hardly here that a connecting link is likely to be found.[138] It may well, therefore, be in the Miseroni workshop that the necessary continuity of tradition should be sought. In a picture of the workshop painted in 1653, four huge cutting-wheels may be seen, with six or seven workmen employed about them, at least one of whom appears to be an engraver.[139] Furthermore, the work executed in the Miseroni workshop included not only the facetting of hardstones, but actual engraved work of great refinement.[140] It is perhaps therefore not fanciful to see in this repository of skills the source from which sprang the abundance of glass-engraving talent which appeared in northern Bohemia in the second half of the seventeenth century. Be that as it may, it appears that the glass-engraving technique was established at the Dobrá Voda works on the Nové Hrady estate of Count Buquoy in 1650,[141] and a goblet once in the Schlossmuseum, Berlin, an obvious forerunner of later Bohemian goblets in shape, is engraved with a coat-of-arms, and dated 1648.[142]

An indispensable precondition for the full development of Bohemian glass-engraving in the second half of the seventeenth century was the evolution of a substantial, water-clear and colourless glass-material, as near as possible to the rock-crystal in the engraving of which at least some of the earliest master-engravers had probably been trained. The 'crystal' formula evolved in the course of the seventeenth century on the basis of the potash-composition which was already current in Bohemia before 1600. (The extent to which the unperfected traditional material could fall short of this ideal is exemplified by the two Waddesdon glasses with wrought decoration, *62* and *63*.) The essential improvement consisted in the addition to the batch of appreciable quantities of chalk or lime. A book of 1679 refers to glasses 'which in transparency and brilliance came near to crystal', made in the Neuschloss glasshouse of Count Kaunitz (near Falkenau). By the date of the book, however, this concern had already been closed down, and the famous glass-merchant and decorator Georg Franz Kreybich, of Steinschönau, related in 1686 that no high-quality glass was made in this general district, and that he was obliged to get his supplies from the Schreiberhau district in the Hirschberger Tal, on the Silesian side of the Riesengebirge.[143] The earliest specific reference to a lime-rich glass, however, seems to occur in southern Bohemia, where at the Helmbach glasshouse, on the Vimperk estate (not far from Budějovice), Michael Müller, an exceptionally able glassmaker, developed a formula

137 Pesatová, op. cit., p.16.
138 Ibid., p.12, citing *Mittheilungen des nordböhmischen Excursions-Klub*, *48*, p.86; see also O. Drahotová, op. cit., pp.32, 35.
139 Ill. Pesatová, op. cit. Fig.3 (detail); R. J. Charleston, 'Wheel-engraving and -cutting ...', Fig.17.
140 Erich von Strohmer, *Prunkgefässe aus Bergkristall*, Vienna, 1947, Figs.31 ff.
141 O. Drahotová, op. cit., p.35.
142 R. Schmidt, *Das Glas*, Abb.144.
143 Ibid., pp.139–40.

in 1683.[144] It is noteworthy that whereas Kunckel in the first edition of his book (1679) gives only a silica-potash recipe for 'crystal glass' (although hinting that he had a better formula which he could not divulge), in the second edition (1689) he gives two recipes which involve the deliberate addition to the batch of considerable quantities of chalk.[145]

From 1660 onwards the development of wheel-engraving in Bohemia is far better documented. In 1661 the formation of a guild principally of glass-painters was authorized at Kreibitz by Count Johan Octavianus Kinsky, and was confirmed by his son in 1669. Some of these painters also qualified as glass-cutters or engravers – such as David Schlegel, Gabriel Preussler and Georg Hille, Masters of the guild in 1661, 1663 and 1670 respectively. The guild included glassmen from the neighbouring towns or villages of Böhmisch Kamnitz, Blottendorf, and Steinschönau, and the earliest recorded Master who was solely a glass-engraver was a certain Georg Thomass of Kamnitz, Master in 1669. Similar guild-organizations were formed on other estates, such as Bürgstein (Sloup), authorized by Count Ferdinand of Kokořov in 1683, or Steinschönau, which achieved a guild of its own in 1694.[146] These organizations for regulating the glass-decorating industry were not wholly successful in one important respect – they could not prevent a proliferation of decorators so extensive that until the glass export-trade became well-organized towards the end of the seventeenth century, there was not work enough to support them all. It was recorded of neighbouring Silesia, which had parallel guild-organizations: 'glass engravers and also cutters are so numerous that the one spoils it for the other and hardly anyone can earn a proper living, and each should teach two or three youngsters, but when they have brought it so far as to be able to do a bit of scratching, they run away from the masters and set up with their own tools and bungle away, each one for himself'.[147] In view of this proliferation of glass-engravers and the dangers of lowered standards inherent in such a system, it is not surprising that much of the engraved work on Bohemian glasses of the late seventeenth century is of a somewhat perfunctory and routine character.[148] The Waddesdon glasses decorated by wheel-engraving (*64* to *66*) do not rise markedly above this level.

144 Ibid.; O. Drahotová, op. cit., pp.40–1.

145 R. Schmidt, *Das Glas*, p.140 – see J. Kunckel, *Ars Vitraria Experimentalis*, notes to the thirty-sixth Chapter of the First Book of Neri (1679 ed., pp.66–8; 1756 ed., pp.53 and 55). Kunckel adds frankly that he is in the second edition of his book prepared to divulge recipes which nothing would have induced him to publish in the first; from this one may infer that in this decade the chalk additive had become fairly well known.

146 Details of the guild-organizations may be found in Jablonec nad Nisou, Muzeum Skla a Bižutérie, *Sklo Borských Výtvarníků* (Introduction by K. Hetteš), Jablonec, 1964, pp.17–22; Pesatová, op. cit., pp.18–20; V. Ryneš, 'Czech Glass Cutters and Carvers at the beginning of the 18th century', *Czechoslovak Glass Review*, 19, No.9, 1964, pp.264–8; id., 'Aus der Galerie der Veredler des böhmischen Glases in Steinschönau . . .', *Ars Vitraria*, I, 1966, pp.121–6 (especially pp.123–5).

147 Cit. A. Polak, op. cit., p.106. This author gives the best extant account in English of the development of the Bohemian glass-industry at this period (pp.105–11).

148 Pesatová, op. cit., p.20, has pointed out that the superficial 'painterly' character of this shallow engraving may be due to the fact that many of the engravers were in fact painters by training.

Bohemian glasses of the eighteenth century decorated in gold

A rich vein in the Waddesdon Collection is furnished by the glasses with decoration formed by gold-leaf sandwiched between two layers of glass in the wall of a vessel (*Zwischengoldgläser*). No less than nine beakers represent this method (*67–75*). The technique of using a gold-leaf etched with a point as embellishment of glass was known in Antiquity, and revived in Italy in the fourteenth century at latest, since which time it has hardly at any time been in abeyance, normally being used as a decoration of sheet-glass. Here, however, the leaf was not sandwiched between two layers of glass as it had been in Antiquity, and as it began to be once more in Germany from the end of the seventeenth century onwards. The first mention of it occurs in Kunckel's *Ars Vitraria Experimentalis* (1679): 'Take two plain glasses, which fit snugly into each other, which also, especially in point of height, are so fashioned that the inner glass does not exceed the outer in height, but both are equally tall; paint the larger glass on the inside with oil-colours in the manner of a semiprecious stone as best you can; let it dry, then scratch in it veins or what you will this way and that with a sharp graduating-needle. Then swill it round with old linseed-oil, pour away the surplus, and let it get almost dry upside-down; then when it is tacky lay on fine leaves of gold or other metal, and press them from the inside with a piece of cotton-wool, and then let it dry out well; then the veins you have etched appear filled with gold. In the meantime, take the second or smaller glass, paint it as thinly as possible by means of a brush with old clear linseed-oil or pure varnish, and cover it thoroughly with gold-leaf or other metal foil, so that from the inside it resembles a gilt beaker; let it too dry off, and put the glasses one inside the other . . .'[149] Kunckel explains that the glasses must fit snugly, and the join at the rim must then be cemented with a lacquer-varnish so that it is imperceptible, and finally be gilt with leaf-gold. Glasses of this kind survive, but the drawback with them must have been that despite all precautions, liquid or air must have got between the layers and produced blisters or other defects. A new method was therefore devised by means of which the inner glass projected above the level of the outer, this upper projection being thick enough to cover the edge of the outer glass. When the gold-leaf decoration was completed, the outer glass was slipped over it and probably cemented: then the two together were facetted vertically on the glass-cutter's wheel. The outer shell, which was open at the bottom, projected beyond the inner glass below, and the cavity was sealed by a disc specially cut to fit. This in turn was decorated with an etched gold leaf, usually washed over with a transparent crimson-pink (rarely green) varnish, and cemented into place (*see diagram overleaf*). The fact that the join was on the outside below the rim

[149] J. Kunckel, *Ars Vitraria Experimentalis* . . ., Part II, Frankfurt and Leipzig, 1679, p.12, Chap.XXVII. 'Ein sonderliches curieuses Trinckglas zu machen': 'Nimm zwey glatte Gläser, welche sich gerade in einander fügen, welche auch, sonderlich was die Höhe betrifft, also beschaffen seyn, dass das innere Glas, dem eusern ja nicht an der Höhe vorgehe, sondern beyde gleich hoch seyn; mahle das grössere inwendig mit Oehlfarben nach Edelgestein-Art auffs beste als du kanst; lass es trocken werden, alsdenn reisse mit einer spitzigen Gradiernadel hin und wieder, Aederlein oder was du wilt darein. Ferner schwäncke altes Leinöhl darinn herumb, lasse es wieder wohl heraus lauffen, und umbgestürtzt fast trocken werden; wann es demnach ein wenig klebrigt ist, so lege Blättlein von Gold oder Metall hinein, drücke sie mit einer Baumwollen inwendig an, und lass es folgends wohl austrucknen, so scheinen die gerissenen Aederlein gold-reich heraus. Indessen nimm das andere oder kleinere Glas, streiche es auch vermittelst eines Pensels mit alten klaren Leinöhl oder einen reinen Fürniss auffs dünste an, und belege es über und über mit geschlagenen Gold oder Metall, so sihet es von inwendig einen verguldeten Becherlein gleich, lasse es auch trocken werden und setze sie in einander . . .'

helped to ensure that liquids did not penetrate between the layers when the glass was used for drinking.

The majority of these glasses are decorated with etched gold-leaf alone (*67–72*), but some have silver-leaf, often combined with crimson and other coloured varnishes (*74*), and others have silver- and gold-foils combined (*73–5*). Very much rarer are the glasses where the metal-foils are combined with a zone of wheel-engraving (*75*).

The *Zwischengoldgläser* span a considerable stretch of time.[150] A beaker shown in the Prague Exhibition of 1891 was dateable by a chronogram to 1728, while another in the Berlin Kunstgewerbemuseum was dated 1753.[151] The extremely individual character of this painstaking and laborious work suggests that it was not done in the way of commerce, but that glasses were made for presentation to important people, as their inscriptions and coats-of-arms (*73*)

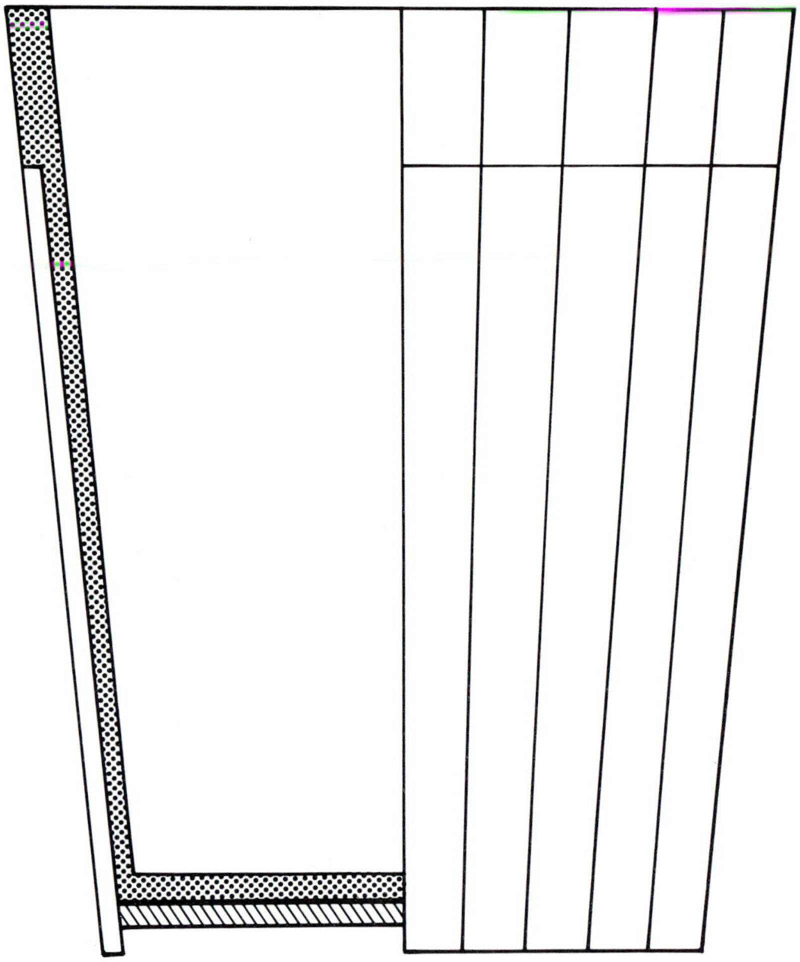

Diagram showing the cross-section of a 'Zwischengoldglas'. The hatched portion represents the inserted base-disc. (Not accurately to scale).

[150] The dating and attribution of the 'Zwischengoldgläser' are most fully discussed by G. E. Pazaurek, 'Die Heimat der Zwischengoldgläser', *Mittheilungen des Nordböhmischen Gewerbe-Museums*, XVI Jahrgang, 1898, Nr.3, pp.53–60.

[151] Illustrated by R. Schmidt, *Das Glas*, Fig.221, left.

often suggest. Apart from the very numerous hunting-scenes (*69–73*), there is a considerable body of glasses with religious motifs (*72*). The greater number of the saints represented on these glasses are reckoned among the patron saints of Bohemia (e.g., St John Nepomuk, St Wenzeslaus, etc.); when identifiable church-views occur, they seem to be of Bohemian buildings.[152] The name 'Woborzill' appearing on a glass of 1740 also suggests a Bohemian origin,[153] as does the fact that the glasses are almost invariably facetted, a speciality of the Bohemian glass-cutters (*Eckigreiber*). The prevalence of religious themes has suggested that one or more religious houses in Bohemia may have sheltered workshops producing these glasses. That most of the inscriptions are in German suggests that the centre or centres of production were located in the German-speaking periphery of Bohemia.[154] The only known inscription which may record the name of an artist mentions 'P. PACIFICUS KLIEGEL', a name which has an unmistakable German ring.

A far more economical way of producing glasses with gold decoration was the time-honoured method of surface-gilding by means of gold-leaf or precipitated gold.[155] This technique is represented at Waddesdon by a large table-service of twenty-nine pieces, at first sight homogeneous in character, but on closer inspection revealing discrepancies which suggest that its components originally came from a number of different services(*76*). The larger decanters (*76, 1–4*) have an idiosyncratic gilt decoration of *rococo* architectural 'ruins' accompanied by tall stylized figures in bicorn or short top hats; the smaller decanters (*76, 5* and *6*, perhaps for spirits) have a gayer and more open decoration of pronounced *rococo* character, with little pastoral figures spreading their arms in characteristic gestures, while the handled bottles (*76, 7* and *8*) have a gilt decoration of pendant swags of a pronounced neo-classical character. In a similar way, there are differences of shape, size and decoration among the wine-glasses (*76, 9–23*) and cordial-glasses (*76, 24–29*). The wine glasses appear to fall into two groups differing in their decoration and roughly corresponding to the two decanter-styles referred to above (*76, 9–18* in the first group, *76, 19–23* in the second): the cordial-glasses, perhaps because of the restricted area available for decoration, have no more than a simple building in a 'landscape' (*76, 24–27*), or an animal in flight (*76, 28* and perhaps *76, 29*). Within these groups there are differences of size and shape (on the wine-glasses, for example, the polygonal bowls vary from nine- to fourteen-sided).

It has long been thought that the glass table-service, in the sense of a set where the components are linked together by close similarities of shape and ornament, and where different types of glass were reserved for different purposes, is a relatively recent development, belonging essentially to the age of the gastronome in France, in the years after the Revolution, but especially in the early years of the nineteenth century.[156] In Bohemia, however, there is good evidence

[152] e.g. that at Haindorf in the Isergebirge, mentioned ibid., p.367.
[153] The 'Woborzill' beaker apparently comes from the Pauline convent at Woborzischt near Przibram (see G. E. Pazaurek, *Die Gläsersammlung des Nordböhmischen Gewerbemuseums in Reichenberg*, Leipzig, 1902, pp.22–5).
[154] The possibility is not excluded that other German centres may have been involved, especially in the production of some of the less standardized glasses.
[155] For the methods of doing this, see R. J. Charleston, 'Enamelling and Gilding on Glass', *The Glass Circle*, I, 1972, pp.21–4.
[156] Elisa Steenberg, *Svenskt Adertonhundratals Glas*, Stockholm, 1952, Chap. III ('Composition of Services'), pp.125–44.

that the making of suites of glasses, linked in shape and decoration but varying in capacity, began before the middle of the eighteenth century.[157] In the monastery of Rajhrad, not far from Brno, in Moravia, there were a number of services dating from the eighteenth century, of which two are of particular interest in the context of the Waddesdon service.[158] One consists of seventeen pieces, among which there are four sizes of stemmed wine-glasses and a footed water-glass: they are all decorated with vertical facets and a simple gold band at the rim. The second service[159] comprises three sizes of stemmed wine-glasses and a footed water-glass: all are vertically facetted and, for good measure, wheel-engraved with the arms of an unidentified abbot. Both sets are attributed to a date about 1760. Although there is no internal evidence to throw light on the functions of these various sizes of glasses, they may reasonably be deduced in the light of information gleaned from pattern-books of some twenty years later.

In the pattern-book (now preserved in the Museum at Jablonec nad Nisou) of the Harrachov glass-works, there are drawings representing a decanter (page 262) much like the larger Waddesdon decanters in shape, together with three stemmed goblets of different sizes, all held together as a set by the unity of their decoration. The wine-glasses in this set were intended for the consumption of champagne, burgundy and tokay respectively.[160] The same picture is repeated in patterns 176–78 of the same pattern-book, showing three diminishing sizes of the same glass, marked for champagne, burgundy and *rosali* (*rosoglio*) respectively.[161] All of these stemmed forms have ovoid bowls, but the same considerations apply to the straight-sided facetted glasses with shouldered facetted stems of the type which make up the Waddesdon set.[162] Patterns 121 and 122 of the Harrachov pattern-book (page 273) show a larger and a smaller glass of this form, the larger (No.121) clearly marked 'Champagnier', and the smaller (No.122) being presumably for burgundy. The same picture is confirmed and extended in yet another pattern-book, now preserved in the Henry Francis Du Pont Museum at Winterthur, Delaware, USA.[163] The date of this book is uncertain, but it may probably be placed on internal evidence somewhat nearer to 1800 than the Harrachov pattern-book which has already been discussed. Here are represented on one page two sets of three glasses in diminishing sizes, one set composed of wine-glasses with ovoid bowls and plain stems (Nos.66–8, referred to as *Römers*), the other of

[157] Jiřina Vydrová, 'Les débuts de la Différenciation des Types de Verre de Table en Bohême', *Annales du 5e Congrès de l'Association Internationale pour l'histoire du Verre, Prague, 1970*, Liège, 1972, pp.205–15. There is mention here (p.206, No.4c) of a set of glasses *en suite*, with engraved armorials of the families of Wimpfen and Thurn, giving a date about 1720; this set consists of a large covered goblet, four smaller goblets, and a small stemmed bowl (ibid., Fig.1; see also Victoria and Albert Museum, *Bohemian Glass*, London, 1965, No.13). There is, however, no indication of what these glasses were used for.

[158] J. Vydrová, loc. cit., p.206, n.4(f) and Fig.3.

[159] Ibid., p.206, n.4(f) and Fig.4.

[160] Ibid., p.210, n.6 and Fig.5. Further drawings show a similar set with the addition of two tumblers, presumably for beer and water respectively (p.214 and Fig.8). For further details of this pattern-book, see Stanislav Urban, 'Kapitoly z dějin novosvětské hutě hrabat Harrachů' ('Chapters from the History of the Nový Svět Glassworks'), *Ars Vitraria, 3*, Jablonec nad Nisou, 1971, pp.30–41 (English Summary, p.207).

[161] For *rosoglio*, see J. Vydrová, loc. cit., p.210. It was a sweet cordial based on raisins.

[162] On this 'Silesian' form, see ibid., p.214.

[163] See Dwight Lanmon, 'The Importation of Glass into the United States, 1780–1820', *Annales du 5e Congrès de l'Association Internationale pour l'Histoire du Verre, Prague, 1970*, Liège, 1972, pp.233–45, particularly p.244 and Figs.2–4, 6 ('Gardiner's Island Glass Catalogue').

glasses with the straight-sided facetted bowls on facetted inverted-baluster stems (Nos.69–71, page 273). In each case the smallest glass is indicated as being for 'liquer' (Nos.66 and 69); of the middle-size, the facetted glass (No.70) is called a 'wine-glass', while the largest facetted glass (No.71) is called a 'larger' wine-glass. In this catalogue the champagne-glass is represented as a tall tapering 'flute' on a square base, a further indication of a relatively late date. On this evidence, therefore, it may reasonably be assumed that the larger glasses at Waddesdon are for wine, the smaller for 'liquers', probably spirituous drinks of a wide variety of types, such as brandy and cordials.

Although it seems likely that the Waddesdon 'service' is in fact composed of parts of a number of similar services of much the same date, it may be that such sets were sometimes put up by dealers rather than in the glasshouses themselves. Dealers almost certainly made up the cases of spirit-'squares', glasses and funnels which still survive, as the occasional label demonstrates.[164] With this contingency to be taken into account, it seems not impossible that the components of this 'service' have been together since the late eighteenth century, although it seems far more probable that they were assembled in the nineteenth.

As for the glasses themselves, there can be little doubt that they are of Bohemian origin, as the appearance of identical forms in the Bohemian pattern-books demonstrates with reasonable conclusiveness. Glasses of the same type have been claimed for Venice,[165] and may be found in Italian collections; it should be borne in mind, however, that this was a period of extreme weakness in Venetian glass-making, and in this phase when the economic balance was tipped decisively against Venice, much Bohemian glass was certainly imported there, as it was to Holland and Spain.[166] Glasses of closely similar shapes and decoration, the latter executed in characteristic Bohemian enamel-colours, tend to clinch the ascription.

Glasses of this kind are peculiarly difficult to date with any precision. The Harrachov pattern-book to which allusion has been made, and drawings from which accompany this text, bears dates in 1787–8, and the Winterthur pattern-book appears to be of late eighteenth/early nineteenth century date. A goblet in private possession, with decoration of the same general character as appears on the Waddesdon glasses, bears the arms of Gregorio Barnaba Chiaramonti, Bishop of successively Tivoli (1782–85) and Imola (1785–1800). He became Pope as Pius VII in 1800 and since his arms show the twelve-tasselled hat of a bishop, it may reasonably be concluded that the glass was made for him between 1782 and 1800[167].

164 An example with the label of a Dutch retailer is in the Henry Francis Du Pont Museum at Winterthur – see Lanmon, loc. cit. in previous note, p.244 and Fig.5. In the pattern-book of a Bohemian glasshouse of about 1830–40, however, there is an illustration of such a cellaret, suggesting perhaps that they may also have been made up in the glasshouses themselves (see R. J. Charleston, 'A Glass Pattern-Book of the Biedermeier Period', *Proceedings of the 7th International Congress on Glass*, Brussels, 1965, paper 261, p.2 and Fig.3). A further example of a set of glasses presumably put up by a dealer – this time a 'nest' of beakers in a paper-lined leather case – is furnished by an *étui* marked 'FRIED. REICHENBACH' in the Museum für Kunsthandwerk in Frankfurt-am-Main (*Europäisches und Aussereuropäisches Glas*, p.207, No.454).

165 e.g. A. Gasparetto, *Il vetro di Murano*, Venice, 1958, Pl.131.

166 See, e.g., Astone Gasparetto, 'The Relations between Venice and Bohemia in the field of glass-making at the beginning of the 18th century', *Glass Review*, 1, 1975, pp.6–10; Olga Drahotová, 'Bohemian Crystal Glass and its Influence on 18th century Venetian Glass', ibid., 2, 1975, pp.14–19.

167 I am indebted to my colleague John Harthan for this identification.

Bibliography

This Bibliography includes only books and long articles of particular importance within their own field. References to other articles are quoted in full in the relevant parts of the Introduction or Catalogue.

GENERAL

E. Dillon, *Glass*, London, 1907.

R. Schmidt, *Das Glas*, Berlin and Leipzig, 2nd ed., 1922.

Victoria and Albert Museum, *Glass: a Handbook for the Study of Glass Vessels of all Periods and Countries, and a Guide to the Museum Collection*, by W. B. Honey, London, 1946.

W. Bernt, *Altes Glas*, Munich, *nd.* (1951 ?).

Corning Museum of Glass, *Glass Drinking Vessels from the Collections of Jerome Strauss and the Ruth Bryan Strauss Memorial Foundation*, Corning, 1955.

I. Schlosser, *Das alte Glas*, Brunswick, 1956.

Kestner Museum, Hanover, *Die Glas-Sammlung*, by Christel Mosel, Hanover, 1957.

Musée Curtius, Liège, *Trois Millénaires d'Art Verrier*, Liège, 1958.

Kunstgewerbemuseum der Stadt Köln, *Glas*, by Brigitte Klesse, Cologne, 1963.

Elka Schrijver, *Glass and Crystal*, I, *from earliest times to 1850*, London, 1963, and II, *from 1850 to the Present*, London, 1964.

George Savage, *Glass*, London, 1965.

Brigitte Klesse, *Glassammlung Helfried Krug*, Munich, 1965. (Klesse, *Krug*, I)

Kunstmuseum, Düsseldorf, *Glas*, by Elfriede Heinemeyer, Düsseldorf, 1966.

C. Isings, *Schönes altes Glas*, Hanover, 1966.

F. Kämpfer and Klaus G. Beyer, *Glass: a World History*, Dresden, 1966.

British Museum, *Masterpieces of Glass*, by D. B. Harden and others, London, 1968.

Kunstmuseum, Düsseldorf, *Meisterwerke der Glaskunst aus Internationalem Privatbesitz*, by A. von Saldern, Düsseldorf, 1968.

Toledo Museum of Art, *Art in Glass*, Toledo, 1969.

Gustav Weiss, *The Book of Glass*, London, 1971.

Museum für Kunsthandwerk, Frankfurt-am-Main, *Europäisches und Aussereuropäisches Glas*, by Annaliese Ohm, Frankfurt, 1973.

Brigitte Klesse, *Glassammlung Helfried Krug*, II, Bonn, 1973. (Klesse, *Krug*, II)

Corning Museum of Glass, *Glass from the Corning Museum of Glass: a Guide to the Collections*, Corning, 1974.

Ada Polak, *Glass, its makers and its public*, London, 1975.

ANCIENT GLASS (a) ROMAN

Louis P. di Cesnola, *A Descriptive Atlas of the Cesnola Collection of Cypriote Antiquities in the Metropolitan Museum of Art, New York*, New York, Vol. III, Part 2, 1903.

C. C. Edgar, *Graeco-Egyptian Glass* (*Catalogue Général des Antiquités Egyptiennes du Musée du Caire*), Cairo, 1905.

A. Kisa, *Das Glas im Altertume*, 3 vols., Leipzig, 1908.

A. de Ridder, *Collection de Clercq, Catalogue*, Vol.6 (*Les Terres cuites et les verres*), Paris, 1909.

D. B. Harden, *Roman Glass from Karanis*, Ann Arbor, 1936.

P. V. C. Baur, 'Glassware', in (ed.) Carl H. Kraeling, *Gerasa: City of the Decapolis*, New Haven, 1938, pp.505–46.

D. B. Harden, 'Tomb-Groups of Glass of Roman Date from Syria and Palestine', *Iraq*, XI, Part 2, 1949, pp.151–9.

Olof Vessberg, 'Roman Glass in Cyprus', *Opuscula Archaeologica*, VII, 1952, pp.109–65.

D. B. Harden, 'Roman Tombs at Vasa: the Glass', *Report of the Department of Antiquities, Cyprus, 1940–1948*, 1955, pp.46–60.

Corning Museum of Glass, *Glass from the Ancient World, the Ray Winfield Smith Collection*, Corning, 1957.

C. Isings, *Roman Glass from dated Finds*, *Archaeologica Traiectina*, Groningen/Djakarta, 1957.

Christoph W. Clairmont, *The Glass Vessels* in *The Excavations at Dura-Europos . . . Final Report IV, Part V*, New Haven, 1963.

D. B. Harden, 'Some Tomb Groups of late Roman Date in the Amman Museum', *Annales du 3e Congres des "Journées Internationales du Verre", Damascus, Novembre 1964*, Liège, *n.d.* (1965), pp.48–55.

Paul Åström, 'Collections of Cypriote Glass', *Opuscula Atheniensia*, V, Lund, 1965, pp.123–58.

D. B. Harden, 'Ancient Glass, I: Pre-Roman', II: 'Roman', III: 'Post-Roman', *The Archaeological Journal*, CXXV, 1969, pp.46–72; CXXVI, 1970, pp.44–77; CXXVIII, 1972, pp.78–117.

ANCIENT GLASS (b) ISLAMIC

C. J. Lamm, *Mittelalterliche Gläser und Steinschnittarbeiten aus dem Nahen Osten*, I Text, Berlin, 1930; II, Illustrations, 1929.

Arthur Lane, 'Medieval Finds at Al Mina in North Syria', *Archaeologia*, LXXXVII, 1938, pp.61–74.

P. J. Riis and Vagn Poulsen, *Hama . . . Les Verreries et Poteries Médiévales*, Copenhagen, 1957.

VENETIAN

R. Schmidt, 'Die Venezianischen Emailgläser des XV. und XVI. Jahrhunderts', *Jahrbuch der Königlich Preuszischen Kunstsammlungen*, XXXII, 1911, pp.249–86.

Österreichisches Museum für Angewandte Kunst, Vienna, *Venezianer Gläser*, by Ignaz Schlosser, Vienna, 1951.

Corning Museum of Glass, *Three Great Centuries of Venetian Glass*, Corning, 1958.

A. Gasparetto, *Il Vetro di Murano*, Venice, 1958.

G. Boesen, *Venetianske Glas på Rosenborg*, Copenhagen, 1960.

K. Hetteš, *Old Venetian Glass*, London, 1960.

G. Mariacher, *Italian Blown Glass from Ancient Rome to Venice*, London, 1961.

Uměleckoprůmyslové muzeum v Praze (Museum of Applied Art, Prague), *Benátské sklo* (*Venetian Glass*) Prague, 1973.

'FAÇON DE VENISE'

J. Barrelet, *La Verrerie en France*, Paris, 1953.

G. Taddei, *L'Arte del Vetro in Firenze e nel suo Dominio*, Florence, 1954.

R. Chambon, *L'Histoire de la Verrerie en Belgique du IIme Siècle à nos Jours*, Brussels, 1955.

E. Egg, *Die Glashütten zu Hall und Innsbruck im 16. Jahrhundert*, Innsbruck, 1962.

Alice Wilson Frothingham, *Spanish Glass*, London, 1963.

F.-A. Dreier, *Glaskunst in Hessen-Kassel*, Kassel, 1969.

A. von Saldern, 'Façon de Venise', in *Reallexikon zur deutschen Kunstgeschichte*, Stuttgart, 69. Lieferung, 1972, Cols. 1024–56.

GERMAN AND BOHEMIAN

E. Schebek, *Böhmens Glasindustrie und Glashandel*, Prague, 1878.

G. E. Pazaurek, 'Die Heimat der Zwischengoldgläser', *Mittheilungen des nordböhmischen Gewerbe-Museums*, XVI. Jahrgang, 1898, Nr.3, pp.53–60.

G. E. Pazaurek, *Die Gläsersammlung des nordböhmischen Gewerbemuseums in Reichenberg*, Leipzig, 1902.

R. Schmidt, *Brandenburgische Gläser*, Berlin, 1914.

R. Schmidt, *Die Gläser der Sammlung Mühsam*, Berlin, 1914. (Schmidt, *Mühsam* I)

R. Schmidt, *Die Gläser der Sammlung Mühsam*, Neue Folge, Berlin *n.d.* (1927). (Schmidt, *Mühsam* II)

K. Hetteš, *The Bohemian Glass*, (?)Prague, n.d. (*c.*1950).

Victoria and Albert Museum, *Bohemian Glass*, London, 1965.

A. von Saldern, *German enameled Glass*, Corning, 1965.

Zuzana Pešatová, *Bohemian Engraved Glass*, Feltham, 1968.

F.-A. Dreier, 'Fünf Silbermontierte Prunkvasen aus Goldrubinglas', *Berichte aus den staatlichen Museen des preussischen Kulturbesitzes*, Neue Folge XIX, Heft 2, Berlin, 1969, pp.78–83.

Prague, Museum of Applied Art, *České Sklo 17 a 18 století/Bohemian Glass of the 17th and the 18th centuries*, Prague, 1970.

Olga Drahotová, *Bohemian Glass*, Prague, 1970.

Glossary

Alabastron
(Probably originally an *unguentarium* (q.v.) made of alabaster). A slender unguent-flask with narrow neck and spreading lip, with small side-handles.

Cristallo
A more or less colourless transparent glass.

Cut-out base
Foot-rim made by pinching out a hollow ring of glass from the wall or base of the vessel (cf. the drawing of 2, etc).

Façon de Venise
A way and a style of making glass practised in the glasshouses working in the Venetian tradition in other centres.

A fili See Thread-decoration.

A filigrana See Thread-decoration.

Gather
A quantity of glass drawn from the furnace by dipping and rotating the iron in the molten glass.

Glory hole
An aperture in the wall of a furnace used primarily for 'warming-in' a glass (reheating for further working).

Humpen
A cylindrical drinking-vessel, normally of considerable capacity (see also *Reichsadlerhumpen* and *Kurfürstenhumpen*).

Kick (or punt)
Indented base of a vessel, pushed in to give it stability on the resultant ring-base.

Knop
A (normally bulbous) element in the formation of a stem, usually hollow-blown.

Kurfürstenhumpen
(Literally, 'Electors' Beaker'). A cylindrical vessel, usually of considerable capacity, enamelled (or less often decorated in diamond-point) with the Holy Roman Emperor seated among his Electors, or riding in their company.

Lattimo
(From Italian, *latte*, milk). Colourless glass opacified with tin-oxide or other substance to give an opaque-white.

Lehr
(From Italian, *l'aere*). At first the uppermost chamber of the Venetian-type three-storey furnace, the lehr could also be a separately-built chamber deriving its heat from a subsidiary fire-box. The glass to be cooled was moved progressively away from the source of heat. In the late seventeenth and eighteenth centuries a tunnel was built from the uppermost chamber of the furnace to the 'Sarrosel Room' (q.v.).

Marver
(From French, *marbre*, marble). A flat slab, originally of marble, latterly of metal, usually iron, on which the gather (q.v.) of glass is rolled ('marvered') to smooth and consolidate it before blowing and further working.

Merese
A flat solid disc interposed between two separately-worked parts of a (normally) stemmed vessel – e.g. between bowl and stem, stem and foot, or components of a stem.

Pad base
Foot-rim made of a separate paraison of glass backed on to the original paraison.

Paraison
A partially inflated gather (q.v.) of glass.

Pontil ('punty')
From the Italian, *pontello*. The solid iron rod to which the glass is attached (by means of a small gather or 'wad' of glass) for further working after being detached from the blowing-iron. When the pontil is in turn detached, the resultant scar is called the 'pontil-mark'.

Printie
A circular cut made by rotating the glass against the rounded edge of a cutting-wheel.

Prunt
A decorative pad of glass applied to the wall of a vessel and often impressed with a mask or a pattern of concentric dots ('raspberry prunt') – cf. *24*.

Raspberry prunt   See preceding entry.

Reichsadlerhumpen
(Literally 'Imperial Eagle beaker'). A cylindrical vessel, usually of considerable capacity, enamelled (or less often decorated in diamond-point) with, among other insignia, the spread-eagle of the Holy Roman Empire, its feathers composed of the coat-of-arms of the constituent parts of the Empire.

A reticella   See Thread-decoration.

A retorti (or *retortoli*)   See Thread-decoration.

Sarrosel Room
When a glass is taken finished to the lehr (q.v.) to cool off, in the practice of the late seventeenth and eighteenth centuries it proceeded down a tunnel, growing progressively cooler, until it reached the Sarrosel Room, where it was removed, fully cooled, from the lehr, for sorting and packing.

Seed
Very numerous small air-bubbles impairing the transparency of the glass.

Stangenglass (or *Stange*, literally, 'a pole')
German word for a tall cylindrical glass of relatively small diameter.

Thread-decoration, usually in opaque-white (*lattimo*, q.v.)
Single threads or collapsed tubes could be used singly (*a fili*), or cable-twists (*a retorti, a retortoli* or *a filigrana*). When two paraisons were blown one inside the other, with the threads running in opposite senses, the resultant mesh-pattern was called *a reticello(a)* or *vetro di trina* (19th century usage).

Unguentarium
A small flask for perfumes or other toilet preparations.

Vetro di trina   See under Thread-decoration.

Zwischengoldgläser
Glasses, usually beakers and goblets, decorated by means of gold (sometimes silver) foil etched with a point to give the desired design, and sandwiched between an inner and an outer wall of glass (*see drawing on page 40*).

CATALOGUE OF GLASS

1. Bowl

Roman (Syrian or Cypriot); 1st century A.D.

Bowl of greenish-colourless glass, lightly weathered

Height, $2\frac{7}{16}$ in. (6·2 cm); diameter of rim, $2\frac{7}{8}$ in. (7·4 cm)

Plain cup-shaped bowl of hemispherical proportions but with the almost straight sides curving gently inwards to the rim. The base is very slightly pushed up to produce a dimple-like 'kick'. The outside of the glass is covered with iridescence, the inside with a patchy layer of brownish enamel-like decomposition marked with black flecks.

Small bowls of this sort are one of the commonest types of early glass in the West, being found as far north as Britain, as well as in Italy and Gaul. They date from the very earliest years of the Imperial period through the first century A.D. Further east they are found at Corinth and Cyprus (Isings, pp.27–30, Form 12, with numerous parallels). In Cyprus indeed they seem to have been common, some thirty-eight examples apparently coming from Idalium alone (Cesnola, *Atlas*, Pl. LXXXIV,1), although the Cypriot form appears on the whole to be somewhat more straight-sided and to have external horizontal cut lines (Harden, 'Vasa', p.49, No.12, Fig.20,j; Vessberg, pp.114–15, Type A.III, Pl. I,25). A closely similar bowl comes from Samothrace (Elsbeth B. Dusenbery, 'Ancient Glass from the Cemeteries of Samothrace', *Journal of Glass Studies*, IX, 1967, p.45, Fig.36). The fact that no examples of this very widespread type appear to be recorded from Syria or Palestine is almost certainly a matter of chance.

Painted with the Nazi inventory number: R2662 (see p.7).

See profile drawing on page 276.

Baron Edmond de Rothschild P/2051 ter**

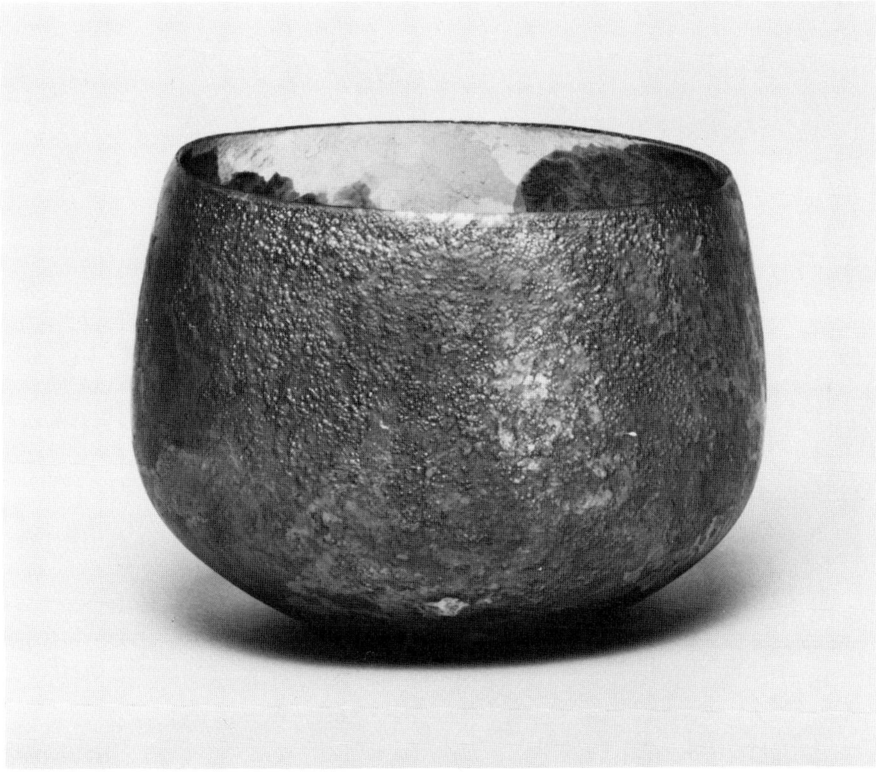

2. Bowl

Roman (probably Cypriot); 2nd century A.D.

Bowl of greenish-colourless glass, lightly weathered

Height, $2\frac{3}{8}$ in. (6 cm); diameter, $3\frac{5}{8}$ in. (9·2 cm)

Bowl with irregularly sloping straight sides, out-curved rim and applied pad-base, forming a low spreading pedestal foot. Underneath are traces of a pontil-mark. A single thin trailed thread has been laid round the bowl below the rim. The glass is patchily weathered with a thin brown enamel-like layer of decomposition.

This type of bowl seems to be most numerous on Cyprus. Analogous bowls in the West seem rare, and appear to date from the late first or second century (Isings, p.106, Type 87, with 'cut-out' base). In Cyprus, the Cyprus Museum in Nicosia has half a dozen or more close analogues (Vessberg, p.117, Type C, Pl. II,12 ff), but unfortunately none appear to come from datable contexts. An even more numerous series came from Idalium (Cesnola, *Atlas*, Pl. LXXXIV,3; Pl. LXXXVI,3) but the type seems not to have been represented at Vasa, where the earliest levels dated from before the middle of the third century. A date in the second century seems reasonable for the Waddesdon bowl.

Painted with the Nazi inventory number: R2660 (see p.7).

See profile drawing on page 276.

Baron Edmond de Rothschild P/2051 ter**

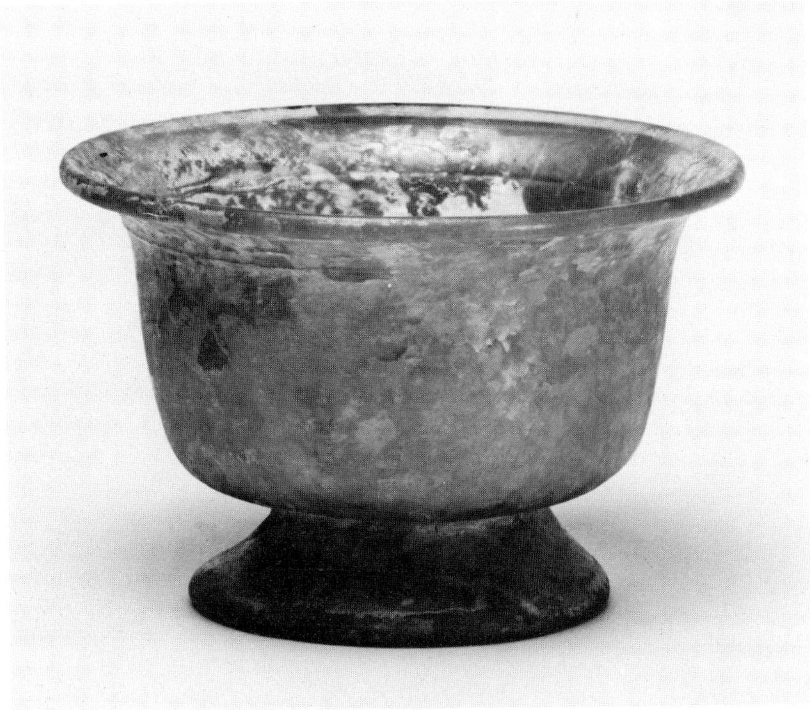

3. Bowl

Roman (probably Cypriot); 2nd to mid-3rd century A.D.

Bowl of pale green glass, lightly weathered

Height, 2 in. (5 cm); diameter, $4\frac{7}{8}$ in. (12·4 cm)

Bowl with straight sides flaring outwards from a high 'cut-out' base-ring to a slightly thickened rim. The surface is covered with iridescence or thin patchy brownish weathering.

The shape of this glass, with its high 'cut-out' foot-rim, seems almost specific to Cyprus. An example in the Cyprus Museum (D 1506, Vessberg, p.117, Type B. II.δ, Pl. II,11) is matched by two from Vasa (Harden, 'Vasa', p.48, Nos.3–4, fig.19,b), and a further example from Idalium (Cesnola, *Atlas*, Pl. LXXXVII,2). The first three examples mentioned here have folded rims (two folded inwards and one outwards), but dishes with the same high foot and straight sides have the plain rim of the Waddesdon bowl (cf. an example from Kazaphani, in the Cyprus Museum, D 968, cit. Vessberg, p.113, Pl.I,9; and two from Vasa, cit. Harden, 'Vasa', p.48, Nos.1–2, Fig.19,a). The Vasa glasses occur (with one exception which may have been reinterred in the early fourth century) in levels which suggest a date before the middle of the third century.

Further examples of this type of bowl occur in the de Clercq Collection (de Ridder, No.403, Pl. XXI, from Cyprus, diameter 14·6 cm), and in the Cairo Museum, from Hawara (Edgar, Pl. II, No.32.462, diameter 8·5 cm; this appears to have an inward-folded rim, and is of colourless glass).

See observations on *4* below.

Painted with the Nazi inventory number: R2653 (see p.7).

See profile drawing on page 277.

Baron Edmond de Rothschild P/2051 ter**

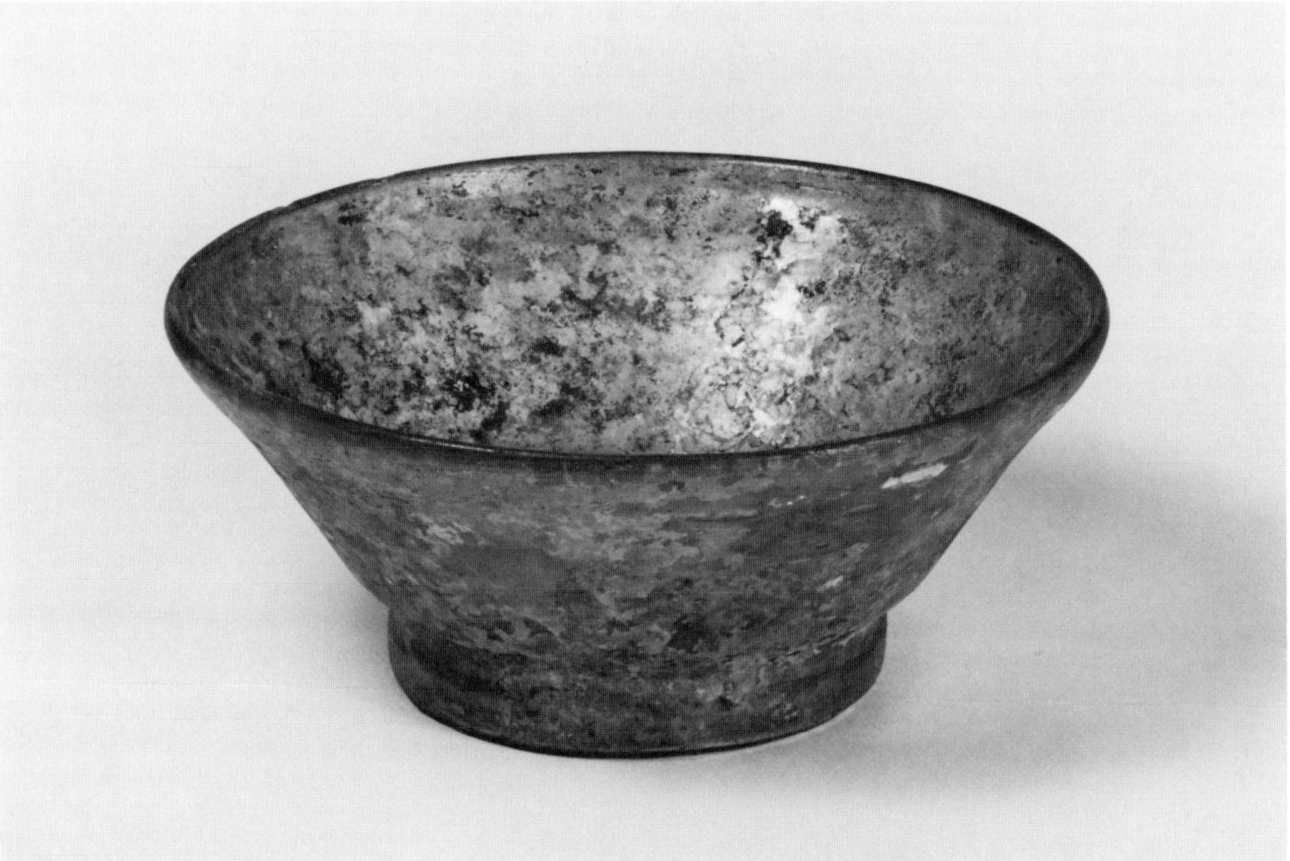

4. Bowl

Roman (Cypriot or Syrian); probably 2nd to mid-3rd century A.D.

Bowl of greenish-colourless glass, heavily weathered

Height, 2⅛ in. (5·5 cm); diameter, 3¾ in. (9·8 cm)

Bowl with almost straight everted sides, curving outwards slightly at the rim, and with applied pad base, slightly 'kicked', and showing traces of a pontil-mark. The glass is heavily weathered almost all over, with a thickish light-brown enamel-like layer of decomposition.

The shape of this bowl seems to find its nearest parallels in a number found in Cyprus, although it is difficult to match precisely. A group of five bowls in the Cyprus Museum in Nicosia comes close (Vessberg, p.46, deep bowl Type B.1.β.1), although the wall of these tends to be rather more curved. The same straight profile of the bowl occurs, however, on a second group which has a taller vertical-sided foot (Vessberg, p.117, deep bowl Type B.II.δ). Here, however, the rim of the bowl is folded, as occurs on two similar bowls from Vasa, also in Cyprus (Harden, 'Vasa', p.48,3–4, Fig.19,b): but a larger type of straight-sided dish on the tall foot has a plain rim, and it seems likely that this detail is interchangeable. This type of glass has affinities with two different classes occurring on the Continent of Europe, sometimes with an almost straight, but concave, profile, sometimes with a convex curved side and outsplayed rim. Both occur at Pompeii (destroyed A.D. 79), but the second type lingers on through the second century (Isings, pp.57–8, Forms 41b, and 42). Three of the Vasa glasses occurred in a context giving a date before A.D. 250 (Harden, loc. cit.). A bowl of similar type, but with slightly more spreading rim, was in the Cairo Museum (Edgar, Pl. II, No.32.450). It provides no dating evidence.

See observations on *3* above.

Painted with the Nazi inventory number: R2661 (see p.7).

See profile drawing on page 277.

Baron Edmond de Rothschild P/2051 ter**

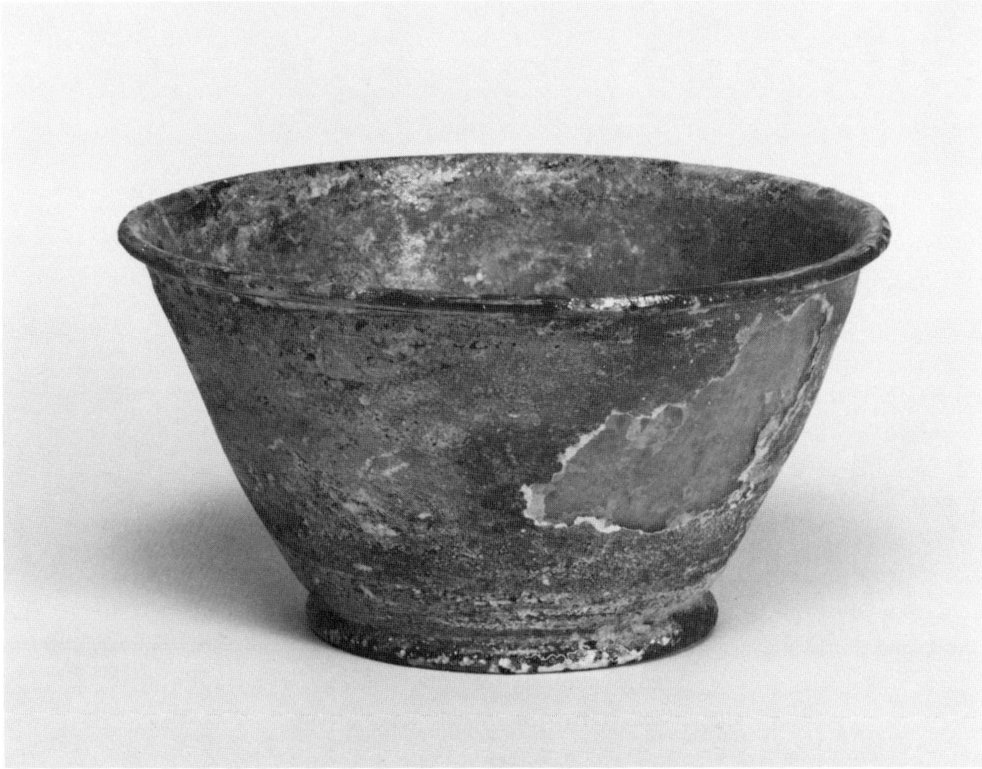

5. Dish

Roman (Syrian or Cypriot); probably 2nd century A.D., or later

Dish of pale green glass, lightly weathered

Height, $1\frac{1}{2}$ in. (3·8 cm); diameter, $7\frac{1}{4}$ in. (18·5 cm)

Low dish with almost flat base, steeply rising sides, and thickened rim. On the base is a pontil-mark. The sides have marked horizontal *striae*. The surface is covered with iridescence and one or two patches of brownish enamel-like weathering.

Dishes or shallow bowls of this form, without any sort of foot-ring, normally seem to belong in the first two centuries A.D. A number of such flat-based dishes, of varying rim profile, are found in first century contexts in the West (Isings, pp.36–9, Forms 18, 19, 22 and 23), although these are more usually of 'mould-pressed' than blown glass (ibid., pp.36–7). Examples have occurred in Cyprus, although not in datable finds (Vessberg, pp.112–13, Type A, Pl. I,1–2). The first of the two pieces cited is considerably distorted, presumably in the making (Cesnola, *Atlas*, Pl. LXXXVI,1), and may constitute *prima facie* evidence for the manufacture of this type in Cyprus itself (cf. Vessberg, pp.158–9). Its occurrence, however, is too rare to allow of any safe generalizations, and since the shape occurs in 'Late' varieties of Roman 'red gloss' pottery in the southern and eastern Mediterranean in the third–fourth centuries, the glass version may also be of this date (cf. R. J. Charleston, *Roman Pottery*, London, 1955, Pl.25,A).

Painted with the Nazi inventory number: R2654 (see p.7).

See profile drawing on page 277.

Baron Edmond de Rothschild — P/2051 ter**

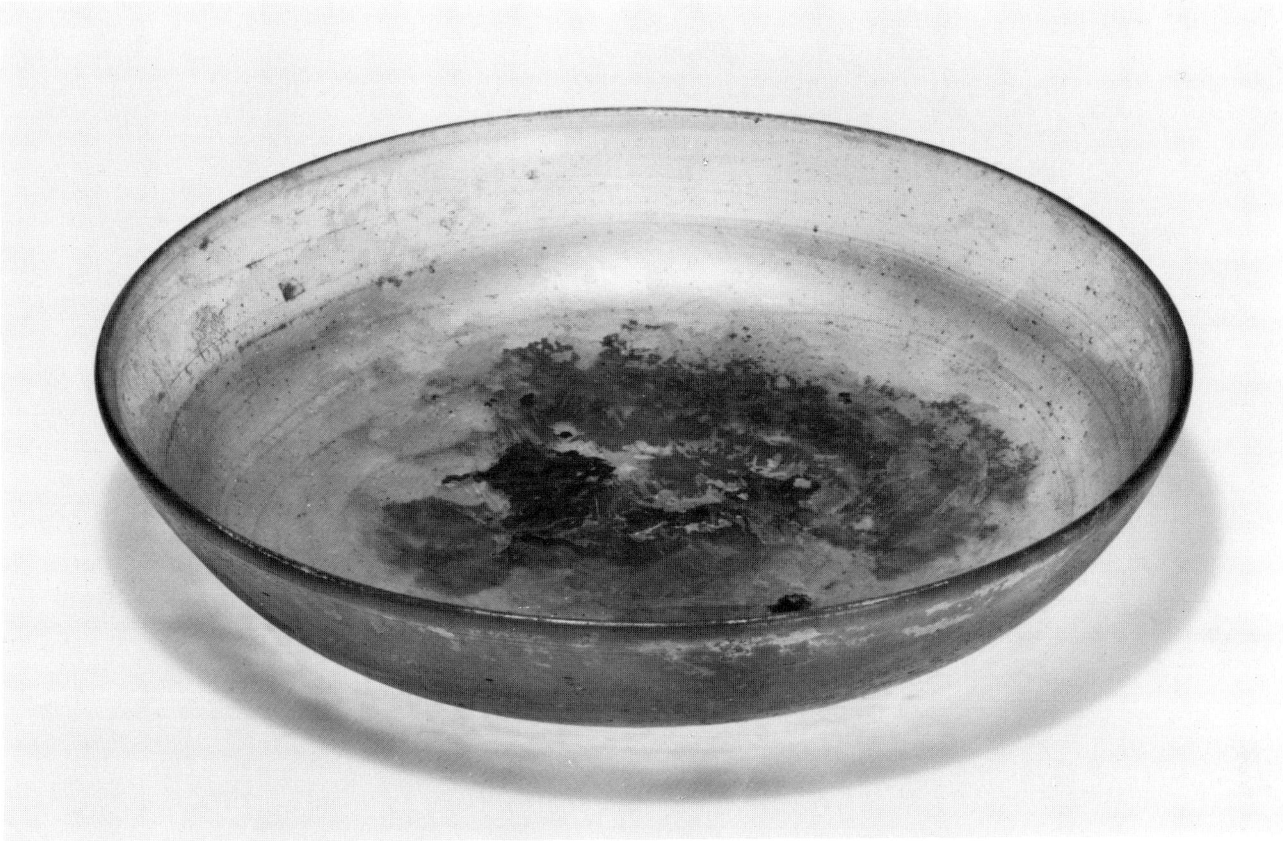

61

6. Toilet-bottle

Roman (Syrian or Cypriot); probably 3rd century A.D.

Unguentarium of greenish-colourless glass, heavily weathered

Height, 6¾ in. (17·2 cm); diameter of rim, 2 in. (5 cm)

Bottle with low truncated-conical body, flat base, and long slightly tapering neck, the spreading horizontal lip being formed by turning the edge of the glass upward and inward and flattening it. The glass is heavily weathered, with a thickish light-brown enamel-like layer of decomposition.

These unguent-bottles are among the commonest glasses made in the Roman world, and the general type was produced in the West from the first century onwards (Isings, pp.41–3, Form 28), and lasted probably into the fourth century (ibid., pp.97–9, Form 82). The particular shape of the Waddesdon bottle seems to occur most frequently in Cyprus, at least six examples occurring in a tomb, very prolific of glass, found in the Limassol Oasis, and dated on the evidence of coins to the late second or early third century (Vessberg, pp.155–6, Pls. XXII–XXIII, Nos.32, 26, 10, 24, 12, 69). Numerous examples were in the Cesnola Collection (Cesnola, *Atlas*, Pl. C, showing several similar flasks from Idalium, numerous other parallels being apparently also in the collection: see also Vessberg, pp.137, 139, Type IV, Pl. VIII, 25, etc.; Harden, 'Vasa', p.56, Type B.i(a), Fig.24,e, Pl. VI,J, and p.59). The type is far from rare, however, in the Syro-Palestinian area. In Palestine itself numerous examples were found in a tomb at W'ar Abu Eṣ Ṣafa, near Jerusalem, dated by the excavator to the first century B.C.–first century A.D. (D. C. Baramki, 'An Ancient Tomb Chamber at W'ar Abu Eṣ Ṣafa . . .', *Quarterly of the Department of Antiquities in Palestine*, IV, 1935, pp.168–9, Pl. LXXX. The date of this burial, although accepted by D. B. Harden in 'Tomb-Groups', p.155, is shown by Isings, pp.43, 98, as second or third century A.D.). Very similar flasks were found at Beirut in tombs containing coins of Maximianus and Constantinus Chlorus, suggesting a date in the early fourth century (M. Chéhab, 'Sarcophages en Plomb du Musée National Libanais', *Syria*, XVI, 1935, pp.62–3); whilst examples of this general type found at Jerash are also probably not earlier in date than the fourth century (Baur, pp.514, 542, Nos.92–4, Fig.26 on p.535). The same general type, however, occurring at Dura-Europos, must date from before the sack of that city in A.D. 256 (Christoph W. Clairmont, *The Glass Vessels*, New Haven, 1963, pp.135–6, Nos.690, 692, 693, Pl.XVI). An example excavated at Acre was found with a coin of Domitian (reigned A.D. 81–96), the probability of a first century date being strengthened by the general character of this cemetery (Maria Teresa Fortuna, 'I vetri soffiati della Necropoli di Akko', *Journal of Glass Studies*, VII 1965, pp.17, 21, Fig.11); another found at Homs, however, comes from a tomb-complex dating between the second and the first half of the third century (Selim Abdul Hak, 'Contribution d'une Découverte archéologique récente à l'Etude de la Verrerie Syrienne à l'Epoque Romaine', ibid., pp.26–7, 29–30, Fig.9).

The type does not appear to have existed in Egypt, no parallels having come to light at Karanis (Harden, *Karanis*, pp.265 ff) or survived in the published collection of the Cairo Museum (Edgar, the nearest approximation being No.32.622, illustrated on Pl. VII).

See profile drawing on page 278.

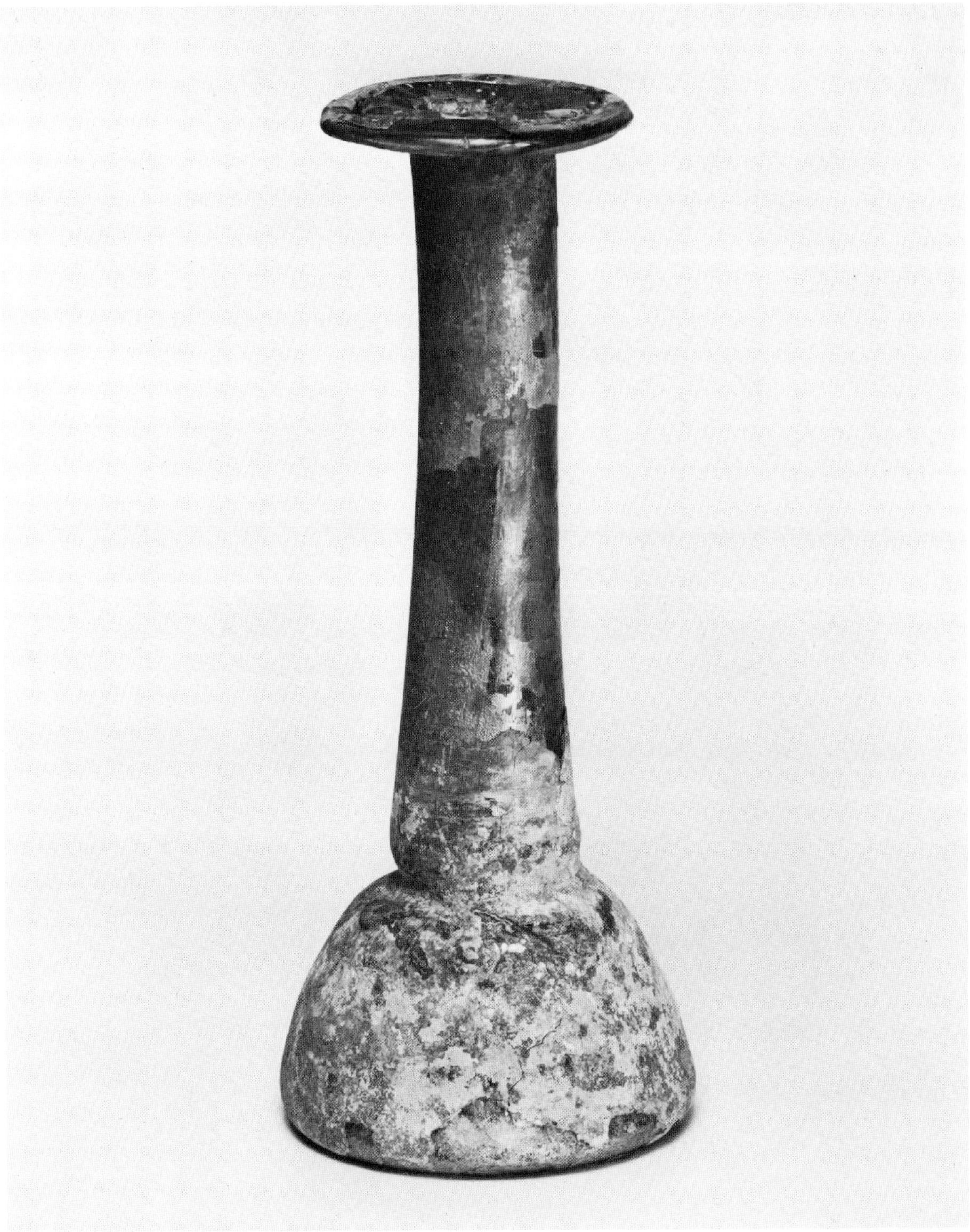

Baron Edmond de Rothschild P/2051 ter**

7. Flask

Roman (Syrian or Cypriot); probably 3rd century A.D.

Flask of greenish-colourless glass, patchily weathered

Height, $6\frac{1}{4}$ in. (16 cm); diameter of rim, $1\frac{3}{4}$ in. (4·5 cm)

Flask with ovoid body drawn into a slender neck which flares in a gentle curve to a slightly thickened rim. The flask stands on a low foot-rim which was probably made by the 'cut-out' method, and which has a pontil-mark in the centre of it. A double circuit of thin threading has been laid round the lower part of the neck. The surface of the glass shows patches of thin brownish enamel-like weathering.

This form of flask, which always appears to have a low foot-rim and often a turn or two of threading round the neck, seems to be restricted to the Eastern Empire. Examples from Cyprus include one from Idalium, in the British Museum (Harden, *Karanis*, p.215, n.1, Fig.4,c); a second of unknown provenance in the Cyprus Museum, Nicosia (Vessberg, p.135, Type B.I, Pl.VIII,1); and two further pieces of comparable profile from the Cesnola Collection (Cesnola, *Atlas*, Pl.XCVIII,5 and 6; cf. Vessberg, loc. cit.). A jug from Cyprus with identical profile and thread-decoration, differing only in the addition of the handle, is in the National Museum, Copenhagen (Vessberg, p.130, Type B.III, Pl. XVI,1). The neck-fragment of a thread-decorated flask found at Karanis is ascribed to this type of flask and dated by implication to the third century or later (Harden, *Karanis*, pp.191 and 215, No.637, Pl. XVIII); a slightly differing form with threading above the foot as well as round the neck was found at Oxyrhynchus (Edgar, No.32.754, Pl. X; cf. an example very similar in form, with several thread-circuits, in the Ashmolean Museum, illustrated B. Goldsmid, 'Glass in the Ashmolean Museum', *Antique Dealer and Collectors' Guide*, February, 1951, p.23). Dr Harden has pointed out that the general treatment of the neck resembles that of Western 'Snake-thread' flasks (e.g., Harden, 'Ancient Glass, II: Roman', Pl. VI,A, a bottle from Idalium, Cyprus). At least one example of the type of flask under discussion is to be found within the confines of the ancient territory of Syria (Museum Haaretz, Tel-Aviv, *Ancient Glass*, by Y. Israeli, Tel-Aviv, 1965, Pl. V,b).

See profile drawing on page 278.

Baron Edmond de Rothschild P/2051 ter**

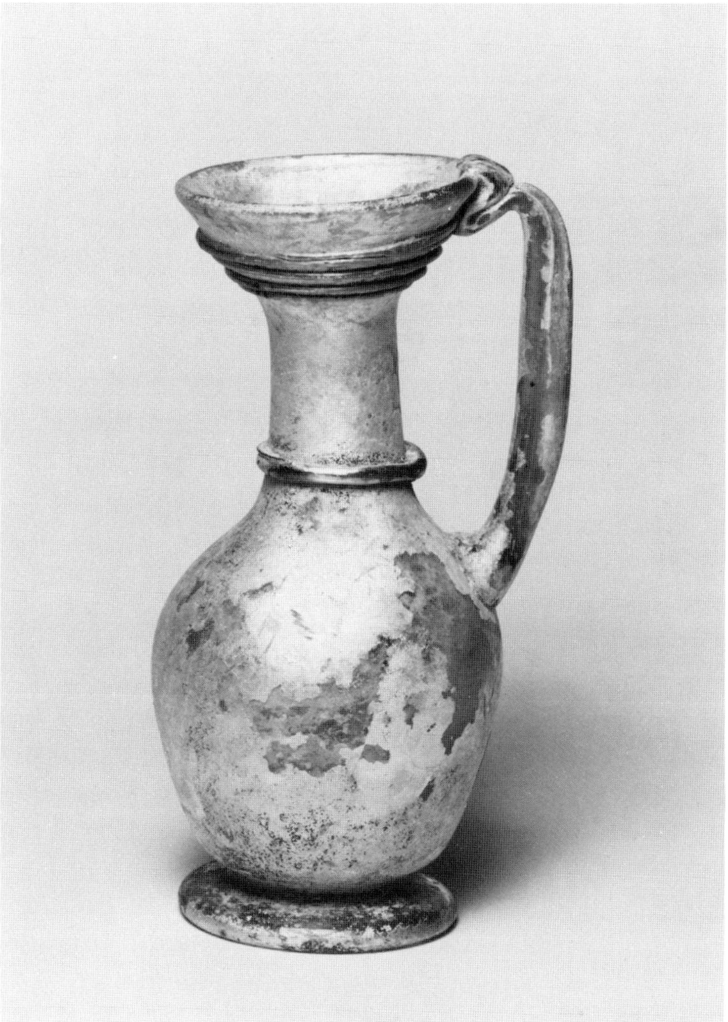

9. Dish

Roman (probably Syrian); probably 4th century A.D.

Dish on low foot, pale green glass, originally heavily weathered

Height, $1\frac{5}{8}$ in. (4·2 cm); diameter, 7 in. (17·8 cm)

Low dish with almost vertical sides, the 'cut-out' foot-ring being pushed up in such a way as to give a convex curve to the area of the base outside it; underneath the sagging base is a pontil-mark. The inside of the dish shows an extensive patch of brown enamel-like weathering which probably originally covered the dish, leaving an iridescent surface where it has scaled off.

Dishes of this type occur both in Cyprus (Vessberg, p.113, Type B.1; Pl.I,3–5; Cesnola, *Atlas*, Pl. LXXXV,1, from Idalium, and others on Pls. LXXXVII and LXXXVIII), and in Syria (de Ridder, Pl.XXV, 495, p.233, from Tyre: Pl.XXVI, 497, p.234, from Tyre, although both these pieces are differentiated by a decorative thread laid round the vertical part of the side). The form seems not to occur at Karanis (Harden, *Karanis*, pp.64–5), but examples excavated in Egypt (illustrated Edgar, Pl.1, No.32,406; 32,407, etc.) are considered by Dr Harden to be probably of Syrian origin (*Karanis*, p.64). A similar shape, but with an elaborately outward-folded rim, occurs in the West (Isings, p.148, Form 118) as well as in Syria (Harden 'Tomb-Groups', pp.151 ff, No.G.22, Pl. XLIX,2 and other parallels in Isings, loc. cit.: see also Harden, 'Amman', pp.51–2, Fig.10, a group from Beit Ras, near Irbid in East Jordan) and Egypt (Harden, *Karanis*, p.71, No.88, Pl.XII). All these seem to be from fourth century contexts. An example was found in Palestine at Karm-al-Shaikh, Jerusalem, in a fourth–fifth century burial (D.C.B., 'Note on a cemetery at Karm-al-Shaikh, Jerusalem', *Quarterly of the Department of Antiquities in Palestine*, I, 1932, pp.3–8, Pl. VII, 13; Harden, 'Tomb-Groups', p.156).

Painted with the Nazi inventory number: R2655 (see p.7).

See profile drawing on page 279.

Baron Edmond de Rothschild — P/2051 ter**

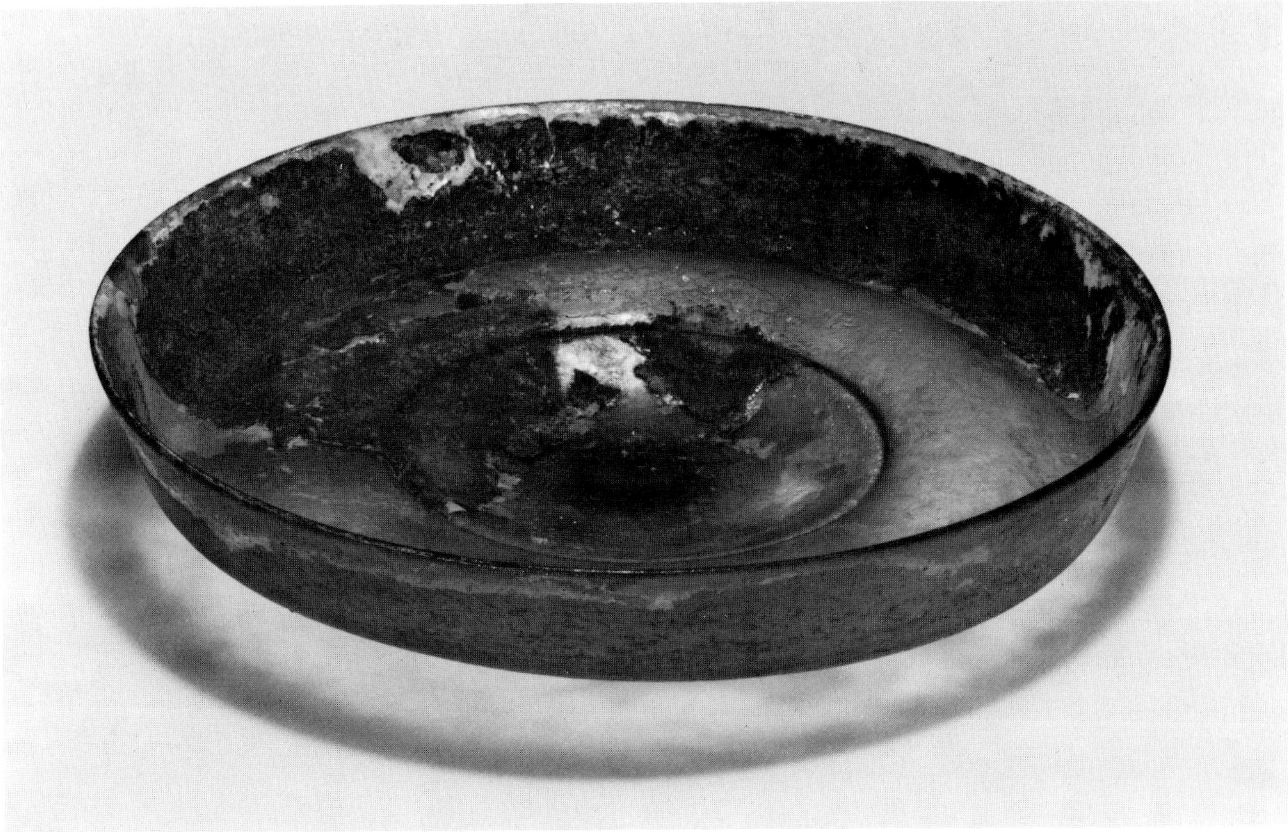

10. Dish

Roman (perhaps Syrian); probably 4th century A.D.

Dish on a low foot, pale-green glass, heavily weathered outside

Height, $1\frac{1}{2}$ in. (3·8 cm); length, $7\frac{7}{8}$ in. (20 cm); width, $5\frac{5}{8}$ in. (14·3 cm)

Low dish of irregular oval plan, with gently sloping sides which are turned rather more sharply upward at the thickened rim. The dish stands on a low oblong foot made from an applied pad, with pontil-mark visible below. The outside of the dish is covered with a thickish brown enamel-like layer of decomposition, showing iridescent where this has flaked off.

The oval dish on a foot-ring is of common occurrence at Karanis, in Egypt (Harden, *Karanis*, p.47), but is rare in the West, although it does occur (Isings, p.117, Form 97b). It does not appear to be recorded in Syria, nor in more than one example in Cyprus (Vessberg, p.113, quoting A. P. di Cesnola, *Salaminia, Cyprus; the history, treasures and antiquities of Salamis . . .*, London, 1882, Pl. XVIII,3). At Karanis the type was probably introduced in the late third century and was fully established in the fourth century. The dish with plain rim is perhaps somewhat earlier than that with folded rim (Harden, loc. cit.). A dish found at Cologne dated from the end of the third century, and an example from Remagen from the late third or fourth century (Isings, loc. cit.).

Painted with the Nazi inventory number: R2656 (see p.7).

See profile drawing on page 279.

Baron Edmond de Rothschild — P/2051 ter**

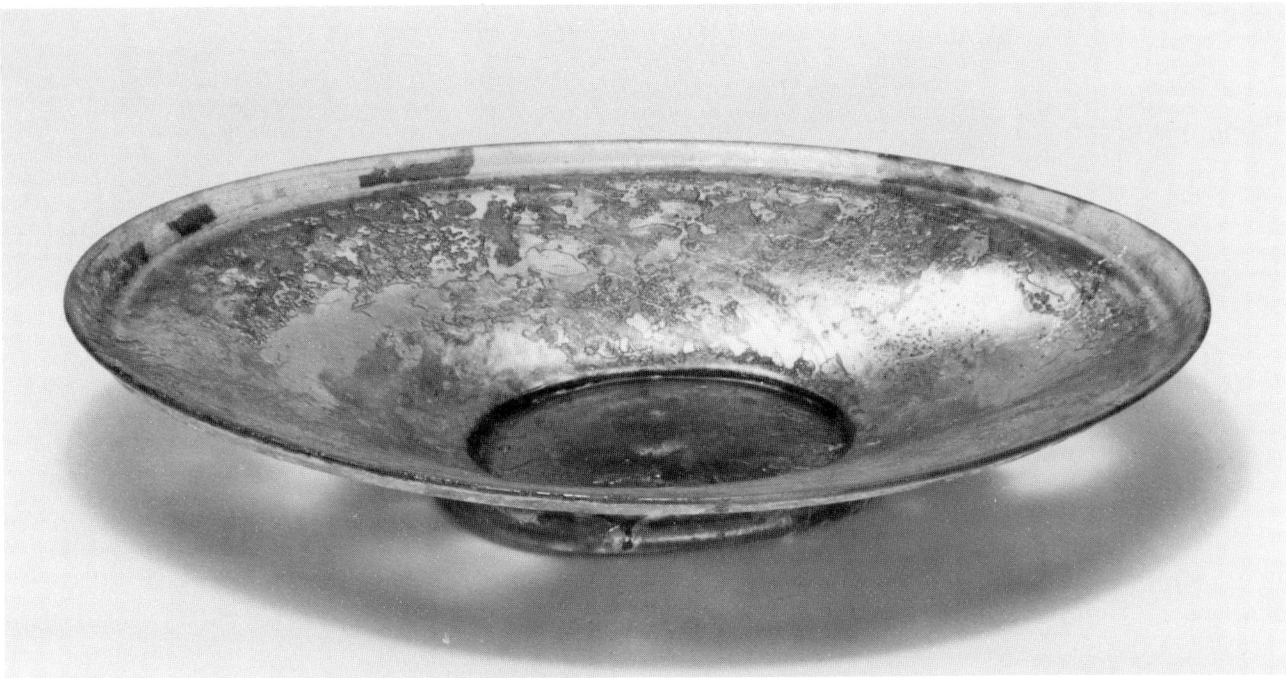

11. Flask

Roman (probably Syrian); 4th century A.D.

Flask of pale-green glass, lightly weathered

Height, $5\frac{1}{4}$ in. (13·3 cm); diameter of rim, 2 in. (5 cm)

Flask with slightly depressed globular body, and short, gently everted trumpet-neck. The base is slightly pushed in to a very low 'kick'. The surface of the glass is patchily covered with a thin light-brown enamel-like layer of decomposition.

Flasks of this form are found throughout the Roman world, from Portugal (Jorge Alarcão, 'Abraded and Engraved late Roman Glass from Portugal', *Journal of Glass Studies*, XII, 1970, p.30, Figs.5–7) and other countries in the West (Isings, pp.123–5, Form 104,b, very numerous examples dating from the second half of the third century through the fourth) to Cyprus, Egypt and Syria in the East. There are examples from Vasa, in Cyprus, dating from the middle or end of the fourth century (Harden, 'Vasa', pp.53–4, Nos.2–8; cf. Vessberg, p.133, Type A.I.γ, two specimens in the Cyprus Museum, Pl.VII,14; also Åström, p.138, Fig.14, Nos.729, 731), while comparable flasks from Karanis and elsewhere in Egypt date from the fourth or even the fifth century (Harden, *Karanis*, pp.191, 215, No.635, Pl. XVIII; Edgar, No.32.585 (with folded rim), Pl. VI). The type is also common in Syria, four variants with minor differences coming from Beit Ras, near Irbid in East Jordan, in a tomb-group dating on internal evidence to the fourth–fifth centuries (Harden, 'Amman', pp.51–2, Fig.10); another came from Ajlun, west-north-west of Jerash, in a group dating mainly from as late as the sixth century, but with types from the fourth and fifth centuries included (ibid., pp.52–3, Fig.12).

Painted with the Nazi inventory number: R2658 (see p.7).

See profile drawing on page 280.

Baron Edmond de Rothschild P/2051 ter**

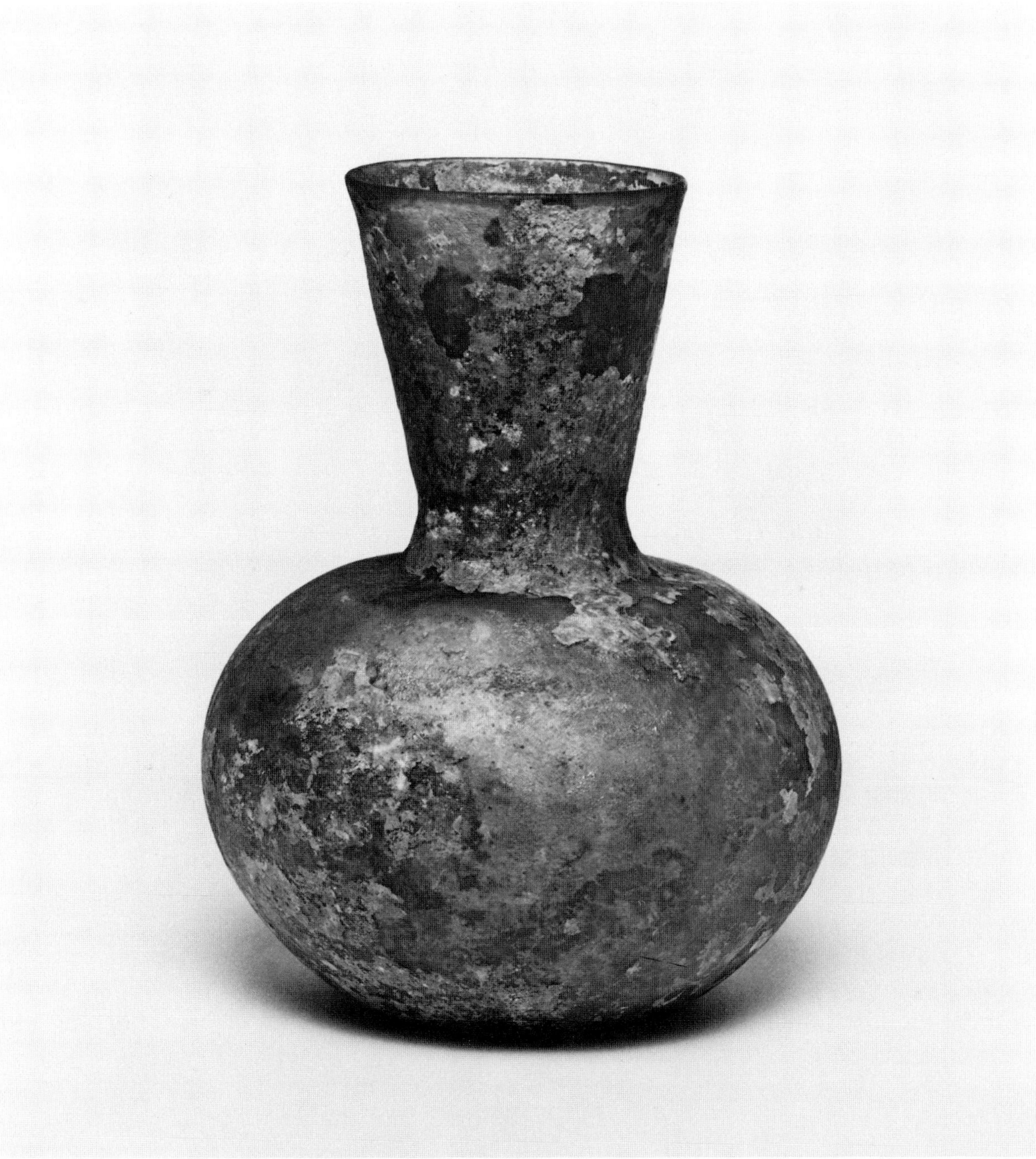

12. Twin-tube toilet-bottle

Roman (Syrian); 4th–5th century A.D.

Twin-*unguentarium* of green glass, with thread-decoration, heavily weathered

Height, 3¾ in. (9·5 cm); width of base, 1½ in. (3·8 cm)

Unguentarium formed of two parallel conjoined tubes tapering towards the top except for an out-turned lip. The twin-bottles are decorated by means of a zigzag of glass thread round the lip, and a dozen circuits of thin glass threading round the combined body. The glass has been completely covered by a layer of thick enamel-like weathering, which has flaked away in places taking the thread decoration with it.

The twin-tube *unguentarium* is a typical form of the Syro-Palestinian area, and does not occur either in the West or in Egypt (Harden, *Karanis*, p.270,g and references cited there); nor in Cyprus (Vessberg, p.143), although at least one example has been found in South Russia (N. Sorokina, 'Das Antike Glas der Nordschwarzmeerküste', *Annales du 4e Congrès des 'Journées Internationales du Verre'*, Liège, n.d. (1968), pp.75–6, Fig.2,25, thought by the author to be, like many of the other glasses from this area, a Syrian import).

The flasks occur in variant forms, often with elaborate tooled overarching or side-handles. They were apparently made by pinching a seam down a bag-shaped paraison to form a partition-wall between the two halves. The Waddesdon flask, distinguished by its absence of handles and by the zigzag thread-trailing below the lip, finds parallels in a flask, presumably from Palestine, in the Metropolitan Museum, New York (F. Neuburg, *Glass in Antiquity*, London, 1949, Pl. XIX,65), and in two specimens from a tomb in Shikmona in Israel, found in company with lamps and coins suggesting a third–fourth century dating (*Bulletin des 'Journées Internationales du Verre'*, *4*, Liège, 1965–6, pp.112–13, Fig.85).

The general type is normally dated to the fourth–early fifth century. An example with simple overarching loop-handle was found in a tomb-group at Beit Fajjar, in Palestine, in a context suggesting a date in the fourth century (S. A. S. Husseini, 'A 4th century A. D. Tomb at Beit Fajjar', *Quarterly of the Department of Antiquities in Palestine*, IV, 1935, pp.175–77, Pl. LXXXV,3; see also Harden, 'Amman', pp.49–50, Fig.8); a specimen with simple side-handles as well as a far more complex example with elaborate handle have been found in a tomb at Ajlun, west-north-west of Jerash, which contained many glasses of later (sixth century) types (ibid., pp.52–4, Figs.12–13). Two fragmentary specimens occurred among the mainly fourth–fifth century glasses at Jerash (Baur, pp.513–4, 546). Many of these tomb-groups, however, contained objects covering a long span of time (Harden, 'Amman', p.50). The absence of this type from the numerous glass fragments found at Dura-Europos (sacked in A.D.256) may support the supposition that the double-*unguentarium* does not occur as early as the third century (see also Harden, 'Ancient Glass, II: Roman', p.63, n.91).

See profile drawing on page 280.

Baron Edmond de Rothschild — P/2051 ter**

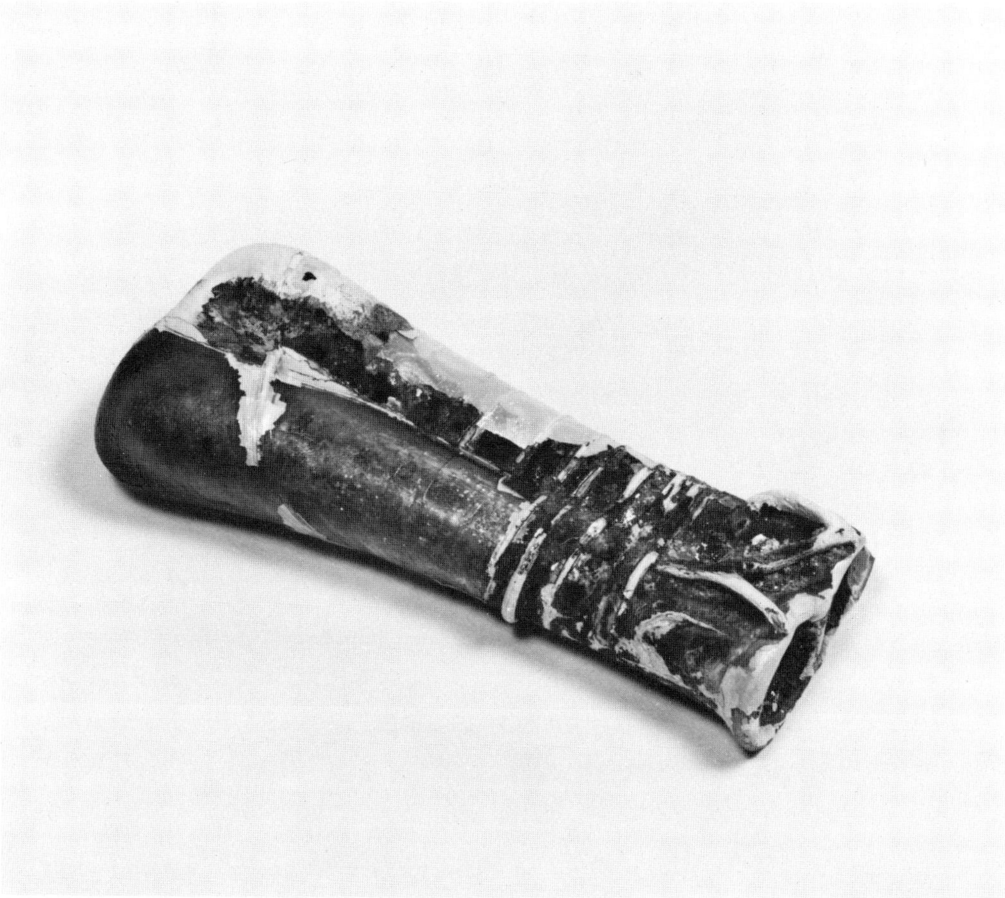

13. Alabastron, with thread-decoration

Perhaps Venetian; late 19th century

Greenish-blue glass with opaque yellow and red thread-decoration

Height, 5 in (12·7 cm); diameter of rim, 1½ in. (3·8 cm)

From the rounded base the alabastron tapers slightly to the shoulder, the neck having a concave profile and spreading out to a rim of somewhat the same diameter as the widest part of the body. The body and neck are almost completely covered with alternating threads of opaque-red and opaque-yellow, dragged into a feather-pattern and marvered flush with the body. A thread of opaque-yellow has been laid round the rim. The alabastron has two applied serpentine side-handles made of the same material as the body.

Several features of this glass point to its being a modern imitation of the 'core' glasses of the Hellenistic period and earlier in the eastern Mediterranean area. In the first place, it is blown, and not formed on a removable core, as were the glasses made before the invention of the blow-pipe: it has a pontil-mark on the base, which suggests the same conclusion. Secondly, its greenish-blue metal is quite translucent, whereas the normally dark-blue material of the earlier glasses was opaque. Thirdly, its shape, and particularly the form of the handles, cannot be matched from the Hellenistic glasses, which normally run very true to form. Lastly, the 'weathered' surface of the glass appears to have been simulated by artificial treatment, a brilliant spot which was missed in the process betraying the true surface-quality of the metal.

It is not known where such imitations were made, but perhaps the most likely centre is Venice, where 'Roman' glasses are made today with comparably counterfeited 'weathering', and where the opaque glasses required for the decorative threading have always been readily available.

Baron Edmond de Rothschild — P/2051 ter**

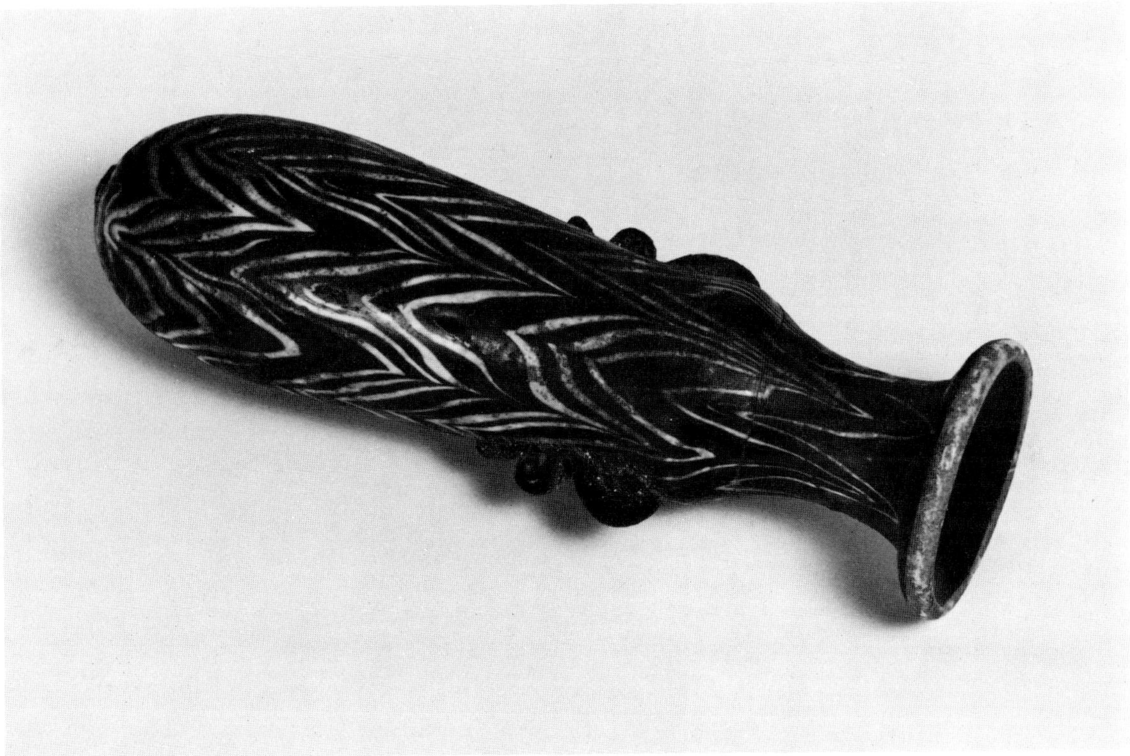

14. Beaker, with enamelled and gilt decoration

Syrian (probably Aleppo); about 1260–70

Brownish-colourless bubbled glass painted in enamel colours and gilt

Height, $5\frac{3}{8}$ in. (13·3 cm); diameter of rim, $3\frac{1}{4}$ in. (8·3 cm)

Slightly tapering cylindrical beaker flaring at the mouth, standing on a low footring made of an overlapping applied coil of glass.

Below the rim is an inscription in gold *nasq* characters outlined in red enamel against a ground of blue enamel. Below this, on the front of the beaker, is the figure of a young man wearing a white turban and a long garment striped in red, white and blue. In his right hand he holds a conical red beaker striped horizontally with green, whilst another – this one striped blue – is set down at his left hand. He is seated between two trees with large green leaves and red and yellow flowers. All the drawing is executed in a thick red enamel over a faint wash of gold.

The reverse of the goblet has suffered from burial and the decoration is not clearly discernible, but seems to show a second figure seated and playing a tambourine.

The inscription reads: عزّ مولانا السلطا ن الملك العالم العادل المجاهدى ('Glory to our Lord, the Sultān, the King, the Wise, the Just and the Warrior (of the Frontiers)' Unfortunately, this gives no clue as to the identity of the Sultan concerned.

This beaker, with its seated figures wearing boldly striped costumes, has a number of parallels, the closest being a beaker originally from the Löwenburg at Kassel, now in the Hessisches Landesmuseum, Kassel.[1] This, however, had an upper border of flying cranes[2] in place of the zone with the inscription on the Waddesdon beaker. Despite this difference, however, the two pieces are closely linked not only by the rendering of the figures, but by the form of the 'trees' which separate them, with their herring-bone arrangement of leaves, the sprays terminating in three flowers;[3] and by the curious fronds terminating in semi-palmettes which seem on both glasses to be used as space-fillers.[4] The Kassel glass is unfortunately not datable by external evidence, but some idea of date is suggested by the presence of a figure with comparable boldly striped garments on a fragment found at Al Mina in north Syria, and therefore dating from before 1268.[5] Confirmatory evidence is furnished by a beaker once in a private collection in Egypt, painted with seated figures playing tambourines and separated by the familiar 'trees'

[1] Lamm, *Mittelalterliche Gläser*, I, pp.329–30; II, Pl.127,3: a better illustration is in I. Schlosser, *Das Alte Glas*, Brunswick, 1956, Fig.44.

[2] This feature is found on a beaker which has been in Europe since the Middle Ages and is now in the Grünes Gewölbe, Dresden (Lamm, *M.G.*, I, p.332; II, Pl.129,2); see also op. cit., I, pp.313–14; II, Pl.120 (fragments from Egypt), especially 120, 15, which also shows a seated figure and the tree-motif discussed below.

[3] This tree-motif recurs frequently – see e.g., Lamm, *M.G.*, I, pp.314–15; II, Pl.121,3,4 (with parts of seated figures visible) 5,8,9,14 (with part of seated figure), 16,18,19, etc. (fragments from Egypt).

[4] Of common occurrence – see Lamm, *M.G.*, I, p.315; II, Pl.121,10,16,18,19, etc. (fragments from Egypt).

[5] Arthur Lane, 'Medieval Finds at Al Mina in North Syria', *Archaeologia*, LXXXVII, 1938, pp.73–4, Fig.13,C. Al Mina is only 80 miles from Aleppo.

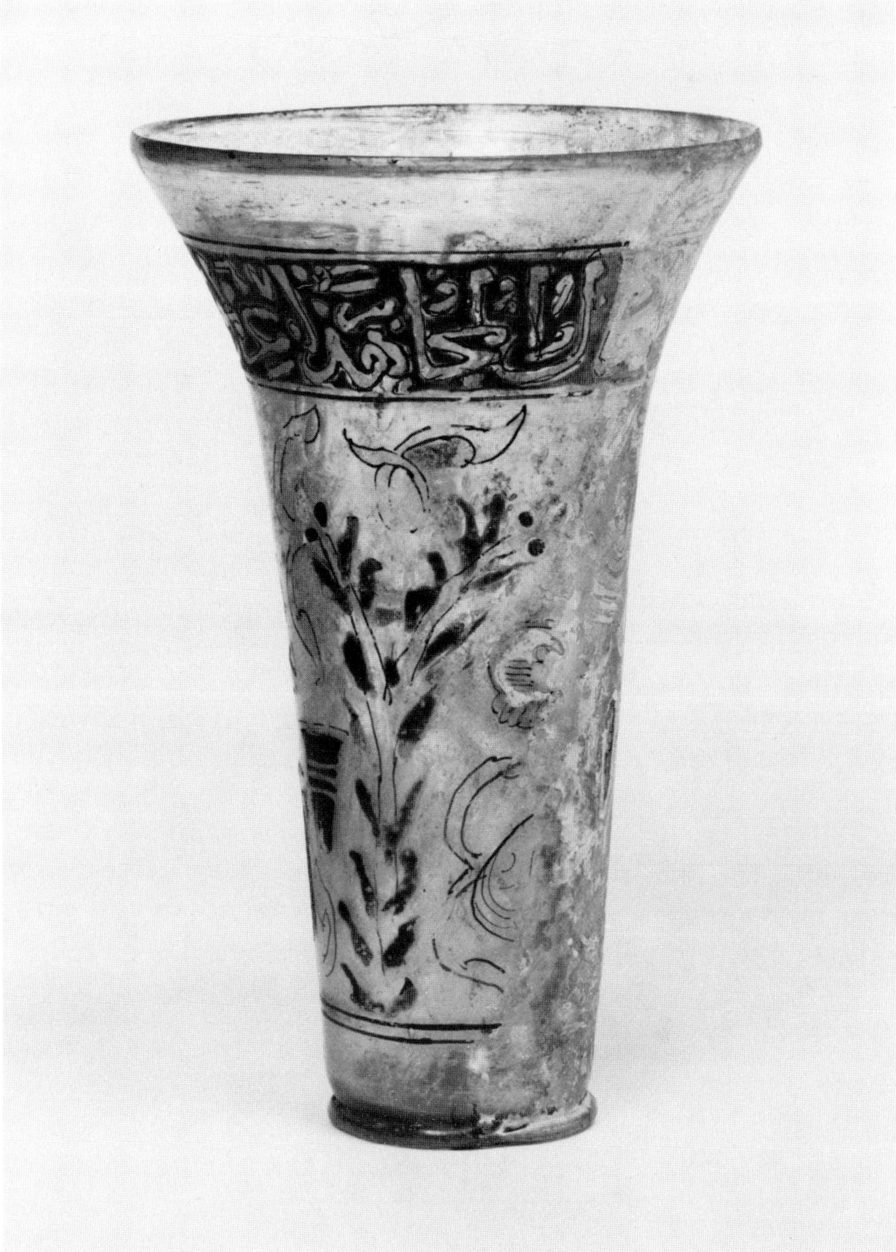

15. Goblet, with coat-of-arms

Venetian; about 1500

Brownish-colourless glass, painted in enamel colours and gilt

Height, $6\frac{1}{4}$ in. (16 cm); diameter of rim, 4 in. (10 cm)

Goblet with bowl of ogee profile and flat base surrounded by a pincered frilled trail, partly gilt; tall pedestal foot with folded rim. The bowl of the glass is painted in red, blue, orange, yellow, green, black and white enamels, and gilt. On front and back are roundels framed in borders of scale-design etched in gold and enlivened with dots of blue, red and green enamel, and bounded by a blue line on the outside, a red line on the inner circumference. Within each roundel is a coat-of-arms – per pale gules and azure, two claw-hammers in saltire or; on a chief of the second three fleurs-de-lis surmounted by a label of four points all of the third; the whole surrounded by a bordure erminois. An ecclesiastical hat of ten tassels sable.

Between the roundels are trophy-like motifs of Renaissance leaf-scrolls bearing up a white spread-eagle between a blue and a red S-shaped cornucopia, each topped with fruits painted in green, yellow and black. Below the rim of the glass runs a border of scale-pattern etched in the gold ground and enlivened with dots of green enamel, between lines of blue enamel above and red enamel below. Round the stem runs a zone of ornament composed of white dots above two sets of two horizontal lines in red and yellow, with vertical blue dashes between them, and below this, white comma-like dashes terminating each in a group of three blue dots.

This goblet was almost certainly made for Filippo Sergardi, a cleric from a Sienese family originating in Montalcino. His father Giovanni became a member of the Sienese Council in 1463, and his brother Niccolò was Console di Mercanzia in 1496, and Console dei Regolatori in 1497.[1] Filippo himself entered the Church, and in 1494 was created Abbot in the diocese of Strigonia, in Hungary. Subsequently he became Chierico della Camera Apostolica in Rome, and was finally despatched by Pope Alexander VI as his Nuncio Extraordinary to the Emperor Maximilian in Germany, where he died.[2] He was also nominated Bishop of Corinth, but unfortunately the date of his elevation is not recorded. The last incumbent of the see is given by one authority[3] as Hugo Rhetut in 1450, but another[4] gives Antonius Sarasco in 1476, and Julius de Blanchis as Archbishop up to 1512. Filippo Sergardi is not mentioned, and it is impossible to determine from the number of tassels on his hat in the coat-of-arms, whether this was his hat as Abbot or as Bishop, this point of protocol not having been rigidly formulated at this early period.

Goblets of this shape are by no means uncommon, and examples may be found, e.g., in the

[1] V. Spreti, *Enciclopedia Storico-Nobiliare Italiana*, VI, Milan, 1932, p.255.
[2] I. Urgugieri, *Le Pompe Sanesi . . .*, Pistoia, 1649, pp.113, 224–5.
[3] P. B. Gams, *Series Episcoporum Ecclesiae Catholicae . . .*, Ratisbon, 1873, p.431.
[4] C. Eubel, *Hierarchia Catholica Medii Aevi*, III, Münster, 1910, p.194; II, p.152.

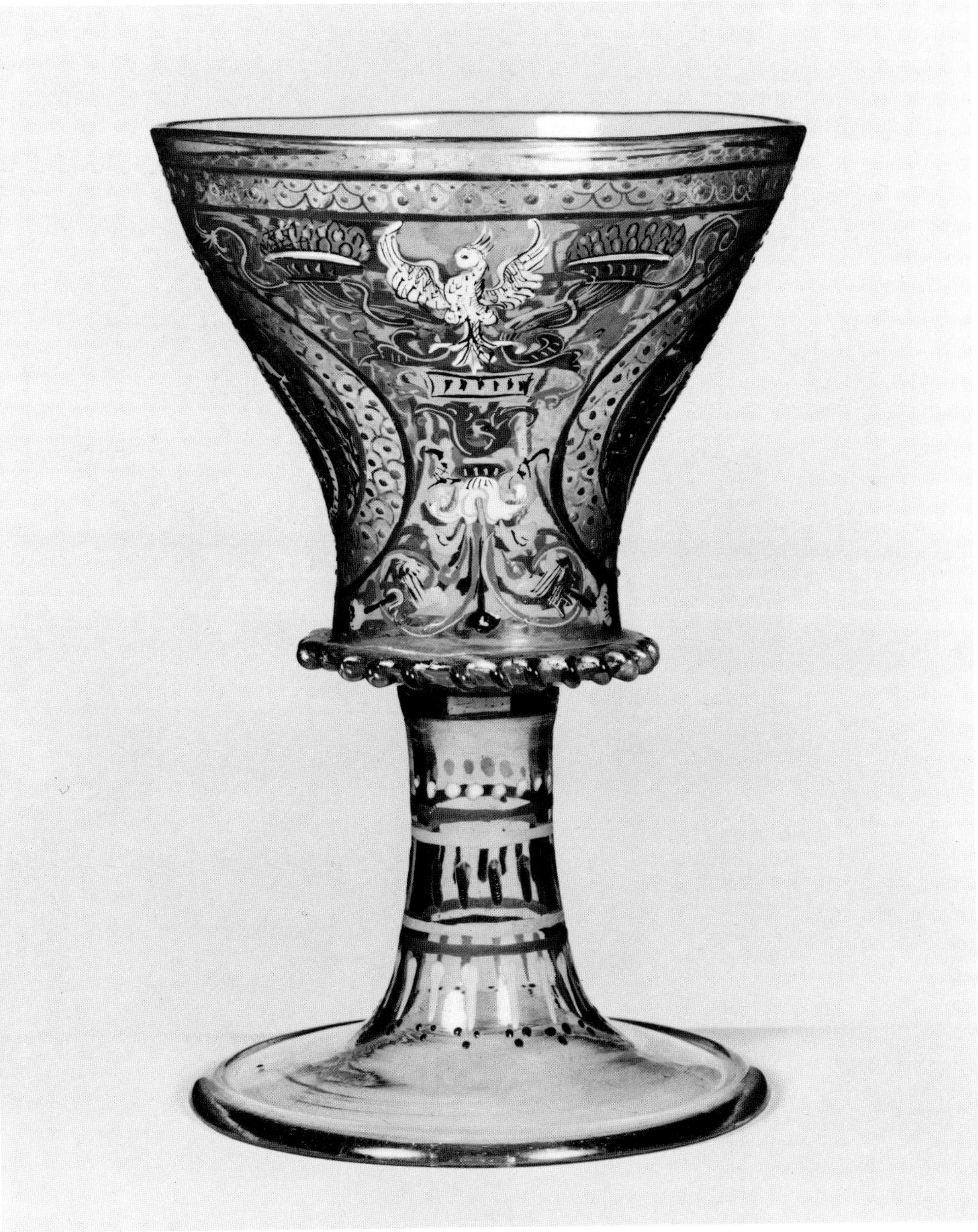

foliate tails hook into a central trophy-like motif surmounted by a white spreadeagle closely resembling that on the Waddesdon goblet, also standing above a hatched bar, while at the base of the 'trophy' is a white inverted calyx reminiscent of that on the Waddesdon glass. Allied to both of these is a handled jug in the Rijksmuseum, Amsterdam,[14] decorated on either side with roundels enclosing figures, and in front with mermen closely allied to those on the Victoria and Albert Museum beaker, and having a 'trophy' between them. Here, instead of the white spreadeagle, we find the two cornucopia motifs which flank the bird on the Waddesdon goblet. It is difficult to resist the conclusion that here we are dealing with glasses emanating from the same workshop, if not from the hand of the same painter.

The motif of a bird with spread wings between two cornucopias is closely allied to, if not derived from, an engraving attributed to Nicoletto Rosex da Modena (*illustrated below*), works from whose hand bear dates between 1500 and 1512.[15]

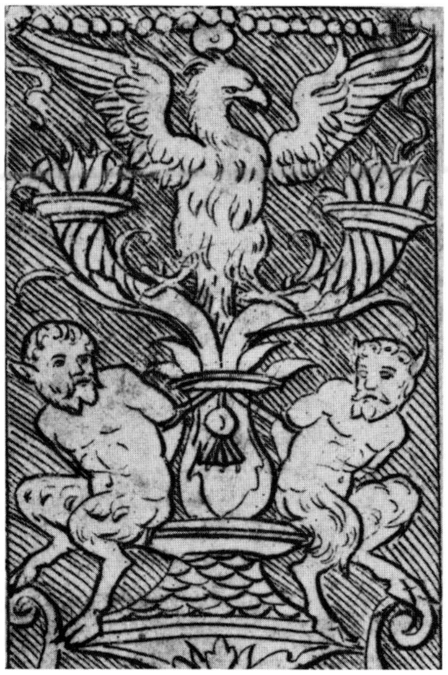

Engraving attributed to Nicoletto Rosex da Modena, c.*1500*
(British Museum)

1922 Waddesdon Inventory (Smoking Room): 'Cup on stem, medallions enamelled with coats of arms. Venetian'.

Miss Alice de Rothschild | W3/40/1**

[14] Franz Carl Lipp, *Bemalte Gläser*, Munich, 1974, Figs.27–9.
[15] Arthur M. Hind, *Early Italian Engraving*, Part II, Vol. V, London, 1948, pp.107 and 138; Vol. VI, London, 1948, Pl.695, No.117.

16. Goblet, with centaurs

Venetian; early 16th century

Greyish-colourless glass, painted in enamels and gilt

Height, 8⅝ in. (22 cm); diameter of rim, 6 in. (15·2 cm)

Goblet with bowl of ogee profile and flat base surrounded by a pincered frilled trail: stem composed of an upper section with central flattened knop, the whole blown into a vertically ribbed mould, separated by mereses from the bowl and from the lower section, which spreads into a plain foot with folded rim.

The bowl of the glass is painted in red, blue, orange-yellow, leaf-green, turquoise, beige-brown, black and white enamels, and gilt. The decoration consists of a frieze of three centaurs in different combinations of colours, each holding a circular shield on his left arm and a large feathered dart in his right hand. Behind each runs a double S-shaped leafy scroll, terminating at one end in a poppy-like flower, at the other in a trefoil. The ground is formed of a green-sward dotted with yellow fleurets and bounded by a turquoise stream. Below the rim runs a border of ovals etched in the gold and outlined by red dots, with a central white dot: the border is edged with lines of dots in red, white and blue enamels. Round the foot runs a border composed of circlets alternately in blue and green, threaded on a yellow ribbon from which diverge pairs of green leaves between each pair of circlets; below this, and separated from it by a red and two yellow lines, is an arcade motif terminating below in alternately green and blue enamel trefoils.

This goblet resembles very closely, in its bowl with frilled base and its rib-moulded stem with central flattened knop, the famous enamelled glass once in the City Museum, Dresden, made for the Bohemian Jörg von Kopidlnansky von Kopidlna, and dated 1511 (Schmidt, *Das Glas*, p.96 and Fig.59). This glass, which differed in having a vertically ribbed foot and also in retaining its cover, was decorated with a ship in a roundel, and with the owner's coats-of-arms at the sides. These were surrounded by leafy scrollwork of a type commonly found in a considerable class of these Venetian enamelled glasses dating from the very end of the fifteenth and the beginning of the sixteenth century; and the ground below was enlivened with black dashes and clumps of 'grass' in very much the manner of the Waddesdon goblet. The same shape is found again in a glass belonging to the Founders' Company, of London, decorated with *putti* and a king riding on sea-horses with interlinked tails (cf. the commentary on *15* here). The foot of this glass has been broken off and replaced in silver-gilt. It was stated by the donor in 1631 to have formed part of the booty brought from Boulogne after its surrender to Henry VIII in 1546. Since the replacement-foot is hall-marked for 1547–8, this seems likely enough (R. J. Charleston, 'Pottery, Porcelain and Glass' in (edd. Ralph Edwards and L. G. G. Ramsey), *Period Guides: Tudor Period*, London, 1956, p.84, Pl.46,b). Two further goblets of this shape, differing only in the formation of the foot, bear virtually identical decorations of fantastic male and female figures terminating below the waist in fleshy leaf-scrolls, and holding suspended either a Medusa head or a classical tablet (*tabella ansata*). One was in the Collection of Baron Maurice de

Rothschild in Paris (Schmidt, 'Die Venezianischen Emailgläser . . .', pp.270–1, Fig.13); the other is in the Victoria and Albert Museum (Mus. No.674–1884, from the Castellani Collection, Sale (1884), Lot 407).

These parallels could be multiplied (see, e.g., Corning, *Three Great Centuries . . .*, No.21, Pl.7, and the notes thereon). One further glass, however, does deserve mention here. It is a blue goblet in the Cleveland Museum of Art, with vertically ribbed foot and a rather pronounced frill round the base of the bowl (Corning, *Three Great Centuries . . .*, No.18, Pl.6). On one side it is decorated in enamels with the late-medieval theme of a pedlar being robbed of his wares by monkeys; on the other with three women on the bank of a stream, and one in the water, being shot at by blindfold cupid. This latter design has been traced to a Ferrarese Tarocchi card dated by A. M. Hind to about 1460–7 (William D. Wixom, 'A Venetian Tazza with Putti', *Cleveland Museum of Art Bulletin*, February, 1962, pp.24–31, Figs.3–4, citing Arthur M. Hind, *Early Italian Engraving*, New York and London, 1938, I.239, E.1.43a, Pl.362). The *Pedlar and Monkeys* design may similarly be traced to late-fifteenth-century Italian engravings (see Hind, op. cit., II, Pls.72–3). Two of the female figures from the Tarocchi card moreover reappear, as Mr Wixom has pointed out, in different positions on the famous blue *coppa nuziale* in the Murano Museum, traditionally attributed to Angelo Barovier (see Gasparetto, Figs.21,21 *bis*, 21 *ter*; also p.17 here). Although this attribution is to be treated with scepticism, this glass is nevertheless acceptable as one of the earliest fifteenth-century Venetian enamelled glasses extant, and the Cleveland goblet may also be accepted as dating from the last quarter of that century. It is therefore evident that the shape of the Waddesdon goblet was already current in coloured metal well before 1500, even if it was not transposed into colourless *cristallo* until somewhat later.

The decorative theme of centaurs was much favoured on the enamelled glasses of this period. A blue jug in the Robert Lehman Collection (now in the Metropolitan Museum of Art, New York), and dating probably from the late fifteenth century, is decorated with a figure of a rather small centaur executing a Parthian shot with a huge bow, amidst leafy scrollwork of the type common to these glasses (Corning, *Three Great Centuries . . .*, No.10, Pl.5). A wine-glass on a low pushed-in pedestal-foot, of a shape which appears to have come into fashion about 1500, in the Town Museum of Prostějov, in Czechoslovakia, is decorated with centaurs who hold a baton in their left hand and an unidentified object in their right (a pelt?). This glass is dated in diamond-point '1518', and was therefore presumably made either in or before that year (Prague, *Venetian Glass . . .*, No.7, Fig.5). Centaurs also appear on the goblet in the Gemeentemuseum, The Hague, cited in the commentary on *15* here. Possible sources, with suitable adaptations, for these themes may be suggested. There is a drawing in the British Museum representing a battle between two centaurs (Mus.No.B.M.1920–2–14–1(23^v) – see *A Book of Drawings formerly ascribed to Montegna*, Introduction by Campbell Dodgson, London, 1923, p.15, Pl. XXIV). It comes from a book of drawings, dated to about 1465, which belonged to a certain Matteo, son of Messer Ruberto Macigni, who became a Venetian citizen in 1500. Matteo himself died in Padua on 31 August, 1582 (ibid., p.7). The attribution of these drawings is uncertain, but an artist working at a small court such as that of Ferrara has been suggested (ibid., p.9). It is therefore by no means inconceivable that these drawings should have been available for copying in Venice about 1500. That the glass-painters in Murano were closely in touch with the major arts in Venice has recently been well demonstrated by Mr T. H. Clarke (*Journal of Glass Studies*, XVI, 1974, pp.22–56).

17. Pilgrim-flask

Venetian; early 16th century

Greyish-colourless glass painted in enamel colours and gilt

Height, $12\frac{7}{8}$ in. (32·7 cm); width, 7 in. (17·8 cm)

Flask with flattened piriform body drawn into a slender neck slightly spreading towards the lip, and mounted on a low pedestal foot with upward-folded rim. The flask has two loops on each side, above and below the greatest diameter of the body.

The body of the flask is decorated on each side with a large roundel formed of concentric zones of imbricated design etched in the gold leaf and enlivened with blue, red, green and white enamel dots. This roundel is encircled by smaller roundels formed of circles of white dots surrounding a central motif of alternately a blue blob with a white centre or a gold disc with a white dot on a larger red dot. The foot is decorated with a border of cross-hatching on gold within two lines of white dots, and two similar borders embellish the neck. In addition, the rim of the foot is painted with dashes of alternate white and red enamel, while on the neck a thick line of blue enamel has been painted between two thinner lines of red enamel, as a border between the two cross-hatched borders. On the narrow sides of the flask are lozenge-shaped panels of gold leaf etched with scroll patterns and enlivened with dots of white, blue and red enamels.

Pilgrim-flasks are by no means uncommon among the Venetian enamelled and gilt glasses of the late fifteenth and sixteenth centuries. An example in the Museo Civico in Bologna bears the arms of the Bentivoglio and Sforza families, and was probably made in connection with the marriage of Alessandro Bentivoglio and Ippolita Sforza in 1492 (Gasparetto, p.83 and Fig.21). This piece has the up-folded foot-rim decorated with a cable border in exactly the manner of the Waddesdon glass.

The pilgrim-flask form occurs in *calcedonio* glass, imitating agate and other comparable figured natural stones, and this material appears to have been one particularly favoured in the years about 1500 (an example is in the E. Wolf Collection, São Paulo, Brazil – see Museu de arte moderna de São Paulo, *4000 Anos de vidro*, 1957, No.27, *ill.*).

Other examples with enamelled coats-of-arms yield less helpful suggestions for dating. One in the Wallace Collection, London, bears the arms of Wilhelm von Rappolstein of Alsace, and of the first Count Liechtenstein, who married von Rappolstein's daughter (George Savage, *Glass*, p.58, *ill.*); the Liechtenstein arms are those borne by the Count between 1523 and 1526 (information kindly communicated by Mr. A. V. B. Norman): another flask, in the Bayerisches Nationalmuseum, Munich, has the arms of von Pfintzing, a Nuremberg family with possessions in Silesia (Walther Bernt, *Altes Glas*, pp.51–2, Pl.31), and probably dates before 1535 (kindly communicated by Dr Rainer Rückert, who points out that the von Pfintzing arms were altered in 1554, thus ruling out the late sixteenth-century date proposed by Bernt): a third, with arms of an unidentified bishop, is in the E. Wolf Collection in São Paulo. Enamelling appears to have continued in Italy as a means of decorating glasses for customers in northern and central Europe after it has fallen out of fashion in Venice itself, and this may to some extent account for the late dating of the first two flasks with heraldic bearings mentioned above. It is probably wise to date the Waddesdon flask by the Italian analogues, such as the Bentivoglio-Sforza example already cited and in accord with the generally early sixteenth-century dating of Venetian glasses with a slight ornamentation of imbricated designs incised through the gold leaf, rows of enamelled dots, etc. A flask with a decoration of this kind which comes very close to the Waddesdon example is in the Leopold Blumka Collection, New York (Corning, *Three Great Centuries . . .*, No.39, *ill*): another is in the Musées Royaux d'Art et d'Histoire in Brussels (Anne-Marie Berryer, *La Verrerie Ancienne aux Musées Royaux d'Art et d'Histoire*, Brussels, 1957, p.19 and Pl. VIII; see also Liège, Musée Curtius, *Trois Millénaires . . .*, pp.123–4, No.255).

1922 Eythrope Inventory: 'Enamelled pilgrim bottle, Venetian 16th century'.

Miss Alice de Rothschild E1/13/3*

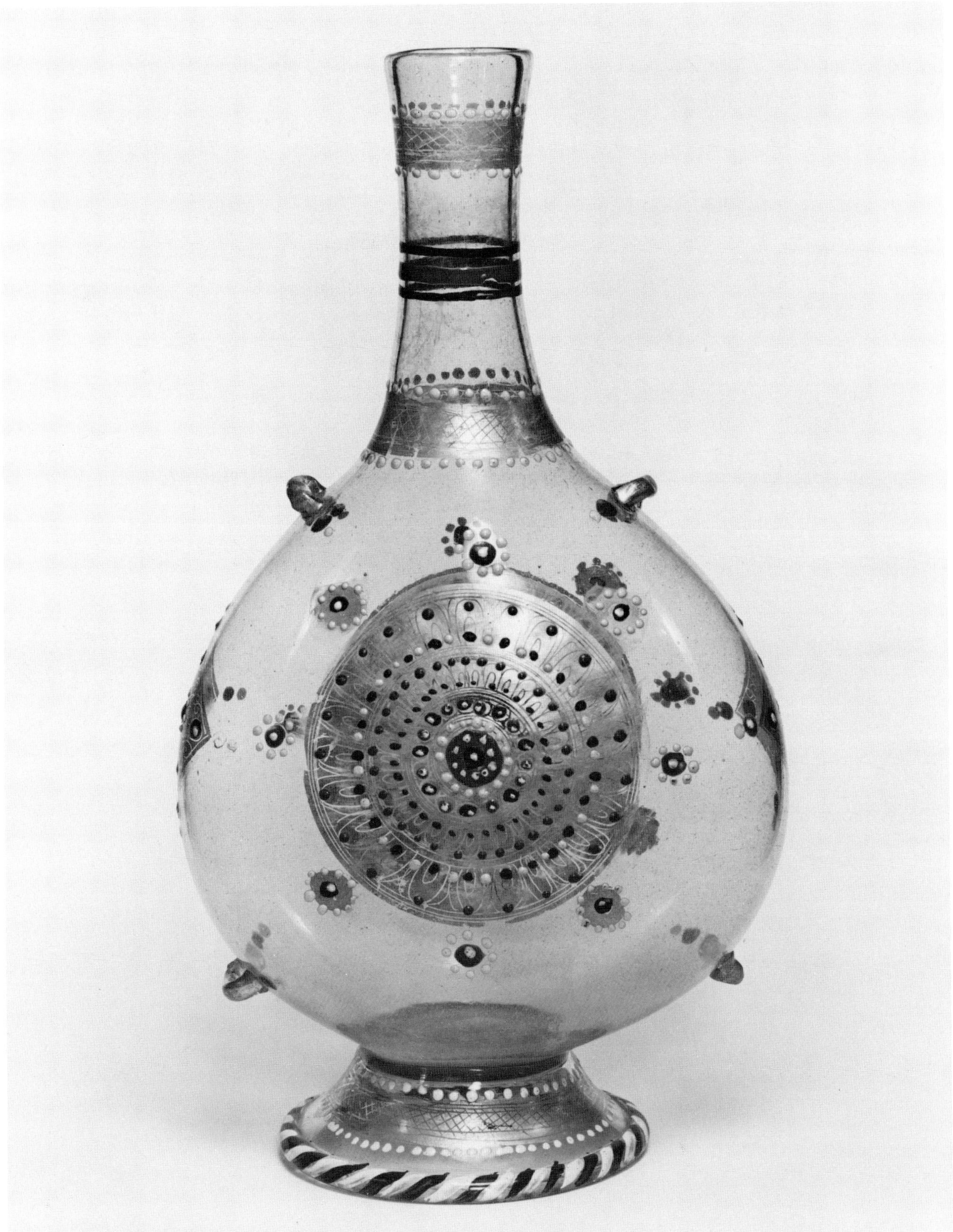

18. Pilgrim-flask

Venetian; early 16th century

Greenish-colourless glass painted in enamel colours and gilt

Height, $14\frac{1}{4}$ in. (36·2 cm); width, 8 in. (20 cm)

Flask with flattened almost circular body drawn in to a slender neck slightly spreading towards the lip, and mounted on a low pedestal foot with upward-folded rim. The flask has two loops on each side, above and below the greatest diameter of the body, and a hole pierced on each side of the foot, to take a suspensory cord passing underneath.

The body of the flask is decorated on each side with a central gilt disc-motif outlined by a circle of alternating blue and red enamel dots, with a single red dot in the centre. At the cardinal points are symmetrical gilt palmette-motifs with dots of blue, red and green enamel; these convert the circular design into a lozenge-shaped one, presumably corresponding to the original dimensions and shape of the gold leaf used. Round this runs a broad circular border of gold etched with circles, each enlivened by a circle of white enamel dots round a central green one; this central zone is edged with single lines of alternating blue and red enamel dots, and these in turn by rows of white dots. Similar borders encircle the neck and foot. The foot-rim and side-loops are gilt.

For the general discussion of this type of glass, see the previous item.

A flask coming close to the Waddesdon example in the lozenge-shaped disposition of the central motif is in the Toledo Museum of Art, Ohio (*Art in Glass*, p.47, *ill.*)

1922 Eythrope Inventory: 'Small enamelled pilgrim bottle. Venetian, 17th century'.

Miss Alice de Rothschild E2/16/2**

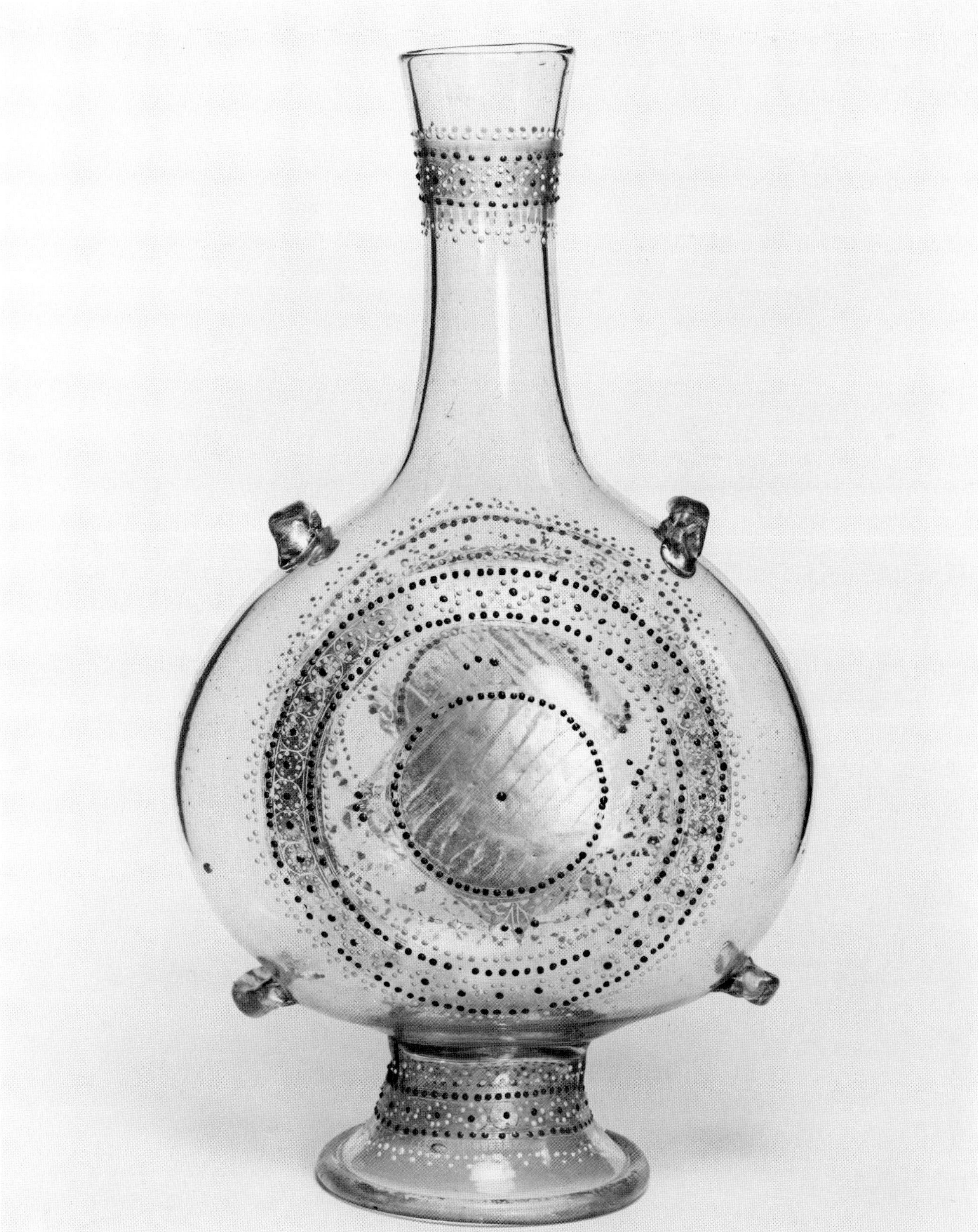

19. Ewer in the form of a lion

Probably Venetian; second half of 16th century

Greyish 'crystal' glass with decoration of incorporated opaque-white (*lattimo*) stripes, blown in a mould, and mounted in silver

Height, overall, $9\frac{7}{8}$ in. (25 cm)

The ewer is mould-blown in a fixed mould having the form of a half-crouching lion holding between its forepaws a cartouche in the form of a shield with scrolled border. The remainder of the paraison has been drawn out into a long tube which is turned upwards and curved outwards to represent the lion's tail. The original crown on the head of the lion has been broken off and replaced in silver. The glass is decorated with narrow parallel stripes (formed of collapsed tubes) of opaque-white glass incorporated in the 'crystal' matrix. The point of junction of these stripes is on the lion's forehead, and the paraison was probably twisted before it was blown in the mould.

The head is open inside the crown, and round this aperture runs a thick coil of glass, presumably originally the lower part of the glass crown.

The ewer has been mounted on an elaborate *repoussé* silver foot of later date. This foot and the silver crown both bear two impressed marks, one a French import-mark used between 1809 and 1819 (Rosenberg, *Der Goldschmiede Merkzeichen*, Berlin, 1922–28, IV, p.206, No. 5903), but also used in Holland 1812–13 (E. Voet, *Nederlandse Goud- en Zilvermerken 1445–1951*, The Hague, 1951, p.41; the other a Belgian import-mark used between 1831 and 1869 (Rosenberg[3], IV, p.12, No.5002).

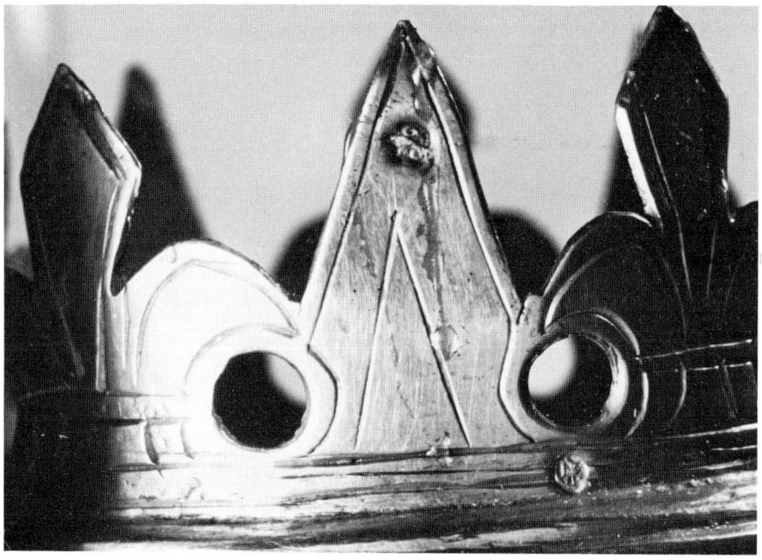

Above: French import mark used between 1809 and 1819. Below: Belgian import mark used between 1831 and 1869 (Enlarged four times)

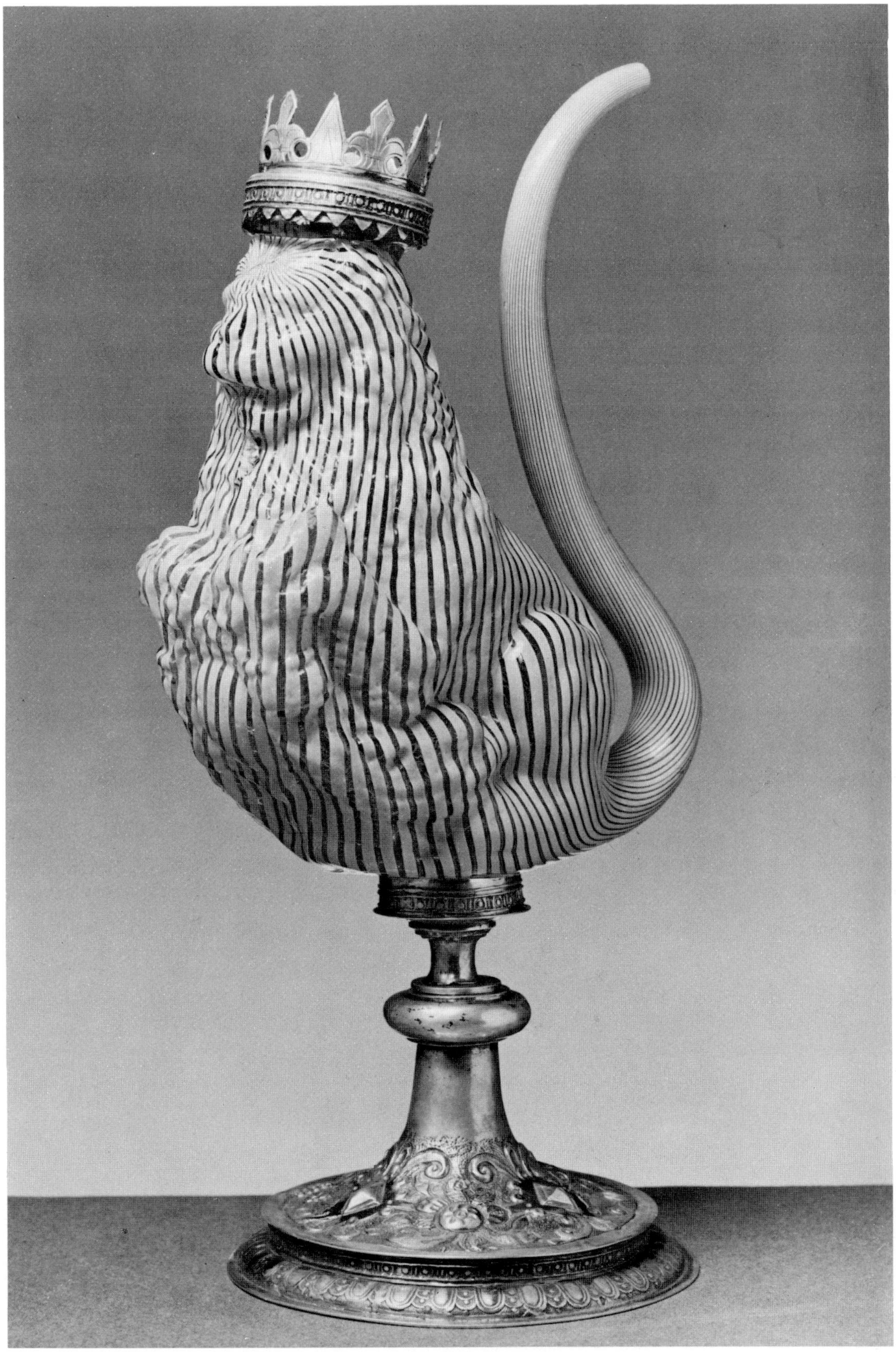

Glass ewers in the form of crowned lions are by no means uncommon, but most of them are differentiated from the Waddesdon example by having a 'shield' formed of strong S-shaped scrolls at the sides and a motif on the front resembling an acanthus-leaf. One of these is in the Museo Arqueologico Nacional, Madrid (Inv. No.13144) and has been furnished with a glass stem fixed to it by a metal join; whether this stem is original, there is no way of knowing, but it certainly seems Spanish in character (the piece is illustrated without its stem by Luis Pérez Bueno, *Vidrios y Vidrieras*, Barcelona 1942, Pl. facing p.93). A second example, mounted on a patterned metal foot like the Waddesdon piece, was once in the Miguel y Badía Collection (J. Gudiol Ricart and P. M. de Artíñano, *Vidrio: Resumen de la Historia del Vidrio; Catálogo de la Colleción Alfonso Macaya*, Barcelona, 1935, Pl. XIV), and is now in the Barcelona Museum (J. G. Ricart, *Los vidrios Catalanes*, Barcelona, 1941, Pl.34, in colour). Exceptionally, this piece is ornamented with coloured canes, sliced in the millefiori manner and embedded haphazard in the crystal matrix. A third ewer, on its original ribbed foot with two depressed spherical knops above, entered the Musée Céramique at Sèvres in 1885 (Legs du Baron Ch. Davillier); perhaps because of its traditional ascription to Spain, and because of the Spanish provenance of the two pieces already cited, it was published as Spanish by Mrs A. W. Frothingham (*Hispanic Glass*, New York, 1941, p.44 and Fig.31), although there is nothing specifically Spanish about the formation of its stem and foot. The glass shows traces of gilding in several places. A fourth of these glasses was in the Emile Gavet Collection (*Catalogue des Objets d'Art . . . composant la Collection de M. Emile Gavet*, Galerie Georges Petit, Paris, 1897, Lot 598, *ill.*). This glass apparently preserved its original hollow-blown stem with masks, and this had the exaggeratedly drawn-out form which seems to be a characteristic of some Barcelona glass (see, for example, J. Ainaud de Lasarte, *Ars Hispaniae, X, Cerámica y Vidrio*, Madrid, 1952, Fig.899). In the Museum Kulturen, in Lund, Sweden, is a fifth of these lions, mounted on a hollow stem above a domed foot. This example differs from the others in having wings, and a tail curled into a partial spiral and not merely curved into an S. The wings suggest that this example may have been adapted to represent the Lion of St Mark, and that it may therefore be of Venetian origin; but the form of the foot and the generally rather loose handling of detail, suggest that it may be relatively late in date (Sven T. Kjellberg, 'Europeiskt Glas,' *Kulturen*, Lund, 1953, Fig.13; from the Lanna Collection – see *Sammlung des Freiherrn Adalbert von Lanna, Prag*, Sale Catalogue Rudolph Lepke, Berlin, 23 March, 1911, No.741, *ill.*). The ewers at Sèvres, at Lund and in Madrid share with the Waddesdon example the feature that all have an aperture in the head, within the crown. The ewer was presumably filled through this and the contents poured out through the tail.

A curious parallel to the lion-ewers already cited is furnished by a trick-glass in the Wallace Collection (XXIII.72/G.IV.B.72). This consists of a wide shallow bowl mounted on a hollow stem mould-blown with masks: in the centre of the bowl is fixed the small hollow mould-blown figure of a lion with a twisted curled tail rather like that of the Lund lion-ewer. A small hole in the underside of the lion admits fluid gradually to its body as the level outside rises. The lion is 'crowned' with a 'raspberry prunt'. There seems to be no particular reason for thinking that this glass is anything other than Venetian in origin. One is reminded of the description given by René François, chaplain of Louis XIII, of the Venetian glass of his time: 'Mourano de Venise a beau temps d'amuser ainsi la soif en remplissant L'Europe de mille et mille galanteries de verre et de chrystal . .; on boit un navire de vin, une gondole; on avale . . . un cloche, un tonneau, un oiseau . . . toute sorte de bestes potables et non potables . . .' (*cit.*, A. von Saldern, 'Façon de

Venise,' in *Reallexikon zur deutschen Kunstgeschichte*, Stuttgart, 69. Lieferung, 1972, Cols.1028–9); and another French seventeenth-century author refers to glass 'façonné en toutes sortes de formes, en nef, en cloche, en cheval, en oiseau, en église' (J. Barrelet, *La Verrerie en France*, Paris, 1953, p.90.)

R. Chambon in 'Verres de la Renaissance soufflés dans un moule fermé' (*Glastechnische Berichte*, Bd.32 K, Heft VIII, 1959, pp.49–56), whilst arguing that most of the fixed-mould glasses of the sixteenth century were likely to have been made in Antwerp, seems willing to concede a Spanish origin for the lion-ewers (*loc. cit.*, pp.54–5).

The Waddesdon ewer differs from all the examples quoted not only in the design of its shield, but in being decorated with *lattimo* stripes, a most exacting technique in combination with mould blowing, and one in which the Spanish glass-workers did not excel: in particular, the trick of nipping all the decorative threads neatly together into one point of junction and cutting off the surplus seems to have been more a Venetian than a Spanish one.

1898 Waddesdon Inventory (Smoking Room, No.1120) 'A white ribbed Venetian glass Jug with guardant lion, silver stem and crown. 10 in.'.

Baron Ferdinand de Rothschild — W1/138/6

20. Lobed glass, with white thread decoration

Probably Venetian; late 16th or early 17th century

Greyish 'crystal' glass with decoration of intersecting opaque-white (*lattimo*) threads

Height, 5⅛ in. (13 cm); width of bowl, 5½ in. (14 cm)

Bowl with rim curving out from a globular body which is mould-blown with a diaper of projecting bosses simulating the appearance of a fruit. The upper part of the bowl has been tooled into eight pointed lobes to give a star-like form at the rim, which is folded. The low pedestal-foot, which is made from a separate paraison folded in at the edge, is separated from the body by a merese of clear glass.

The mesh design is formed by a series of cables of opaque-white glass threads incorporated into the body of the glass.

Glasses of this multi-lobed form are by no means rare in Venetian glass, whether plain or, as here, decorated with *lattimo* threads. A six-lobed glass on a tallish stem in the Gambier-Parry Collection at the Courtauld Institute Galleries, London, shows much of the classic elegance characteristic of the mid-sixteenth century, but the slender inverted baluster stem survived into the seventeenth century, as is attested by the Ligozzi drawings, probably dating mainly from the first quarter of the seventeenth century, although some may well be earlier (Zecchin, 'Disegni Vetrari . . .', pp.57 ff): a date in the second half of the sixteenth century would certainly be reasonable for the Gambier-Parry glass (see R. J. Charleston, 'Glass in the Gambier-Parry Collection', *Burlington Magazine*, March, 1967, p.165 and Fig.77). The same is true of an eight-lobed glass with *lattimo* embellishment in the Museum of Decorative Arts, Prague (Prague, *Venetian Glass . . .*, No.102, Fig.26). The seventeenth century, however, produced a squatter type of glass, as the Ligozzi drawings also show, and a *lattimo*-decorated six-lobed glass in the Düsseldorf Museum illustrates this tendency (*Catalogue*, No.248). Slightly taller is a *lattimo*-decorated four-lobed glass in the Musée Curtius, Liège (*Trois Millénaires . . .*, No.278), and this tendency is even more pronounced in another *lattimo*-striped and four-lobed

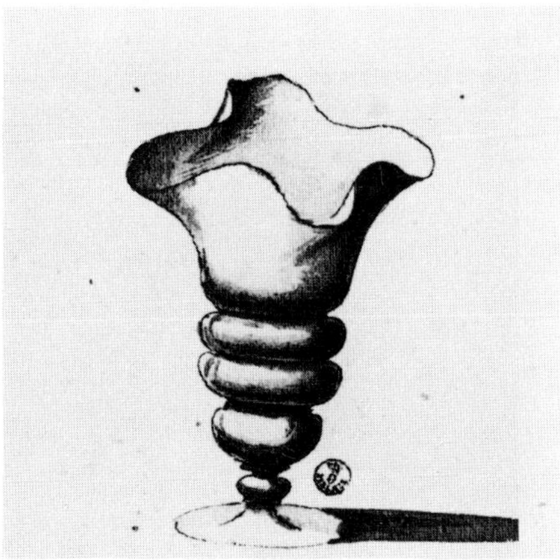

Drawings for glasses, attributed to Giacomo Ligozzi (c.1547–1626) (Gabinetto Disegni e Stampe degli Uffizi, Florence)

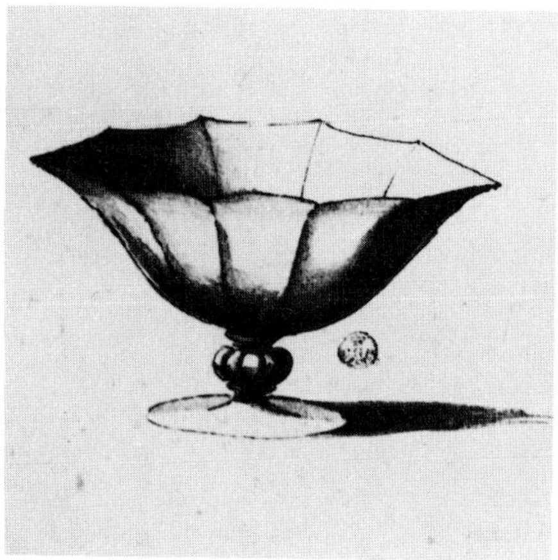

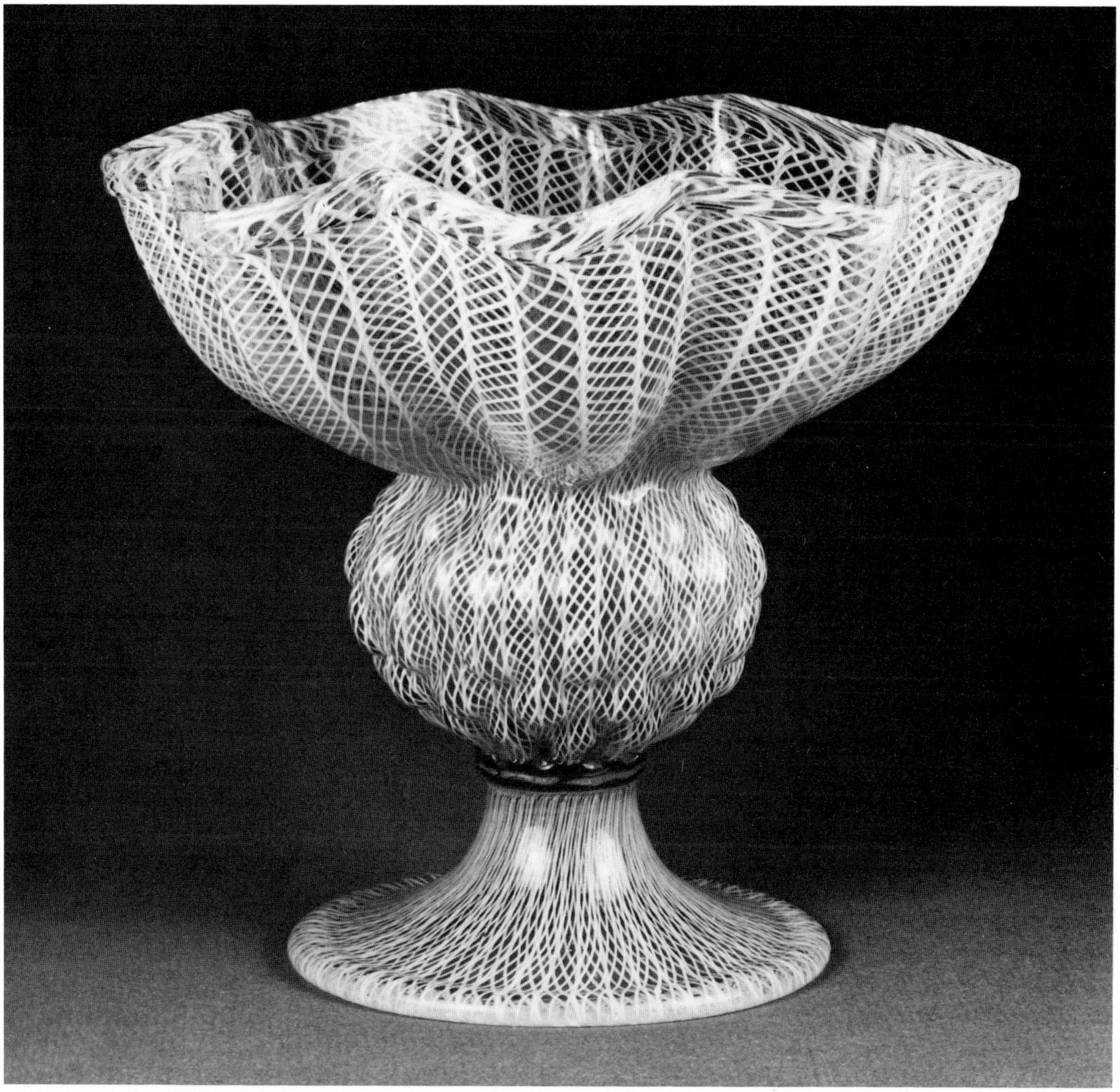

glass in the Jerome Strauss Collection (Corning, *Three Great Centuries . . .*, No.71, *ill.*). This has several bulges below the bowl proper, and begins to approximate to the bowl-form of the glass *22* here; it is strikingly like the Ligozzi drawing.

1898 Waddesdon Inventory, made after Baron Ferdinand's death (Smoking Room, No.1111): 'A circular scalloped shaped Venetian glass tazza, on stem, white lined. 5 in.'.
1922 Waddesdon Inventory (Smoking Room): 'Lobed cup, trellis pattern, Venetian 17th century'.

Baron Ferdinand de Rothschild WI/141/7*

21. Dish, with white thread decoration

Venetian; probably 17th century

Brownish 'crystal' glass with decoration of intersecting opaque-white (*lattimo*) threads forming a mesh-pattern (*a reticello*)

Diameter, 21¾ in. (55·5 cm)

Broad-brimmed dish with conical central projection, and folded rim, decorated by a mesh-pattern of intersecting opaque-white lines with air-bubbles trapped in the mesh.

The dish is broken and restored.

Dishes of this type are by no means rare, but the Waddesdon example is of uncommon size. A dish previously in the Schlossmuseum, Berlin, is recorded as having had a diameter of 65 cm (A. Boesen, *Venetianske Glas på Rosenborg*, Copenhagen, 1960, No.28), and one in the Veste Coburg has a diameter of 61 cm.; a third in the Museum of the History of Science, Florence, has a diameter of 60 cm (M. Celeste Cantù and M. Luisa Righini Bonelli, *The Accademia del Cimento (1657–1667)*, (?) Florence, n.d., p.12). Smaller examples are in the Museo Vetrario, Murano (53 cm – see Boesen, loc. cit.); the Kunstmuseum, Düsseldorf (52 cm – see *Catalogue*, No.263); Rosenborg Castle, Copenhagen (52 cm – see Boesen, loc. cit.); the R. Lehman Collection, now in the Metropolitan Museum of Art, New York (48·3 cm – see Corning, *Three Great Centuries . . .*, No.72, *ill.*); and the Museo Civico, Turin (48·1 cm – see Luigi Mallè, *Vetri-Vetrate-Giade . . .*, Turin, 1971, No.44).

The dating of these dishes, thanks to their simple basic shape, offers peculiar difficulties. It may perhaps reasonably be assumed that the largest dishes occur relatively late in the series. That at Rosenborg was probably brought back from Venice, with a number of smaller dishes, by King Frederick IV of Denmark in 1709 (Boesen, op. cit., pp.63–5): that in Florence, if indeed it formed part of the collections of the Accademia del Cimento, which is not certain, must have been made before the dissolution of that body in 1667.

1922 Waddesdon Inventory (Smoking Room): 'Large round salver, with spiral design; Venetian 17th century'.

Miss Alice de Rothschild — w1/141/9*

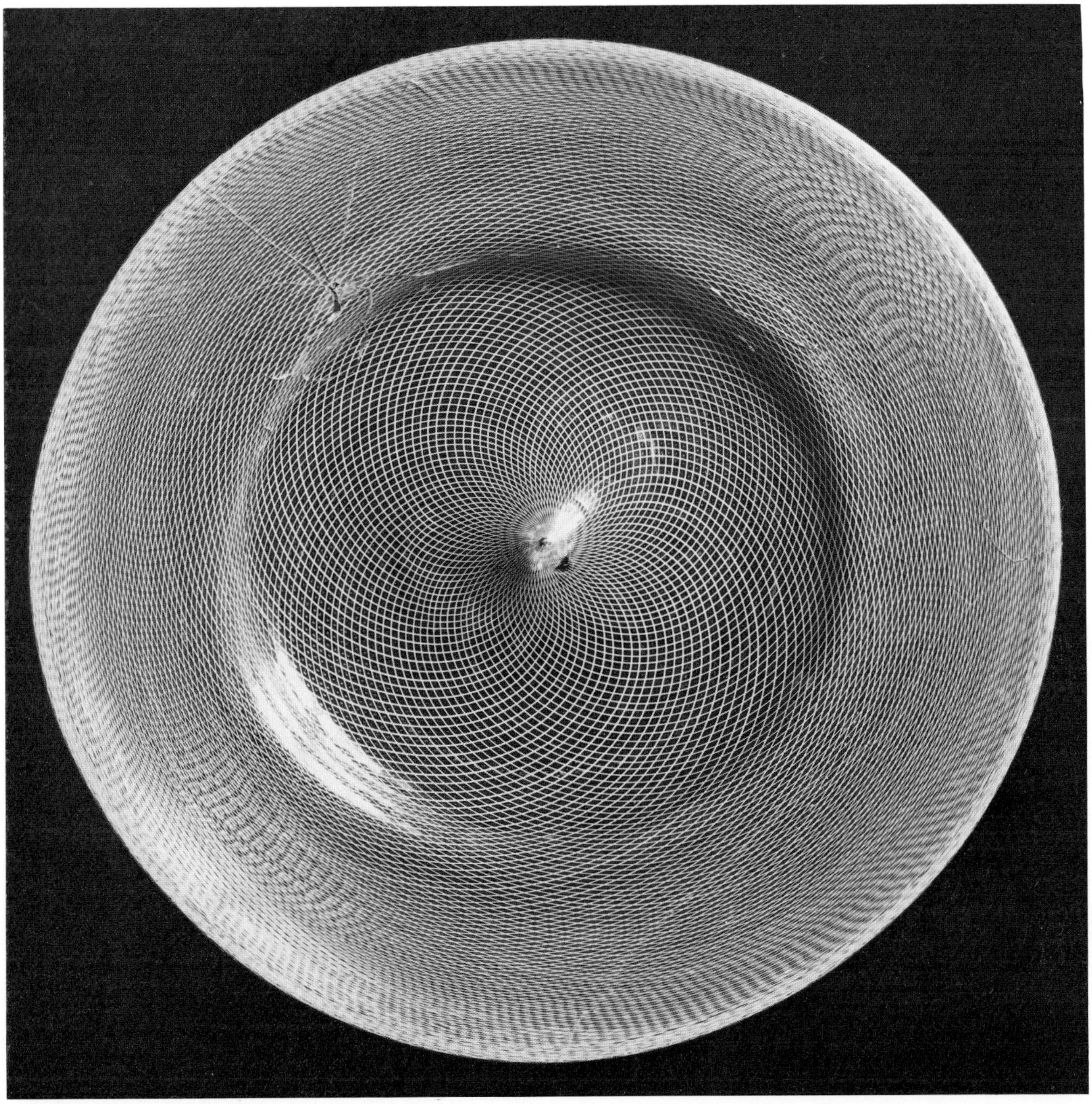

22A. Goblet, with white thread decoration

Façon de Venise (Netherlandish or German); second half of 16th or beginning of 17th century

Greenish-grey 'crystal' glass with decoration of incorporated opaque-white (*lattimo*) threads, mounted in silver-gilt

Height, 10⅜ in. (26·5 cm); diameter of rim, 4¼ in. (10·8 cm)

Goblet with cup-topped bowl, formed of a series of rings diminishing in diameter towards the base: a central depressed knop is separated by a disc of plain glass from the tall pedestal-foot with upward-folded rim.

The glass is decorated throughout by a series of vertical white bands, alternately plain (*a fili*) and twisted (*a retorti*) incorporated in the surface of the glass.

The goblet was presumably broken, and has been mended with a silver-gilt mount which grips the central knop and the base of the bowl by means of bands of acanthus-leaf lappets, and follows approximately the contour of both components of the glass, being decorated above and below the central constriction with embossed gadroons on a ground of punched dots. There are three S-shaped broken scrolls affixed as 'wings' between the upper and lower parts of the mount, perhaps echoing a contemporary type of glass-embellishment in the form of wings.

A number of goblets of this class, with *lattimo*-striped decoration and multilobed form, survive in European and American collections. Most probably originally had covers, usually also of multi-lobed form matching the glass, with a button-finial, usually in plain metal. Examples of this kind are in the Krug Collection, Mülheim (Klesse, *Krug*, I, No.90, *ill.*); in the Victoria and Albert Museum, London (Mus.No.1914–1855, Honey, *Glass*, p.59 and Pl.30,B); in the Butler Collection, sold at Christie's in May, 1964 (Savage, *Glass*, Fig.40); in the Kunstgewerbemuseum, Cologne (Klesse, *Glas*, No.179, with metal foot); in the Museum of Arts and Crafts, Prague (E. Poche, *Umĕleckoprůmyslové Museum v Praze*, Prague, 1955, Fig.26, with metal mounts); and in the Museum für Kunsthandwerk, Frankfurt-am-Main (*Catalogue*, Nos.197 and 198, *ill.*), these last three having four or five lobes above the cover and below the bowl respectively. Other goblets of the same kind have less closely matched covers. One is in the Krug Collection (Klesse, *Krug*, II, No.505, *ill.*, the finial restored; also Düsseldorf, Kunstmuseum, *Meisterwerke*, No.63, *ill.*, with further literature): a second is in the Jerome Strauss Collection (Corning, *Three Great Centuries . . .*, No.60, *ill.*). The cover of this piece has been declared to be a replacement (Klesse, *Krug*, I, p.139), but the formation of the canes on glass and cover match: on the basis of such a small sample it is probably unwise to be dogmatic.

The remaining glasses in this series lack covers. Examples are in the Museum für Kunsthandwerk, Frankfurt-am-Main (*Catalogue*, No.195, *ill.*, with alternating plain and cable *lattimo* stripes in 'wrythen' formation, the stem apparently mended with a metal mount having ring-like projections); in the Museum für Angewandte Kunst, Vienna (Vienna, *Venezianer Gläser*, Pl.34); in the Musées Royaux d'Art et d'Histoire, Brussels (Chambon, *L'Histoire de la Verrerie en Belgique*, Pl.XIV, No.46); two in an Antwerp private collection (ibid., Nos.47 and 48, the latter with a metal foot and stem having wing-like projections).

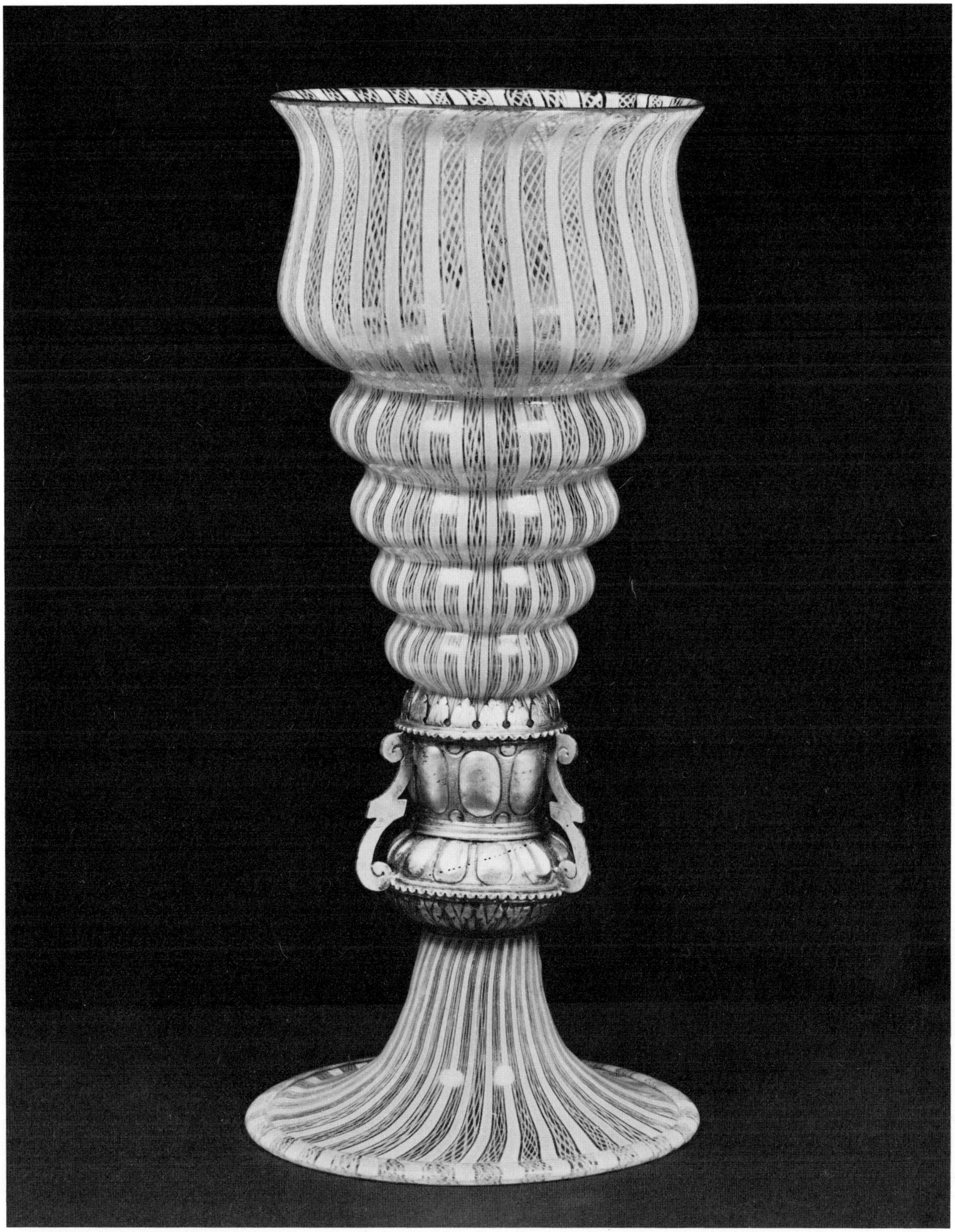

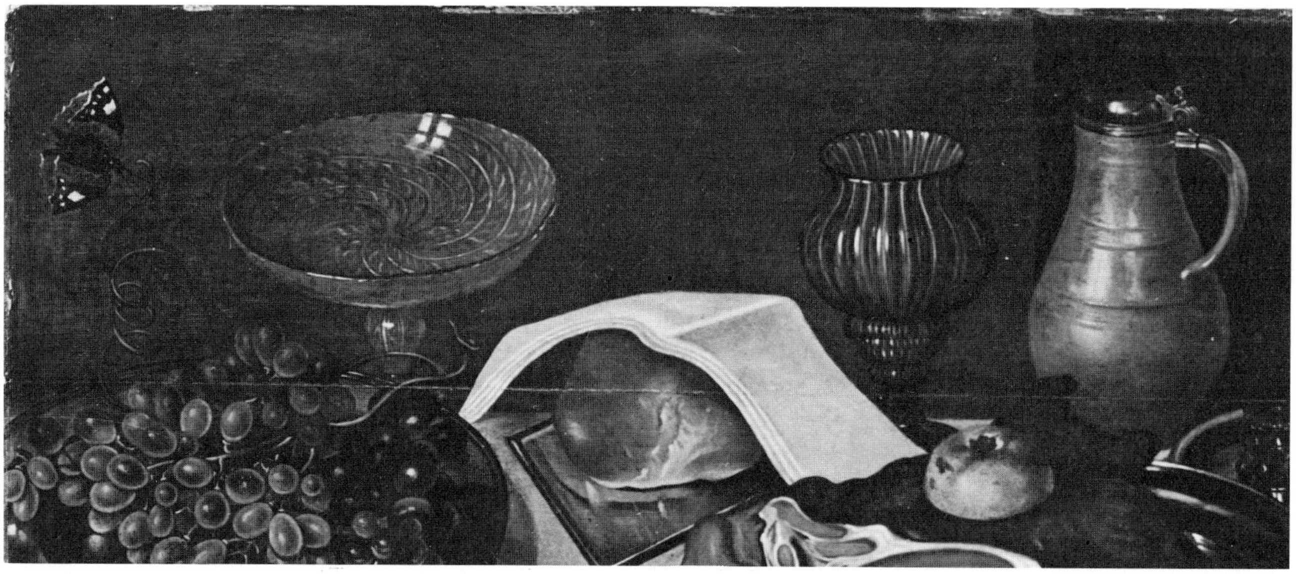

Still-life (detail) by an unidentified Netherlands Master; about 1580 (Present owner unknown; photograph by courtesy of the Van Abbe-Museum, Eindhoven)

Glasses of this type have been variously attributed, but the attributions have seldom been argued. That some at least were made in the South Netherlands seems extremely probable. The illustrated catalogue of the Colinet glassworks at Beauwelz, in present-day Belgium, shows a glass of this shape without cover, under the heading 'Verres Coquilles', which seems to imply the waved outline of diminishing rings. It is described as 'Grant verre a col en coquille verre couleurs fils tirés pied francais';[1] a later hand has added: 'le faisont aussy sans piet francois mais plat'.[2] The Colinet catalogue is datable to 1550–5.[3] A glass of this type is shown in a still-life painting by a Netherlands Master, dated to about 1580 (Ingvar Bergström, *Dutch Still-Life Painting*, London, 1957, Fig.21): and a goblet with bowl of much the same shape, but without *lattimo* stripes,

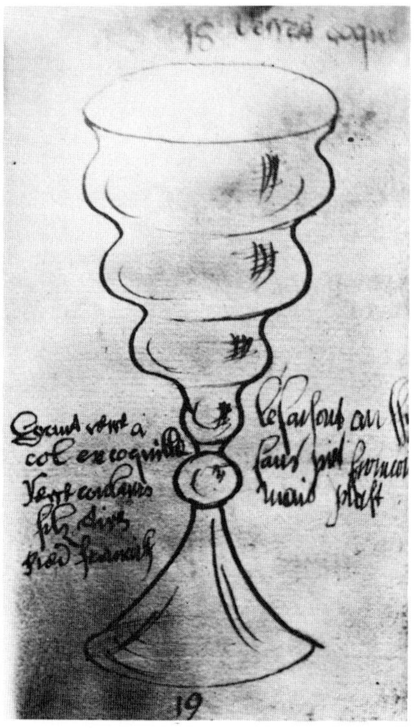

Drawing from the Colinet Catalogue (Collection of M. Raymond Chambon)

[1] 'Tall glass with shell bowl coloured glasses drawn threads French foot'. This elliptical description is difficult to translate. The sense of 'en coquille' is perhaps illuminated by the meaning of 'coquille' used of bread-making: 'to swell, or form blisters'. 'Pied francais' in the Colinet Catalogue normally means a pedestal foot. No examples of this type of glass in coloured metal, or with coloured threads, appear to be recorded (except the probably Bohemian enamelled glasses described below).

[2] 'They also make it without the French foot but flat'. The model with flat foot is shown as No.19 on the same page of the pattern book. See also n.1.

[3] See Chambon, op.cit., pp.86–8; id., 'Note relative à deux chefs-d'œuvre en verre fabriqués en Belgique au milieu du XVIe siècle', *Silicates industriels*, XVI,7 (August-September, 1951), pp.1–3; id., 'Influence vénitienne sur la production verrière de la Belgique à la fin du XVe et au XVIe Siècle', *Cahiers de la Céramique*, 18, 1960, pp.120–34, esp. Fig.25a and b, where drawings of these *verres coquilles* are reproduced.

and mounted on a 'cigar-stem' with wings, is shown in a *Still-life with Fruit* by Osias Beert (1580–1624) in the possession of William Hallsborough, Ltd, London. It seems, therefore, that this form of glass was both known and made in the Netherlands in the second half of the sixteenth and probably also the early years of the seventeenth century.

None of the glasses with Netherlands affiliations seems to have a cover.

The shape was also familiar in the Bohemian/German area. A number of coloured (and occasionally colourless) enamelled glasses of Bohemian type, with dates in the last two decades of the sixteenth century, are known in this shape (e.g., A. von Saldern, *German Enameled Glass*, pp.250–1, No.5, in blue glass, dated 1597; p.252, No.6, in colourless glass, dated 1598; Fig.302, in blue glass dated 1582; Lipp, *Bemalte Gläser*, Fig.177, d.1582; cf. also the blue glass dated 1601 in Düsseldorf, Kunstmuseum, *Alte Gläser*, Fig.26), and the Bohemian glass-makers at this date were certainly able to exploit the *lattimo*-striped technique (see K. Hetteš, 'Venetian Trends in Bohemian Glassmaking in the 16th and 17th Centuries', *Journal of Glass Studies*, V, 1963, pp.39 ff.). The attribution of the Victoria and Albert Museum glass to the Kassel glasshouse is unsubstantiated, although reasonable enough, since the works there certainly made *lattimo*-striped glasses and derived both workmen and styles from the Netherlands (Schmidt, *Das Glas*, pp.127–9, Fig.73; Franz-Adrian Dreier, *Glaskunst in Hessen-Kassel*, Kassel, 1969, Section III, Figs.7–9). The footless bowl of a glass of this type, but with a bulging profile of almost constant diameter, in the Münster Landesmuseum für Kunst und Kulturgeschichte, is mounted in a gilt metal foot in the form of a rampant lion, and celebrated the conclusion of the Peace of Nuremberg in 1650. It is attributed to Nuremberg, but the glass was presumably already of some age when mounted in this way (*Gazette des Beaux Arts*, LXXIX, (February, 1972), p.36, Fig.121). A glass in the Cecil Higgins Museum, Bedford, has a form and decoration close to those of the Waddesdon goblet, but is in addition painted with butterflies in black enamel (*Schwarzlot*) of a type much used in Nuremberg during the second half of the seventeenth century by members of the School of Johann Schaper (see Schmidt, *Das Glas*, pp.215–18, etc.). This glass too was presumably of some age when it was over-painted in this way.

Lastly, a number of these glasses have been attributed to Venice. Although it is perfectly likely that such glasses should have been made there, perhaps primarily for customers in the northern countries, there seems to be no proof of this. Two sheets, however, among the Ligozzi drawings in the Gabinetto dei Disegni in Florence, show glasses of this general profile, although differing in other details (sheets 97326, *recto*, No.202; and 97336, *verso*, No.255, with short stem and dragon-shaped 'wings'). These designs seem to have been intended for execution in the Florence glasshouses of the Grand Dukes, so this general shape may indeed have been made in Italy in the years on either side of 1600 (see Zecchin, 'Disegni Vetrari . . .', pp.57–9; see also observations on *20* in the present Catalogue).

It may well be that the glasses with elaborate, more or less matching, covers are either of German origin or made for the German market.

The cover of the Waddesdon glass does not belong to it, and its description follows on the next page.

22B. Cover, with white thread decoration

Probably Venetian; first half of 16th century

Greyish 'crystal' glass with decoration of applied cables of opaque-white (*lattimo*) threads

Height, $3\frac{1}{2}$ in. (9 cm); diameter, $4\frac{1}{2}$ in. (11·5 cm)

Flanged cover, rising in the centre in steeple form to a button finial surmounted by two discs of plain glass (probably incomplete). The cover is decorated by means of fourteen cables, each of eight opaque-white threads laid on vertically, and standing proud of the surface of the cover. Round the edge of the cover is laid a border consisting of a central cable between two plain bands of opaque-white, marvered into the surface of the glass.

A cover of this type probably came from a beaker (cf. Corning, *Three Great Centuries . . .*, No.66, *ill.*) or a goblet (cf. Prague, *Venetian Glass*, No.97, Fig.26) of slightly earlier type. Goblets with closely corresponding cable-decoration in relief are in the Museum of Applied Arts, Prague (Hetteš, *Old Venetian Glass*, Fig.56) and in a German private collection (Düsseldorf, Kunstmuseum, *Meisterwerke*, No.107, *ill.*). A matching goblet and cover are in the Jerome Strauss Collection (Corning, *Strauss Collections*, No.119, *ill.*).

1922 Eythrope Inventory: 'Tall cup and cover, trellis pattern, silver-gilt stem. Venetian 17th century'.

Miss Alice de Rothschild — E1/14/1**

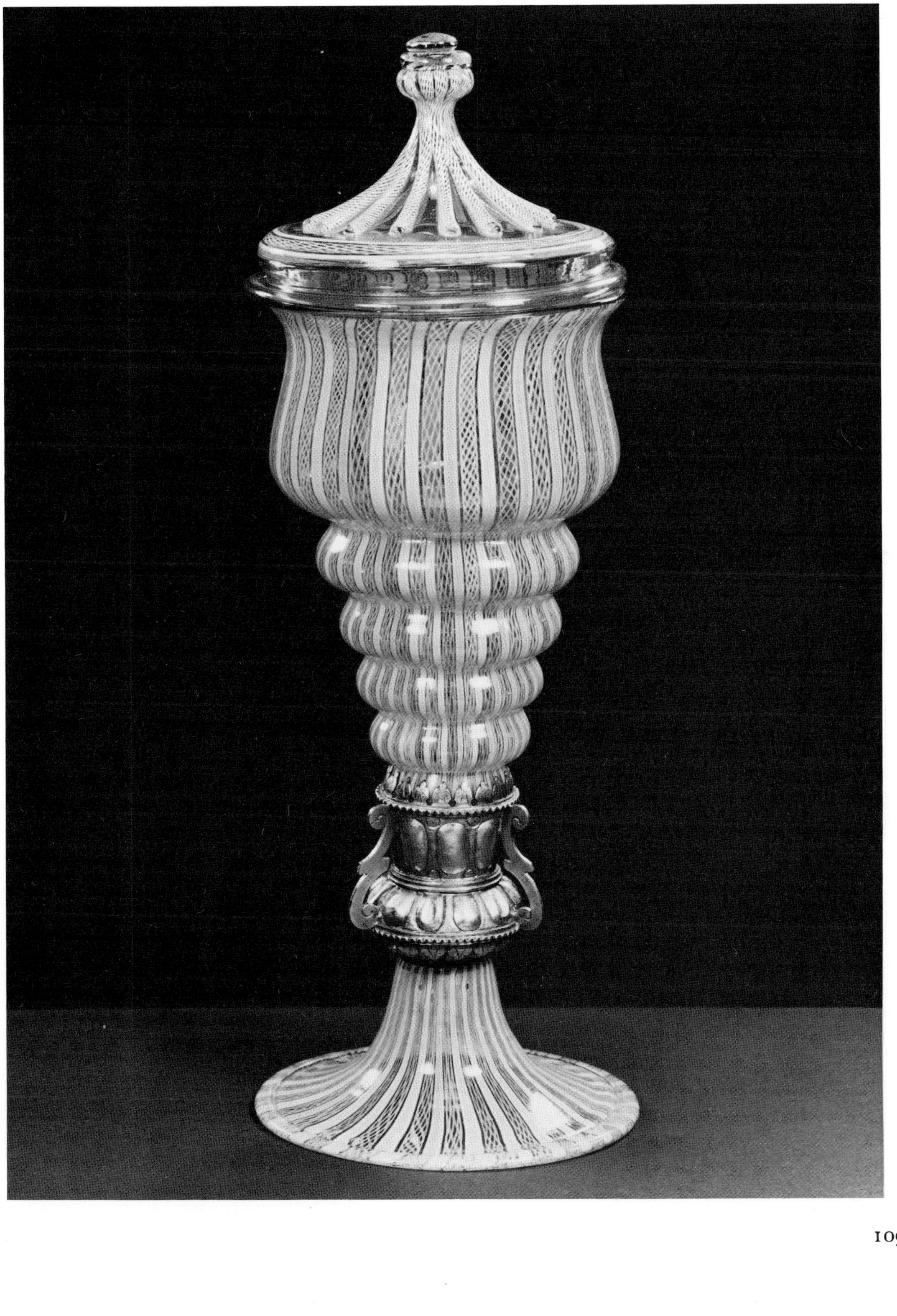

23. Beaker and cover, with white thread decoration

Façon de Venise (origin uncertain); late 16th or 17th century

Yellowish-green 'crystal' glass with decoration of opaque-white (*lattimo*) threads intersecting in a mesh pattern (*a reticello*)

Height, with cover, 14 in. (35·5 cm); without cover, 11 in. (28 cm); diameter of rim, $3\frac{1}{8}$ in. (8 cm)

Cylindrical beaker with slightly offset spreading pedestal foot, made from a separate paraison, the join covered by a waved collar. The domed cover has a wide flange and a finial composed of two hollow knops, joined to the cover by a thin and a thicker merese.

The mesh design is formed by blowing one paraison decorated with incorporated parallel opaque-white threads 'wrythen' in one direction, and a second with the threads twisted in the opposite sense. One is then blown within the other to form a single paraison. The slight projection of the threads ensures that an air-bubble is trapped at each intersection.

This beaker displays several curious characteristics. Apart from the unusual and awkward form of the offset foot, it is made of a markedly green-toned glass, and differs from the glasses made in the usual *a reticello* manner in that the *lattimo* threads appear to project on the inner surface of the inner paraison, and to lie (as normally) on the outer surface of the outer shell. From the yellowish-green tone of the metal and the ungainly form of the foot one might be tempted to suppose that this glass was of Spanish (probably Barcelona) origin, but the shape is unfamiliar in Spanish glass of the sixteenth–seventeenth centuries (although footless cylindrical covered jars are known in plain green glass), and seems to be of German inspiration (*Stangenglas*). Furthermore, the Spanish glass-blowers seem hardly to have ventured into the field of making the more complex types of *lattimo*-striped glasses: the most they seem to have achieved was to use twisted cables (*a retortoli*) rather than plain threads or tubes in vertical or 'wrythen' arrangements. On the other hand, the glass seems unlike the generality of German or Austrian *lattimo*-striped glasses.

1922 Waddesdon Inventory (Smoking Room): 'Tall beaker and cover, trellis design. Venetian, 17th century'.

Miss Alice de Rothschild — W1/140/9*

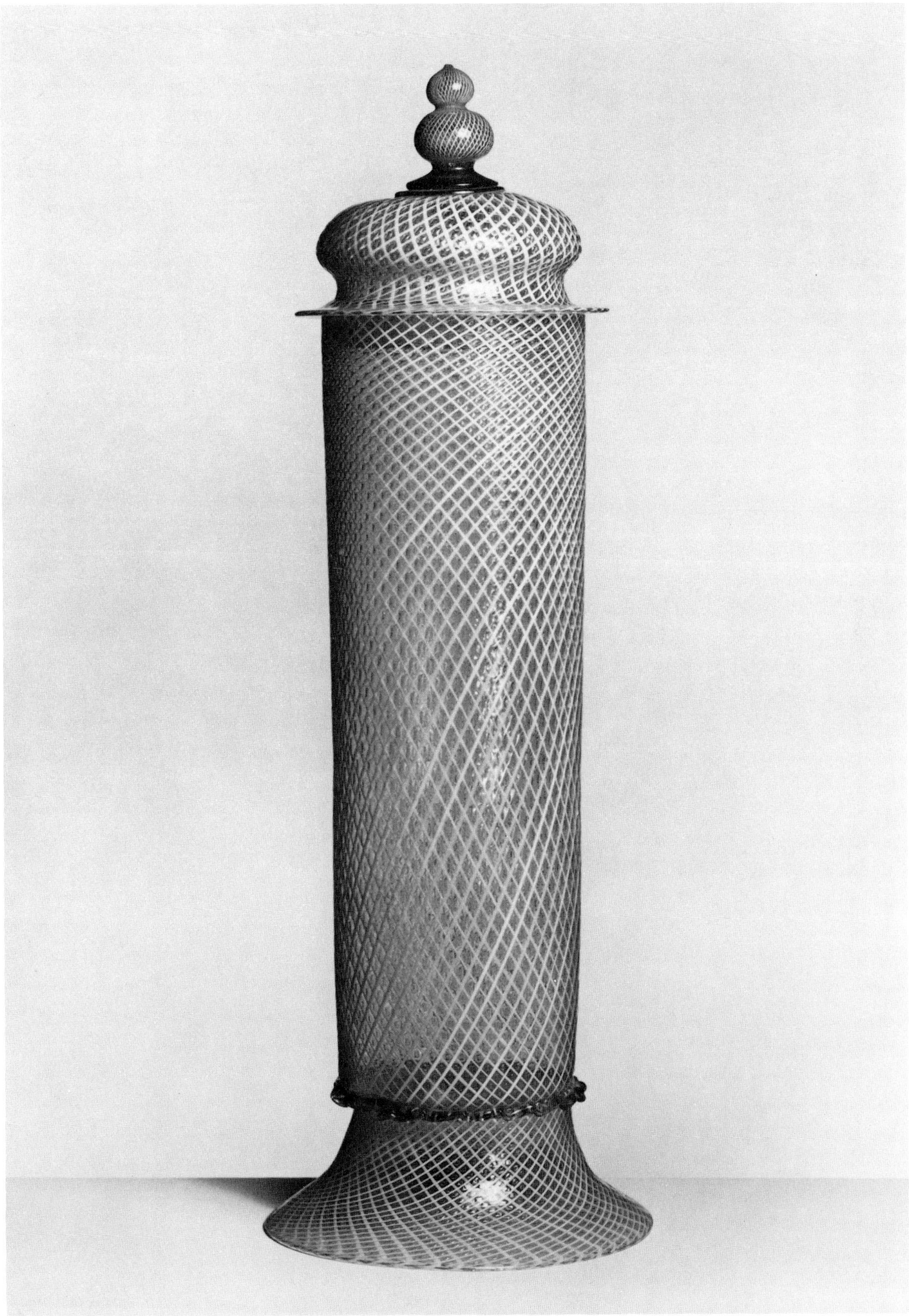

III

24. Goblet-vase, with applied and gilt decoration

Façon de Venise (probably Netherlandish); about 1600

Greyish-colourless glass with applied ornaments in the form of indented bands, and masks alternating with small 'prunts', all gilt

Height of vase, 7 in. (18 cm); diameter of rim, 3⅜ in. (8·5 cm)

Bowl of convex lower and concave upper profile, being widest at the shoulder and then drawn in sharply to a wide neck. Round the shoulder and slightly below it run two horizontal notched cordons interrupted by three larger alternating with three smaller applied 'prunts', the former moulded with lions' masks, the latter with a 'raspberry' motif, and all six heightened with gold. The lower portion of the stem is a hollow inverted baluster blown into a mould with two lions' masks separated by quatrefoils and swags, the whole heightened with gilding. Plain folded foot.

Glasses of this type were probably made in many of the main glass-making centres of Europe. Thus, characteristic examples, recognizable by the manner of their decoration, may be attributed to Spain (e.g. a piece enamelled in Barcelona style in the Rijksmuseum, Amsterdam, illustrated Schrijver, *Glass and Crystal*, London, 1963, Pl.17,a; example with embedded *millefiori* canes and mould-blown pineapple stem in the Victoria and Albert Museum, illustrated by Alice Wilson Frothingham, *Barcelona Glass in Venetian Style*, New York, 1956, Fig.5); or to Innsbruck (Erich Egg, *Die Glashütten zu Hall und Innsbruck im 16. Jahrhundert*, Innsbruck, 1962, Figs.27–35). With many pieces, however, it is difficult to be sure whether they are Venetian, perhaps for the export-market, or *façon de Venise* glasses made elsewhere.

The Colinet catalogue of 1550–5 (see commentary on *22a* here) illustrates a glass of this character under the heading 'Verres Cibores' ('Ciborium Glasses'). It is accompanied by the rubric: 'Verres cibores a panse pour vin ou bier en verre cracquelé ou non':[1] to this a later hand has added 'avecq ou sans coverte' ('with or without cover'). The Colinet glass is shown with hollow-blown lion-mask stem, and lion-masks applied as prunts on a zone of horizontal threading,[2] all as on the Waddesdon glass. Glasses which show the same combination of features include pieces in the Musées Royaux d'Art et d'Histoire, Brussels (Liège, *Trois Millénaires* . . ., No.307, *ill.*); in the Kestner Museum, Hanover (*Glas-Sammlung*, No.3, Pl.4, with cover but without threading); in the Victoria and Albert Museum, London (Mus.No.C.209–1936, Honey, *Glass*, p.64, Pl.31,C, of ice-glass); and in the Gambier-Parry Collection in the Courtauld Institute Galleries, London (Charleston, 'Glass in the Gambier-Parry Collection', p.165, this glass having the turquoise-tipped prunts characteristic of Netherlands, possibly Antwerp, glass-making – see Chambon, *L'Histoire de la Verrerie en Belgique*, Pls. VIII–X).

1 'Bellied ciborium glasses for wine or beer in ice-glass or plain.'

2 R. Chambon, 'Influence vénitienne sur la production verrière de la Belgique', *Cahiers de la Céramique, 18,* 1960, pp.120–34, especially Figs.14–17.

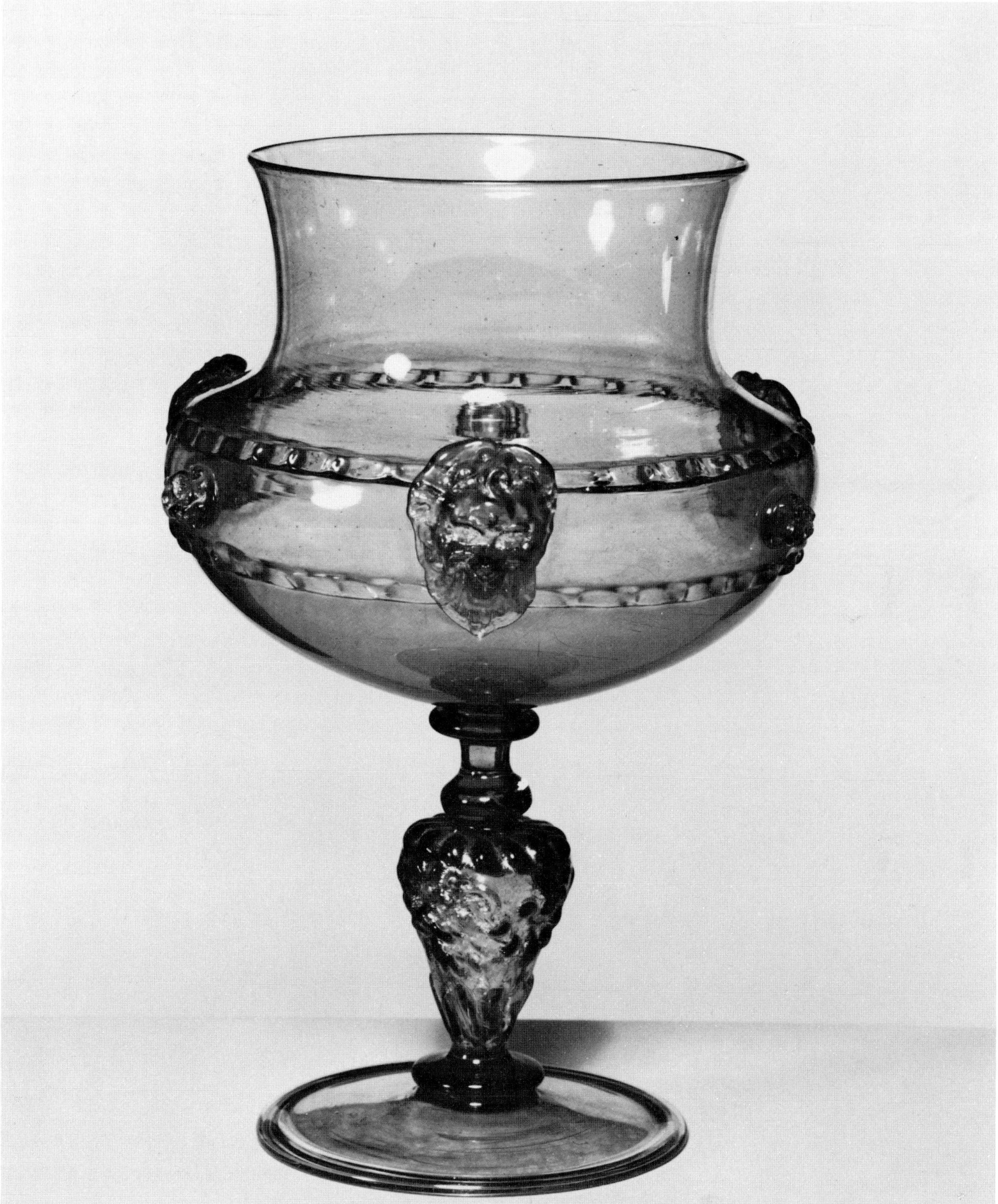

These 'goblet-vases' seem to have been used for flowers as well as for drinks (see Corning, *Glass Vessels in Dutch Painting*, Corning, 1952, p.25 and Pl. VII).

The cover of this glass evidently does not belong to it, but must have come from a glass of very similar character.

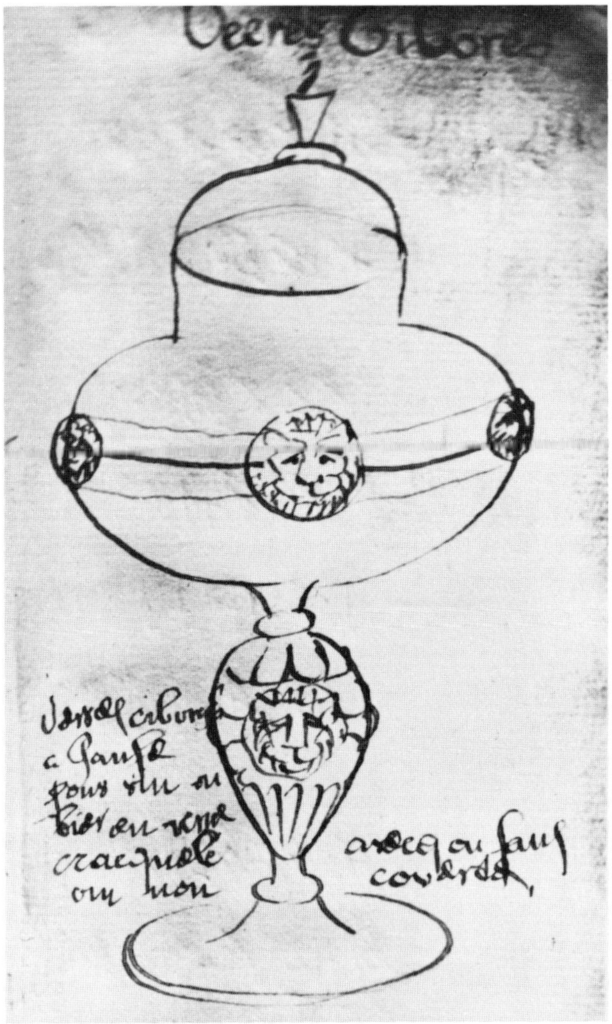

Drawing from the Colinet Catalogue, c.1550–55
(Collection of M. Raymond Chambon)

1922 Waddesdon Inventory (Smoking Room): 'Cup and cover on stem, three raised masks of animals. Venetian, 17th century'.

Miss Alice de Rothschild — W1/141/8*

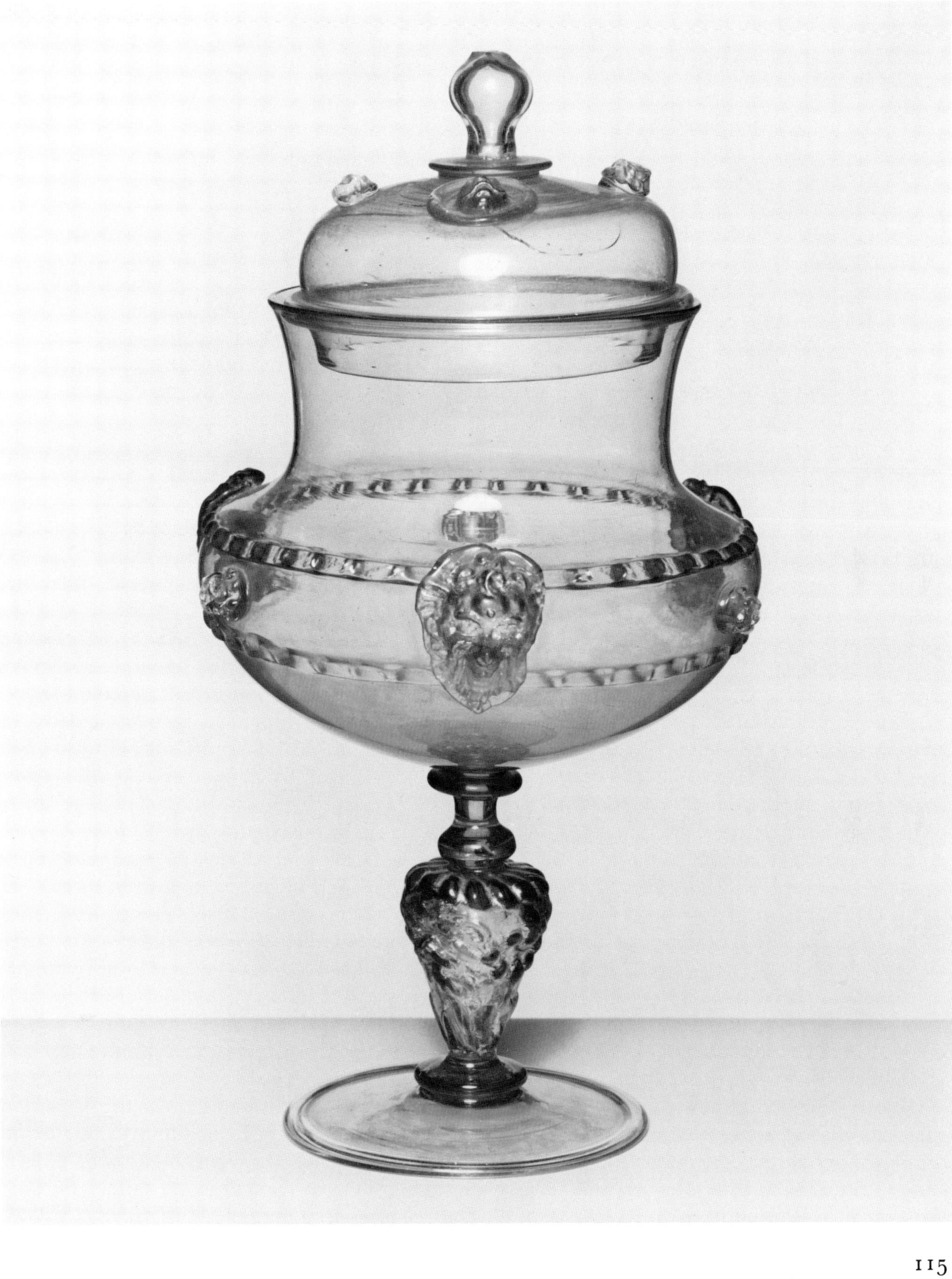

25. Lamp, in the form of a fantastic bird

Venetian or *façon de Venise* (perhaps Netherlandish); 17th century

Mauvish bubbled 'crystal' glass, mould-blown and with tooled blue embellishments

Height, $7\frac{5}{8}$ in. (19·5 cm); diameter of foot, $3\frac{3}{8}$ in. (8·5 cm)

The body of the lamp is formed of a double pecten-shell blown in a mould, supported on a hollow inverted baluster stem with vertical mould-blown ribbing, and a plain foot. The body terminates above in a vertical bird's head with open beak (probably to take the wick) and throat with three constrictions, its eyes and crest in blue glass. On either side of the body are pincered 'wings' of blue glass, and at the back is an applied spout probably for filling the lamp, finished with a thread of blue glass pulled out into a point above. A hole in the same position on the other side of the body indicates that something has been broken away at that point, probably a disc-like draught-shield or reflector.

A number of lamps of this character are known, often with their disc-like draught-shields or reflectors broken off. Two lamps in the Poldi-Pezzoli Museum in Milan (Inventory Nos. 1231 and 1238) retain their discs, the body of one being dumb-bell-shaped, the other divided into four tubes in the manner of a *Kuttrolf*. An example in the Museo Vetrario, Murano, has the body mould-blown in the form of a bunch of grapes above a ribbed knop-stem (Mariacher, *Italian Blown Glass*, Pl.56,c). Another, preserving its disc, the body encircled by a trailed thread, is in the Amatller Collection, Barcelona (Luis P. Bueno, *Vidrios y Vidrieras*, Barcelona, 1942, Pl. facing p.81). Another lamp with this feature, also retaining its disc, has a double-headed bird (Gasparetto, Fig.106), and yet another piece in the Museum of Arts and Crafts, Prague, has two bodies as well as two heads (Prague, *Venetian Glass*, No.141, Fig.34, there identified as a cruet; but the broken-away attachment for the disc betrays its true function).

In Rosenborg Castle, Copenhagen, is a comparable lamp, but with dragon's head and yellow embellishments, probably brought home from Venice by King Frederick IV of Denmark in 1709 (see Boesen, *Venetianske Glas på Rosenborg*, pp.63–5, and No.20, with further literature).

A second bird-lamp with body in the form of a scallop-shell is in the Museo Vetrario, Murano (unpublished), and another was in the Castellani Collection (*Collection Alessandro Castellani, Objets d'Art Antiques, du Moyen Age et de la Renaissance*, Rome, 1884, Lot 502 (sold 28 March): 'Curieux vase dont la panse est formée d'un oiseau en verre incolore et dont la tête et les ailes sont rehaussées d'émail bleu et jaune. Le corps de l'oiseau simule une coquille à double valve – Haut., 21 cent.; larg., 11 cent.'). A fourth, from the Cartwright Collection at Aynhoe Park, was sold at Sotheby's on 14 July, 1975, Lot 302.

Glasses with bodies blown in the form of a scallop-shell are not unfamiliar in the Venetian sphere of glass-making – compare, for example, a two-handled vase in opaque-white glass in the Melvin Billups Collection (Corning, *A Decade of Glass Collecting*, Corning, 1962, No.48, *ill.*); a vase with blue handles in the Wallace Collection (Savage, *Glass*, Fig.35); and a vase from the Soulages Collection in the Victoria and Albert Museum, London (Mus.No.5511–1859).

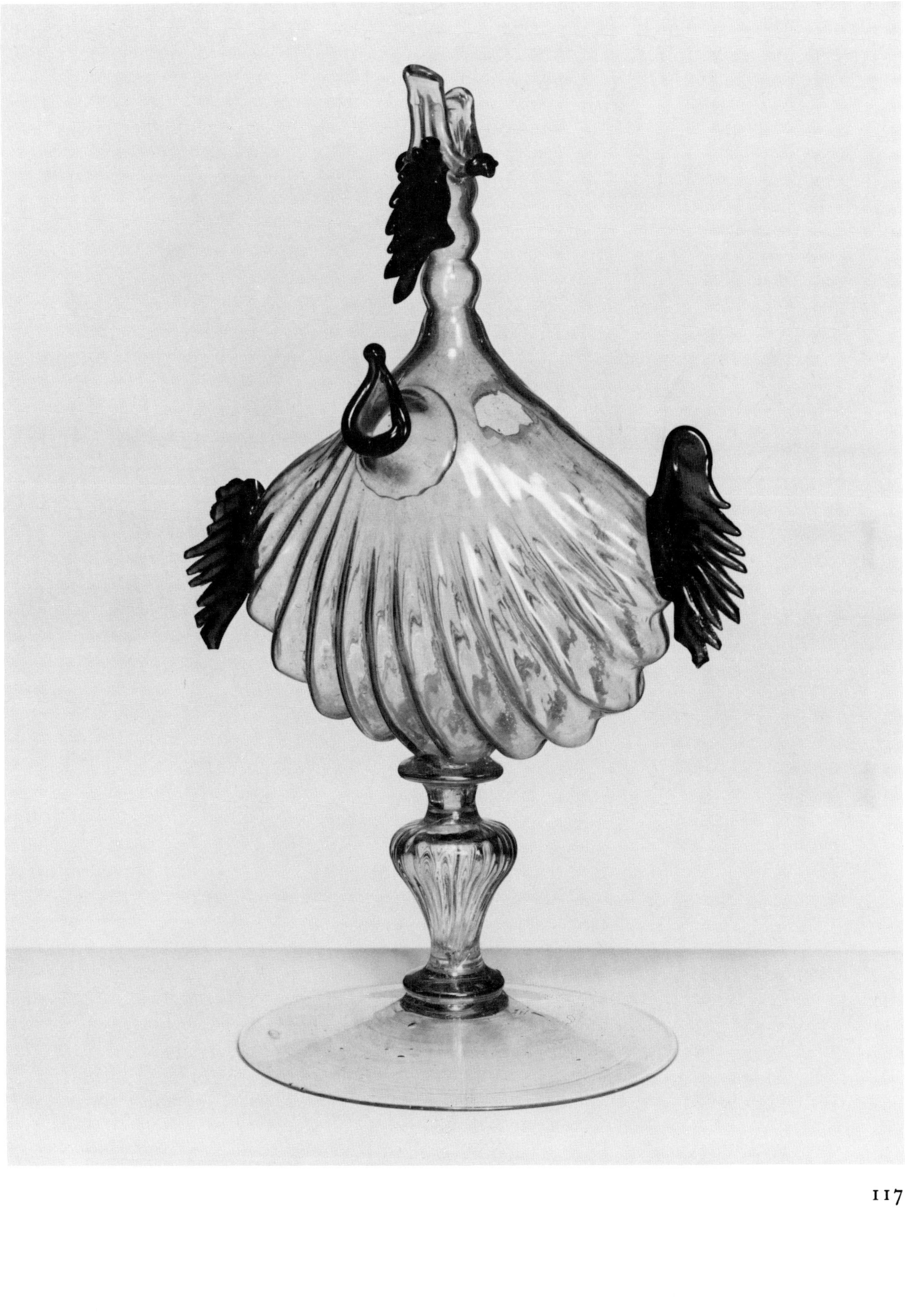

Curiously enough, and a possible pointer to a long-lived tradition, flasks mould-blown in the form of scallop-shells and bunches of grapes – the two forms represented among these lamps – were common enough in Roman times (see, e.g., M. S. Dudley Westropp, 'Molded Glass,' *Antiques*, December, 1928, Fig.1; British Museum, *Masterpieces*, No.65, *ill.*, etc.)

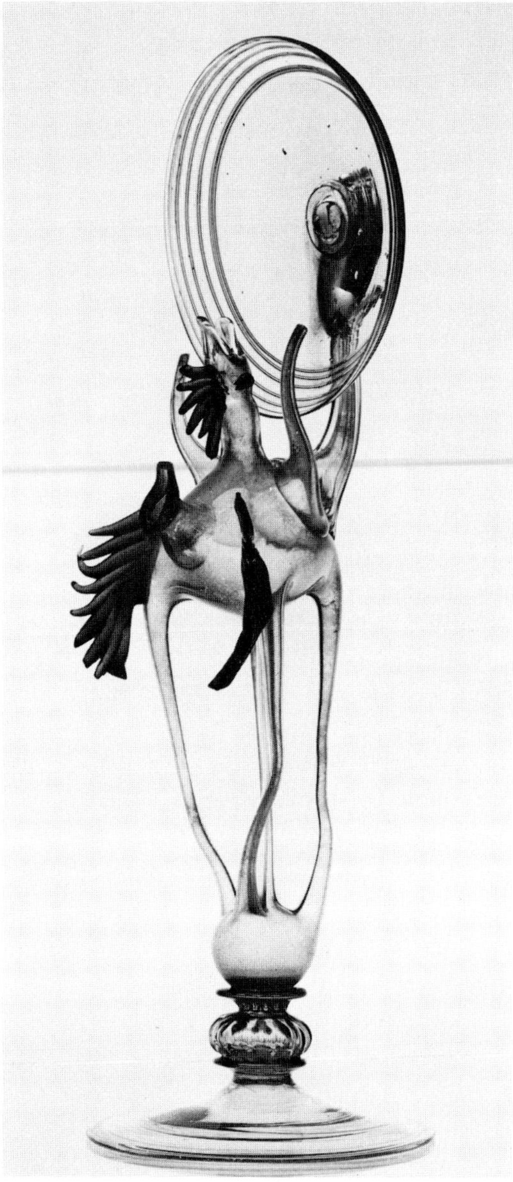

Lamp shown to illustrate the form of reflector used
Probably Venetian; 17th century (*Museo Poldi-Pezzoli, Milan*)

1922 Waddesdon Inventory (Smoking Room): 'Ewer in form of a bird, Venetian 17th century'.

Miss Alice de Rothschild — W1/141/4*

26. Goblet, with 'serpent' stem

Façon de Venise (probably Netherlandish); third quarter of 17th century

Yellowish-colourless 'crystal' glass with wrought decoration.

Height, $10\frac{1}{8}$ in. (25·8 cm); diameter of rim, $4\frac{1}{8}$ in. (10·5 cm)

Goblet with low spreading, almost straight-sided, bowl separated by a merese and a short straight section of stem from an elaborately wrought flat central stem-section. This is composed of an involuted coil of plain glass incorporating spirals of opaque-white and red, the whole with the outline of an inverted heart, at the upper end of which the coil terminates in two 'dragons' heads' with recurved pincered plumes. The outer edge of this complex is ornamented with two pincered 'wings' of plain colourless glass, and the whole is separated from the plain folded foot by a straight stem-section terminating below in a flattened knop.

The material of this glass is rather full of 'seed'.

Glasses of this general type are frequently found in public and private collections in Europe and America, and are usually attributed to the Netherlands. *Verres à serpents* are mentioned in documents of the seventeenth century there, as in contracts between the Bonhommes, owners of glasshouses at Liège, Brussels and elsewhere in the Low Countries, and their workmen (between 1667 and 1673). These make it clear that the work was laborious and time-consuming, two dozen such glasses being the equivalent of 120 ordinary wine-glasses. The Colinet family of Barbançon also made *verres à serpents*.[1] The numbers of these glasses made in the Netherlands is suggested by the official figures of glass sales in Brussels between 1659 and 1663, when nearly eight thousand *verres à serpents* were sold.[2]

A number of these glasses are datable, notably a goblet with diamond-point decoration attributable to the engraver Peter Wolff, active in Cologne between 1666 and 1677, in the Kunstgewerbemuseum, Cologne (Klesse, *Glas*, pp.14 and 111, No.227); another signed by him was in the J. P. Guépin Collection, Amstelveen (Düsseldorf Kunstmuseum, *Meisterwerke*, No.110, *ill.*). A goblet with diamond-point inscription dated 1672 is in the Museum of Applied Arts in Prague (Národni Galerie V Praze, *Sbírky Užitého Umĕni V Umĕleckoprůmyslovém Muzeu*, Introduction by Jarmila Brožová, Prague, 1965, No.37, *ill*), and another, dated 1673, is in the Victoria and Albert Museum, London (Honey, *Glass*, p.131, Mus.No.C.425–1936). A flute with comparable stem, dated 1681, is in Rosenborg Castle, Copenhagen (G. Boesen, *Gamle Glas*, Copenhagen, 1950, Pl.13, right). A goblet of the general type of this group is shown in a still-life painting by J. Hulsdonck (1582–1647) datable to about 1640 (A.-E. Liederwald, op. cit. in n.1.).

[1] R. Chambon, *L'Histoire de la Verrerie en Belgique*, pp.114, 320. Further discussion of these flat-stemmed glasses and other types of 'Serpent-stemmed' goblets, may be found in A. von Saldern, 'Spiral-Schlangenpokale à la Façon de Venise', *Jahrbuch der Hamburger Kunstsammlungen*, 17, 1972, pp.67–80; id., 'Façon de Venise', Col.1039; A.-E. Liederwald, *Niederländische Glasformen des 17. Jahrhunderts*, Doctoral thesis, Freiburg im Breisgau, 1964, pp.28ff.

[2] F. Pholien, *La Verrerie et ses Artistes au Pays de Liège*, Liège, 1899, p.81. This figure, however, is in contrast to some 426,000 *verres ordinaires* and 43,000 *verres à la vénitienne*.

Glasses with 'serpent stems' are sometimes ascribed to the glasshouse at Kassel, but on somewhat slender evidence (see F.-A. Dreier, *Glaskunst in Hessen-Kassel*, Kassel, 1969, Section III and Fig.11). They may also have been made in Cologne (see Klesse, op. cit., p.14, and Düsseldorf Kunstmuseum, *Meisterwerke*, No.110), where sections of canes of the sort used in these stems have been excavated (Klesse, op. cit., No.226).

Despite their generic likeness, the goblets of this class reveal considerable differences of stem- and bowl- formation. Glasses closely similar to the Waddesdon piece are to be found, e.g., in the Germanisches Nationalmuseum, Nuremberg (Kaempfer and Beyer, *Glass*, p.276, No.81); in the Museum Boymans-van Beuningen, Rotterdam (Isings, *Schönes Altes Glas*, Pl.85, left); and in the Kunstgewerbemuseum, Cologne (Klesse, op. cit. No.228).

1898 Waddesdon Inventory (Smoking Room, No.1119): 'A pair of Venetian glass gobelets with double serpent stems and circular bases. $10\frac{1}{4}$ in.'.
1922 Waddesdon Inventory (Smoking Room): 'Tall cup on winged stem, Venetian 17th century'.

Baron Ferdinand de Rothschild — W1/139/2*

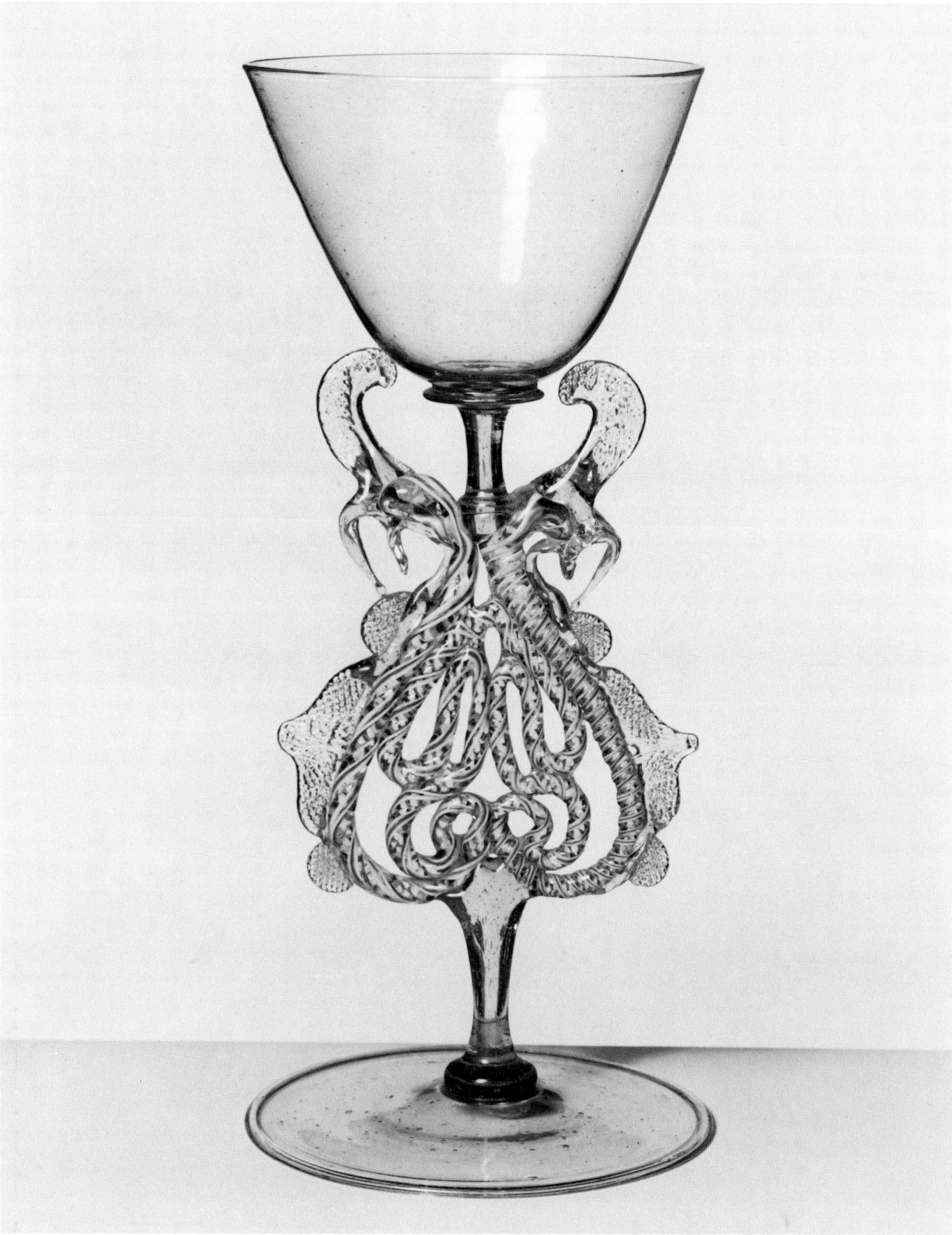

27. Goblet, with 'serpent' stem

Façon de Venise (probably Netherlandish); third quarter of 17th century

Yellowish-colourless 'crystal' glass with wrought decoration

Height, $10\frac{1}{4}$ in. (26 cm); diameter of rim, $3\frac{5}{8}$ in. (9·3 cm)

Goblet with round-funnel bowl separated by a merese and a short straight section of stem from an elaborately wrought flat central stem-section. This is composed of an involuted coil of plain glass incorporating spirals of opaque-white, the whole with the outline of an inverted heart, at the upper end of which the coil terminates in two 'dragons' heads' with recurved pincered plumes. The outer edge of this complex is ornamented with two pincered 'wings' of plain colourless glass, and the whole is separated from the plain folded foot by a straight stem-section terminating below in a flattened knop.

The material of this glass is rather full of 'seed'.

For a discussion of the attribution and dating of this type of glass, see *26*.

Glasses closely comparable with the present goblet may be found in the Kunstgewerbe-museum, Cologne (Klesse, *Glas*, Nos.217 and 220, *ill.*); in the Royal Scottish Museum, Edinburgh (*Journal of Glass Studies*, VII, 1965, p.125, No.23); in the Musées Royaux d'Art et d'Histoire, Brussels (Chambon, *L'Histoire de la Verrerie en Belgique*, Pl.XXIII, 72); and in the Corning Museum of Glass (Mus. No.CMG 50.3.56), etc.

1898 Waddesdon Inventory (Smoking Room, No.1119): 'A pair of Venetian glass gobelets with double serpent stems and circular bases. $10\frac{1}{4}$ in.'.
1922 Waddesdon Inventory (Smoking Room): 'Tall cup on winged stem, Venetian 17th century'.

Baron Ferdinand de Rothschild — W1/140/1*

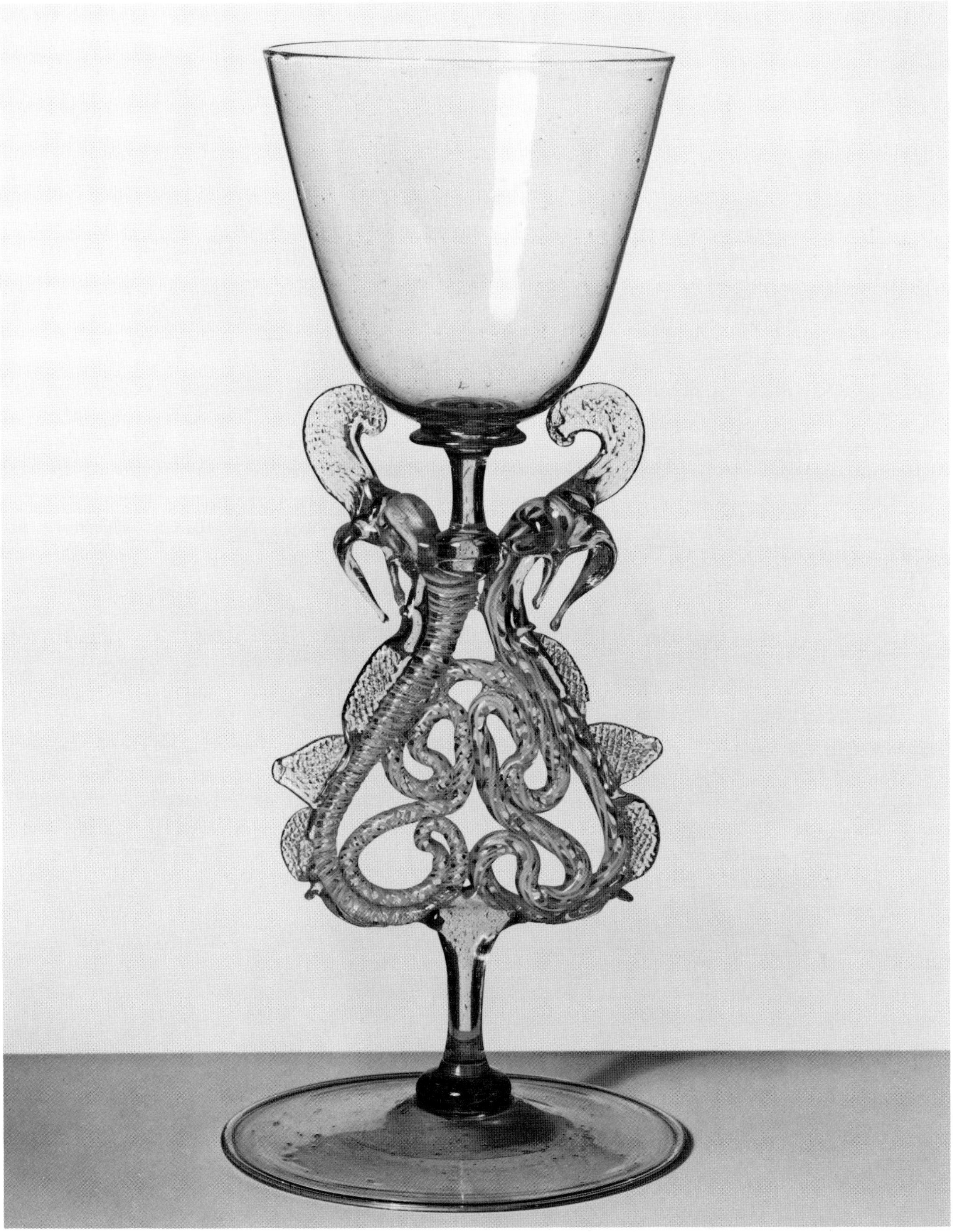

28. Vase with decoration of linked dragons

Probably Venetian; late 16th or early 17th century

Blue glass with mould-blown and gilt decoration

Height, $9\frac{1}{2}$ in. (24 cm); diameter of rim, 3 in. (7·5 cm)

Ovoid vase with gadrooned shoulder drawn in to a tapering conical neck, which flares out above an applied waved collar to a fairly wide folded rim. A pair of opposed scrolled S-shaped handles runs down to the shoulder from points of attachment on the waved collar. The body of the vase bears on either side a mould-blown decoration of a pair of addorsed grotesque winged dragons linked together at their neck and the ends of their lower scroll-like wings and limbs. The whole of this ornament, as well as the handles and the gadrooning of the shoulder, is covered with gold. Round the lower part of the neck is painted in gold a border of looped trefoils, while a similarly painted cable-border runs round the upper part of the neck below the rim. The vase stands on a depressed hollow knop with vertical mould-blown ribbing, above a plain folded foot (broken).

This vase is virtually certainly from the same ill-fitting four-piece mould as a goblet-vase of colourless glass in the Corning Museum of Glass (Mus.No.60.3.87, illustrated in E. Barrington Haynes, *Glass through the Ages*, Harmondsworth, 1966, Pl.16,a; also Houston Museum of Fine Arts, *The Art of Glass* (Exhibition 19 September to 15 November, 1970), *Catalogue*, No.28).

R. Chambon, in 'Verres de la Renaissance soufflés dans un moule fermé' (*Glastechnische Berichte*, Bd.32 K, Heft VIII, 1959, pp.49–56) illustrates a covered vase (without handles) in *lattimo*-striped glass which appears to bear an identical decoration (Fig.4), and cites (pp.52–3) two further examples from the old literature. Chambon argues, on the basis of a goblet blown in a fixed mould and bearing the arms of the City and Marquisate of Antwerp (loc.cit., Fig.1), that most glasses with this type of decoration were probably made in Antwerp. It is to be remarked, however, that nearly all the glasses cited and illustrated by him are not goblets but vases, ewers and bottles, forms which the moulds with upper and lower borders of gadrooning virtually dictate. The glass made in Antwerp, furthermore, was of Venetian type, and a mould which appears to correspond with at least one group of these mould-blown pieces is recorded in an inventory of 1578 taken in the glasshouse at Murano of Giovan Antonio de Zanchi dal Castello (L. Zecchin, 'Una fornace Muranese all' Insegna dell' Angelo', *Vetro e silicati*, XI, No.62, 1967, p.26). It is described as 'una forma de fiaschi a meduse' ('a mould for bottles with Medusa-heads'), and mould-blown objects of closed shapes are known with mask-decorations (e.g. I. Schlosser, *Das Alte Glas*, Brunswick, 1956, p.83, Fig.63). That fixed moulds were familiar in the Italian sphere of glass-making generally is demonstrated by the examples shown in the drawings by Jacopo Ligozzi (d.1637) in the Gabinetto Disegni e Stampe degli Uffizi, in Florence (e.g.folios 97309, *verso*, No.114; 97312, *verso*, No.125; 97291, No.59; 97294, No.72). These drawings were probably designed to be reproduced, as far as this could ever have been technically feasible, in the glasshouse set up in the Boboli Gardens by the Grand Duke

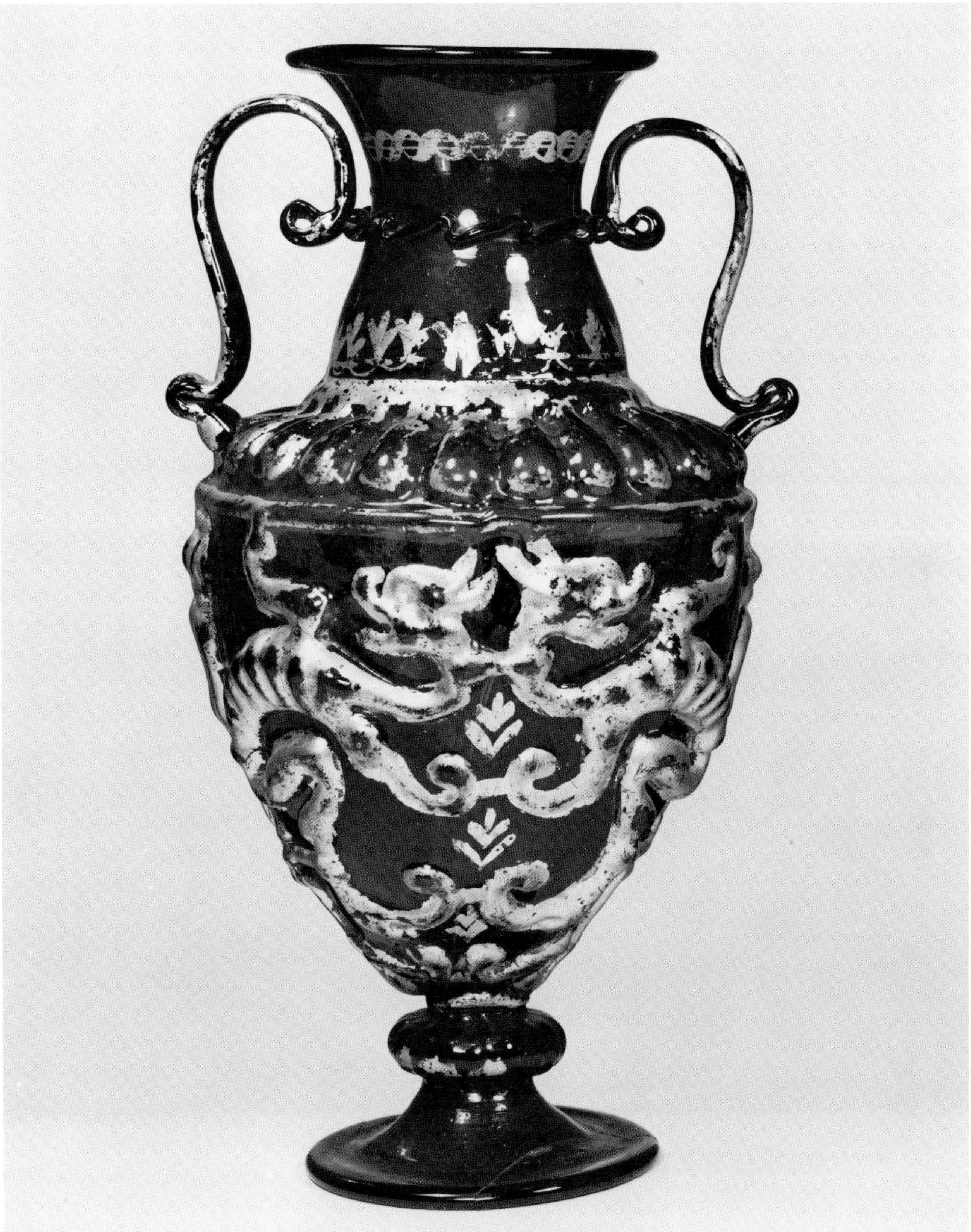

Cosimo II in 1618 (see G. Taddei, *L'Arte del Vetro in Firenze e nel suo Dominio*, Florence, 1954, pp.52–3; Luigi Zecchin, 'La Toscana e il vetro muranese', *Giornale Economico*, July, 1954, pp.16–25; id., 'Disegni Vetrari del Seicento', in *Sulla Storia dell'Arte Vetraria*, *Giornale Economico*, 1959–61, pp.57ff). Indeed, a document in the Florentine archives makes it clear that the preparation of complicated moulds, sometimes in three sections, and occasionally specially designed by local sculptors, was an important ancillary activity of the glasshouse (see L. Zecchin 'La Vetreria Medicea di Palazzo Pitti', *Giornale Economico*, No.3, March, 1961, p.295). Ingegnere Zecchin suggests that these moulds were made primarily for the use of the local glassmen, but since in general the influences on the Florentine glassmen were predominantly Venetian, the use of moulds there may only reflect further the Venetian tradition.

The normal vase-form of these fixed-mould pieces is that of the 'goblet-vase', with or without cover, standing on a low foot with a depressed hollow knop, the collar-neck usually spreading slightly towards the rim. The shape of the Waddesdon vase, however, is echoed in a vase with *lattimo* stripes formerly in the Walter F. Smith Collection, Trenton, New Jersey. This has a mould-blown decoration of two lions and two double-headed eagles, but is without handles (*Journal of Glass Studies*, IV, 1962, p.144, No.27; cf. Chambon, loc.cit., p.52).

From the Hamilton Palace Collection, Sale 3 July 1882 (eighth day), *Catalogue*, London and Paris, 1882, Lot 836: 'A Two-handled Vase, of blue glass, with dragons and ornaments in relief and gilt', bought by E. Joseph for £84.

1898 Inventory (Smoking Room, No.1119) 'A dark blue German glass two-handled vase, with gilt embossed double eagles and gilt decorations. 9½ in. (repaired)'.
1922 Waddesdon Inventory (Smoking Room).

Baron Ferdinand de Rothschild — W1/139/4*

Opposite: Side view of 28

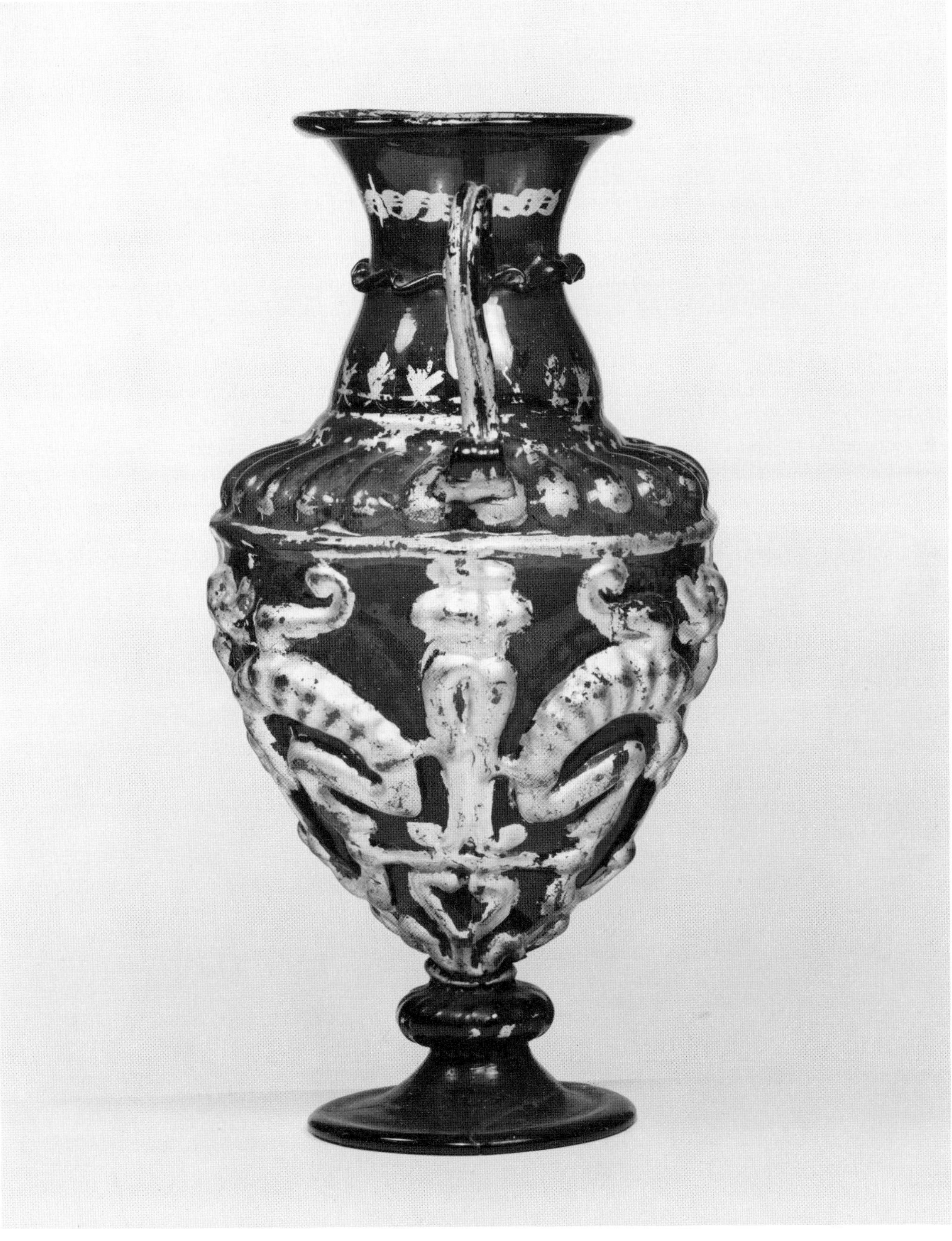

29. Bowl, in gilt metal mounts

Probably Italian; 17th century

Blue slightly bubbled glass, mounted in gilt base metal

Height, $5\frac{1}{2}$ in. (14 cm); length, 8 in. (20·5 cm)

Boat-shaped bowl, oval on plan, drawn in at the shoulder, with slightly flaring rim pinched in on the short axis, and indented at one end. On the base is a raised button to afford anchorage to the mounts.

The bowl is mounted on a low pedestal foot surrounded by a decorative border of large fleurs-de-lis alternating with smaller acanthus-like leaves. The fleurs-de-lis are joined by four straps to a collar running round the narrowest part of the vessel and formed of fleurs-de-lis alternating with blunt-ended lappets. From the straps on the short axis rises a pair of angled S-shaped scrolled handles with a dorsal beaded rib terminating in an acanthus-leaf.

The attribution and dating of this class of glasses are discussed in the Introduction (pages 21–5).

A design for a glass closely resembling the present bowl, but on a taller metal stem, is included among the Ligozzi drawings in the Gabinetto Disegni e Stampe degli Uffizi, Florence (fol. 97330, *verso*, No. 222).

A closely similar boat-shaped bowl, in ribbed emerald-green glass, is in the Victoria and Albert Museum, London (Mus.No. 19–1867).

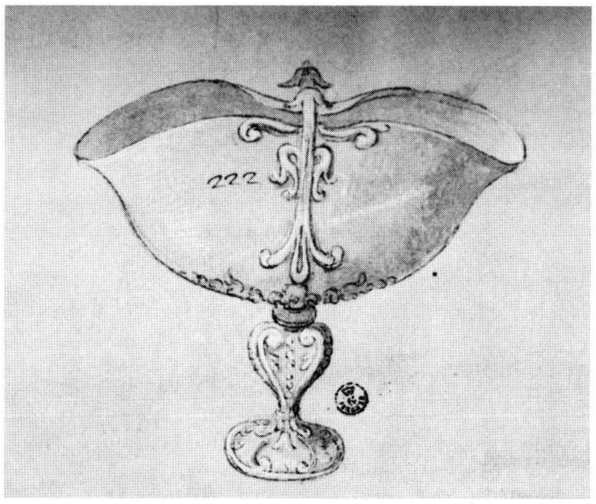

Drawing attributed to Giacomo Ligozzi (c.1547–1626)
(Gabinetto Disegni e Stampe degli Uffizi, Florence)

1922 Eythrope Inventory: 'One blue oblong bowl, metal gilt mounts'.

Miss Alice de Rothschild — E2/35/4i*

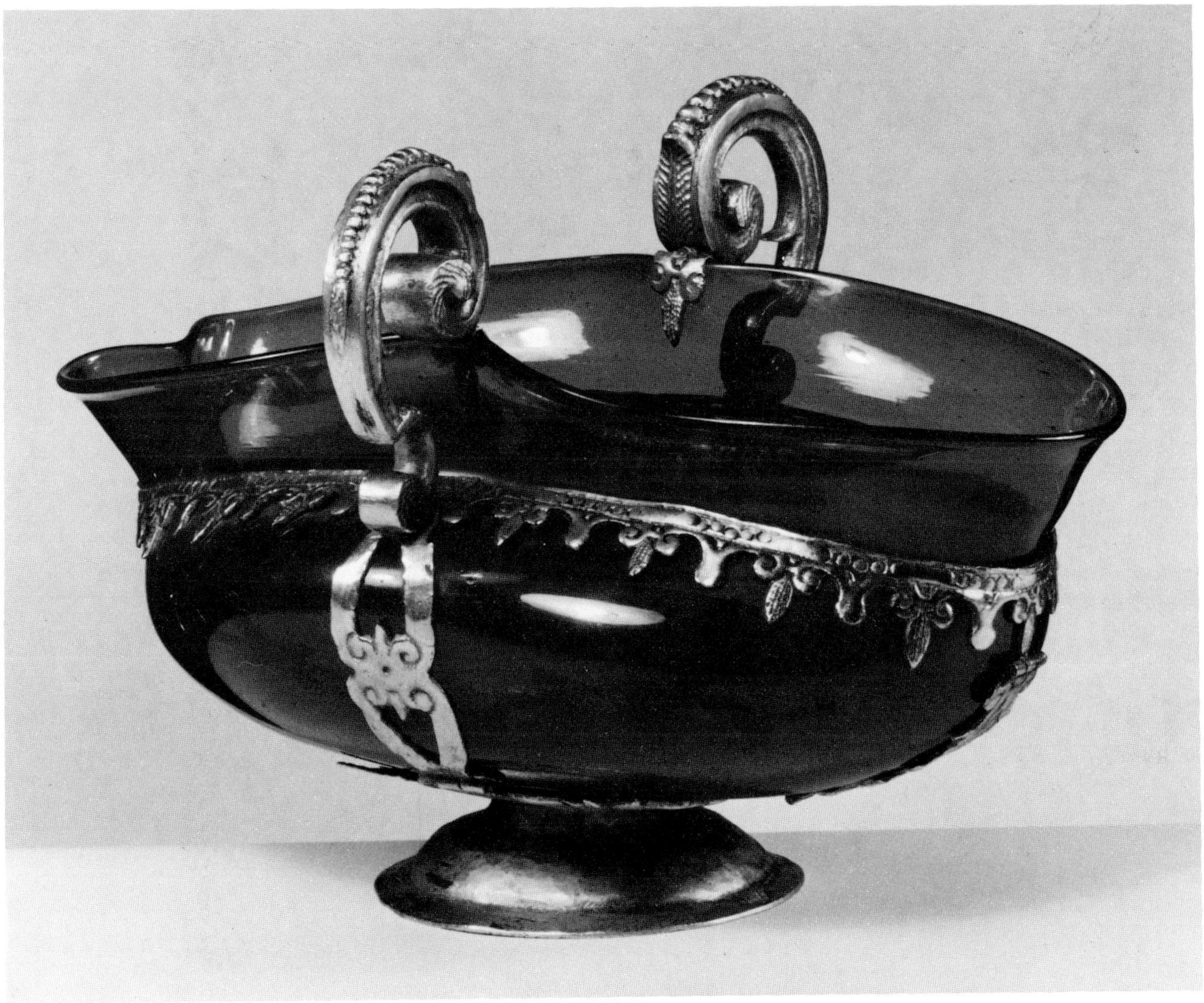

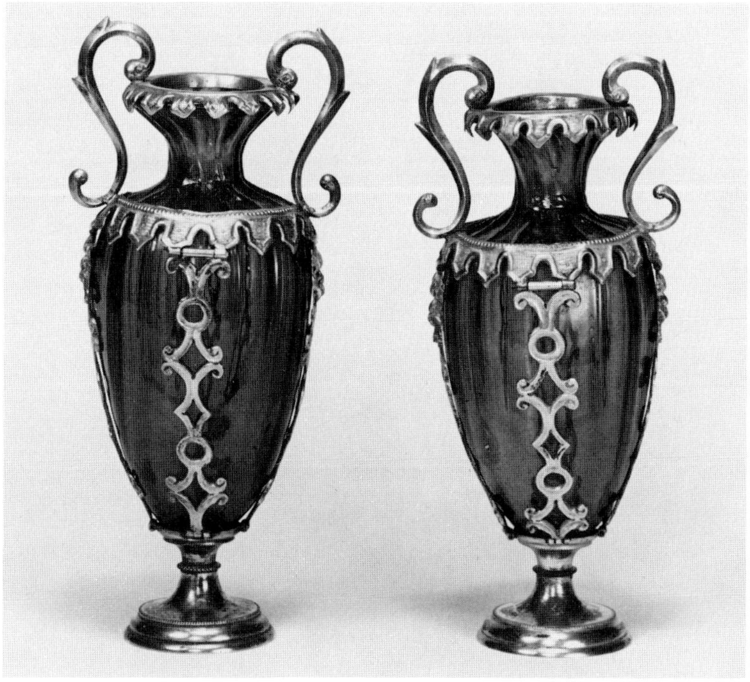

30. Pair of vases, in gilt metal mounts

Probably Italian; 17th century

Blue bubbled glass with mould-blown ribbing, mounted in gilt base metal

Height, $6\frac{7}{8}$ in. (17·5 cm) and $7\frac{3}{8}$ in. (18·8 cm); diameter of foot, $1\frac{7}{8}$ in. (4·8 cm)

Vases of inverted baluster form with flaring neck, decorated with slightly 'wrythen' vertical mould-blown ribbing. Each vase is mounted on a gilt-metal pedestal foot with turned mouldings and lines of horizontal beading. These are connected with a collar of lappets surrounding the shoulder of the vases by means of four openwork straps, two of which terminate above in masks: above these rise scrolled S-shaped handles which connect the shoulder-collar with a similarly decorated collar running round the rim of the vase.

The attribution and dating of this class of glasses are discussed in the Introduction (pages 21–5).

A comparable pair of vases was sold at Sotheby's on 10 May 1971, Lot 23, described as 'Venetian or South German; 17th century'.

1922 Eythrope Inventory: 'Pair of Venetian glass bottles, blue, metal gilt mounts with handles. 17th century'.

Miss Alice de Rothschild E1/12/6*

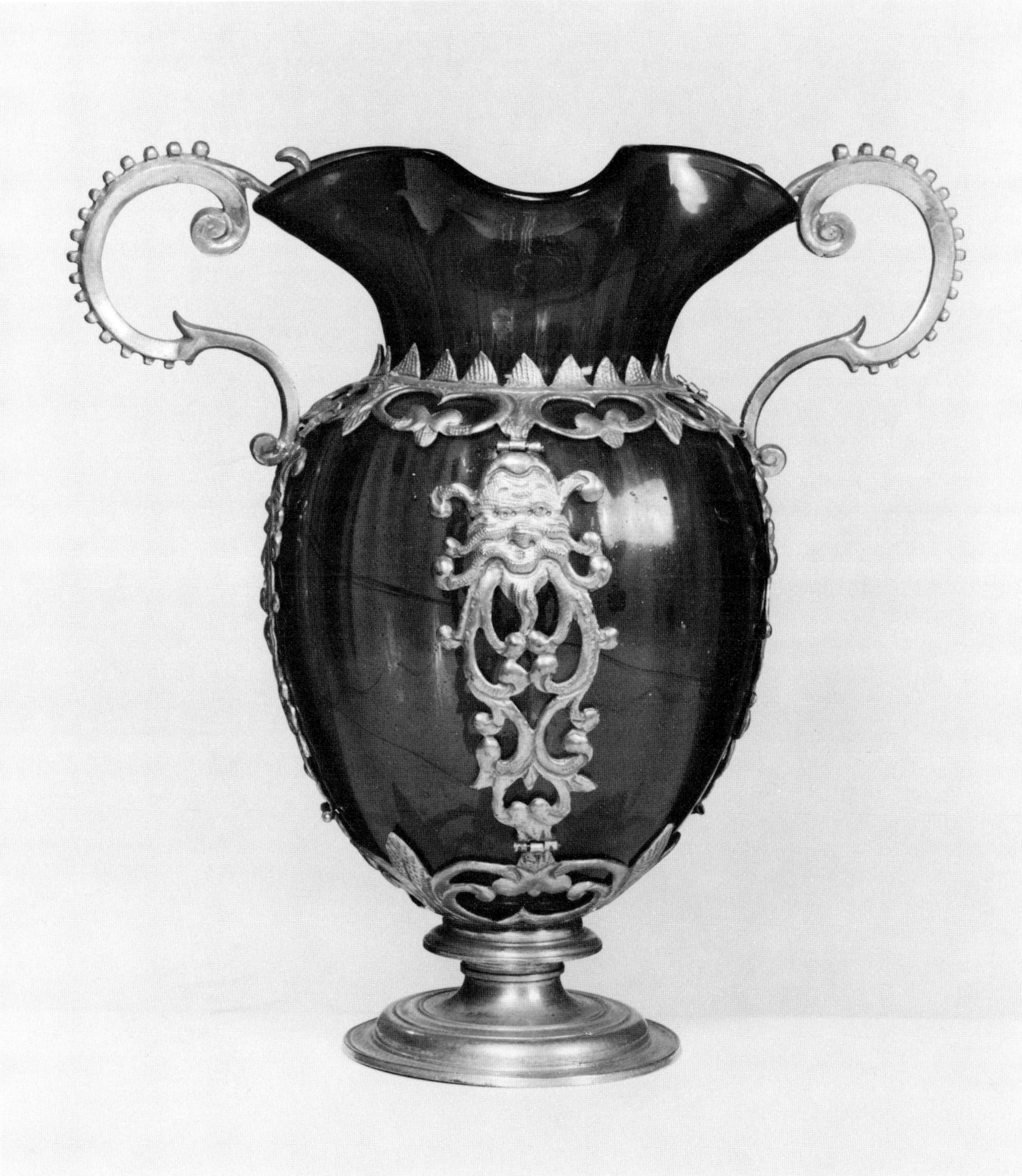

Alessandro Castellani, Objets d'Art, Antiques du Moyen Age et de la Renaissance . . ., Rome (28 March 1884), Lot 482; 'Trois vases en forme de balustre à côtes, en verre bleu uni, garnis de montures en cuivre découpé et doré avec anses ornées de têtes de génies. Venise, XVIe siècle – Haut., 26 cent.; larg., 22 cent.' The 1898 Waddesdon Inventory (see below) lists 'a set of three dark blue vases with gilt mounts and double handles (2 and 1)', and it seems possible that this pair and *33* formed this set. The colours agree, and the nature of the mounts seems compatible, although the dimensions do not exactly correspond (26 cm = $10\frac{1}{4}$ in.).

1898 Waddesdon Inventory (Smoking Room, No. 1120) – see above.
1922 Waddesdon Inventory (Smoking Room): 'Pair of vases, blue, metal gilt mounts with heads. Venetian 17th century'.

Baron Ferdinand de Rothschild — W1/141/5*

33. Vase, with gilt metal mounts

Probably Italian; 17th century

Green-blue bubbled glass with mould-blown ribbing, mounted in gilt base metal

Height, $10\frac{3}{4}$in. (27·5 cm); diameter of foot, $3\frac{1}{4}$ in. (8·2 cm)

Ovoid vase drawn in to a narrow trumpet-shaped neck, decorated with vertical mould-blown ribbing which has been almost worked away on the body of the vase. On the base a projection has been left which affords an anchorage to a gilt-metal pedestal foot with turned mouldings, above which is a decorative frieze of openwork false gadrooning. This is connected to a collar of openwork lappets running round the shoulder by means of four openwork straps terminating above in cherubs' heads. Above the side-straps rise scrolled S-shaped handles with dorsal beading, attached by pins to a collar running round the rim of the neck, this being decorated with pendant trefoils and with acanthus-leaf motifs engraved on the upper surface.

The attribution and dating of this class of glasses are discussed in the Introduction (pages 21–5).

A closely similar piece was in the Spitzer Collection (*Catalogue des Objets d'Art . . . composant l'importante et précieuse Collection Spitzer*, Sale, Paris, Chevallier, 17 April to 16 June, 1893, Lot 2083 'Vase monté en cuivre doré – (Fin du XVIe. siècle). Il est de verre vert. La panse ovoide repose sur un pied en cuivre doré. Elle est accompagnée de frettes décorées de tête de chérubins et de deux anses en forme de volutes également en cuivre doré. H. Om.21.' (*ill. Atlas*, Pl. I).

Perhaps from the Castellani Collection (see notes on *32*). 1898 Waddesdon Inventory (Smoking Room): 'A set of 3 dark blue vases with gilt mounts and double handles (2 and 1)'. 1922 Waddesdon Inventory (Smoking Room): 'Vase, blue ground, metal gilt mounts, cherubs heads. Venetian 17th century'.

Baron Ferdinand de Rothschild — W1/139/10*

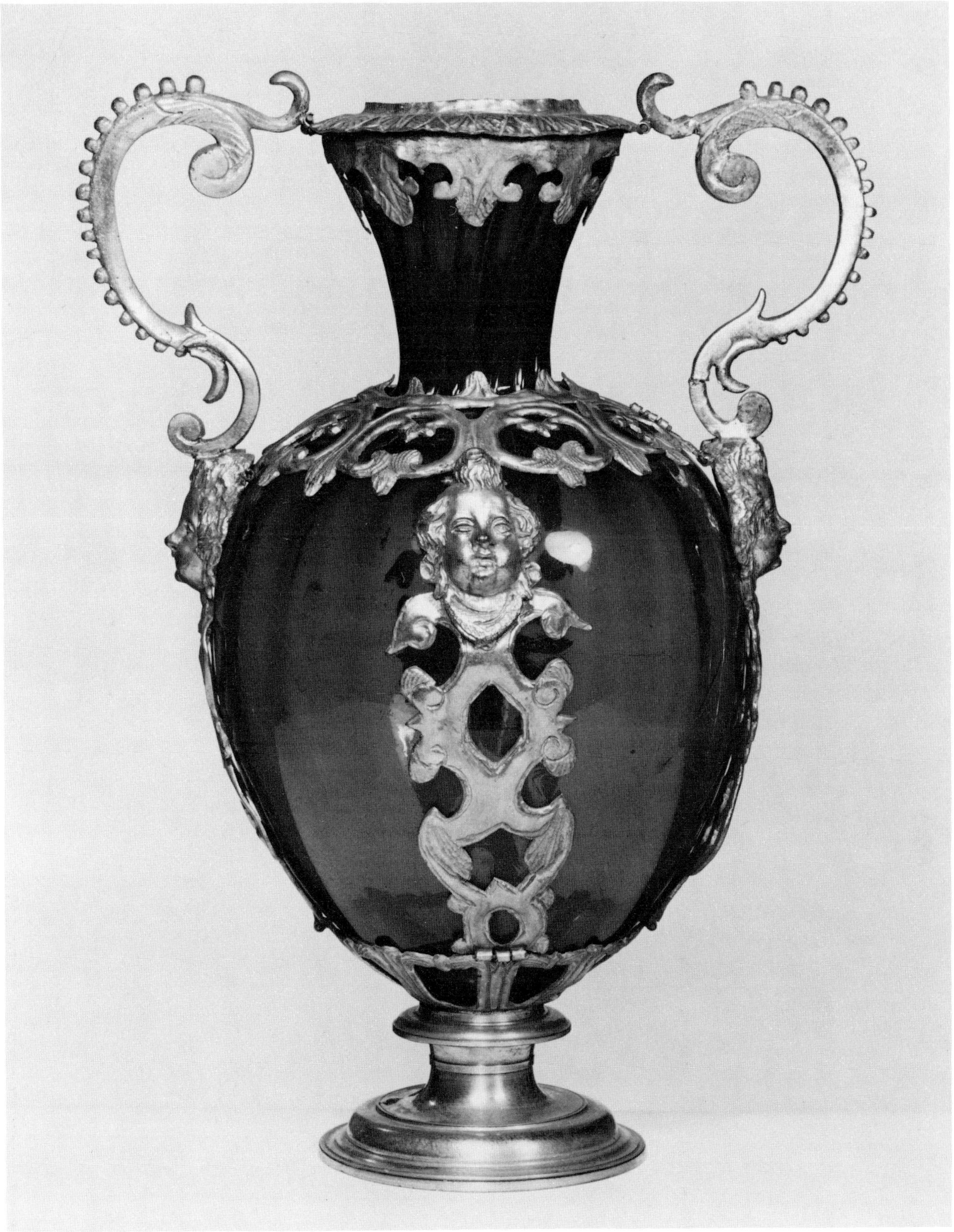

34. Vase, with gilt metal mounts

Probably Italian; 17th century

Green bubbled glass with mould-blown ribbing, mounted in gilt base metal

Height, 10 in. (25·5 cm); diameter of foot, $3\frac{7}{8}$ in. (9·8 cm)

Vase of almost globular form, drawn sharply in to a low trumpet-shaped neck, decorated with vertical mould-blown ribbing which has been almost worked away on the body of the vase. On the vase a projection has been left which affords an anchorage to the gilt-metal pedestal foot with turned mouldings, above which is a decorative frieze of openwork false gadrooning, with acanthus-leaf embellishments. This is joined to an elaborate openwork collar occupying the almost horizontal shoulder of the vase, by means of four openwork straps terminating above in classical heads in full relief. From the tops of those at the sides rise scrolled handles formed like the head of a crozier, and from these run horizontal struts which in turn are fastened by pins to a broad flange-like collar running round the rim of the vase, decorated on its upper surface with engraved acanthus-leaves and surmounted by a tall crown-like openwork gallery.

The attribution and dating of this class of glasses are discussed in the Introduction (pages 21–5).

Mounts with comparable heads in full relief are to be seen on an emerald-green ribbed flask in the Victoria and Albert Museum, London (Mus.No.7445–1861; Savage, *Glass*, Fig.125).

1922 Waddesdon Inventory (Smoking Room): 'Vase, green, metal gilt mounts with heads. Venetian 17th century'.

Miss Alice de Rothschild — W1/141/6*

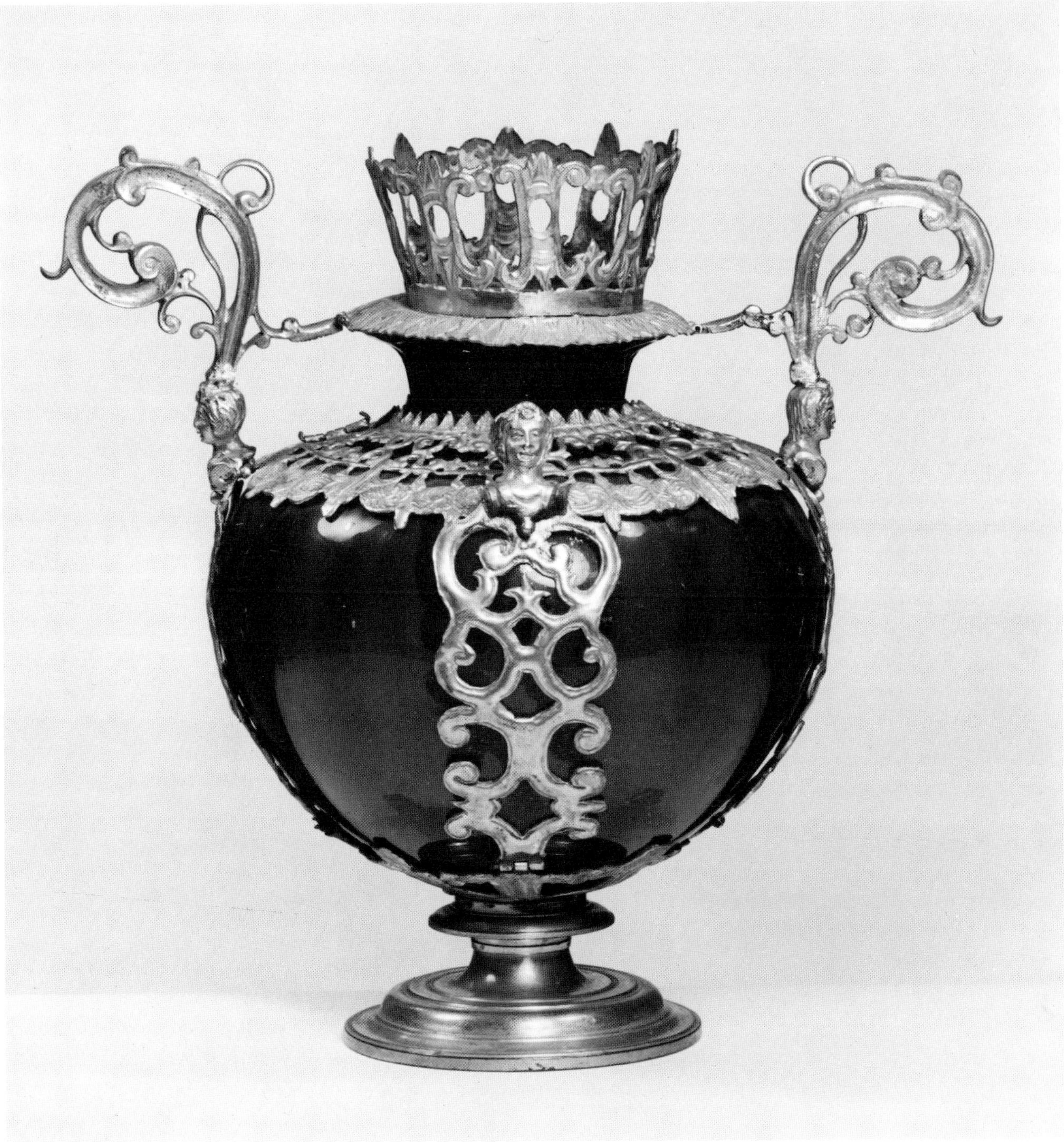

35. Four altar-candlesticks, with gilt metal mounts

Probably Italian; 17th century

Green bubbled glass with vertical mould-blown ribs, the glass components strung on a central rod, and mounted in gilt metal

Height, $20\frac{7}{8}$ in. (53 cm); $21\frac{1}{4}$ in. (54 cm); 21 in. (53·3 cm); and 22 in. (55·8 cm)

Each candlestick is made up of four glass components – at the top, a shallow dish as grease-pan; below this, a long slender baluster and a shorter piriform member above a foot-section in the form of a hemispherical dome surmounted by a depressed knop of much smaller diameter. The stem-components are separated by reel- and capstan-mouldings which terminate above and below in a fringe of alternately wide and narrow fleurs-de-lis, all in gilt metal. The grease-pan and foot are mounted in profiled gilt-metal rims with similar fleur-de-lis borders, the foot-rim being supported on three spreading S-scrolled bracket feet decorated with beading. The rim of the grease-pan is gilt on the upper surface.

The attribution and dating of this class of glasses are discussed in the Introduction (pages 21–5).

Although these candlesticks are described as comprising two pairs, and were so lotted in the Hamilton Palace Sale (see below), they probably originally formed an altar garniture with a central cross. A cross of this kind, mounted on a base identically similar to the bottom three members of the bases of the Waddesdon candlesticks, is in the Poldi-Pezzoli Museum, Milan (Inv.No.1224). It is of blue glass and stands 64 cm ($25\frac{1}{2}$ in.) high (illustrated overleaf). That the cross and its base really belong together seems indicated by the continuation of the fleurs-de-lis motif as an edging of all four members of the cross, while the pair of scrolls at its base, and the angel's head motif are decorative devices constantly found on the mounts of this class of glasses.

A drawing of one of these candlesticks, probably when it was in the Hamilton Palace Collection, is in the Victoria and Albert Museum (Mus.No.2522, Press mark C.120.b). It is marked 'Mr. Holgate, No.128', and is attributed to Albert J. Holgate.

From the Hamilton Palace Collection (*Catalogue of the Collection of Pictures, Works of Art and Decorative Objects, the property of His Grace the Duke of Hamilton . . .*, Christie, Manson and Woods, Eighth Day's Sale (3 July 1882), Lot 855: '*A Pair of Altar Candlesticks*, of old green Venetian glass, mounted with chased metal gilt, on tripod feet – 23 in. high'. Lot 856 was 'A PAIR OF DITTO', and both pairs were bought by E. Joseph for £84 each.

The candlesticks are shown in Baron Ferdinand's Red Book (i.e. before 1898).

1898 Waddesdon Inventory (Smoking Room) No.1112: 'Four tall green Venetian glass candlesticks with baluster stems, metal gilt mounts and three scroll feet, 23 in high (damaged)'.
1922 Waddesdon Inventory (Smoking Room): 'Two pairs of tall Candlesticks, green, with metal gilt mounts. Venetian, 17th century'.

Baron Ferdinand de Rothschild — W1/138/3*

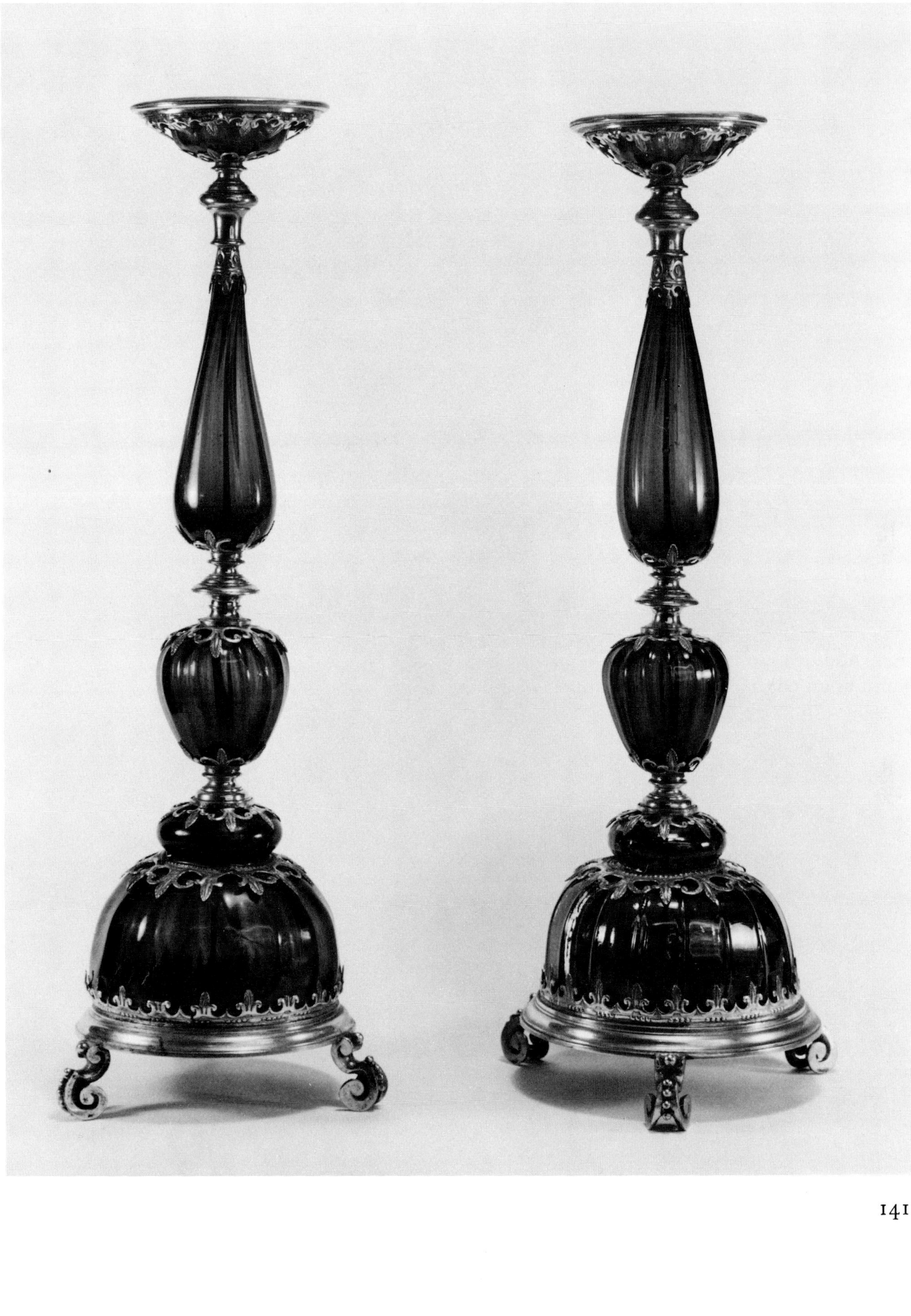

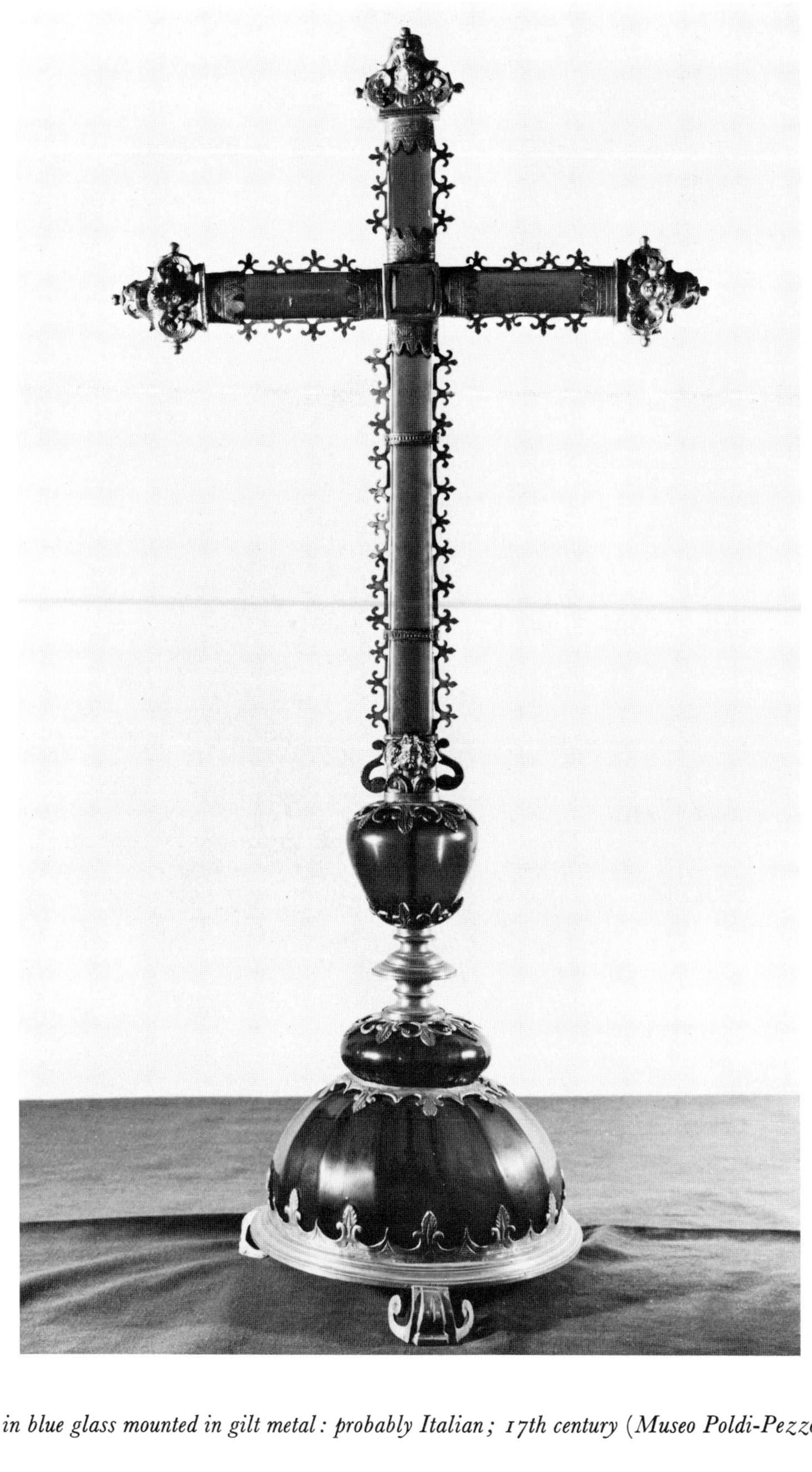

Altar-cross in blue glass mounted in gilt metal: probably Italian; 17th century (Museo Poldi-Pezzoli, Milan)

36. *Stangenglas*, enamelled with the arms of Johann Schmidt

Bohemian or South German; dated 1602

Almost colourless glass with a greenish-grey tinge, painted in enamel colours and gilt

Height, 11½ in. (29·2 cm); diameter of rim, 3½ in. (9 cm) Illustrated overleaf.

Cylindrical beaker on tall pedestal foot with folded rim, the base pushed into a low 'kick'. The beaker is decorated with coats-of-arms front and back: at the front, party per pale, or three roses gules and sable a gryphon rampant or; crest of a mermaid proper, crined or, holding three roses gules in the sinister hand: on the reverse, gules two shields barruly of azure and argent in chief, a plough proper, and a fleur-de-lis or. Below the rim, between borders of white tangent curves and rows of white dots originally on a gold ground, is the inscription in white: 'Johannes Schmidt Appotecker Vonn Straubing' ('Johannes Smith, apothecary of Straubing'). Below the coats-of-arms is a border consisting of a row of white dots, originally on a gold ground, between two lines of white tangent curves. Between the coats-of-arms is the date in white '1602', with cross-motifs in white above and below.

The coat-of-arms on the reverse of the beaker is that of the town of Straubing, on the Danube near Regensburg in Bavaria (Otto Hupp, *Wappen und Siegel der Deutschen Staedte*, 6. Heft, *Königreich Bayern*, *Kreis Ober- und Niederbayern*, Frankfurt, 1912, pp.83, 85–6).

Three roses form part of the arms of several branches of the Schmidt family in Austria and Germany (see J. B. Rietstap, *Armorial Général*, Pls. Vol. V, Paris, 1921, Pls. CCLXIX–CCLXX).

In 1598 Maria, widow of the apothecary Löffler, married the apothecary Johann Faber in the 'Lower Apothecary's Shop' (*Untere Apotheke*), later the 'Lion Apothecary's Shop' (*Löwenapotheke*), situated on the Ludwigsplatz in Straubing. *Faber* is the Latin for 'Smith' (*Schmidt*), and the Johann Faber who married Maria Löffler was without doubt the Johannes Schmidt of the Waddesdon *Stange*. Johann Schmidt, 'Citizen and Apothecary', and his wife Maria are mentioned in 1599. The wife died in 1614, but the date of Johann Schmidt's death is not known. By 1621 his shop is recorded as being in the occupation of one Thomas Höller.[1]

R. Schmidt, *Das Glas*, pp.175–6, cites a number of glasses with dates in the second half of the sixteenth century and the early seventeenth century which bear the arms of Bavarian families. Since three vessel-glasshouses are recorded in the Bayrischer Wald in 1566 (Reichenberg, Schönau and a third on the Lebrach), Schmidt concluded that these glasses might well have been made in this area (see, however, Introduction, page 27).

From the Bernal Collection (*Catalogue of the celebrated Collection . . . of Ralph Bernal*, London, Christie and Manson, 22nd Day's Sale, 17 April 1855, Lot 2850: 'A beaker, with arms of Johannes Schemdt [sic], 1602 – 11½ in. high').

1922 Eythrope Inventory: 'Small tumbler, enamelled coat of arms'.

Miss Alice de Rothschild EI/13/7*

[1] Information kindly supplied by Dr Scherl, of the Historischer Verein für Straubing und Umgebung, Straubing.

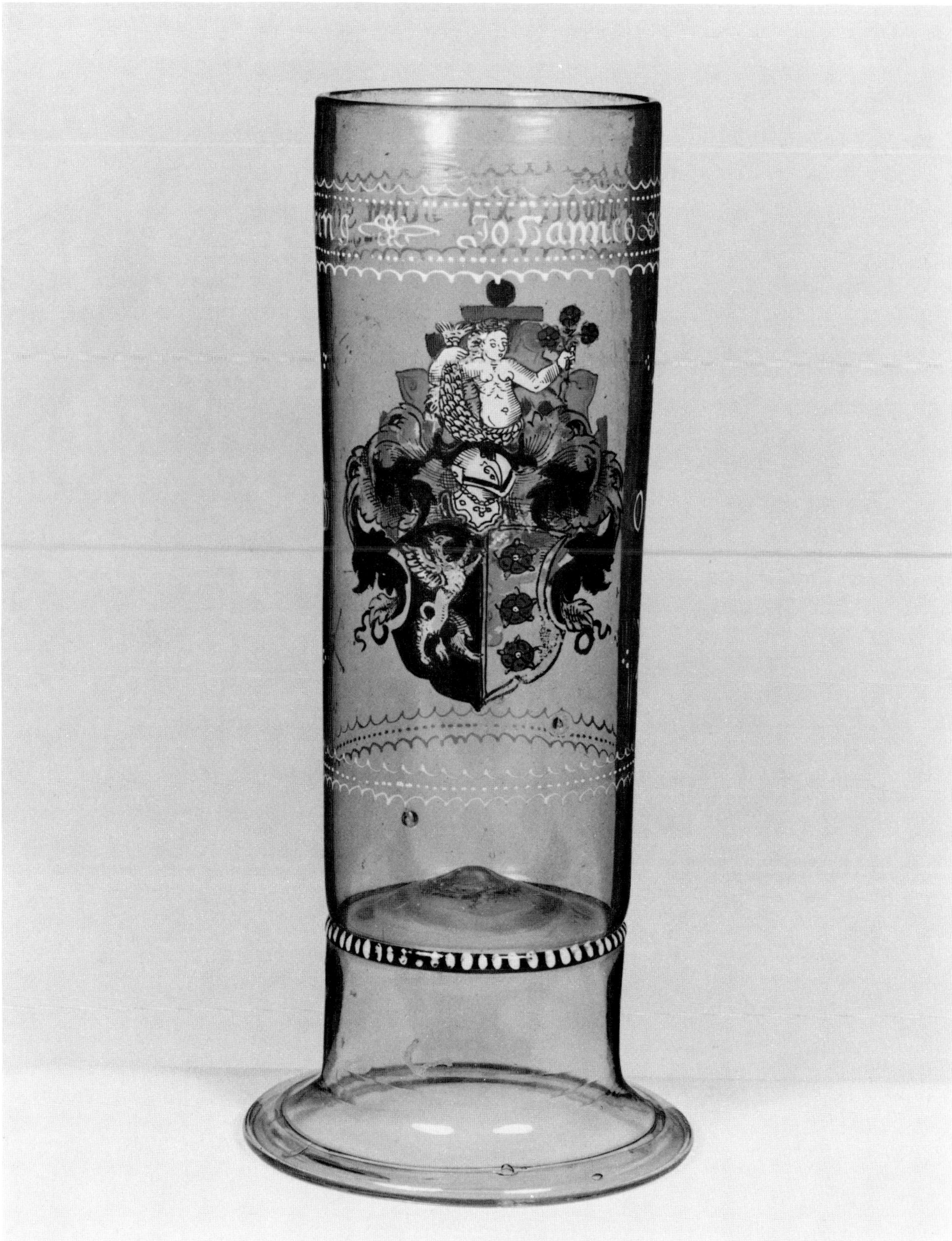

The front of the Stangenglas, *36*

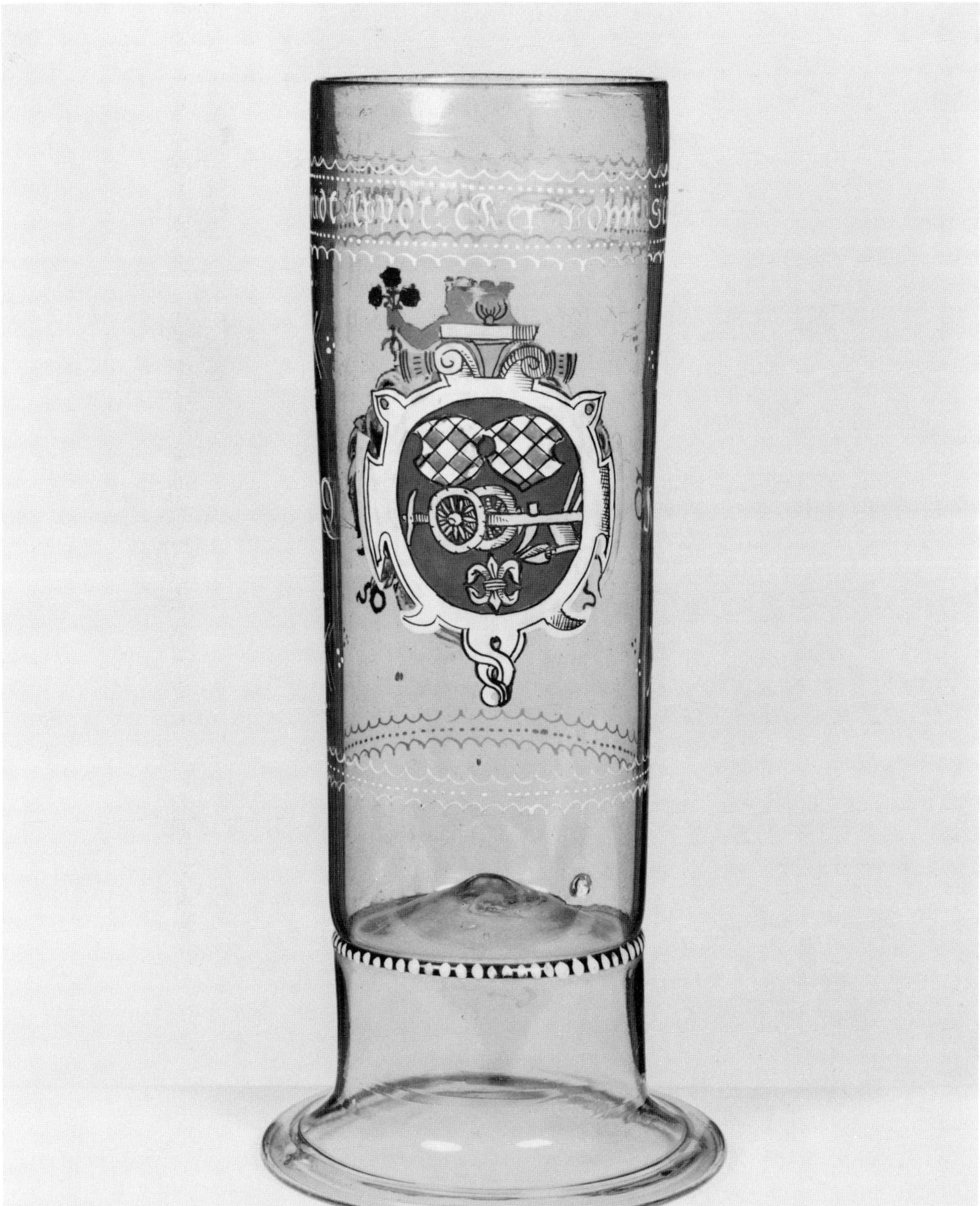

The back of the Stangenglas, *36*

37. Beaker, painted with the 'Ages of Man' (*Lebensalterhumpen*)

Bohemian; dated 1575, but perhaps somewhat later

Humpen of brownish-colourless glass, somewhat bubbled, painted in enamel colours and gilt

Height, $8\frac{7}{8}$ in. (21·5 cm); diameter of rim, $4\frac{3}{4}$ in. (12 cm)

Cylindrical beaker, with applied foot-rim in the form of a thick flattened coil of glass applied round the base of the beaker, which is pushed up into a low conical 'kick'. Painted in black, brownish-red, white, yellow, green and two tones of blue with representations of the Ages of Man, each figure in a round-headed compartment with the arch composed of an inscription in black on white, as follows:

1. '10 IAR EIN KIND' ('10 years, a child'), with a boy trundling a hoop, his dog beside him and his hobby-horse at his feet.
2. '20 IAR EIN IVNGLING' ('20 years, a youth'), with a young man wearing a feathered hat, and a sword, holding a prunted beaker in his left hand, his right hand raised, and his hound (?) at his feet.
3. '30 IAR EIN MAN' ('30 years, a man'), with a bearded warrior holding a spear in his right hand, a charging bull behind him.
4. '40 IAR WOLGETAN' ('40 years, the prime of life'), with an armoured bearded man holding a (?) halberd in his right hand, wearing sword and dagger, with a lion standing behind him, its forepaw on a ball.
5. '50 IAR STILLESTAN' ('50 years, standstill'), with a bearded man wearing a furred gown and holding a scroll, a fox at his feet.
6. '60 IAR GETS ALLTER AN' ('60 years, onset of old age'), with a bearded man leaning on a pedestal, a bear at his feet.
7. '70 IAR EIN GREIS' ('70 years, a greybeard'), with an elderly bearded man seated and resting his left elbow on a cushion placed on a table, a white dog curled up on a cushion at his feet. To his left is a table with a tankard placed on it, and the date '1575' inscribed in black on the front.
8. '80 IAR NIMMER WEIS' ('80 years, never sensible'), with a greybeard leaning on a stick in his right hand, supporting himself on a (?) fence with his left, a white cat at his feet.
9. '90 IAR DER KINDER SPOT' ('90 years, a laughing-stock to the children'), with an old man seated in an arm-chair, a blue donkey at his feet.
10. '100 IAR GENAT DIR GOT' ('100 years, God have mercy on you'), with an old man walking beside Death with raised arm holding an hour-glass, a goose at their feet.

Below the rim is a border composed of white, red and green dots disposed diagonally on a band of gold leaf; this is bounded above and below by double lines of white and blue dots.

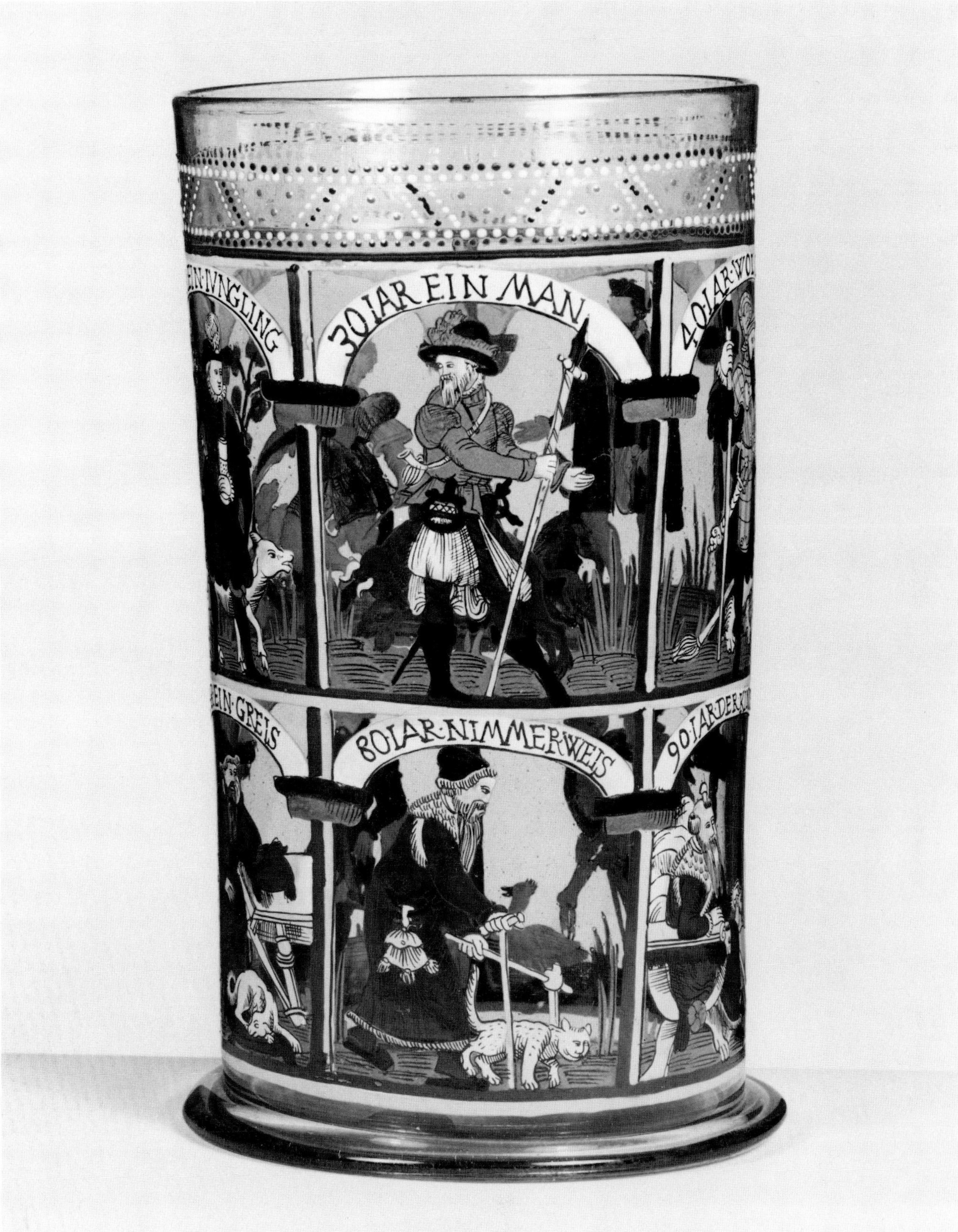
30 IAR EIN MAN
80 IAR·NIMMER·WEIS

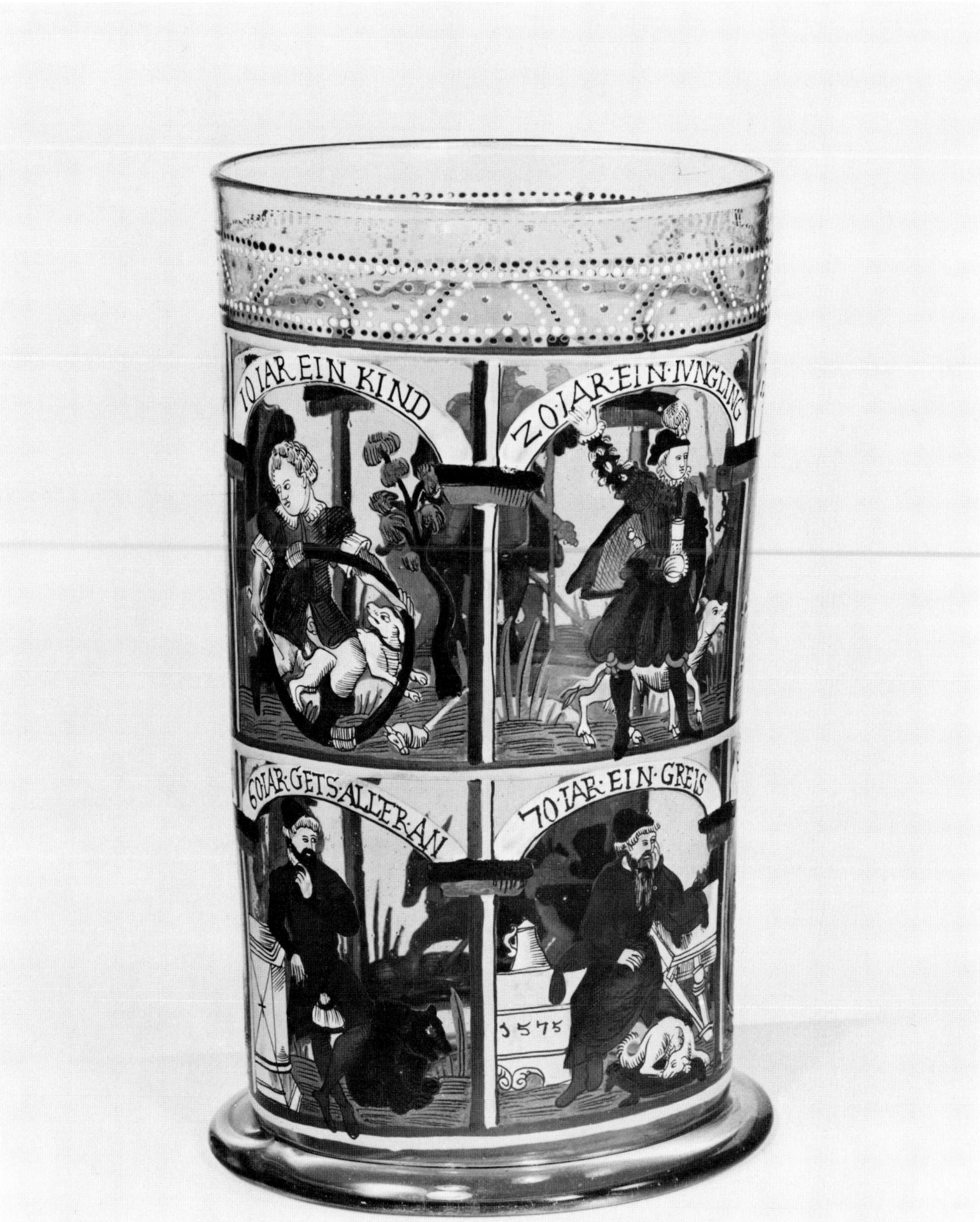
10 IAR EIN KIND
20·IAR·EIN·IVNGLING
60·IAR·GETS·ALTER·AN
70·IAR·EIN·GREIS
1575

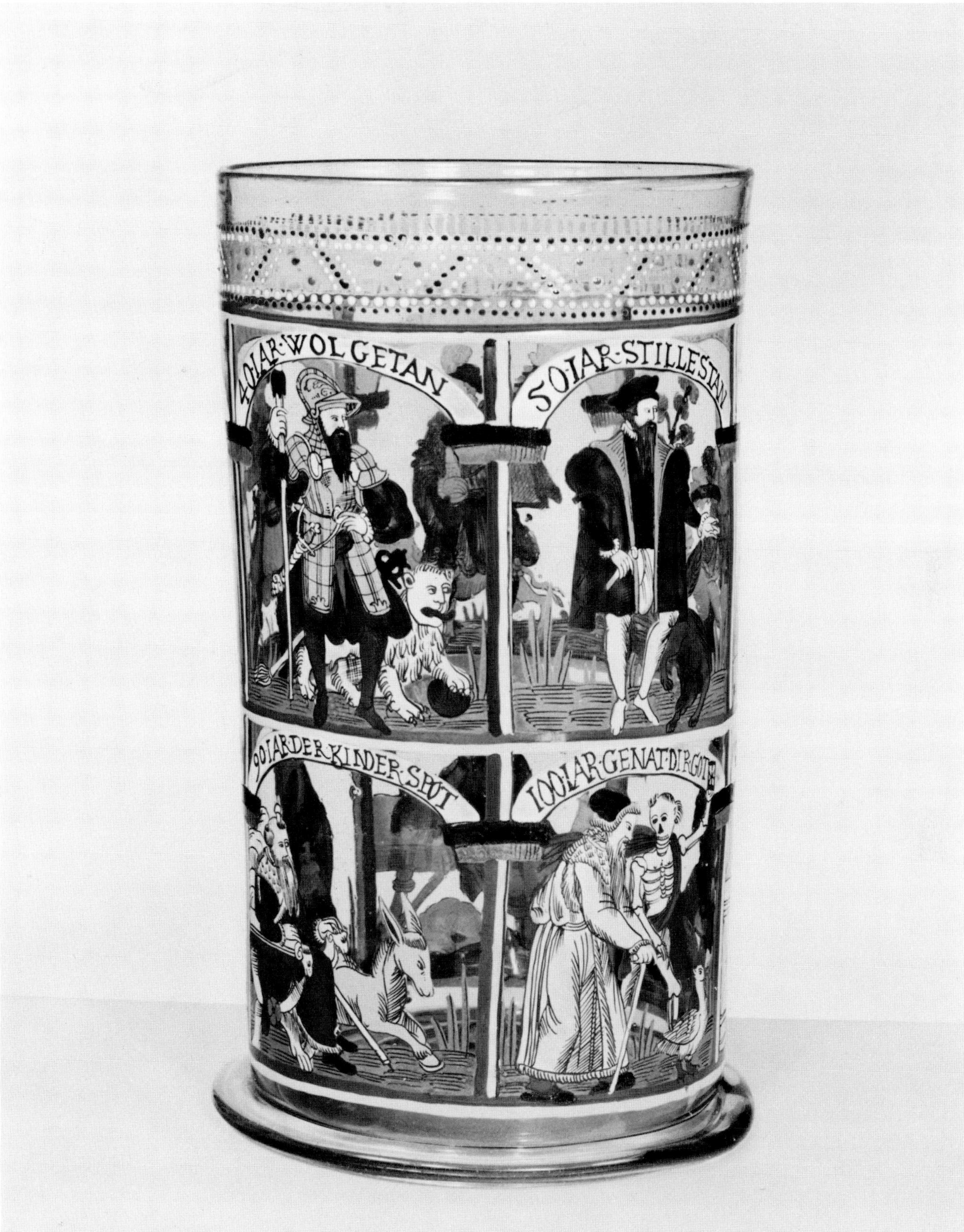
40 IAR·WOL GETAN
50 IAR·STILLESTAN
90 IAR DER·KINDER·SPOT
100 IAR·GENAT·DIR GOT

Beakers painted with the theme of the Ages of Man are by no means uncommon,[1] although they are not so numerous as those decorated with, e.g., the *Reichsadler* (cf. *43* and *44* here) or the Electors of the Holy Roman Empire (cf. *45* here). The present example, however, differs from most others in its versions of the themes represented, which are normally derived from woodcuts by the Master I. R. (active about 1575).[2] A *Humpen* in the Art Institute of Chicago, however, is evidently derived from the same source.[3]

Von Saldern draws attention to a house in Retz, Lower Austria, decorated about 1576 after the woodcuts by the Master I.R. and others by Virgil Solis. Evidently this theme was popular about the time of the date on the Waddesdon *Humpen*, but the general style of this glass suggests that in fact it was made nearer 1600.[4] A similar *Lebensalterhumpen* at Corning bears the date 1578 but for the same reason has been ascribed to the late sixteenth or early seventeenth century.[5]

From the Hollingworth Magniac Collection (*Catalogue of The Renowned Collection of Works of Art of the late Hollingworth Magniac, Esq.*, London, Christie, Manson and Woods, fourth Day's Sale, 6 July, 1892, Lot 413: 'A WIEDERKOM, enamelled in ten compartments with the Ages of Man from 10 to 100 years, and dated 1575 – 8¾ in. high').

1922 Waddesdon Inventory: 'Beaker enamelled with figures in ten compartments. German 1575'.

Miss Alice de Rothschild — W1/139/5*

[1] See, e.g., A. von Saldern, *German Enameled Glass*, Corning, 1965, pp.96–9; R. Schmidt, *Das Glas*, Berlin and Leipzig, 2nd ed., 1922, p.180, Fig.100; W. Bernt, *Altes Glas*, Munich, n.d. (1951), Fig.37; I. Schlosser, *Das Alte Glas*, Brunswick, 1956, Fig.88.

[2] A. von Saldern, op. cit., pp.97–9, and Fig.133; R. Schmidt, *Das Glas*, p.173, Fig.95.

[3] A. von Saldern, op. cit., p.96, Fig.131 (Mus.No.27.1012).

[4] cf., e.g., A. von Saldern, op. cit., Figs.63 (dated 1595), 152 (dated 1599), 167 (dated 1586), 235 (dated 1602). A glass very close in every way, especially in the disposition of its subjects (the Apostles) under arcading, is the *Humpen* ibid., Fig.88, attributed to the late sixteenth century.

[5] A. von Saldern, op. cit., p.315, No.47.

38. Beaker, painted with the 'Ages of Man' (*Lebensalterhumpen*)

Bohemian; late 16th or early 17th century

Humpen of dark greenish-grey glass, slightly bubbled, painted in enamel colours and gilt

Height, $11\frac{3}{8}$ in. (29 cm); diameter of rim, $4\frac{1}{2}$ in. (11·5 cm)

Cylindrical beaker with slightly concave profile, with applied foot in the form of a thick flattened coil of glass applied round the base of the beaker, which is pushed up into a low conical 'kick'.

Painted in black, red, white, blue, yellow and green with representations of the Ages of Man, each figure in a round-headed compartment with the arch composed of an inscription in black on white, as follows:

1. 'X Jar ein Kindt' ('10 years, a child'), with a boy riding a hobby-horse, a bird in his left hand, and a toy windmill stuck in his belt: to his right a Unicorn.
2. 'XX Jar ein Jüngling' ('20 years, a youth'), with a bearded, hatted man holding a hawk on his left wrist, a hound at his feet and a sword at his left hip.
3. 'XXX Jar ein Man' ('30 years, a man'), with a bearded, hatted man holding a large black and yellow striped standard in his right hand, and having a recumbent ox behind him.
4. 'XL Jar Wolgethan' ('40 years, the Prime of Life'), with an armoured bearded man holding a baton on his right hip, his left hand on the hilt of his sword; behind him a lion.
5. 'L Jar Stille Stan' ('50 years, standstill'), with a man wearing a furred shoulder-cloak, holding a glove in his right hand, his left outstretched, while at his feet sits a fox.
6. 'LX Jar gehets allter an' ('60 years, onset of old age'), with a grey-bearded man wearing a furred cap and a long furred cloak, and holding a purse in his left hand, a blue fox behind him holding a bird in its jaws.
7. 'LXX Jar ein Greis' ('70 years, a greybeard'), with an elderly man wearing a blue cloak trimmed with black fur, and holding a parchment in his right hand, a hound at his feet.
8. 'LXXX Jar nimmer weis' ('80 years, never sensible'), with an old, bearded man wearing a long furred coat and holding a stick in his left hand, a rosary in his right, with a blue monkey crouching at his feet.
9. 'XC Jar der Kinder Spott' ('90 years, a laughing-stock to the children'), with an old white-bearded man in a furred cloak, walking on crutches, a blue donkey at his feet.
10. 'C Jar genad Dir Gott' ('100 years, God have mercy on you'), with an old man seated in a blue cloak which Death is drawing aside with one hand, while in the other he holds an hour-glass: beside them is a goose.

Below the rim is a border consisting of a band of gold leaf etched with a point in alternating scale-design, enlivened with dots of red enamel and groups of three diagonal lines enlivened with dots of white enamel, the whole between horizontal single rows of blue and white dots.

On the question of *Lebensalterhumpen* in general, see *37*. The present beaker bears the same designs in all its panels as an example in the Corning Museum of Glass (A. von Saldern, *German Enameled Glass*, p.315, No.47). They are derived from woodcuts by the Master I.R., of which the prototype for the child of ten is illustrated ibid., Fig.133. The Corning glass resembles the Waddesdon piece in all respects except the feature of a row of lily-of-the-valley heads above the top border: it is dated 1578, but may well in fact be later (see *37*).

The *Humpen* carries an old paper label with the number '3295', its Lot number in the Bernal Sale – see *Catalogue of the celebrated Collection . . . of Ralph Bernal*, London, Christie and Manson, 24th Day's Sale, 19 April, 1855: 'A vidercom, with the ten ages of man, in arched divisions – 11 in. high' (bought by Sir H. H. Campbell for thirteen guineas).

1922 Waddesdon Inventory: 'Tall beaker enamelled with figures in ten compartments. German, 17th century'.

Miss Alice de Rothschild — w1/140/8*

38

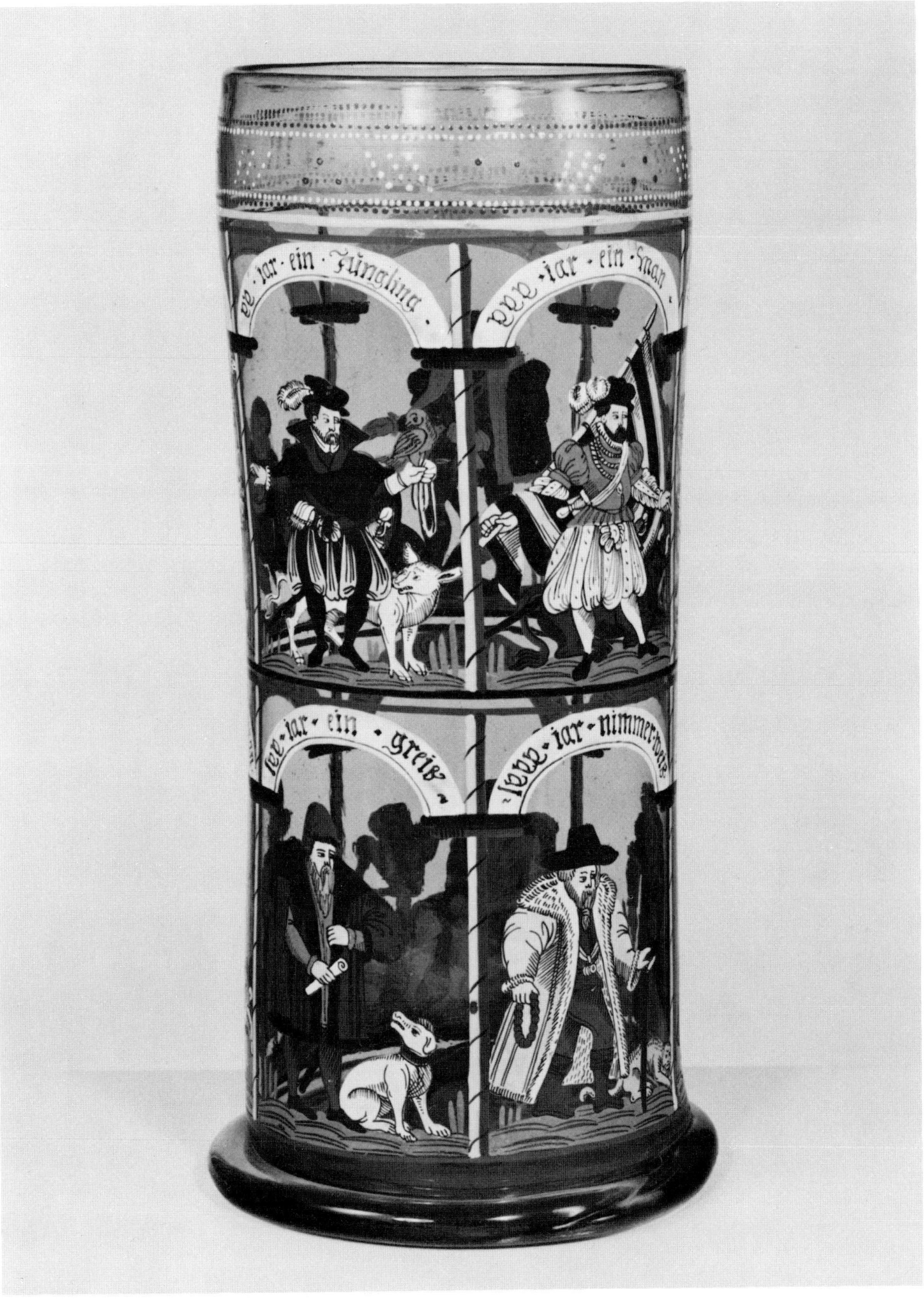

39. Jug and cover, in the form of an owl

Bohemian; dated 1605

Greyish-colourless glass painted in enamel colours and gilt

Height with cover, 8¾ in. (22·3 cm); without cover, 7½ in. (19 cm); diameter of foot, 3½ in. (9 cm)

The body of the vessel, of an inverted baluster shape, forms the body of the owl, and has a separately applied pedestal foot with folded edge. The vessel has a cylindrical neck over which fits a domed cover forming the head of the owl, with eyes and beak applied and gilt. This is painted blue with vertical black markings, suggesting feathers, and black concentric rings round the eyes. The body is painted with blue wings, a green tail and breast, and yellow legs, all picked out in black. In the middle of the breast, where the glass has been slightly pressed in, a heart-shaped patch has been left unenamelled and then gilt. On the reverse is the inscription 'EIN EVLE BIN ICH ALLE VOGEL HASSEN MICH 1605' ('an owl am I. All birds hate me. 1605'), each word successively painted in yellow, red, white, blue and green enamel. Round the neck of the body runs a double row of white dots with a plain gold band between.

A second jug of the same type, but without its cover, is in the Museum of Applied Arts in Prague (I.CVI.57). It is dated 1597 and is inscribed: 'ICH BIN EIN VOGL VND HEISSE DI EVLE WEHR HASET DER HABE IM DIE BEULE 1597' ('I am a bird and am called the owl. A pox on him who hates me! 1597'). Like the Waddesdon example, it has its heart carefully delineated on its breast.

A third example, in the Fritz Biemann Collection, Zurich, also lacking the head, is the earliest of the three. It bears in alternating yellow and red the inscription 'ICH BIN EIN/VOGEL UND HEI/SSE EULE WER/MICH HASSET/DER HABE IN (=ihm?)/ DIE BEULE. 1582', essentially the same as that on the Prague example. The Biemann piece has the foot replaced in metal.[1]

This form is extremely rare in glass, but is not uncommon in pottery and metalwork, the pottery examples being mostly earlier, ranging from 1540 to 1561 in the dated examples, although the type lingered on into the eighteenth century. The shape was made in tin-glazed earthenware ('faience') in Central Europe in the sixteenth century, and in faience and stoneware in the seventeenth and early eighteenth centuries.[2]

One of the early faience jugs has painted on the back a panel with the story of William Tell,[3] and it has commonly been believed that these jugs were given away as prizes for prowess in archery, probably shooting at the popinjay. There is a record of a silver owl given as a prize

[1] K.-A. Wirth, article 'Eulengefäss' in *Reallexikon zur Deutschen Kunstgeschichte*, Stuttgart, 63. Lieferung, 1970, Cols 322–40, Fig.18; id., 'Von silbernen und silbermontierten Eulengefässen', *Anzeiger des Germanischen Nationalmuseums*, Nuremberg, 1968, pp.42–83, Fig.9.

[2] W. B. Honey, 'Augustin Hirschvogel and the Tyrolese Owl-jugs', *Burlington Magazine*, LXIX, September, 1936, pp.111–20. This is much extended by K.-A. Wirth's articles cited in n.1.

[3] W. B. Honey, loc. cit., Pl. I,E.

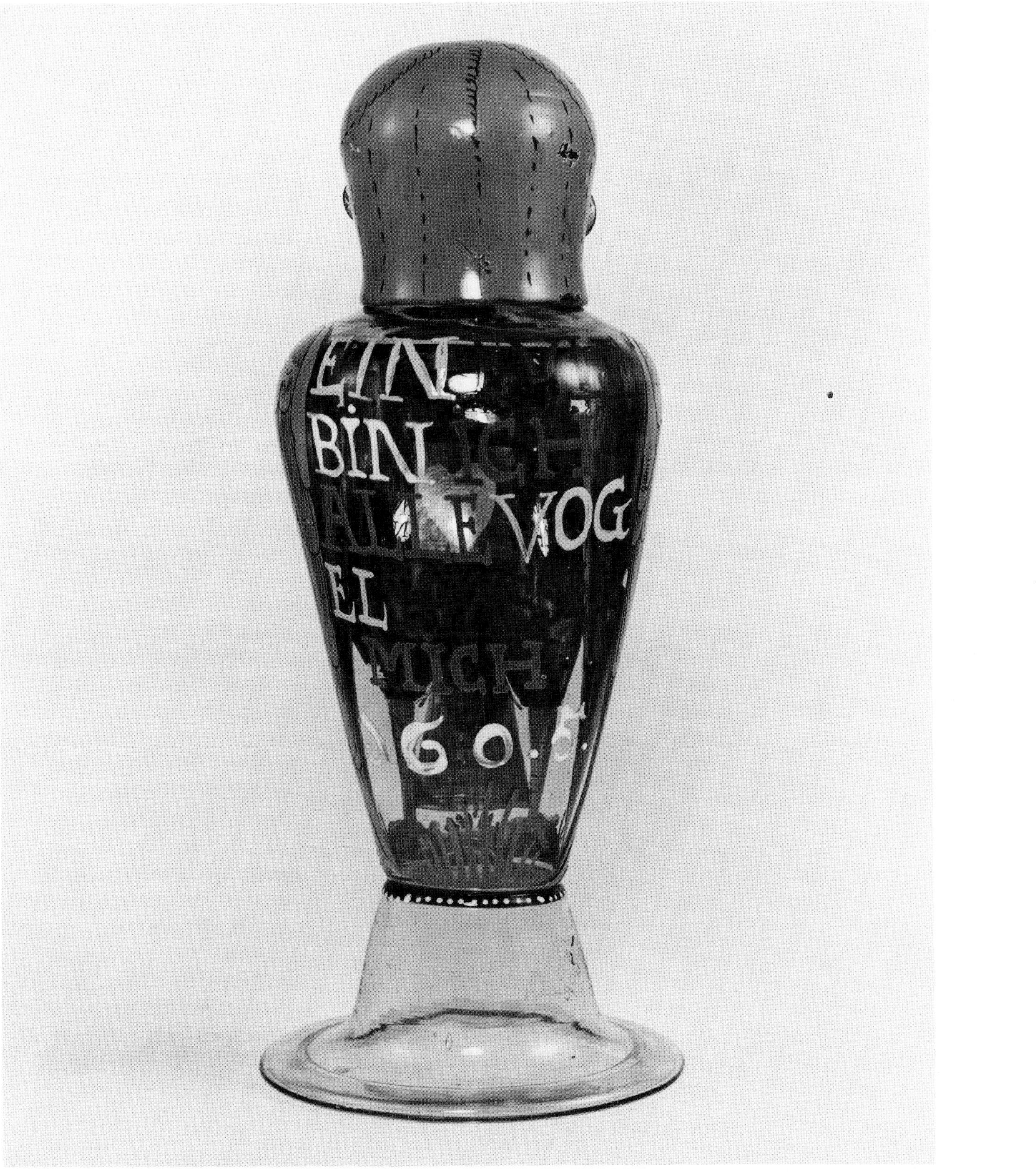
EIN
BIN ICH
ALLE VOG
EL
MICH

40. Jug, painted with a hunting scene

Bohemian; dated 1605

Blue glass painted in enamel colours, with pewter mount

Height with cover, $8\frac{1}{8}$ in. (21 cm); without cover, $7\frac{1}{8}$ in. (18·5 cm); diameter of rim, $2\frac{1}{4}$ in. (5·5 cm)

Jug with tall cylindrical neck and ovoid body, with pushed-in base. Grooved strap-handle, doubled back at the lower sticking-part. Round the base of the neck is laid a glass thread (partially broken away) which has been picked out with dots of white enamel.

The jug is painted with a red fox chased by two hounds, one white and one blue. Between each pair of animals is a curved frond of leaves rendered in green and black. Round the neck is the date '1605' painted in yellow between double lines of red and blue, with a white quatrefoil between each pair of figures.

The mount bears the mark of a five-petalled rose under a crown, with the date 1627. The mark of a crown and rose, in many variations, was one denoting high quality in the pewter (Erwin Hintze, *Die Deutschen Zinngiesser und ihre Marken*, Leipzig, I–VII, 1921–31, *passim* ('Rosenmarken für Feinzinn')).

Hunt-subjects are frequently to be found on the Bohemian enamelled glasses of the late sixteenth and early seventeenth centuries. Of such scenes rendered on the characteristic blue jugs of this shape, the following may be mentioned – a hare-hunt dated 1601 in the Musées Royaux d'Art et d'Histoire, Brussels (Berryer, *La Verrerie Ancienne*, Brussels 1957, Pl. XVII; Isings, *Schönes Altes Glas*, p.72, and Pl.69); a stag-hunt dated 1597 in the Moritzburg zu Halle (O. H. Werner, *Schönes Glas in der Moritzburg zu Halle*, Halle, 1957, No.60, Fig.15); and an undated fox-hunt in the Metropolitan Museum of Art, New York, from the Mühsam Collection (Schmidt, *Mühsam*, II, No.85, ill. p.27). The fox-hunt subject is represented again on a blue dish of the same period in the Österreichisches Museum für Angewandte Kunst, Vienna (A. von Saldern, *German Enameled Glass*, Fig. 317); and on a colourless undated jug in the Corning Museum (ibid., p.253, No.7). The blue jugs cited above bear their dates on the body of the vessel, but the manner of writing the date round the neck is repeated on a number of jugs decorated with other themes (e.g. A. von Saldern, *German Enameled Glass*, Fig.160, dated 1593, with satirical subject of a fox; Fig.307, dated 1597, with a woman's portrait, etc.).

The sources of the decoration of the hunt-jugs are to be found among the prints of Virgil Solis, Hans Weiditz, Stradanus, etc., although exact correspondences of theme are often difficult to establish (see A. von Saldern, op.cit., pp.113 ff).

1922 Eythrope Inventory, 'Enamelled jug, black ground, hound and hares, silver cover. German 17th century'.

Miss Alice de Rothschild E1/12/7*

41. *Humpen*, made for Leonhart von Milckau

Probably Bohemian; dated 1608

Height, $12\frac{1}{4}$ in. (31 cm); diameter of rim, $4\frac{7}{8}$ in. (12·5 cm)

Greenish glass painted in enamel colours and gilt

Cylindrical beaker of irregular, slightly concave profile, with applied foot-rim in the form of a thick flattened coil of glass trailed round the base of the beaker, which is pushed up into a low pointed 'kick'.

Painted in black, red, white, yellow, green and blue enamels with a knight in armour on a prancing black horse with one white fetlock. The rider bears as a charge on his shield and as his crest a two-tailed lion sable, langued gules, crowned or, bearing a curved stave gules. Above him is inscribed in white: 'LEONHART VON MILCKAU 1608'. On the reverse is the representation of FORTUNA standing on a winged globe in the midst of a blue sea, carrying a blue sail over her head, with an inscription in white: 'cui fortuna favet sponsa petita manet' ('He whom Fortune favours finds the bride he sought'). Below the rim is a border formed of a diaper of green dots on a plain band of gold, incised with a point in an imbricated design, and bounded above and below by lines of white and blue dots.

The von Milckau family was a Saxon one, with its seat at Milckau, near Rochlitz on the Mulde (O. Posse, *Die Siegel des Adels der Wettiner Lande*, IV, Dresden, 1911, pp.114–5, Pl.56). On the early seals (fifteenth–sixteenth centuries) the rampant lion sometimes faces right, and normally has a single tail, as on the arms of the family at a later date (J. B. Rietstap, *Armorial Général*, Plates Vol. IV, Paris, 1912, Pl. CCXII). In the 1605 edition of J. Sibmacher, *New Wapenbuch*, Pl.152, the lion has two tails (von Milckau, of Meissen), as on this glass.

The first member of the family to be mentioned in the archives was Hermann von Milckau in 1329, in which year the family already possessed Gross- and Klein-Milckau. Of Meissen origin, the family also had possessions in that area of Saxony, including Alberode and Alten-Schönfels (a town with a castle, south-west of Zwickau). Some time after 1500 the von Milckau also had the 'High House' (*Hohe Haus*) at Rochlitz, built originally to protect the crossing of the Mulde, but this was sold to the Elector of Saxony in 1588 by Hiob von Milckau. The building was still visible in Wilhelm Dilich's view of Rochlitz, drawn about 1628 (information kindly supplied by Dr Joseph Horschik, of Dresden – see also Wilhelm Dilich, 'Federzeichnungen kursächsischen und meissnischen Ortschaften aus den Jahren 1626–1629', herausgegeben von P. E. Richter und C. Krollmann, 3 vols, Dresden, 1907; Walter Fischer, 'Schloss und Stadt Rochlitz in Sachsen', *Mitteilungen des Landesvereins Sächsischen Heimatschutz*, Dresden, XXIV, Heft 1–4, 1935, p.6). Christoph von Milckau, Lord of Alberode, died in 1550, leaving two sons – another Christoph, of Alberode; and Leonhart, of Alten-Schönfels. Leonhart von Milckau was Privy Councillor to the Elector of Saxony and Vice Chancellor ('kursächsischer Geheimer Rat und Vizekanzler' – see (ed.) E. H. Kneschke, *Neues allgemeines Deutsches Adels-Lexikon*, VI, Leipzig, 1930, pp.292 ff, kindly communicated by Professor Dr Kurt Schwabe, President of the Saxon Academy of Science, Leipzig).

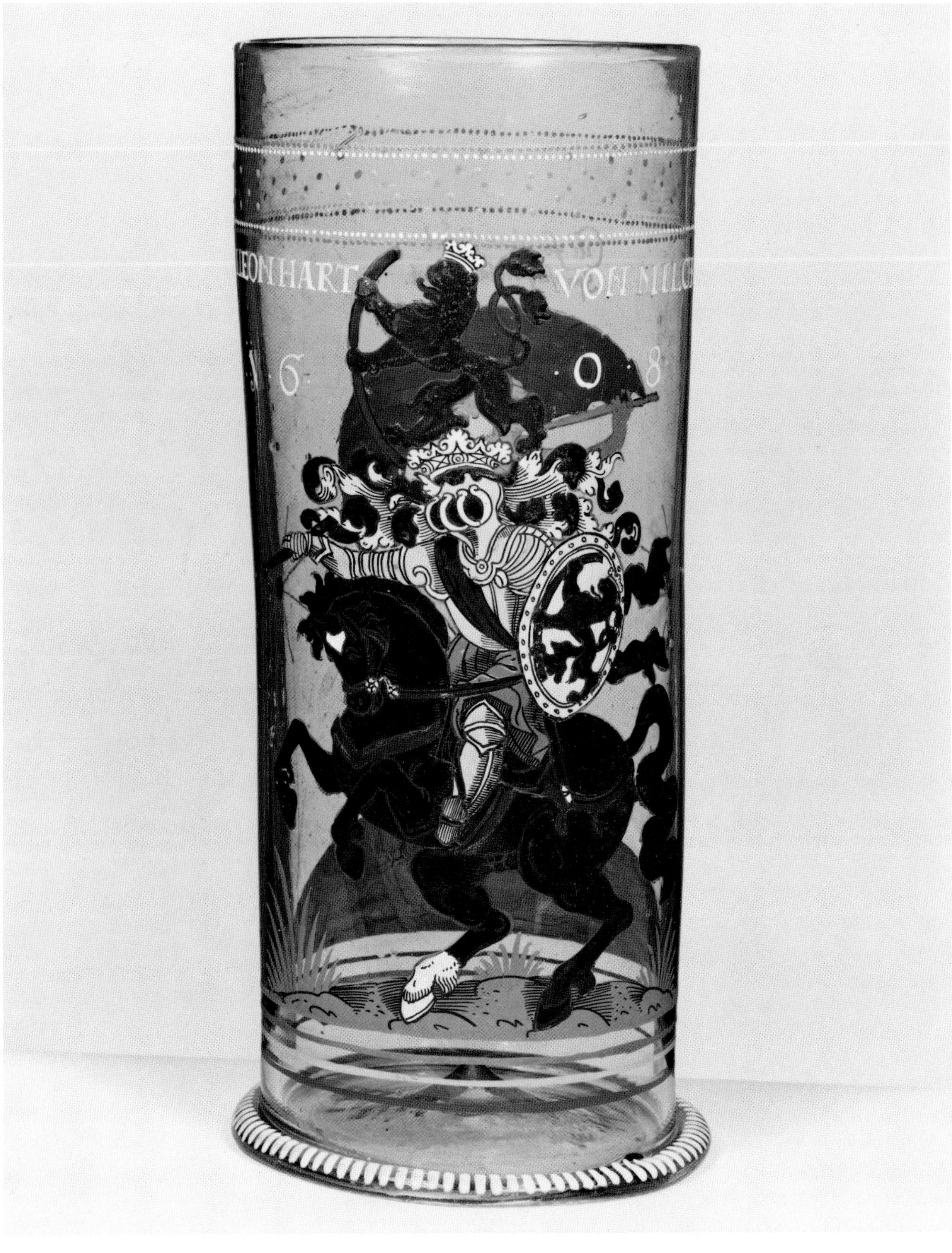
LEONHART
6
0
8

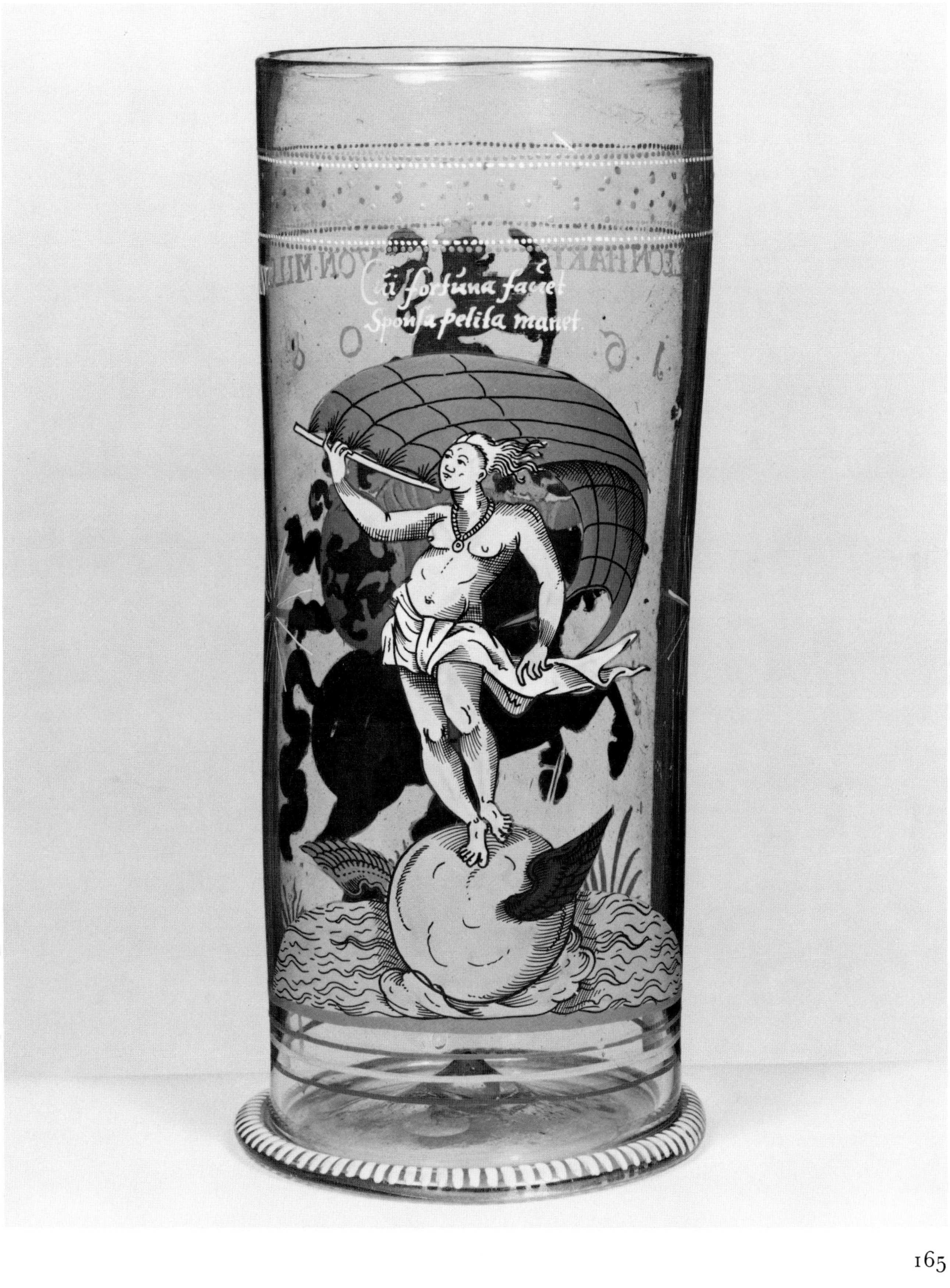
Cui fortuna favet
Sponsa petita manet.

It is just possible that at this early date (1608) the glass might have been made in Saxony, but it was more probably procured from Bohemia (see Axel von Saldern, *German Enameled Glass*, pp.200–01).

From the Bernal Collection – see *Catalogue of the celebrated Collection of Ralph Bernal*, Christie and Manson, 24th Day's Sale, 19 April, 1855, Lot 3086: 'A vidercom, with knight bearing the arms of Leonhart Von Milckau, 1608; Fortune, on the reverse – $12\frac{1}{2}$ in. high', bought by Sir H. H. Campbell for five guineas.

1922 Waddesdon Inventory (Smoking Room): 'Tall beaker enamelled with equestrian figure with coat of arms and figure of fortune. German, 1608'.

Miss Alice de Rothschild W1/140/10*

42. *Humpen*, painted with an equestrian figure

German (probably Franconian); dated 1631

Height, $6\frac{3}{4}$ in. (17·2 cm); diameter of rim, $3\frac{3}{8}$ in. (8·5 cm)

Greenish somewhat bubbled glass, painted in enamel colours and gilt

Cylindrical beaker, slightly drawn inwards towards the rim, with foot-rim in the form of a thick flattened coil of glass applied round the base of the beaker, which is pushed up into a low conical 'kick'.

Painted in black, brown, red, white, yellow, green and blue enamels, with the representation of a cavalier in blue breeches and yellow doublet with green sash, mounted on a brown horse. He holds a long switch in his right hand. Inscribed on the reverse in white: 'Ann Gottes mundt Steht all mein grunndt ANNO DOMINI 1*63*1*' ('My whole life's reason is founded upon the Word of God'). Below this inscription is a spray of lilies-of-the-valley between scrolled leaves of green, blue, red, and yellow. Below the rim is a border of white loops and dots, with a band of plain gilding. The foot-rim is enlivened with radiating white dashes.

Mounted figures are a frequent theme upon German enamelled glasses (A. von Saldern, *German Enameled Glass*, pp.118–19). *Humpen* similar in date and style to that at Waddesdon are a glass in the Städtisches Museum des Kunsthandwerks, Leipzig, with mounted figures of Gustavus Adolphus of Sweden and Johann Georg, Elector of Saxony, dated 1634 (F. Kämpfer and K. G. Beyer, *Glass*, p.281, No.127; A. von Saldern, op.cit., Fig.181); a *Willkomm* with arms of Wolf Sigmund von Lückau, dated 1641 (*Sammlung des Freiherrn Adalbert von Lanna, Prag*, Sale Catalogue Rudolph Lepke, Berlin, 23 March, 1911, No.836, Pl.66; A. von Saldern, op.cit., Fig.182); another with the arms of Brandenburg, also dated 1641 (A. von Saldern, op.cit., Fig.183). All these glasses are probably Franconian (see A. von Saldern, op.cit., pp.191–2), and all have the lily-of-the-valley motif.

From the Bernal Collection (*Catalogue of the celebrated Collection of . . . Ralph Bernal, Esq.*, Christie and Manson, 25th Day's Sale, 20 April, 1855, Lot 3272: 'A small vidercom, with a cavalier on horseback, and inscription – 1631', bought by the Rev. W. Pigott for £1.18.0).

1922 Waddesdon Inventory, 'Goblet enamelled with figure of horseman. German, 1631'.

Miss Alice de Rothschild — W1/139/1

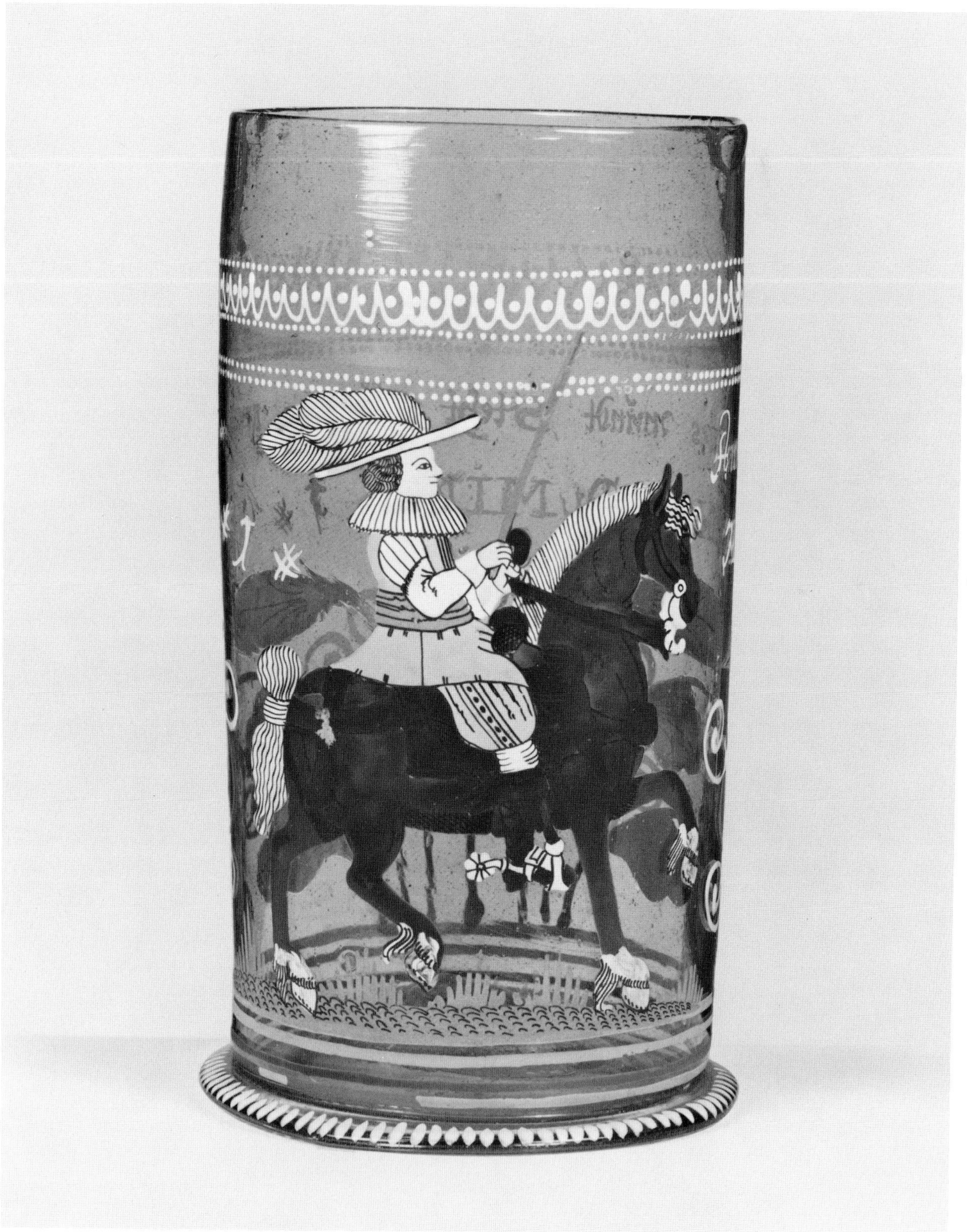

42

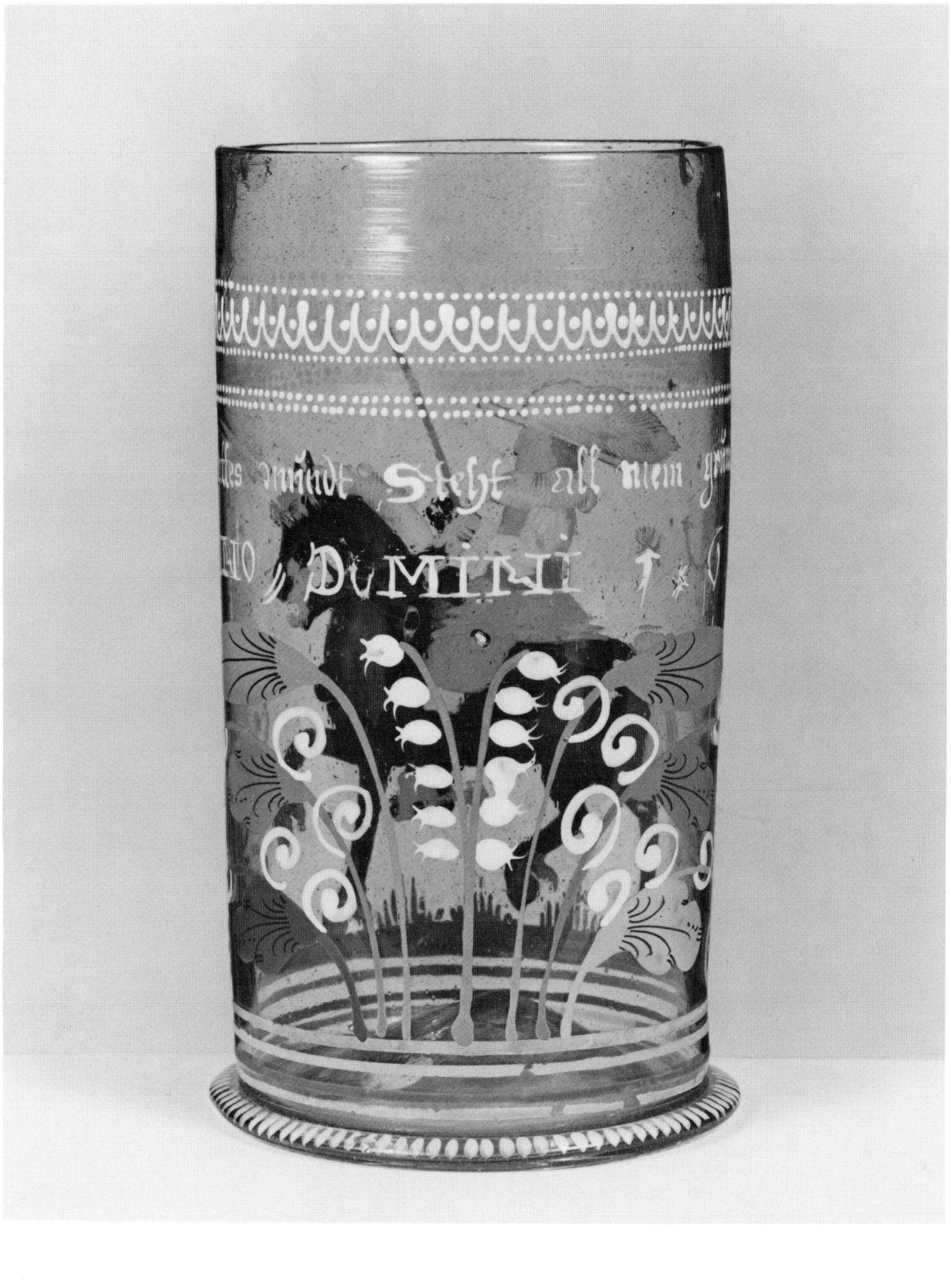

43. Beaker, painted with the arms of the Holy Roman Empire (*Reichsadlerhumpen*)

Franconian; dated 1645

Brownish-colourless glass, painted in enamel colours and gilt

Height, $5\frac{3}{4}$ in. (14·5 cm); diameter of rim, $2\frac{7}{8}$ in. (7·5 cm). Illustrated overleaf.

Cylindrical beaker slightly drawn in towards the rim, with projecting foot-rim pincered out of the paraison in the form of a flat fold round the base angle of the beaker, which is pushed up into a low 'kick'.

Painted in black, brownish-red, blue, yellow, white and green enamels. On the front is rendered a spread-eagle with double heads wearing the Imperial Crown, and on its breast a blue orb with golden cross: on its wings are displayed the coats-of-arms of the constituent parts of the Holy Roman Empire. On the reverse is the inscription in white enamel: 'Gott behütt unnd erhaldt/dass gantze heÿlige Röm/ische reich mitt seinn gliedern allen zv/gleich. ANNO DOMI/NI 1645'. ('God protect and preserve the whole Holy Roman Empire along with all its parts together. Anno Domini 1645'). Below the rim of the beaker is a formal border composed of intersecting arcs of white dots punctuated at the intersections by groups of one red and four blue dots, all on a ground of gold leaf; above and below are borders composed of three and two lines respectively of white dots. The flange of the foot is enlivened by radiating white strokes.

The beaker has been broken and mended.

The *Reichsadlerhumpen* occurs in innumerable variants, and comparisons are not particularly rewarding. It may suffice to draw attention to one or two of the peculiarities of the Waddesdon glass. The first is the rendering of the jewelled bands of the orb as a series of running S-scrolls. This feature recurs on a *Humpen* in the Corning Museum, dated 1608 (A. von Saldern, *German Enameled Glass*, p.269, No.17, there attributed to 'Probably Bohemia'). The second distinguishing feature of the Waddesdon glass is the form of the inscription, more elaborate than the normal 'Das heilige Römische Reich mit samt seinen Gliedern' ('The Holy Roman Empire with all its components'). This more extended form of the inscription occurs again on a *Humpen* in the Frankfurt Museum (see Museum für Kunsthandwerk, *Europäisches und Aussereuropäisches Glas*, No.301, dated 1685 and attributed to Bohemia) and on another in the Kunstmuseum, Düsseldorf (*Catalogue*, No.181, dated 1655 and attributed to Franconia; see also J. Jantzen, *Deutsches Glas aus Fünf Jahrhunderten*, Düsseldorf, 1960, p.25, No.56). Except for its border, the last-named glass has many affinities with the Waddesdon *Humpen*. A third *Humpen* with virtually the same inscription, dated 1659, is in the Corning Museum, and is attributed with great probability to Franconia (A. von Saldern, op. cit., p.277, No.24). This piece shares in addition the unusual feature of a pinched-out foot-rim rather than an applied coil, as does a further Franconian beaker dated 1674 in the same Collection (ibid., p.278, No.25). This piece has an inscription even longer and more fulsome than that on the pieces already cited. A Franconian origin for all seems extremely probable. Another Franconian glass with similar characteristics

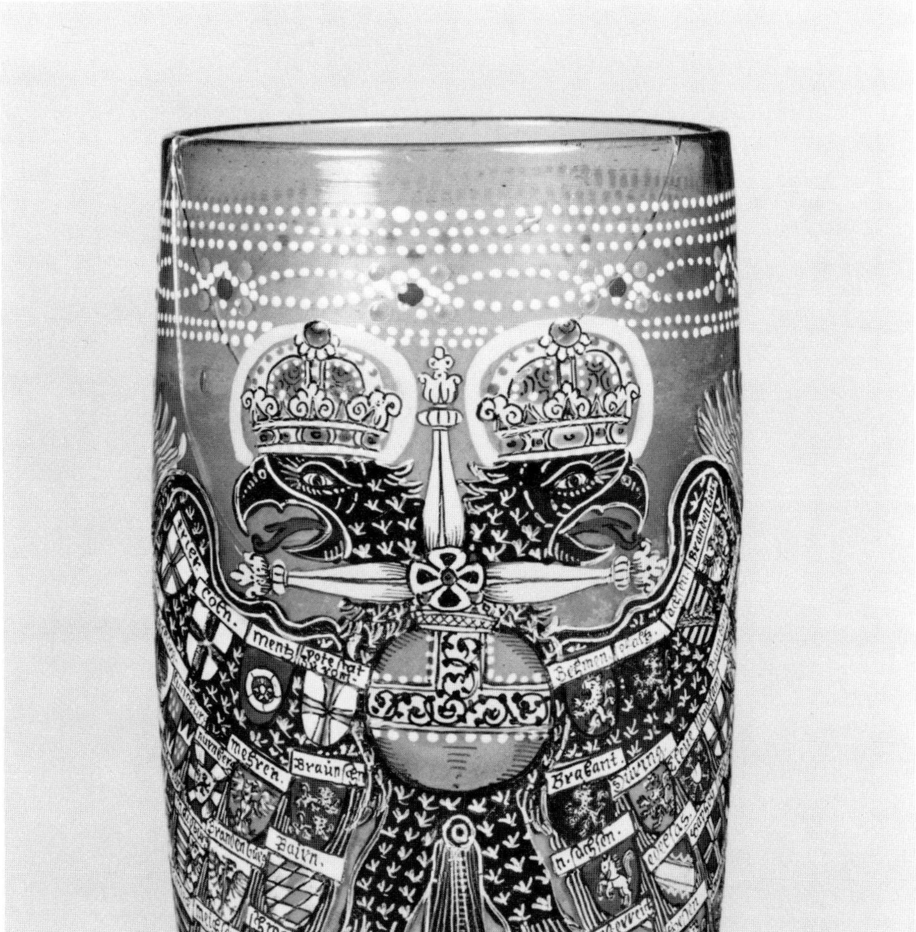

and a border resembling that on the Waddesdon beaker is a *Humpen* dated 1625 in the Museum at Frankfurt (*Europäisches und Aussereuropäisches Glas*, No.304).

1922 Waddesdon Inventory: 'Goblet enamelled with numerous coats of arms. German, 1645'.

Miss Alice de Rothschild — W1/138/8*

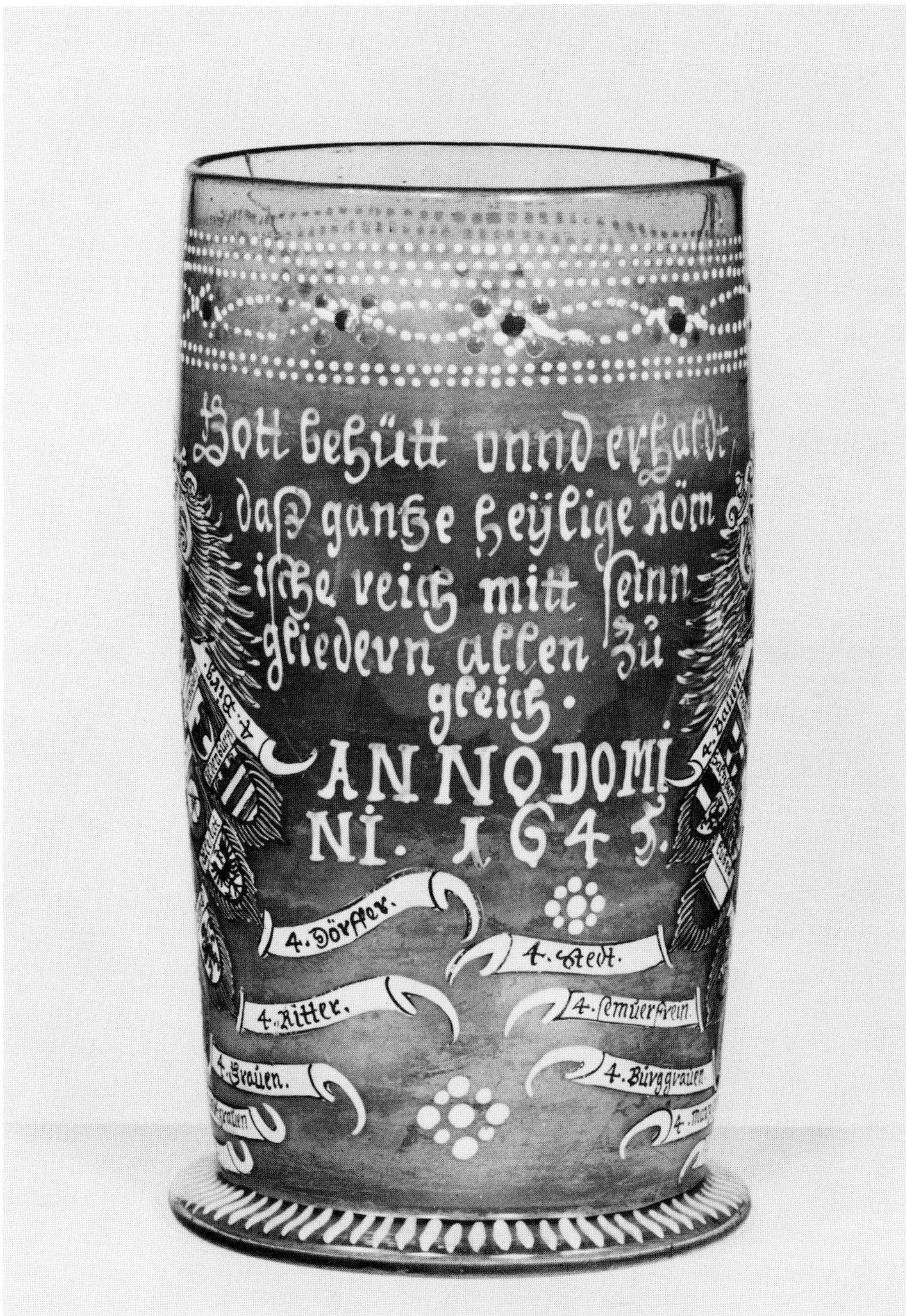

The back of 43

44. Beaker, painted with the arms of the Holy Roman Empire (*Reichsadlerhumpen*)

Bohemian; dated 1661

Greenish-yellow bubbled glass, painted in enamel colours and gilt

Height, $11\frac{3}{8}$ in. (29 cm); diameter of rim, $4\frac{7}{8}$ in. (12·5 cm)

Cylindrical beaker slightly drawn in at the rim, with applied foot in the form of a thick flattened coil of glass applied round the basal angle of the beaker, which is pushed up into a low 'kick'.

Painted in black, brown, blue, yellow, white and green enamels with, on the front, a double-headed spread-eagle wearing the Imperial Crown, and on its breast a blue orb with golden cross; on its wings are displayed the coats-of-arms of the constituent parts of the Holy Roman Empire. On the reverse is the inscription 'Dass Heÿlige Römische Reich mit samt seinen gliedern' ('The Holy Roman Empire with all its members'), and the date 1661, all rendered in white enamel: above and below, a fleuret composed of a central lozenge of gold leaf with a dot of red enamel at its centre and a dot of blue at each of its corners (or vice versa) outlined in white, with white scrollwork between. Below the rim, a border formed of a band of gold leaf edged with dots of white enamel. The flange of the foot is enlivened with radiating white strokes.

Somewhat comparable *Humpen* are in the Corning Museum, dated 1614 and 1632 respectively (A. von Saldern, *German Enameled Glass*, pp.270–1, Nos.18 and 19). Dr von Saldern expresses considerable reserve about these two pieces and hints that both may be nineteenth-century reproductions. He points out that both render the inscription over the shield of Bohemia as 'Böhemb', instead of the more usual 'Behem'. This feature is shared by the Waddesdon glass and by a further *Humpen* in the Museum des Kunsthandwerks in Leipzig, dated 1654 (see Kaempfer and Beyer, *Glass*, Pl.130 and p.281). Assuming the genuineness of this group, it is perhaps possible to discern a gradual falling-away in quality between the earliest, dated 1614, and the Waddesdon glass, dated 1661, especially in the increasingly perfunctory rendering of the feathers, and the progressively stylized distortion of the tongue, which is reduced to a meaningless scroll on the Waddesdon and Leipzig glasses. It should be noted that the spelling 'Böhemb' also occurs on a *Humpen* of 1651 in the Helfried Krug Collection, at Mülheim (A. von Saldern, op. cit., Fig.36; Klesse, *Krug*, I, p.158, No.123, attributed to Saxony). This glass, very exceptionally for such a late date, bears the figure of Christ crucified on the breast of the eagle, in place of the orb. The tongue of the eagle closely resembles that of the Waddesdon piece, and the other examples in this group. It seems slightly improbable that a latter-day forger would have varied his *Reichsadlerhumpen* to the extent displayed by this little group. It would be easier to stick to a single formula, even if the dates were varied. The plain dotted borders of this group are also to be noted. The fact that the pedigree of the Waddesdon glass goes back to before 1855 (see below) is a further indication of its genuineness. The copying of German enamelled

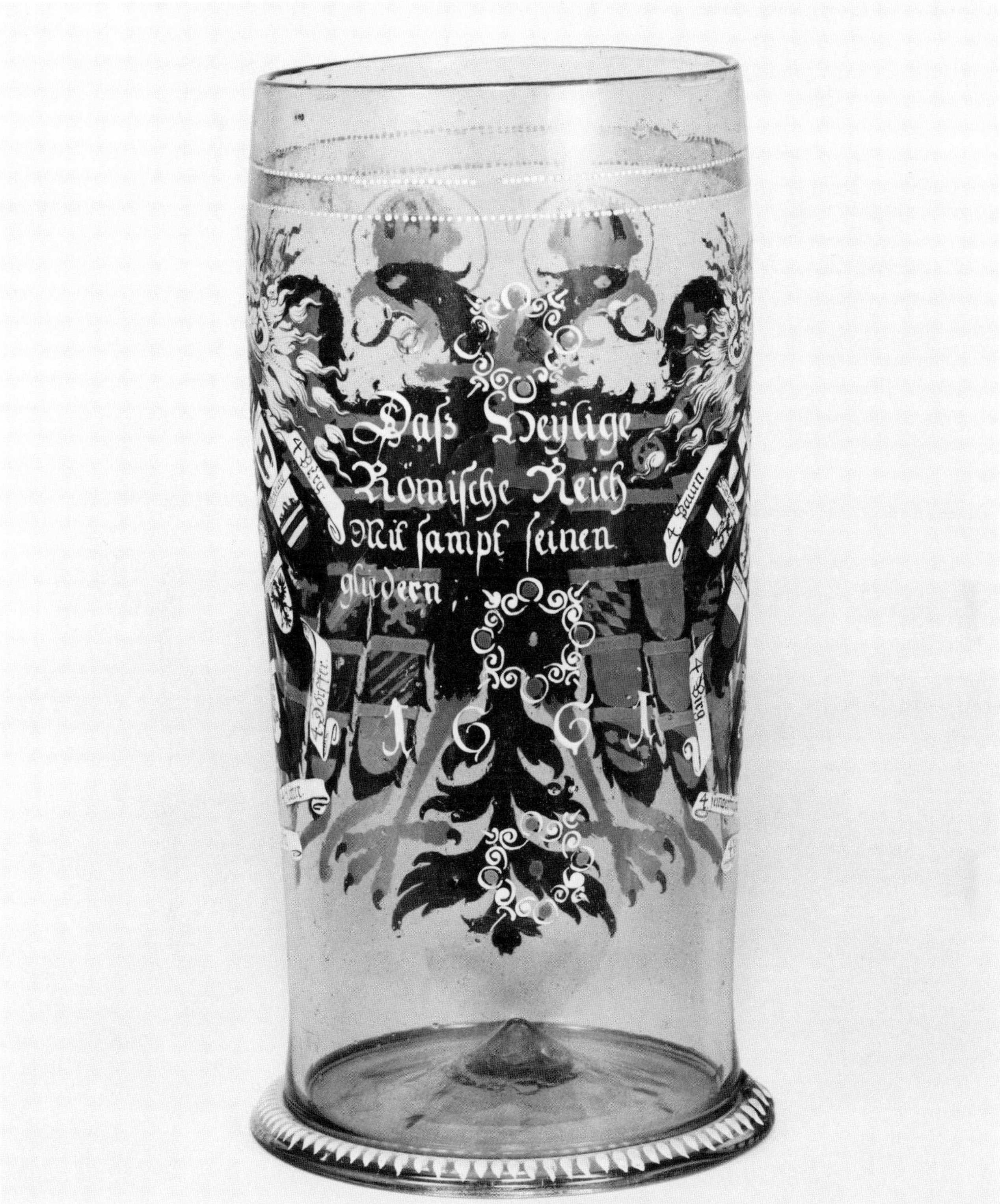

The back of 44

45. Beaker, painted with the Emperor and Electors of the Holy Roman Empire (*Kurfürstenhumpen*)

German; dated 1625, but probably a 19th century copy

Greenish-yellow glass with some bubbling, painted in enamel colours and gilt

Height, $12\frac{7}{8}$ in. (32·7 cm); diameter of rim, $5\frac{3}{4}$ in. (14·6 cm)

Cylindrical beaker with applied foot-rim in the form of a thick flattened coil of glass applied round the basal angle of the beaker, which is pushed up into a low 'kick'. Painted in black, red, blue, yellow, white and green enamels, with two zones of Electors on horseback, their titles inscribed above them in white enamel. In the upper zone they ride from right to left; in the lower, from left to right. In the centre of the upper zone is the Emperor, wearing the Imperial Crown and holding sword and sceptre ('Röm.Keÿser'), with the three spiritual Electors, Mainz ('Chur-Mentz'), Cologne ('Chur-Cölln'), and Trier ('Chur-Trier'), all wearing Electoral bonnets and each holding a paper in his right hand. In the lower zone are three of the four temporal Electors, the Elector Palatine ('Chur-Pfaltz') holding an orb, the Elector of Saxony ('Chur-Sachssen'), holding a sword, and the Elector of Brandenburg ('Chur-Brandenburg'), holding a sceptre: in the middle of the lower zone is a black double-headed spreadeagle with yellow legs, beaks and nimbus, below the date 1625 in white. Between each pair of figures is a flower and a quartered gold lozenge framed in white scrollwork with red dots; and the zones are bounded above and below by blue and red lines. Below the rim is a plain band of gold leaf between single rows of white dots.

This beaker belongs to a class of *Kurfürstenhumpen* showing the Emperor and his Electors on horseback in two registers. Examples are in the Kestner Museum, Hanover (*Glas-Sammlung*, No.54, Pl.16, dated 1628 and attributed to Upper Franconia), the Kunstgewerbe Museum, Cologne (*Glas*, No.235, undated, attributed to Bohemia), and the Corning Museum of Glass (A. von Saldern, *German Enameled Glass*, pp.292–5, Nos.35, dated 1634 and tentatively attributed to Franconia; 36, dated 1648 and tentatively attributed to Silesia; and 37, dated 1658). A further *Humpen* in the Corning Museum (A. von Saldern, op. cit., p.291, No.34), however, is virtually identical in its painting to the Waddesdon example, the only real differences being in the details of the border, and in the fact that on the Corning piece the Emperor is placed directly above the Imperial Eagle. The Corning glass is also dated 1625, and resembles in this a second example in the same collection (A. von Saldern, op. cit., p.427, No.150) which is manifestly a nineteenth-century copy. It stands 38 cm high (without cover) to the 29·2 cm of the previously mentioned glass. The smaller glass is described as being of clear metal with a slight grey tinge and bubbles. The author of the Corning Catalogue felt some doubt as to its authenticity, although giving it the benefit of the doubt in the absence of any 'convincing proof . . . of it being of recent date' (A. von Saldern, op. cit., p.291). The existence of a third piece similarly painted and also dated 1625 can only increase scepticism about both pieces, since it is rather unusual to find glasses of the same type bearing the same date. Moreover, these two

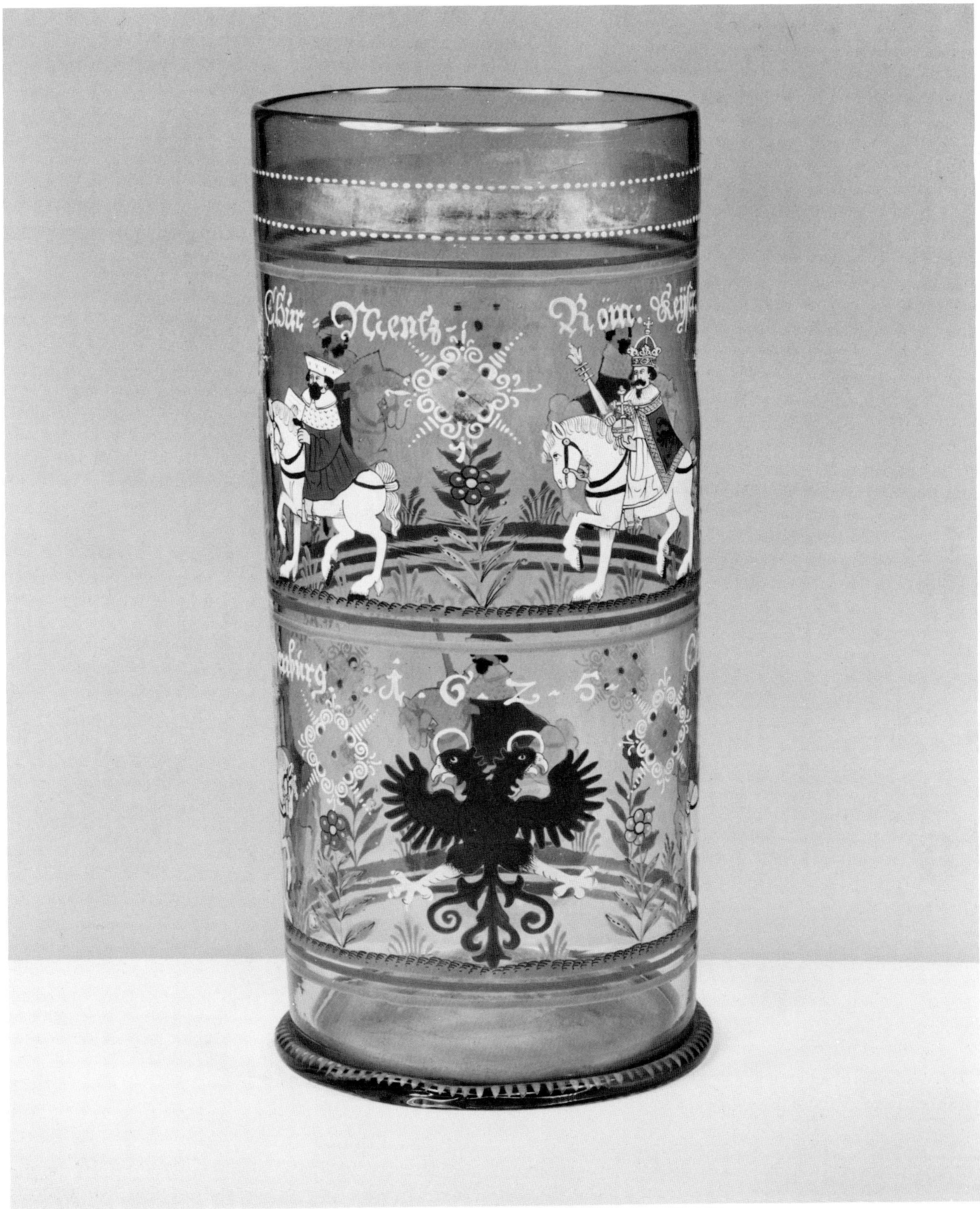
Chür · Mentz ·
Röm:
1 · 6 · Z · 5 ·

pieces, and the admittedly nineteenth-century piece at Corning, all show the same feature that the Electors in the upper register progress from right to left, while those in the lower register ride in the opposite direction (see, however, a *Humpen* attributed to Thuringia, but undated, illustrated in Franz Carl Lipp, *Bemalte Gläser*, Munich, 1974, Pl. III, C, where the Electors in the top register ride from right to left, and those in the lower register from left to right). On all the *Humpen* of this general type cited above, dating from the 1620s to the 1650s, all the Electors ride in the same direction. On the whole, the earlier glasses of the seventeenth century stick faithfully to a particular formula, those of the second half of the century tending to depart from these norms and to display a certain independence and fantasy (cf. A. von Saldern, op. cit., Figs.69 and 70, the latter dated 1701; Johannes Jantzen, *Deutsches Glas aus fünf Jahrhunderten*, Düsseldorf, 1960, p.25, No.58, Pl.27, dated 1691). The possibility remains that either the Waddesdon or the Corning glass may have served as the model for the other, as well as for the avowed nineteenth-century piece at Corning. Without an actual juxtaposition of the two pieces, and in the absence of more convincing criteria than have so far been established in the study of German enamelled glasses, it is probably wise not to reach too definite conclusions, but to preserve a healthy scepticism.

1922 Eythrope Inventory: 'Large goblet, German 1625'.

Miss Alice de Rothschild E1/13/4

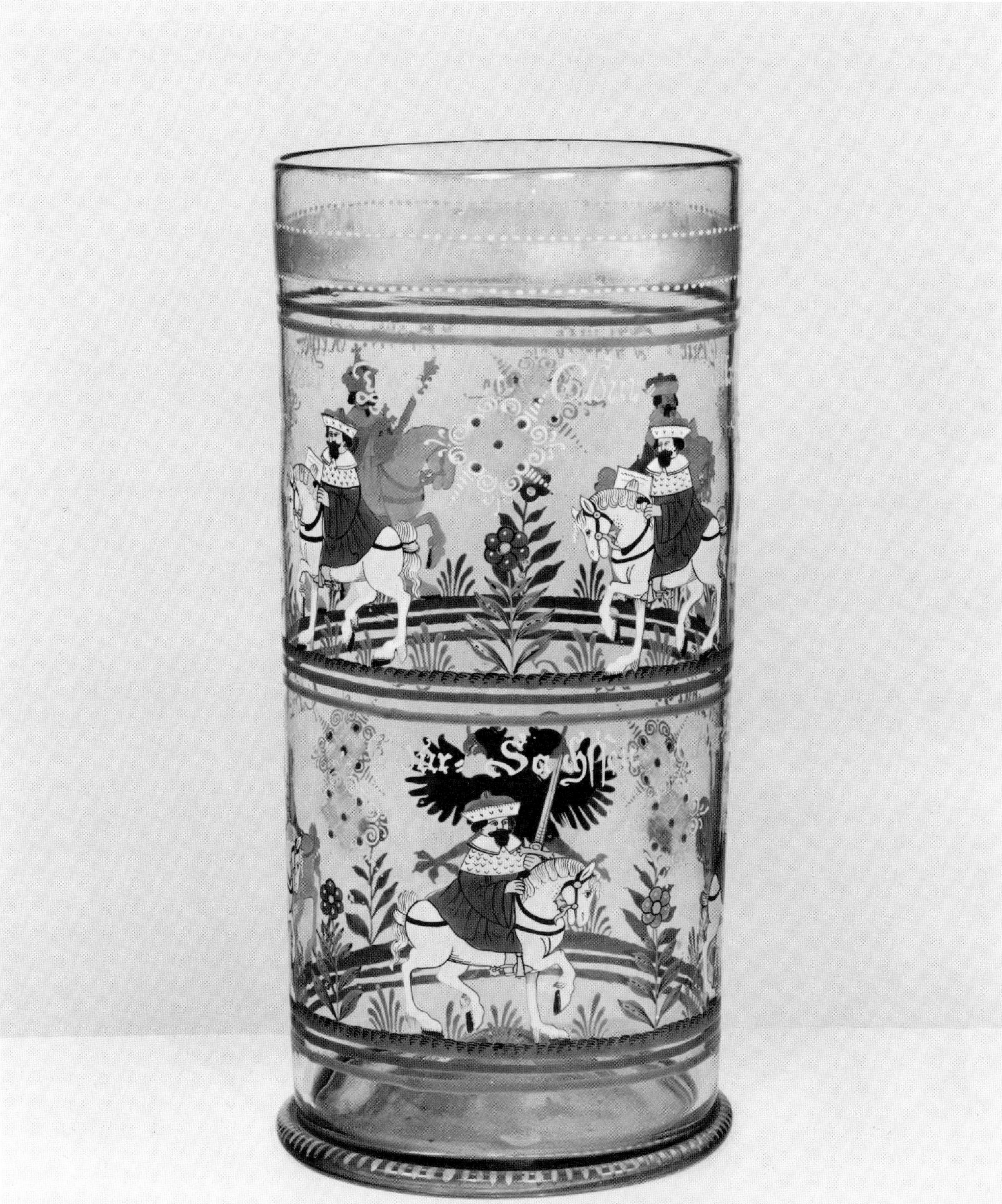

Another view of 45

46. *Humpen*, painted with a satirical scene

German (probably Franconia); about 1720–30

Greenish-colourless glass, lightly bubbled, painted in enamel colours

Height, with cover, 11⅝ in. (29·5 cm); without cover, 7⅝ in. (19·5 cm)

The cylindrical beaker has a spreading foot pinched out of the paraison, and a high conical 'kick'. The domed cover is surmounted by two hollow bulbs integral with the cover itself, and a finial in the form of a solid knob.

The glass is painted in blue, green, yellow, red, beige-brown, white and black enamels, with a predominance of the dark greyish-green. On the front is a hatless but bewigged gentleman in a long green coat with gold buttons, holding a hare by a rope halter round its neck, beside a field of corn: to the right of this a lady wearing a yellow-edged red dress over a green underskirt, raises her left hand and turns her head over her left shoulder: to her left is Cupid as a man in a black tricorn hat and green clothes, shooting his arrow towards the back of the gentleman. Above them runs the inscription: 'Solte meine Kunst wass nützen/Ehr und Leben Zubeschützen/Nur hieher wo Platz genug/Wer schnell fliehet ist recht klug/Herr es ist Zwar scharff gewehret/dass man frembdes feld verherret,/doch allhier auf unserm Sande [*sic.*]/sind was seltnes, Strick und bande/Solcher Thiergen hauft Vergnügen/Sagt man seÿ im feld Zuliegen/Trotz den Bauern die es wehren/Wann sie ihr Geschlecht vermehren/

Wann Cupido der Schelm auf meinen fall nur dichtet/und auf mein keusches Hertz den Pfeil und bogen rich(t)et,/So lauf mit schnellen fuss ich nicht den Wäldern Zu,/diss Kersten feld mein Schatz verheist mir Schutz/und Ruh/Jungfer darff ich meinen Hassen/In ihr feld hin Lauffen lassen/'

('Should my Art accomplish anything in protecting honour and life (it would be) only here, where there is space enough; and who flees hither is wise indeed.

'Sir, there is a hot defence indeed against him who overruns a foreign field; nevertheless here on our sand (?land) are, strangely, a rope and bonds.

'A little animal like this heaps up pleasure, they say, when lying in the fields; despite the peasants' attempts to stop her, she multiplies her kind.

'If Cupid, the rogue, did but versify about my lot, and direct his bow and arrow against my chaste heart, I would not run away with swift foot to the woods – this barley-field, my treasure, promises me protection and peace.

'Maiden, may I let my hare run into your field?')

The glass itself offers no further clue to the interpretation of this cryptic rigmarole.

Round the rim of the beaker runs a plain blue band.

The cover is painted with four sprays of large and small flowers and leaves, the flowers blue, red and yellow; on the bulbs of the finial are painted running leaf-scrolls in green below and yellow above, with fleurets in blue and red respectively.

Miss Alice de Rothschild WI/139/1*

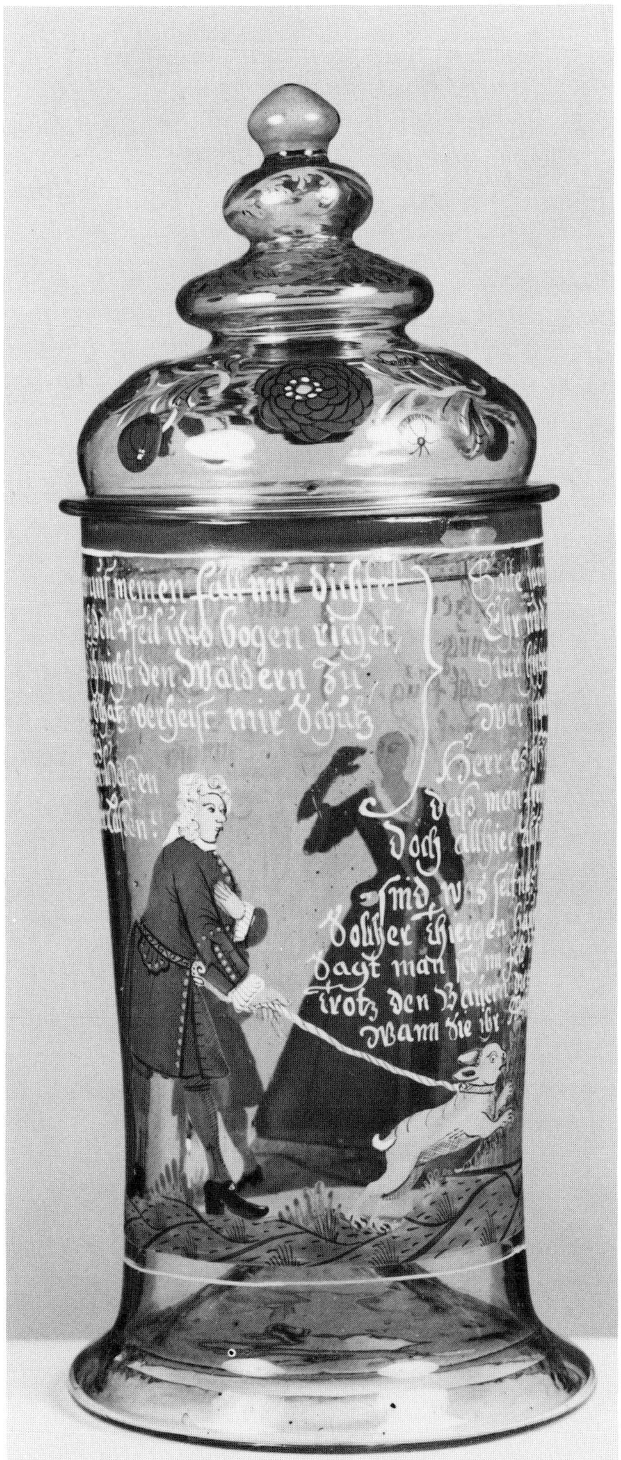
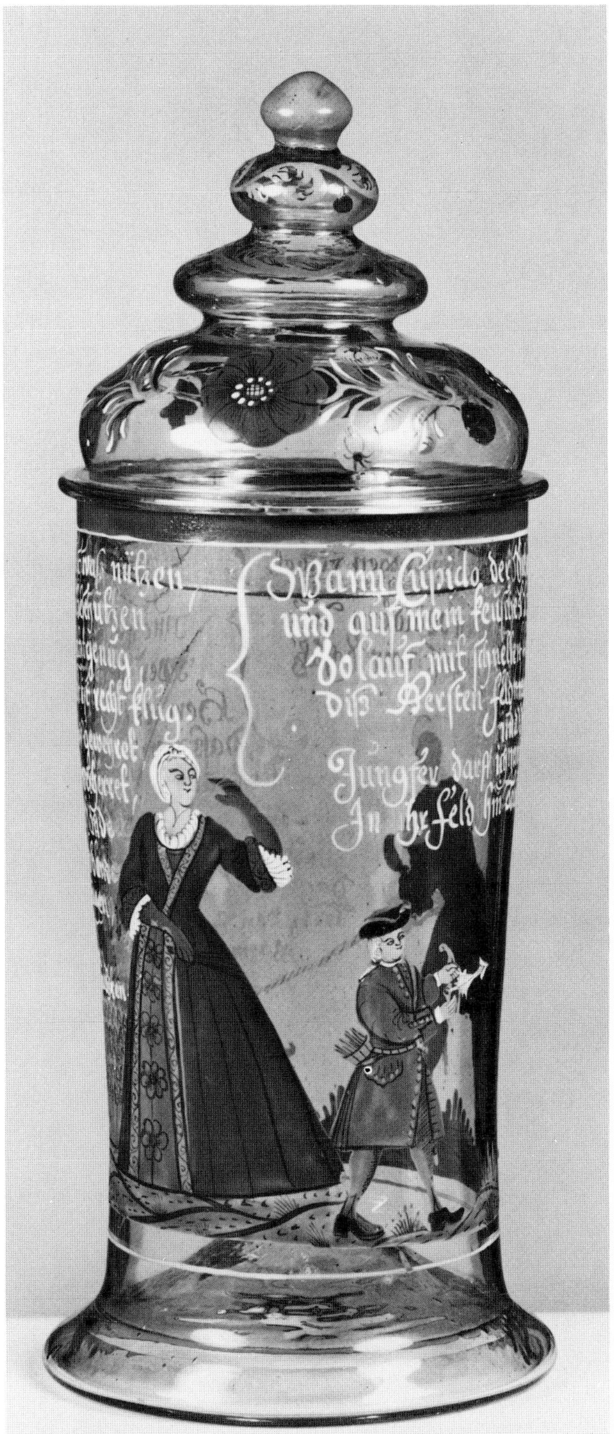

Two views of 46

47. *Humpen*, painted with the 'Ochsenkopf' mountain

Franconian; dated 1748

Bluish-green lightly bubbled glass, painted in enamel colours

Height without cover, 9 in. (23 cm); diameter of rim, $4\frac{1}{2}$ in. (11·5 cm)

Cylindrical beaker with foot pinched out of the paraison and base pushed in to a low conical 'kick'.

Painted on the front in green, black, red and blue with a representation of the 'Ochsenkopf' mountain in the Fichtelgebirge, with stags' heads projecting out of foliage, an ox's head just below the summit, and four rivers, inscribed 'Eger', 'Naab', 'Saal' and 'Maÿn' issuing from the foot of the mountain. On the reverse is a coat-of-arms under a coronet (gules bordered argent, three shuttles or; supporters, lions rampant or, langued gules). Inscribed in white on the front:

'Ich alss Teütsches Paradeiss lass Vier Ströhme von mir/flüssen, welche in Viertheil der Welt sich gantz Schiessreich/Voll ergiessen, von Gold, Silber, Kupfer, Eisen, ist mein/gantzes Eingeweit/Adam graues/alter langet/nicht an meine/frühlings/Zeit' ('I, the German Paradise, let four streams flow from me which pour in flood into the four quarters of the Earth. My bowels are all of gold, silver, copper, iron. Adam's old age does not reach so far back as my Spring-time'). On the reverse, above the coat-of-arms, is inscribed in white: 'Der Spul wischt durch des fadens bahn, noch schneller fahr des/Lebe(n)s stärcke, ach dächte man oftt daran, und würckte schöne/Glaubens Wercke, dann wie man hier die Arbeith thut/so folgt der/Lohn Böss/oder gut/VIVAT dem/Weber' ('The bobbin swishes with the rush of the thread; swifter still is the passage of life and strength: ah! if only Man thought of this often and wrought the goodly works of Faith, then as man here his work fulfils, so follows his reward, bad or good. Long life to the Weaver!'). Below is the inscription in white: 'VIVAT Stephan Entzing Müller/Anno 1748' ('Long live Stephan Entzing Müller/Anno 1748'). Just above the foot, bounded above and below by yellow lines, is the inscription in white: 'Ein jeder guter freünd der mich nur stehen siehet, der dencke nur darbeÿ das ihm der Weitzen/blühet. Er soll mich trincken aus und reinglig setzen hin, damit er sagen kan, dass ich ein Wilkom bin/' ('Let every good friend who sees me standing here, think only that his luck is in. Let him drink me up and set me cleanly down, so that he may say I am a "Welcome"').

At the sides are bunches of flowers and leaves. The upper border is composed of white and red dots and yellow paired leaves.

The cover, which is of an almost colourless glass, although contemporary, does not belong to this beaker. It is painted with a garland of red, blue, and white flowers with greyish-green leaves, between two yellow lines. White dashes enliven the finial and the folded rim (cracked).

A numerous class of beakers and tumblers is decorated with this symbolic representation of the 'Ochsenkopf' ('Ox-head') mountain, abounding in wild life, well-watered, and full of mineral wealth (A. von Saldern, *German Enameled Glass*, pp.182–90). These glasses range in date from 1656 (R. Schmidt, *Das Glas*, p.192, *ill.*; A. von Saldern, op. cit., Fig.334) to 1768

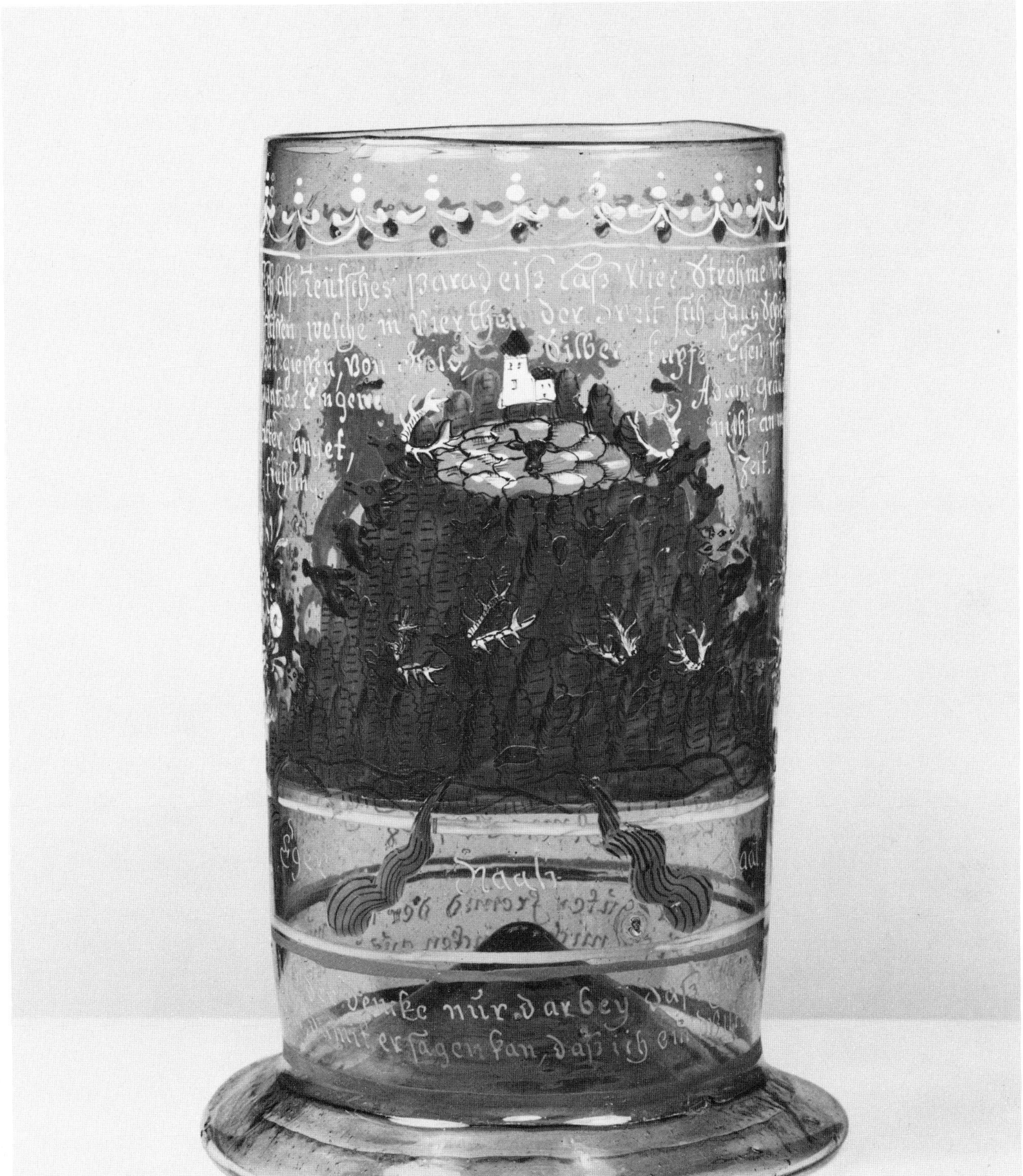

(R. Schmidt, op. cit., p.193), and display a variety of motifs on the reverse, and a great many variations of the inscription, often as wordy as that on the Waddesdon example (T. Ostertag, *Das Fichtelgebirgsglas*, Erlangen, 1933, pp.34–5).

No graphic prototype for the 'Ochsenkopf' motif has yet been identified, but an illustration in a book on the Fichtelgebirge published in 1817 repeats the same formulation of the scene, and perhaps draws on a common source with the glasses (A. von Saldern, op. cit., p.189 and Fig.341).

Glasses comparable with that at Waddesdon are in a private collection in Frankfurt (dated 1664 – see W. Bernt, *Altes Glas*, p.54, Fig.41); in the Österreichisches Museum für angewandte Kunst, Vienna (dated 1737 – see I. Schlosser, *Das Alte Glas*, Fig.90); and in the Corning Museum of Glass, New York (dated 1748 – see A. von Saldern, op. cit., p.398, No.106). A *Humpen* with virtually the same inscription as the Waddesdon glass is in the Museum für Kunsthandwerk, Frankfurt-am-Main (dated 1717 – see *Europäisches und Aussereuropäisches Glas*, No.323). Covers appear to be exceptional on these beakers.

A probable centre of manufacture of these glasses is Bischofsgrün, to the north-east of Bayreuth. A *Humpen* dated 1710 with an inscription to Johann Christoph Blasser (Blaser) is quoted by R. Schmidt as being in Schloss Schwarzburg, and an exceptional goblet with the 'Ochsenkopf' motif, dated 1737, shows by its inscription that it was dedicated by this same Johann Christoph Blasser 'of Bischofsgrün' to his cousin, Captain of the militia of Schweinfurth (R. Schmidt, op. cit., p.193, Fig.106, *right:* A. von Saldern, op. cit., Fig.340). A Johann Christoph Glaser, almost certainly the same man, pledged his services in 1739 or 1740 to the Duke of Sachsen-Weimar, to make 'true porcelain, as in Meissen . . .' (R. Schmidt, op. cit., p.193). This seems to echo a passage in J. Willen's book on the Fichtelgebirge, significantly entitled *Das Teutsche Paradeiss in dem vortrefflichen Fichtelgebirge*, and published in 1692. Willen writes '. . . In Bischofsgrün one should see . . . a number of fine houses belonging to . . . the three glasshouse-masters, especially the well-found glasshouse, in which . . . all sorts of common, but also beautiful porcelain- and crystal-glass are made, ingeniously prepared, engraved, painted and gilt . . .' (A. von Saldern, op. cit., pp.183, 185). The Glaser are mentioned as one of the three glass-making families in the town, and they had been there since before the end of the sixteenth century (ibid.). It seems therefore clear that at least some of these glasses were made in the Glaser glasshouse in Bischofsgrün. There are many different types of glass with the 'Ochsenkopf' motif, however, and they may have been made in any of a number of centres in Franconia (see A. von Saldern, op. cit., p.183).

A number of enamelled glasses exist which commemorate the craft of the weaver. An example dated 1691 is in the Corning Museum of Glass, New York, and shows a similar coat-of-arms, with lions as supporters (A. von Saldern, op. cit., p.369, No.84). Another piece in the Kestner Museum, Hanover, dated 1661, has the shield supported by angels (*Glas-Sammlung*, p.61, No.76). Others are cited by A. von Saldern, op. cit., Appendix, p.448: also Düsseldorf, Kunstmuseum, *Catalogue*, No.179.

1922 Waddesdon Inventory (Smoking Room): 'Tankard and cover enamelled with coat-of arms, view of tower and stags heads. German, 1748'.

Miss Alice de Rothschild — W1/141/2

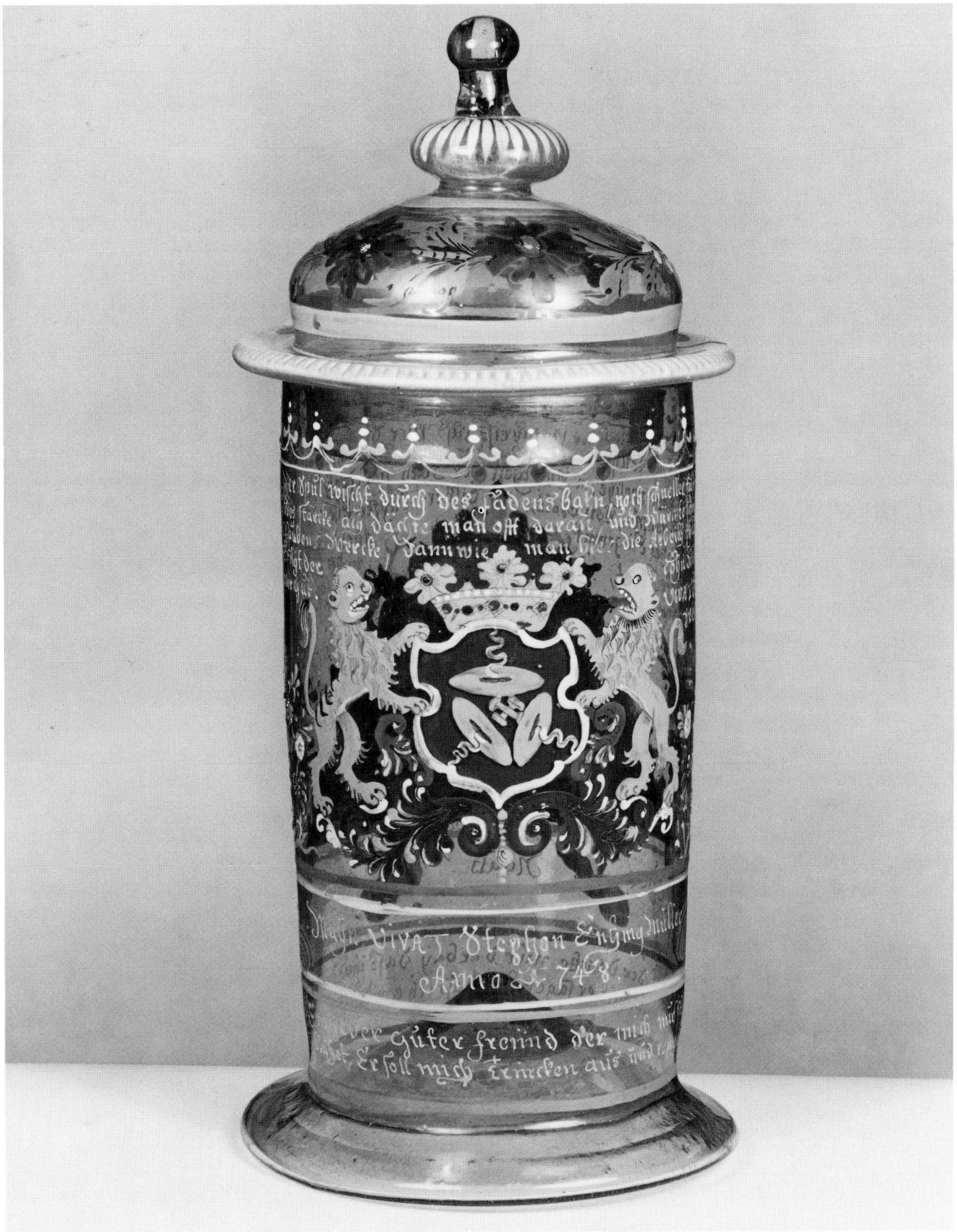

48. *Passglas*, enamelled with 'The Seasons'

Central Germany (Thuringia or Saxony); about 1700–1720

Almost colourless bubbled glass painted in white and black enamels

Height, $10\frac{3}{8}$ in. (26·5 cm); diameter of rim, $3\frac{1}{8}$ in. (8 cm)

Cylindrical beaker with wide doubled foot and base pushed in to a low 'kick'.

The glass is divided by horizontal lines into four zones or *Pässe* (see Introduction, page 30 and A. von Saldern, *German Enameled Glass*, pp.145–7). Each is decorated in the front with the half-length figure of a girl bearing an attribute of one of the Seasons. At the top is Spring carrying a trefoil in her left hand; behind is the inscription 'Frühling es Freu's' ('Enjoyment is in Spring'): next comes Summer carrying a flower-pot in her left arm and in her right hand raising a flower to smell it ('Sommer ver Neuert' – 'Summer renews'): next comes Autumn smelling a fruit from a basket clasped in her left arm ('Herbst thut Geben' – 'Autumn is prodigal'): lastly comes Winter, wearing a fur cap and warming her hands over a brazier ('Winter zu leben' – 'So that one may live in Winter'). Each figure is enclosed in a compartment bounded to right and left by a curved spray of hooked leaf-forms. The foot is decorated with radiating strokes of white enamel.

Glasses of this type are traditionally ascribed to Central Germany, either to Saxony or Thuringia. A *Passglas* in the Museum at Frankfurt-am-Main bears a representation in white of a *taler* with an inscription relating to the Elector Georg III (1680–91) or Georg IV (1691–94) of Saxony. Much white enamel is used on this *Passglas* (Ohm, *Europäisches und Aussereuropäisches Glas*, No.329, *ill.*), and this seems to be characteristic of the class. Apart from the Waddesdon example, a beaker in the Jerome Strauss Collection, State College, Pennsylvania, USA, is painted exclusively in white (A. von Saldern, op. cit., Fig.256), and another, also in the Frankfurt Museum, has white leafy borders at top and bottom (Ohm, op. cit., No.337, *ill.*). This glass has in the top *Pass* a girl, in the second a stag, and in the third a hare, and this formula is repeated on a *Passglas* in the Germanisches Nationalmuseum, Nuremberg (A. von Saldern, op. cit., Fig.258), and on another dated 1719 which was sold at Sotheby's on 2 June, 1969 (K. Middlemas, *Continental Coloured Glass*, London, 1971, pp.39 and 44, *ill.*); this had a fourth *Pass* at the bottom, painted with a hog. Each of these three glasses had in the uppermost *Pass* the figure of a girl closely resembling those which embellish the Waddesdon beaker. A fourth glass, sold at Sotheby's on 12 February 1973, Lot 4 (*ill.*), dated 1716, had in each *Pass* half-length representations of a man and his wife, with appropriate inscriptions, the woman in each case again holding a posy to her nose in the manner of the Waddesdon glass. These beakers seem in shape and decoration to belong to one group, being perhaps the products of a single workshop. Whether this was in Saxony or Thuringia, it is not at present possible to say with certainty.

From the Hollingworth Magniac Collection (*Catalogue of the Renowned Collection of Works of Art of the late Hollingworth Magniac, Esq.*, London, Christie Manson and Woods, Fourth Day's

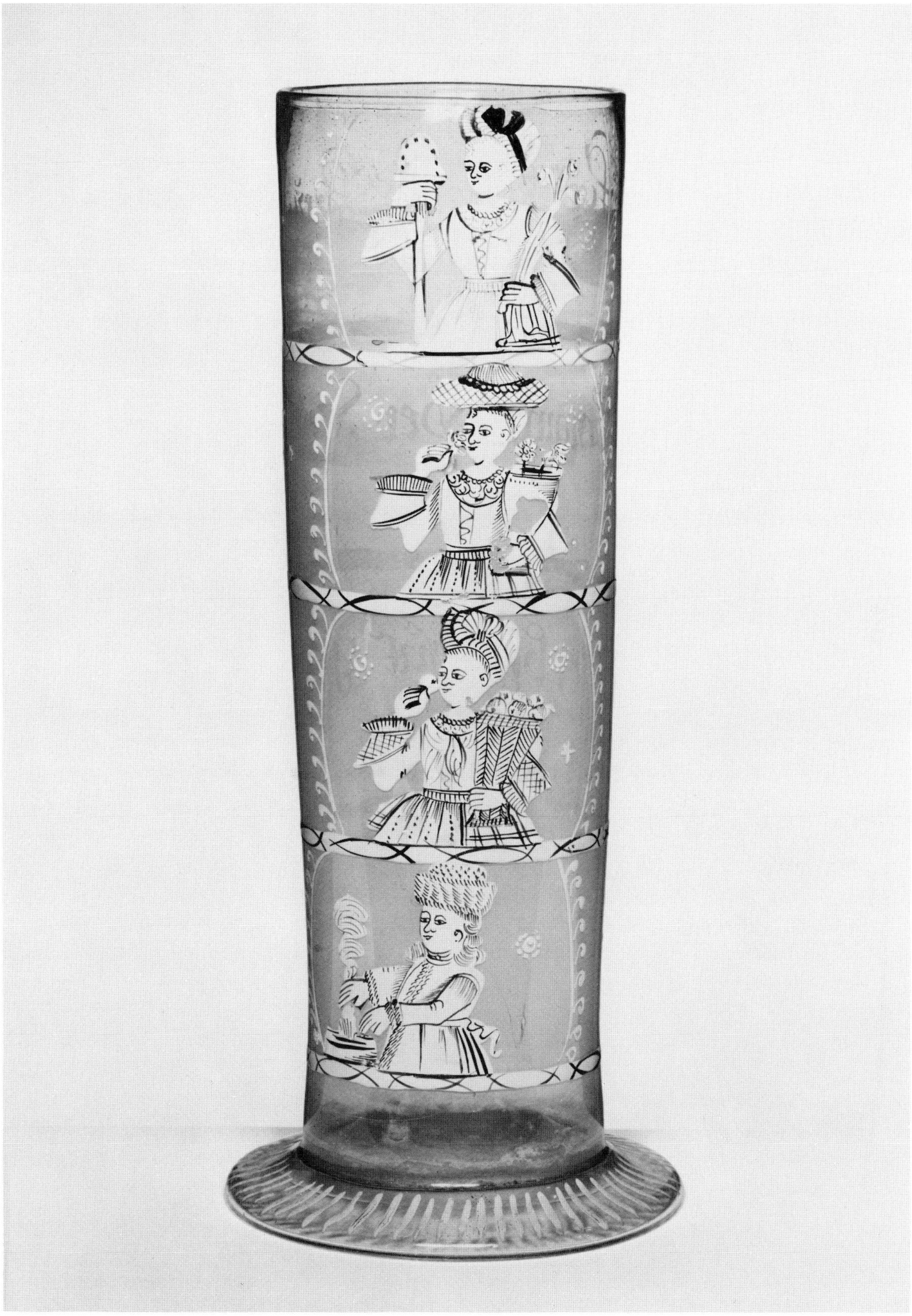

Sale, 6 July, 1892, Lot 423: 'A WIEDERKOM, enamelled with black and white figures emblematical of the Four Seasons, and inscriptions – 10½ in. high 7¾ in. diameter' (this is certainly mistaken, and may refer to the circumference).

1898 Waddesdon Inventory (Smoking Room, No. 1119): 'A cylindrical German glass vase, painted with 4 female domestic figures in white, with inscription on the back. 10½ in.' 1922 Waddesdon Inventory (Smoking Room): 'Tall beaker, opalesque, four half-length figures in costume. German, 17th century'.

Baron Ferdinand de Rothschild WI/140/11

49. Beaker and cover

German; dated 1598, the beaker probably a later replacement

Green glass with mould-blown decoration, mounted in gilt copper

Height with cover, 9 in. (23 cm); without cover, $6\frac{3}{8}$ in. (16·2 cm); diameter of rim, $3\frac{3}{4}$ in. (9·5 cm)

Beaker of flaring form decorated with mould-blown lozenge-diaper design; the pontil-mark on the base has been ground off. The domed flanged cover of ogee profile has a plain spherical knob-finial; this cover too has been mould-blown with a mesh-pattern, but of a different character.

The beaker has been mounted with an elaborate spreading openwork foot of gilt metal, pierced with a lattice design and supported on three figures of lions. Round the rim of this foot are scratched the letters 'A.G.B(?).H.G.Z.M.V.B.H.Z.Y.D.V.E.RK.H.Y.M.A.E.Y.T.P(?R). A.E.T.1598'. The knob of the cover has been mounted with an additional flat-topped seal-like finial of gilt metal bearing an incised coat-of-arms, with the last two digits of the date '98'.

The beaker, which is of an unusual greengage-colour and modern appearance, presumably replaced the original when it was broken at some time.

The coat-of-arms appears to be that of the Counts von Manderschei[d]t – quarterly (1) *or, a fess dancetté gules* (Manderschei[d]t); (2) and (3), *or, a lion rampant sable armed and langued gules, debruised by a label of five points of the last* (Blankenheim); (4), *argent fretty gules* (? Daun) – see J. B. Rietstap, *L'Armorial Général*, Pls. Vol. IV, Paris (1912), s.v. Manderscheid; Pls. Vol. I, Paris, 1903, s.v. Blankenheim. J. Sibmacher, *New Wapenbuch*, Nuremberg, 1605, gives two coats-of-arms for the von Manderschei[d]t family – the first quarterly, first and fourth as (1) above, second and third as (4) above, with Blankenheim in pretence; the second quarterly, first and fourth as (1) above, the second and third as (2) and (3) above, with another quartered shield in pretence. The town of Daun has as its civic arms the fourth quartering of the von Manderschei[d]t coat given above (Brockhaus, *Enzyklopädie*, 4, Wiesbaden, 1968, s.v.) and since the Lords of Daun (first Herren and then Grafen) had Falkenstein as their seat, we may be fairly sure that this was their coat-of-arms, since Falkenstein was one of the estates to which the von Manderschei[d]t laid claim (O. Gundlach, *Bibliotheca familiarum nobilium*, Neustrelitz, 1897, p.649). They were certainly intermarried with the von Daun family (Frank, Baron Freytag von Loringhoven, *Europäische Stammtafeln*, IV, Marburg, 1957, Table 137).

The Counts von Manderschei[d]t had a seat at the Niederburg in Manderscheid (Kreiss Bernkastel-Wittlich in Rhein-Pfalz). Manderscheid became a *Grafschaft* in 1457, and to the family was added in 1469 the *Grafschaft* of Blankenheim (Kreis Schleiden in Nordrhein-Westfalen). To the Blankenheim estates belonged the properties of Gerolstein, Kronenburg and Dollendorf in the Eifel range (Brockhaus, *Enzyklopädie*, 2 and 12, Wiesbaden, 1967 and 1971 s.v. Blankenheim and Manderscheid).

Hermann, Graf zu Manderscheid-Blankenheim (d.1604) married in 1567 Juliane (1529–

95), daughter of Philipp II von Hanau-Münzenberg as her second husband. He may well have been the owner of the glass, in which case the fourth to ninth letters of the inscription on the foot may stand for '*H*ermann *G*raf *Z*u *M*anderschei[d]t *V*nd *B*lankenheim' (Frank Baron Freytag von Loringhoven, op. cit., III, Table 83).

1922 Eythrope Inventory: 'Green beaker and cover, metal mounts. Venetian, 17th century'.

Miss Alice de Rothschild EI/13/1*

The coat-of-arms on the finial of 49

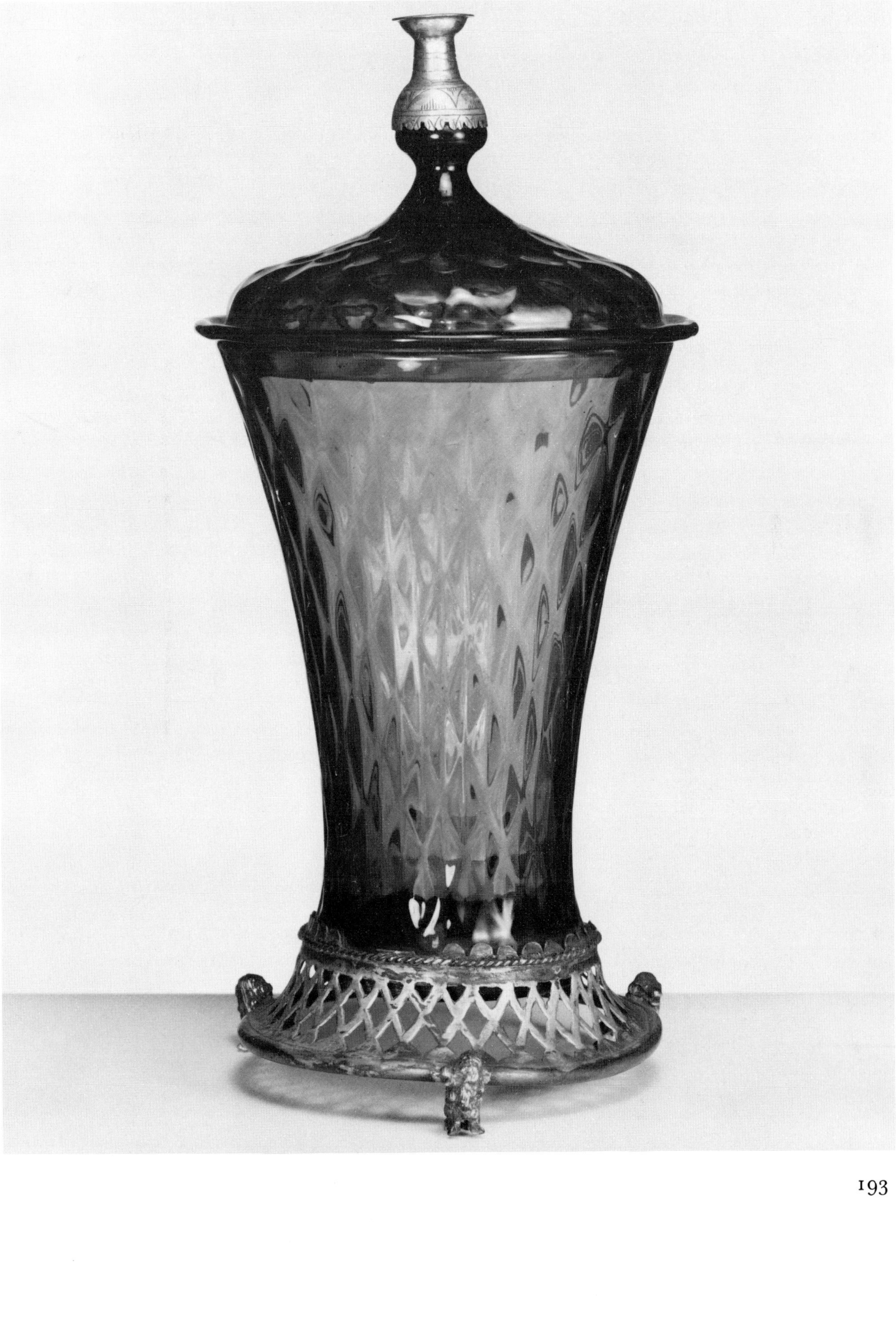

50. Flask in the form of a sceptre, blue glass mounted in pewter

German or Bohemian; about 1678

Blue glass partly blown in a ribbed mould, with applied spiral threading and pincered collar

Length overall, 17¾ in. (45 cm)

Hollow-blown flask, the bulb at the end being made by blowing a 'second gather' into a ribbed mould. Above this is an applied collar with a fringed edge which has been drawn out by tooling. Round the shaft of the 'sceptre' is an applied spiral thread, and on the head is an applied coil which has been roughly tooled into a square form.

The mouth of the flask has been mounted in pewter, with a screw-cap. Round the mount is the incised inscription: 'GABRIEL RITTER CIBIN Ao 1678'. Cibin is the name of a mountain-range and a river in the Transylvanian Alps, south of Hermannstadt (Hungary).

This 'sceptre' is probably to be regarded in the same category as flasks made in the form of pistols, or drinking-vessels in the form of boots. Both these are recorded in the inventory of the Munich glasshouse between 1677 and 1681, so it is perhaps not unduly fanciful to see in the expression *Stab* in the same inventory a form resembling the present 'sceptre' or 'mace' (R. Berliner, loc. cit., p.123; cf. p.121 for pistols and boots. The *Stab* is recorded as being made in crystal or *Waissel*, an inferior quality of colourless glass). A similar sceptre in the Victoria and Albert Museum (568–1872) is made in opalescent-white glass with mould-blown ribbing and gilt metal mounts (cf. observations on *60–1* here). This example, bearing on the mount the arms of Imhoff, of Nuremberg, was also acquired from a Nuremberg dealer, circumstances which argue for the South German origin of the piece. An analogous piece, also in blue glass, with slight enamelling and gilding, was in the Spitzer Collection (*Résumé du Catalogue des Objets d'Art . . . Collection Spitzer*, Sale, Paris, Chevallier (17 April–16 June, 1893), Lot 2084, described as 'Gourde – Venise. Commencement du 16e. siècle': illustrated *Atlas*, Pl. XLIX).

The treatment of the 'collar' on this flask is strikingly reminiscent of that used on the *Ringelbecher*, a typically German conical drinking-glass on a low pedestal foot, the bowl embellished with spiral threading and three free-hanging glass-rings (see, e.g., Klesse, *Krug*, II, No.450, with references to other examples; Berryer, *La Verrerie Ancienne*, Brussels, 1957, p.66, Fig.26, etc).

Perhaps from the Bernal Collection – see *Catalogue of the Celebrated Collection of . . . Ralph Bernal*, Christie and Manson, 23rd Day's Sale, 18 April, 1855, Lot 2935; 'A vessel, in the form of a sceptre, of deep blue, with openwork ornaments on the top, the bottom mounted with pewter – 12 in. long', bought by Durlacher of Bond Street for £1.8.0. It will be noted that the dimensions are discrepant.

1922 Waddesdon Inventory (Smoking Room): 'Mace, blue, Venetian, 17th century.'

Miss Alice de Rothschild — W3/40/3**

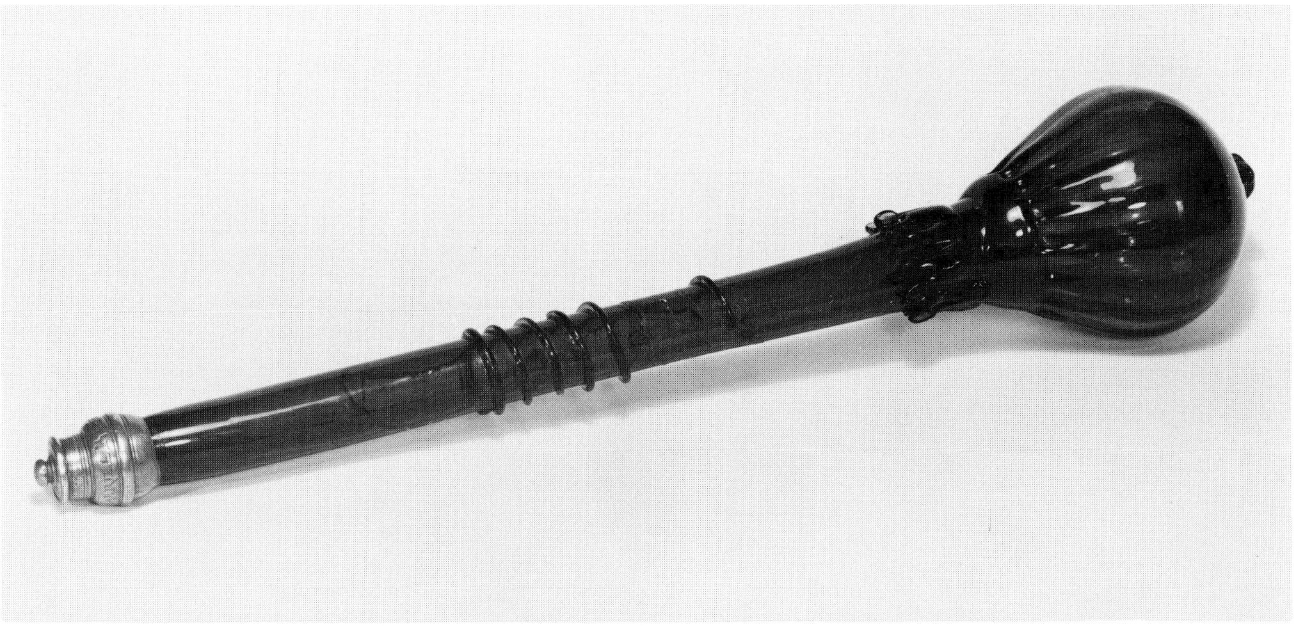

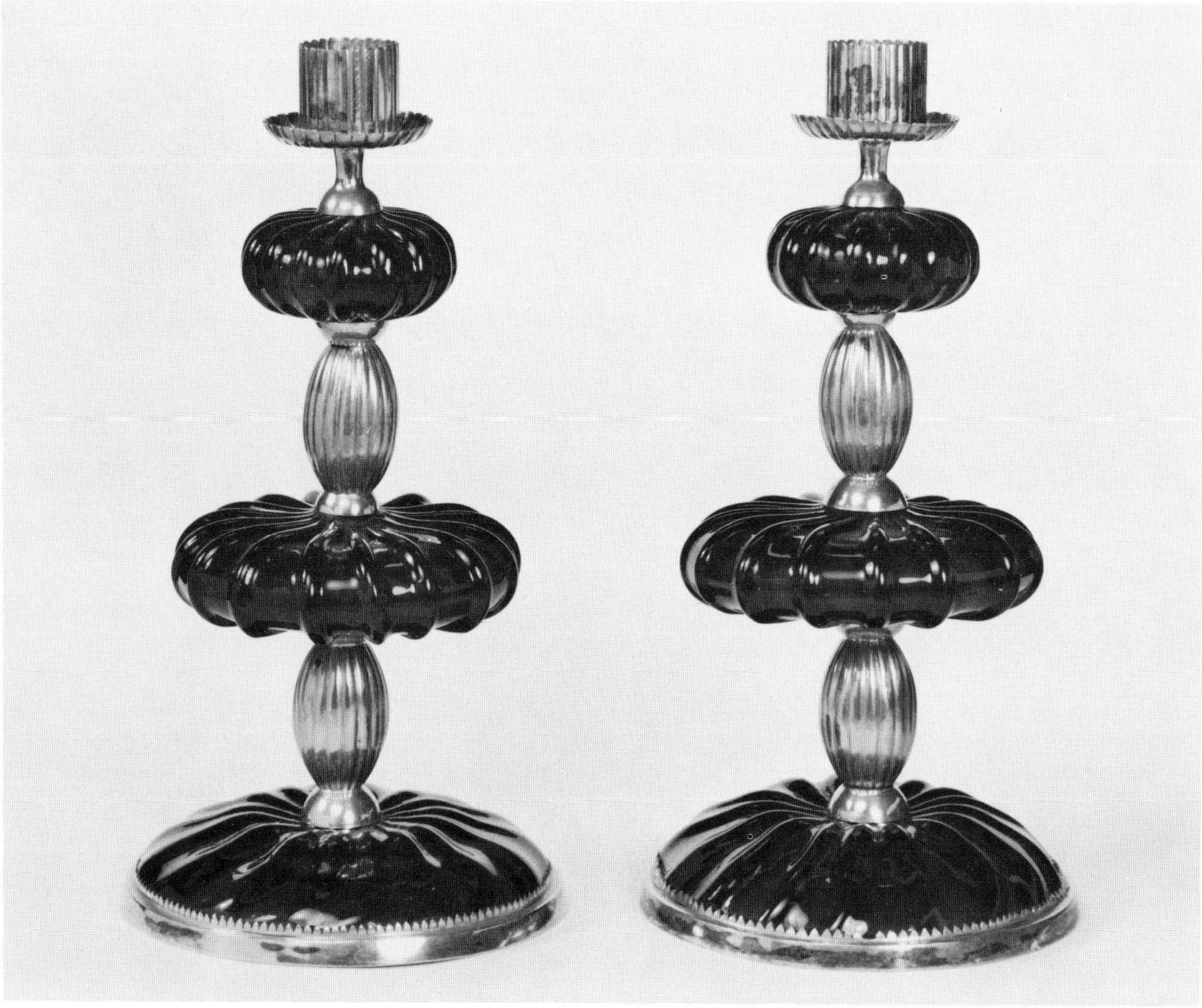

51. Pair of candlesticks, ruby-glass, mounted in silver-gilt

South German (probably Munich or Freising); late 17th century

Bubbled gold-ruby glass with mould-blown ribbing, mounted in silver-gilt.

Height, $7\frac{5}{8}$ in. (19·5 cm); diameter of foot, 4 in. (10 cm) and $3\frac{7}{8}$ in. (9·8 cm)

Each candlestick is formed of a foot and two stem-components of glass blown in a sixteen-rib mould, and mounted with stem-components and candle-holder of silver-gilt, all threaded on a central rod.

The glass stem-components are wide depressed knops, while the foot is a low dome. The glass

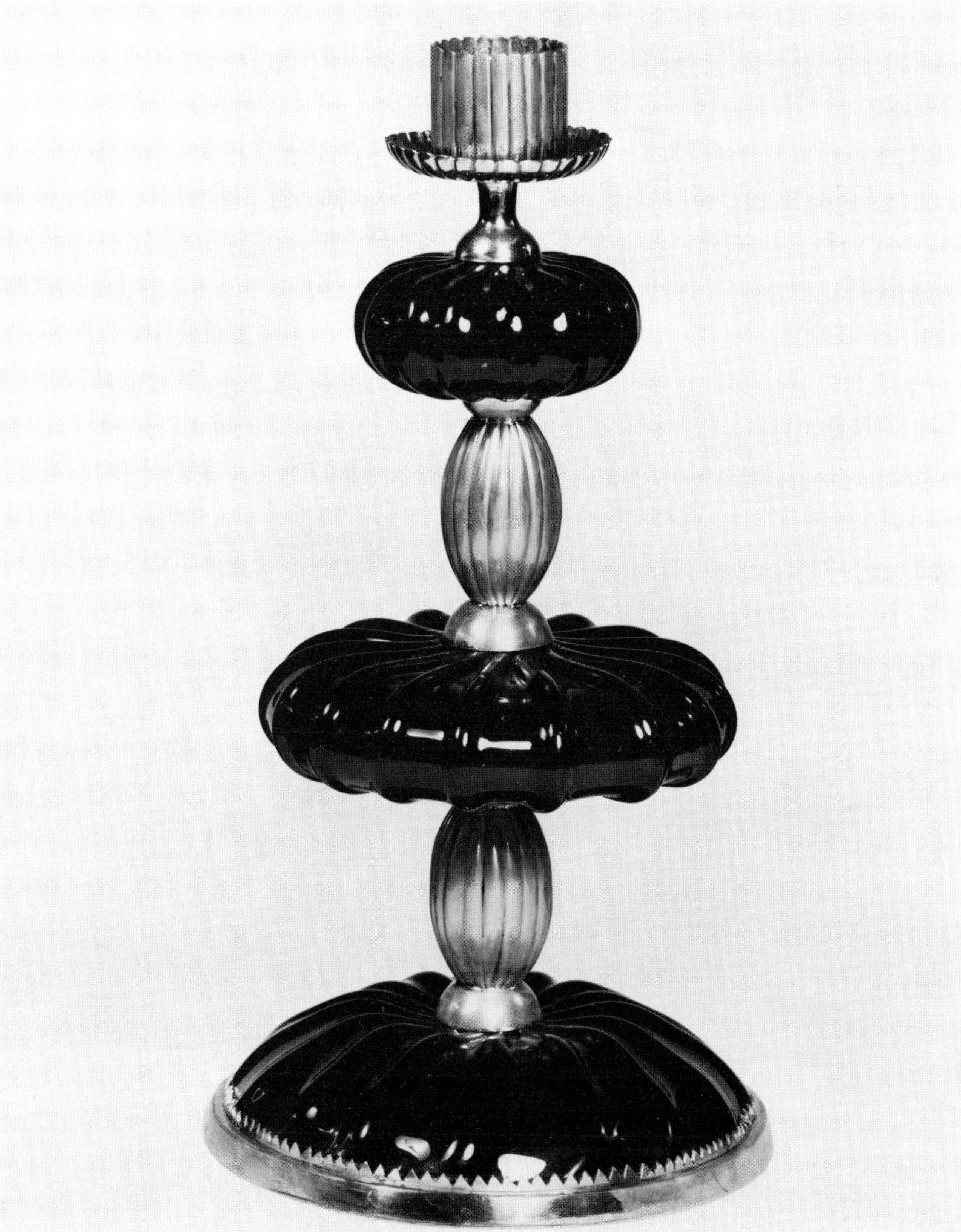

foot is held in an upright silver-gilt rim by means of a fillet of small triangular lappets. The two silver-gilt stem-components are ribbed ovoid forms between domical collars, the ribbing being repeated on the cylindrical candle-holder and grease-pan. The glass has been ground in a number of places to take the metal parts.

On the the foot-rim of each is the impressed mark '13' in an oval cartouche.

The mould-blown ribbing of these candlesticks is found on a number of pieces which may reasonably be attributed to the glasshouses of Munich or Freising (see Introduction, pp.32–33).

A (probably cut-down) ribbed pilgrim-flask (cf. *52* here) in the Ernesto Wolf Collection, São Paulo, bears ribbed mounts of very similar character (see São Paulo, Museu de Arte Moderna, *4000 anos de vidro*, 1957, Cat. No.38 and *ill.*). Exactly similar foot and stem components are found on a magnificent garniture of ribbed ruby-glass vases with Augsburg mounts, in the Kunstgewerbemuseum, Berlin (see F.-A. Dreier, 'Fünf Silbermontierte Prunkvasen aus Goldrubinglas', Abb.1).

The mark '13' in an oval is common to a number of centres, although not to Augsburg or Nuremberg, the number indicating a degree of purity in the silver used. Most of these places, however, are in South Germany – Gmünd in Swabia, Villingen in Baden, Neustadt an der Haardt, near Speyer, Pfalz, and Schweinfurt in Bavaria itself (Rosenberg,[3] II, Nos.2187–9; III, Nos.4528, 3666, 4840).

1922 Eythrope Inventory: 'Pair of ruby glass and silver mounted candlesticks'.

Miss Alice de Rothschild E2/32/13**

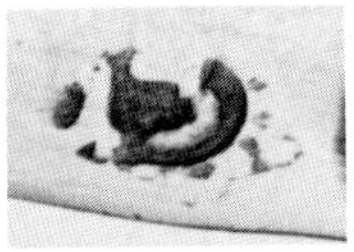

Mark on the silver-gilt foot of 51, enlarged four times

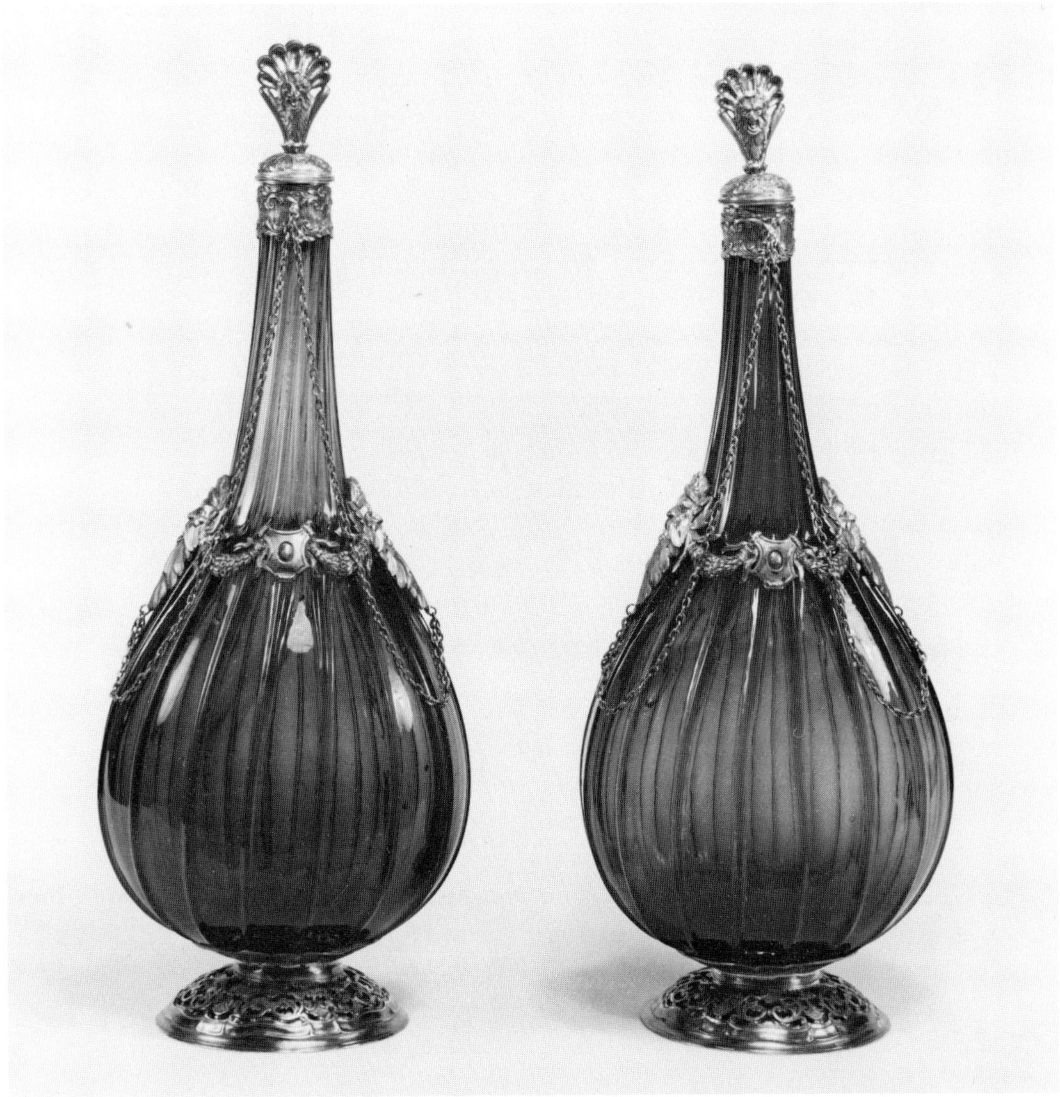

52. Pair of flasks, ruby-glass mounted in silver-gilt

Probably South German (Munich or Freising); late 17th century (the mounts probably 19th century)

Bubbled gold-ruby glass with mould-blown ribbing, mounted in silver-gilt.

Height, with stopper, $14\frac{3}{8}$ in. (36·5 cm); without stopper, $12\frac{1}{8}$ in. (30·5 cm) and $12\frac{1}{4}$ in. (31 cm); width, $5\frac{1}{4}$ in. (13·3 cm) and $5\frac{1}{2}$ in. (14 cm)

Flattened piriform pilgrim-flasks with long tapering necks. Each has been made on a 'double-gather' blown into a sixteen-rib mould.

Each flask has been mounted in silver-gilt with an openwork foot of scrollwork and leaves within a waved border, the openwork protected by a plain plate screwed to the inside of the foot. Round the shoulder of the flask, at the top of the 'double-gather', is a decorative collar formed of bearded wreathed masks at the sides, connected by ribboned swags of leaves and flowers to shield-shaped scrolled cartouches at front and back. Four chains affixed to the beards of the masks connect this decoration with a collar surrounding the rim of the flask. This collar is decorated with scrollwork borders in relief on a ground of punched dots in lines, and the stopper is formed of a double-sided bearded mask with shell headdress, mounted on a cap decorated with scrollwork and shells.

The mounts are unmarked.

The mould-blown ribbing on this pair of flasks (cf. *51*, *53* here) is found on a number of pieces which may reasonably be attributed to the glasshouses of Munich or Freising (see Introduction, pp.32–33). The records of the Munich glasshouse refer to 'gerippte Massflaschen' ('ribbed quart bottles' – see R. Berliner, loc. cit. on p.32, p.122).

A flask of this type in the Schott Collection at Mainz shows clearly below the body of the glass the button to which the mount was intended to be fixed (Eberhard Schenk zu Schweinsberg, *Form und Schmuck des Hohlglases an Beispielen der Sammlung Schott*, *Schott Schriften*, 2, Mainz, 1966, No.19: also Düsseldorf Kunstmuseum, *Meisterwerke*, p.98, No.266). An example with mounts by the Augsburg Master 'T.B.' is illustrated by R. Schmidt, *Brandenburgische Gläser*, Pl.16,3, and F.-A. Dreier, 'Fünf Silbermontierte Prunkvasen aus Goldrubinglas', p.81, Fig.5 (see note on *51* here). An almost identically similar bottle in both shape and mount was once in the Baron L. de Rothschild Collection: it too had an Augsburg mount (South Kensington Museum, *Special Loan Catalogue*, London, 1862, No.5047). Other examples are numerous – see e.g., Moritzburg Halle, Staatliche Galerie, *Schönes Glas in der Moritzburg zu Halle*, by Otto Heinz Werner, Halle, 1957, No.109, Abb.27 (later mount); Cologne, *Glas*, No.285 (attributed to Augsburg); Frankfurt-am-Main, Museum für Kunsthandwerk, *Europäisches und Aussereuropäisches Glas*, No.471 (a pair, with gilt bronze mounts; attributed to (?) Munich); Kämpfer and Beyer, *Glass*, No.141, in Grünes Gewölbe, Dresden (with wheel-engraving between the ribs).

Probably from the David Falcke Collection – see *Catalogue of the Magnificent Collection of . . . Mr. David Falcke*, Christie and Manson, London, 12th Day, 3 May, 1858, Lot 1773: 'A PAIR OF VERY FINE FLUTED BOTTLES AND STOPPERS, of old ruby glass, beautifully mounted with masks and arabesque ornaments, of metal gilt – $14\frac{1}{2}$ in. high.'

1922 Waddesdon Inventory (Smoking Room): 'Pair of ribbed bottles – ruby, silver-gilt mounts, Venetian 17th century.'

Miss Alice de Rothschild — W1/138/4*

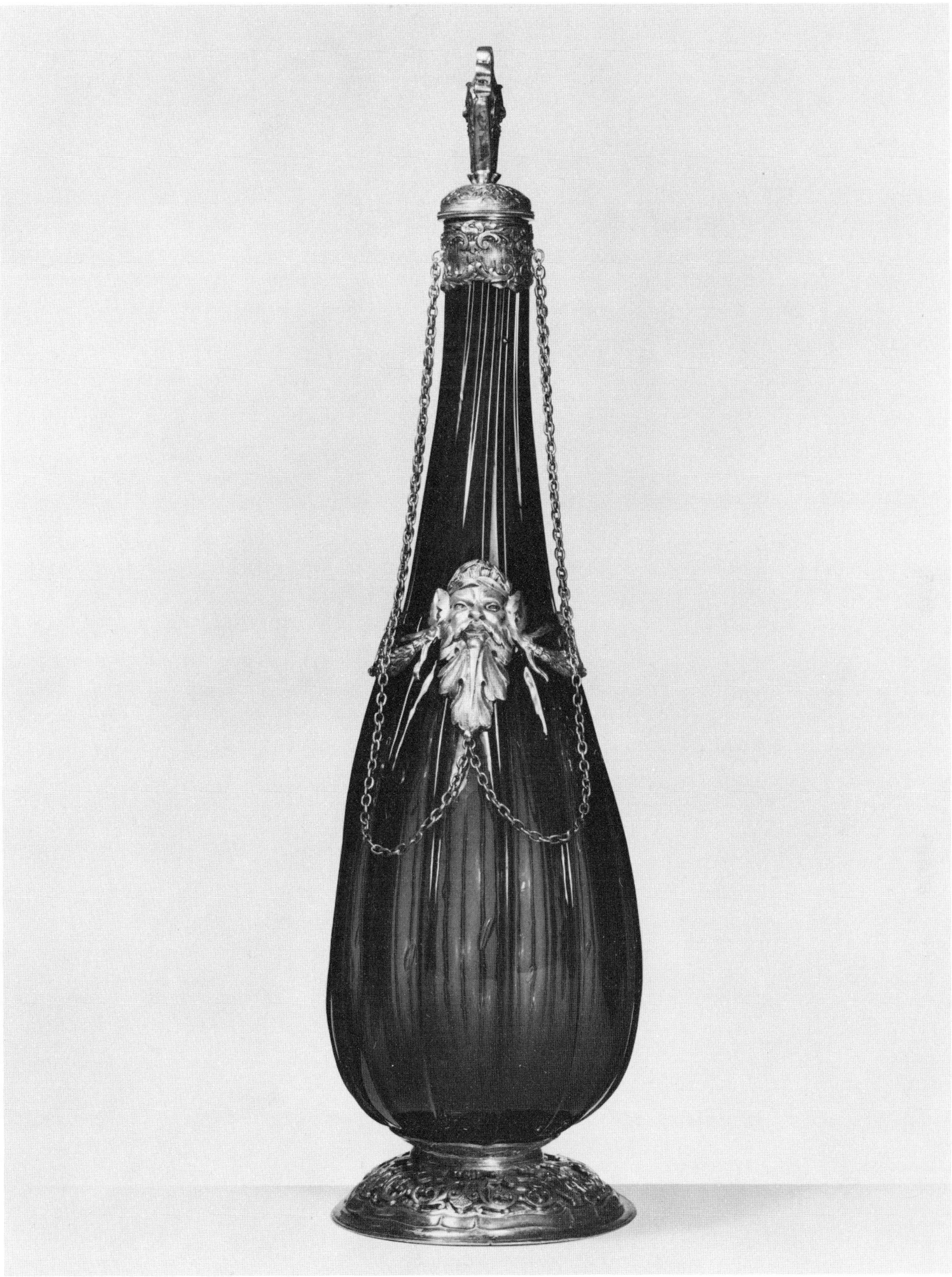

53. Bowl, ruby-glass mounted in silver-gilt

South German (probably Munich or Freising); late 17th or early 18th century

Gold-ruby glass blown into a ribbed mould, mounted in silver-gilt

Height, $2\frac{1}{8}$ in. (5·5 cm); diameter, $3\frac{1}{8}$ in. (8 cm)

Roughly hemispherical bowl with sixteen vertical mould-blown ribs, slightly sloping towards the rim. The bowl is mounted in a silver-gilt gadrooned foot connected by two flat straps with a pair of S-shaped scrolled handles. Underneath the foot is the pineapple mark of Augsburg, together with the maker's mark 'TB' in an oval, and a long zigzag assay mark. The glass is showing signs of incipient 'sickness'.

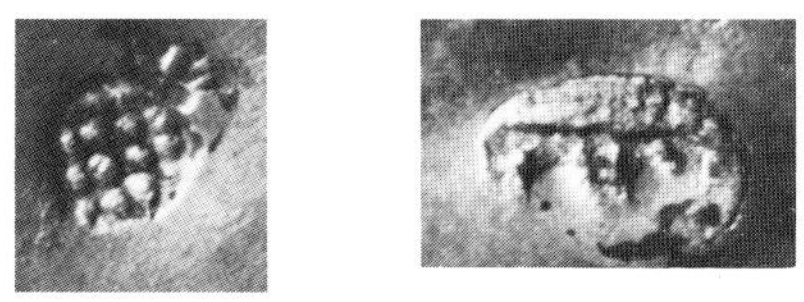

Mark of Augsburg (left) and maker's mark TB (right), enlarged four times

The mould-blown ribbing on this bowl (cf. *51–2* here) is found on a number of pieces which may reasonably be attributed to the glasshouses of Munich or Freising (see Introduction, pp.32–33).

An oval ruby bowl in the Kestner Museum, Hanover, also with sixteen ribs, shows clearly the vestigial foot intended to give an anchorage to a silver-gilt mount (Kestner, *Glas-Sammlung*, No.105, Fig.32). A similar small bowl with gadrooned silver-gilt foot is in the Victoria and Albert Museum (1872–1855). A goblet made from a similar ribbed bowl, in the Staatliche Kunstsammlungen, Kassel, also has mounts by the Augsburg Master T.B. This piece and two analogous ribbed vases in the same collection, all have scrolled handles closely resembling those of the Waddesdon piece (F.-A. Dreier, 'Fünf Silbermontierte Prunkvasen aus Goldrubinglas', pp.80–1, Figs.3–4). A plain bowl with similar silver-gilt handles and foot was included in a silver-gilt travelling medicine-set sold at Christie's on 26 June, 1974, Lot 100. Twelve pieces in the set were marked by a member of the Jäger family, of Augsburg, in the period 1698–1705.

The Master T.B. (identified as Tilmann Bornemann, d.1694, or Theodosius Besmann, d.1724, in the second edition of Rosenberg's *Der Goldschmiede Merkzeichen*, but in the third edition stripped of identity) made a speciality of mounting ruby glass. Apart from *54* in this *Catalogue*, a number of pieces are mentioned by Dreier, loc. cit., and Rosenberg (R^3, I, Nos.708–9). The forms of the Augsburg city mark occurring in conjunction with the TB mark on the mounts of the ruby glass pieces fall within the date-range 1685–1705 (R^3, I, Nos.185, 189, 204, 207); one object is actually dated 1695, which excludes Tilmann Bornemann. Additional pieces with the TB mark are a beaker with characteristic 'South German' wheel-engraving in

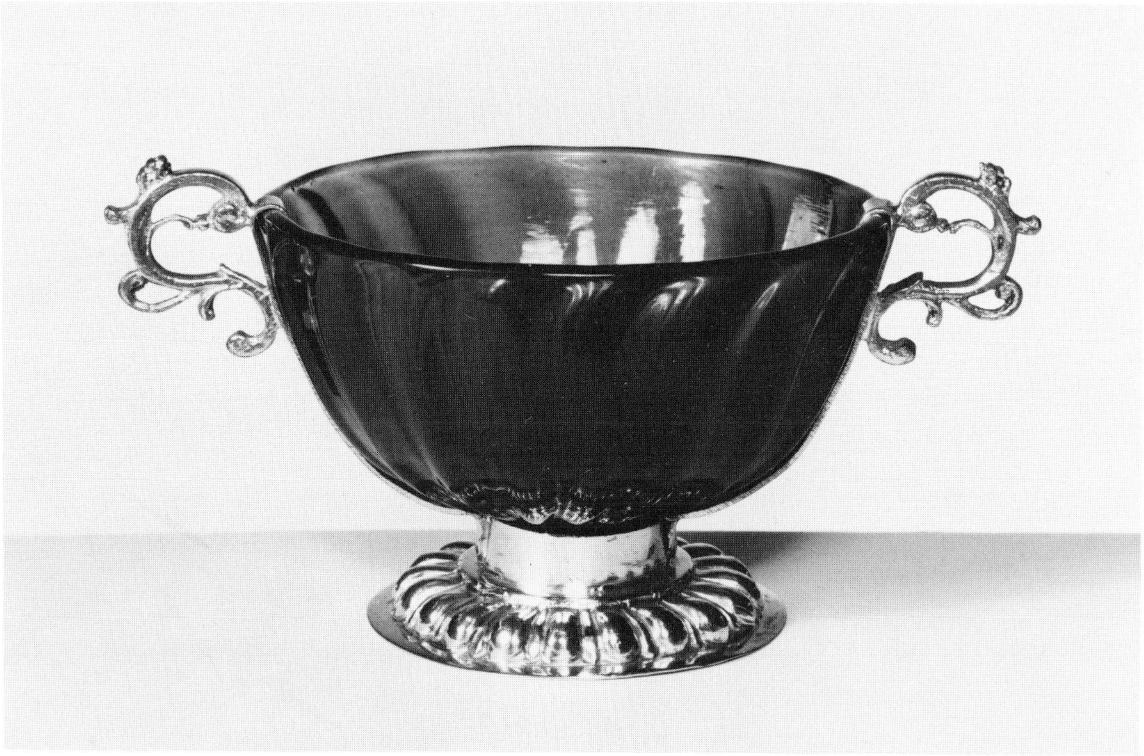

the Nordiska Museum, Stockholm, traditionally said to have been brought to Sweden in 1719 (Marshall Lagerquist, 'Armfeltska Gåvan till Skogaholm', *Nordiska Museets och Skansens Årsbok*, 1959, p.138, No.20); an almost identically mounted beaker in the R. Strasser Collection, New York (R. Strasser, 'Excellence in Continental Glass', *Antiques*, Jan. 1972, pp.158–9); and a slightly differently mounted beaker of the same type with 'South German' engraving, in the British Museum (British Museum, *Masterpieces*, No.258); a further beaker with ribbed glass knob on the cover and typical 'South German' engraving was in the Kaiser Friedrich Museum at Görlitz (R. Schmidt, *Brandenburgische Gläser*, Pl.17,7), and yet another beaker, surmounted by the figure of an old man in silver, in the Gotisches Haus at Wörlitz (ibid., Pl.17,9). See also further references under *54*.

1922 Eythrope Inventory: 'One ruby glass cup, silver mounted'.

Miss Alice de Rothschild E2/14/1**

54. Jug, ruby-glass mounted in silver-gilt

South German (perhaps Munich or Freising); late 17th or early 18th century

Plain bubbled gold-ruby glass, mounted in silver-gilt

Height with cover, $5\frac{5}{8}$ in. (14·3 cm); diameter of rim, $2\frac{3}{8}$ in. (6 cm)

Jug with globular body drawn into a tall, wide cylindrical neck. Plain solid loop-handle slightly tooled at the lower sticking-part, and foot-rim made of a flattened coil of glass which gives purchase to a silver-gilt rim held to it by a fillet of acanthus-like lappets. The shallow-domed flat-topped silver-gilt cover is decorated with a band of oblique gadrooning and has a plain spherical knob. It is hinged to the handle by means of a scrolled thumb-piece, the hinge itself being fixed to a curved plate of acanthus-leaf shape secured by two pins to a plain strap running under the handle.

The cover bears the pineapple mark of Augsburg (Rosenberg, *Der Goldschmiede Merkzeichen*[3], I, pp.30–1, Nos.194A, 199, 203, dated between 1695 and 1705), and the master's mark 'TB' within an oval cartouche.

Mark of Augsburg enlarged four times

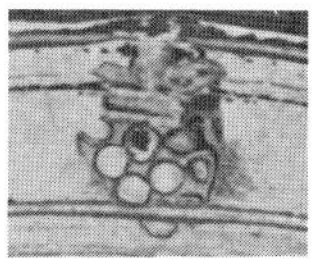

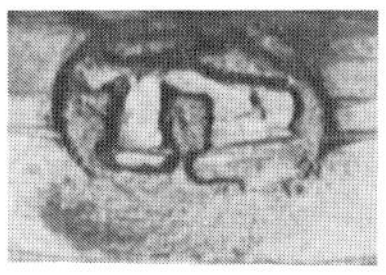

Maker's mark TB enlarged four times

For a note on the goldsmith's mark, see *53* here. The presence of this maker's work on a piece of this character is an argument for attributing the jug to the same centre of manufacture as that of the rib-moulded pieces, *51–53* here. A jug of the same shape and with a mount by the same maker, is in the Toledo Museum of Art, Ohio (Toledo Museum, *Art in Glass*, 1969, Fig. on p.64). This jug is wheel-engraved with the subject of Joshua and Caleb carrying the grapes, certainly not in a Potsdam style of engraving. A third jug with the 'TB' mark, this time plain, was in the Mühsam Collection and is now in the Art Institute, Chicago, U.S.A. (R. Schmidt, *Die Gläser der Sammlung Mühsam*, Berlin, 1914, p.46, No.183; *Brandenburgische Gläser*, Pl.18,1). Another jug of the same form is in the Victoria and Albert Museum, without mount (C.393–1936), and another with a Nuremberg-marked mount was in the Frau Margarete Reichenheim-Oppenheim Collection (R. Schmidt, *Brandenburgische Gläser*, Pl.17,8). Yet another jug of the same shape, and of 'sick' glass, in the Victoria and Albert Museum (C.891–1935) is decorated with typical 'South German' engraving, and is mounted by the Augsburg silversmith 'D.B.'.

Probably from the Hollingworth Magniac Collection, see sale *Catalogue*, Christie, Manson & Woods, London, Fourth Day, 6 July 1892, Lot 410: 'A ruby glass jug with handle – $5\frac{5}{8}$ in. high'.

1922 Eythrope Inventory: 'Ruby glass jug, silver mounts. German, 17th century'.

Miss Alice de Rothschild — E1/13/2*

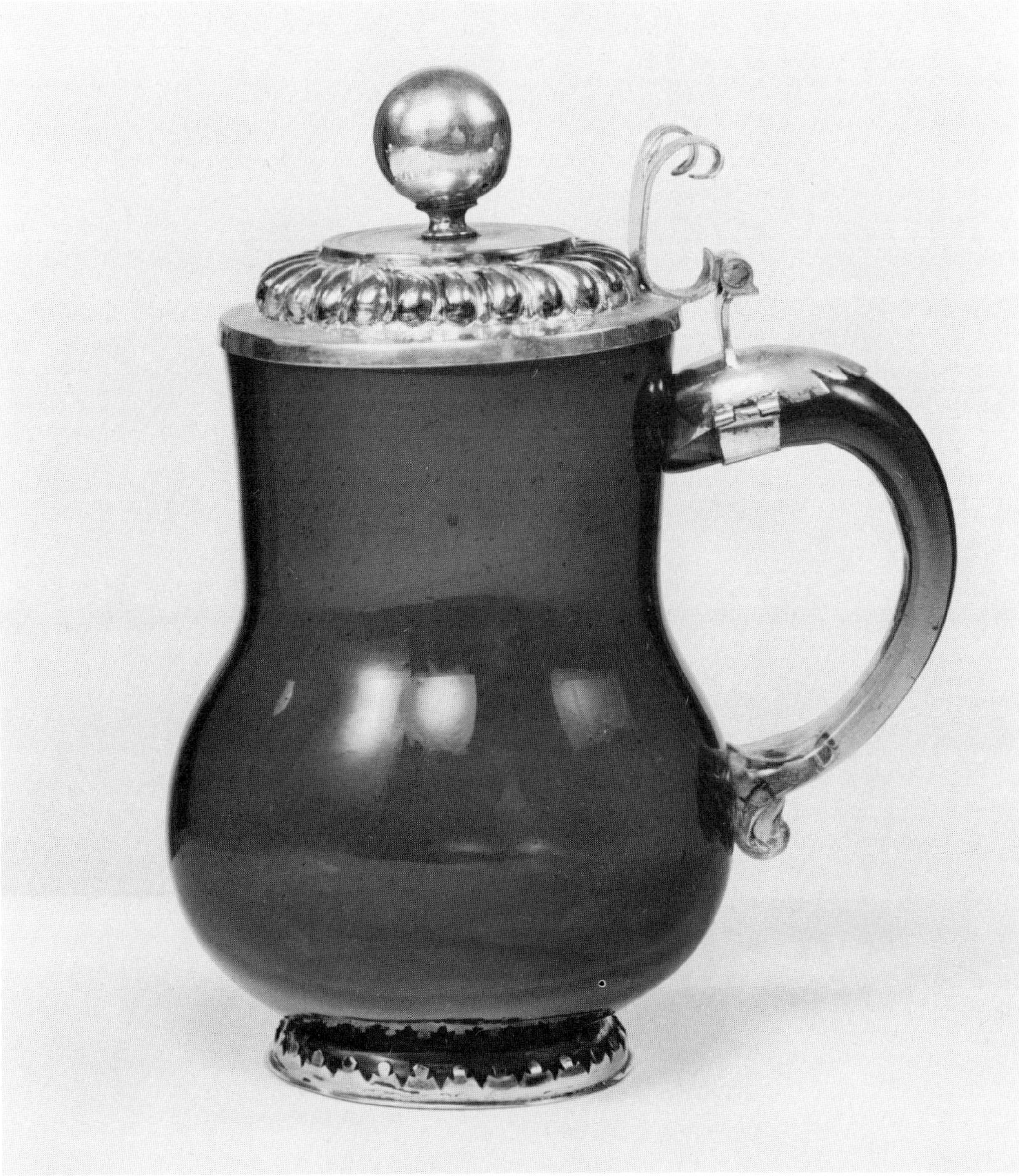

55. Beaker and cover, ruby-glass mounted in silver-gilt

Probably South German (perhaps Munich or Freising); late 17th or early 18th century

Plain, slightly bubbled gold-ruby glass, mounted in silver-gilt

Height, with cover, $8\frac{5}{8}$ in. (22 cm); without cover, $6\frac{7}{8}$ in. (17·5 cm); diameter of rim, $4\frac{5}{8}$ in. (11·8 cm)

Almost cylindrical beaker with sides sloping slightly outwards, and the rim gently flared. Under the foot the pontil-mark has been ground away to leave a circular lens-like 'printie' and this has been decorated by a surround of smaller unpolished circular cuts alternating with groups of thin wheel-engraved strokes. Almost flat cover with edge ground flat to take the mount; in the centre or top, a button of glass has been left to give a seating for the metal knob. The upper surface of the cover is slightly decayed and the inside shows signs of incipient 'glass disease'.

The foot of the beaker is held in a silver-gilt mount formed by a vertical collar edged with acanthus-leaf forms which merges into a convex gadrooned moulding: this in turn runs into a flat foot-rim. The edge of the cover is similarly held in a band edged with acanthus-leaf lappets, and the knob is formed of a plain sphere above a frilled circular plate. The mounts bear the pineapple mark of Augsburg (Rosenberg, *Der Goldschmiede Merkzeichen*[3], I, p.30, Nos.179–182), and the maker's mark $\substack{M\\W}$ in an oval cartouche. These are stamped on the foot-rim and on the rim of the cover.

Mark of Augsburg enlarged four times

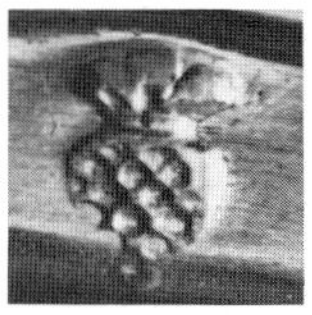

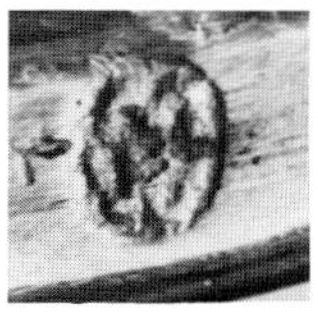

Maker's mark enlarged four times

The mark 'MW' is attributed by Rosenberg (R[3], I, p.159, No.726) to Mattheus or Markus Wolff (married 1685, d.1716). This Master appears not to be otherwise recorded as a maker of mounts for ruby glass.

For other beakers of this form, but mounted by the Master 'T.B.', see notes on *53* above. A further beaker, mounted in gilt metal on three ball feet, but without cover, is in the Grünes Gewölbe, Dresden (Kämpfer and Beyer, *Glass*, Pl.142, p.283, attributed to Potsdam). Despite the attribution to Potsdam of this beaker and the one in the Strasser Collection, a South German ascription seems more plausible in view of the appearance on these beakers of Augsburg mounts, and in one instance of typical 'South German' wheel-engraving also (see *53*).

1922 Waddesdon Inventory, 'Beaker, ruby, with Augsburg silver mounts. Venetian 17th century'.

Miss Alice de Rothschild — W1/139/8*

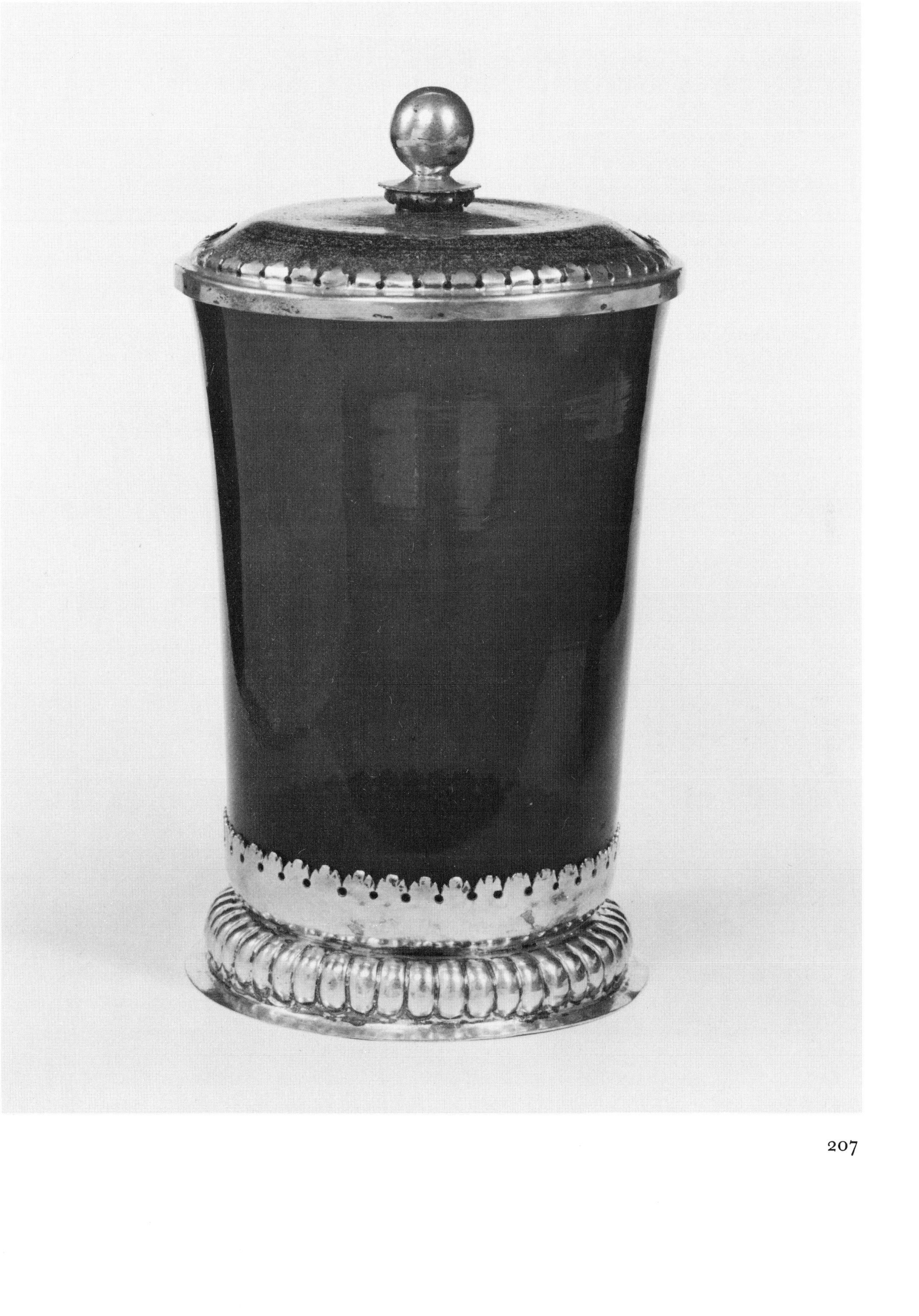

56. Teapot, ruby-glass mounted in silver-gilt

South German (probably Munich or Freising); late 17th or early 18th century

Plain, slightly bubbled gold-ruby glass, mounted in silver-gilt

Height with cover, 4½ in. (11·5 cm); width from spout to handle, 6¼ in. (16 cm)

Teapot of depressed globular shape, with low collar-neck and almost straight spout rising from a broad flattened base. The foot is formed by an applied flattened coil of glass which provides a seating for the silver-gilt foot-rim, held to it by a fillet of acanthus-leaf lappets. From this rises an acanthus-like plate which forms the lower point of attachment of the plain silver-gilt strap handle: the upper end is attached to a collar round the neck and has soldered to it a hinge for the flat cover. This is decorated with a band of gadrooning and has a plain spherical knob and a scrolled thumb-piece. At the top of the spout is a mount, secured by an acanthus-leaf border, with a hinged spout-cover.

The handle bears the impressed pineapple mark of Augsburg (Rosenberg, *Der Goldschmiede Merkzeichen*³, I, p.30, Nos. 186–7, 194, 196), together with the makers' mark 'MB' in an oblong cartouche, and a long zigzag assay mark.

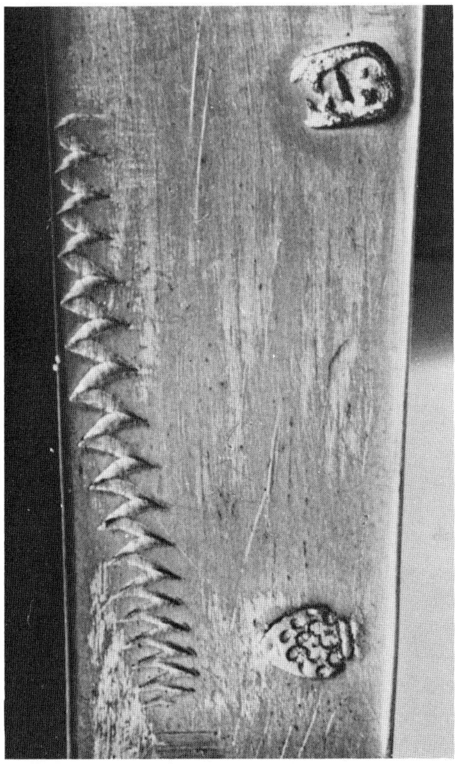

Marks and assay mark on the handle enlarged four times

The mark 'MB' in an oval is not attributed to any master by name (R³, I, No.723, occurring with Augsburg marks spanning the period about 1680–1705). A small engraved ruby bowl in Frankfurt, however, has an 'MB' mark ascribed to Matthäus Baur II (R³, I, No.711, with Augsburg mark for 1695–1705: this mark, however, is enclosed in a rectangle rather than an oval, and the mark on the Waddesdon example may well be the same). The Frankfurt bowl is decorated with characteristic 'South German' engraving (Frankfurt-am-Main, *Europäisches und Aussereuropäisches Glas*, No.473). The 'MB' mark recurs with the Augsburg city mark on a curious pear-shaped covered 'cup' once in the Mühsam Collection, now in the Metropolitan Museum of Art, New York, (R. Schmidt, *Mühsam*, I, p.46, No.182; F.-A. Dreier, 'Fünf Silbermontierte Prunkvasen aus Goldrubinglas', p.80, Fig.2). This 'cup' has the mould-blown ribbing of the group represented here by *51–3* and furnishes further grounds for the attribution to Munich or Freising.

Other teapots of the same general shape, but with glass handles, bear the mark 'CF', perhaps for Christian Frank (married 1666, d.1704: R³, I, No.649). One is in the Kunstgewerbemuseum, Cologne (*Glas*, No.284, the mark attributed tentatively to Dresden), and another was

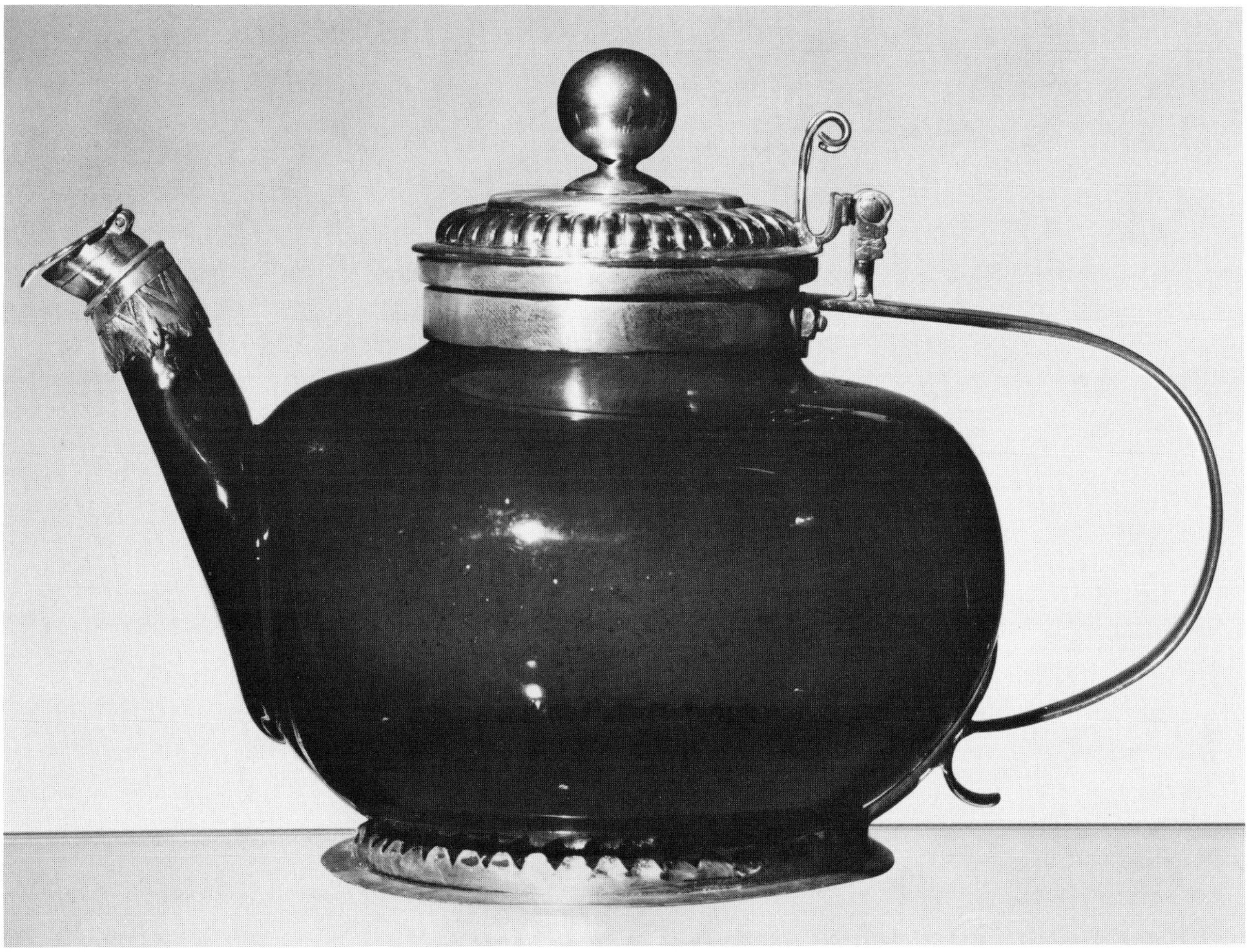

sold at Sotheby's on 2 June 1969, Lot 20 (Keith Middlemas, *Continental Coloured Glass*, London, 1971, *ill.* p.42). An unmarked example in the Kunstgewerbemuseum, Berlin, is illustrated in R. Schmidt's *Brandenburgische Gläser*, Pl. 17,6.

1922 Eythrope Inventory: 'Ruby teapot, silver mounts . . . German 17th century'.

Miss Alice de Rothschild E1/13/2*

57. Dish, ruby-glass mounted in silver-gilt

German; late 17th century

Plain gold-ruby glass, mounted in silver-gilt

Height with foot, $6\frac{7}{8}$ in. (17·5 cm); diameter, $5\frac{1}{2}$ in. (14 cm)

Shallow dish of ogee profile, with low foot. The dish is mounted on an elaborate cast and *repoussé* silver-gilt foot, in the central section of which are the figures of three Apostles standing in niches. The mount, which is unmarked, is probably of Italian or Spanish origin, and of 19th century date.

1922 Eythrope Inventory: 'Ruby glass shallow cup on tall silver-gilt foot'.

Miss Alice de Rothschild E2/32/14**

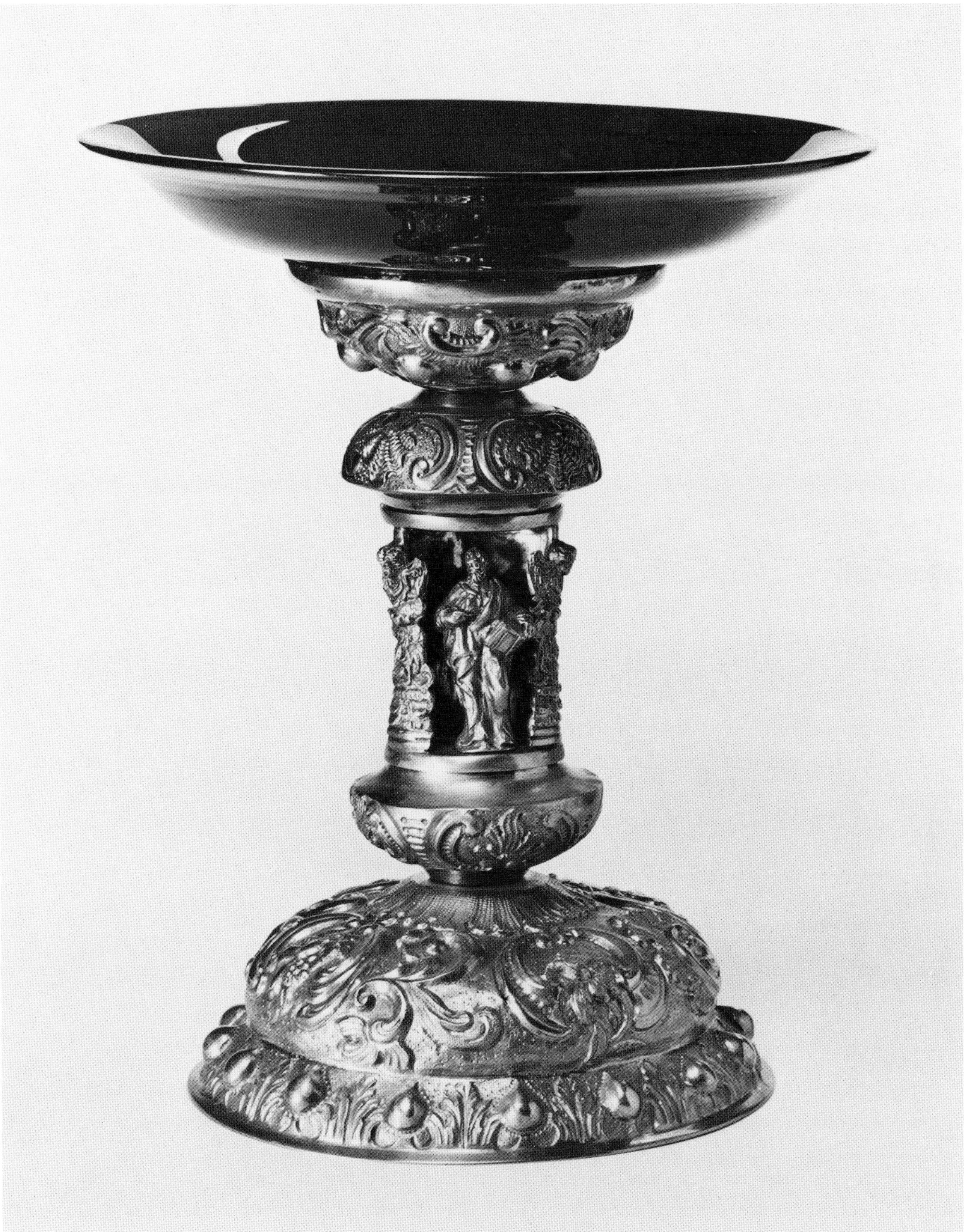

58. Pair of miniature beakers, ruby-glass with gilt metal covers

German; second quarter of 18th century

Plain gold-ruby glass, with covers of *repoussé* gilt metal

Height with cover, $3\frac{1}{8}$ in. (8 cm) and 3 in. (7·6 cm); without cover, $1\frac{7}{8}$ in. (4·8 cm); diameter of rim, $1\frac{3}{8}$ in. (3·5 cm)

Almost cylindrical beakers with bases polished flat and rims trimmed on the wheel. On the base of one is a scratched mark of two parallel strokes.

The beakers have *repoussé* covers of gilt metal in incipient rococo style with finials in the form of a shell mounted on a gadrooned knop. The covers are unmarked.

1922 Eythrope Inventory: 'Two small ruby pots and covers'.

Miss Alice de Rothschild E2/35/49**

59. Jug, ruby-glass with silver-gilt mounts

Illustrated on page 215

Perhaps Bohemian; early 18th century

Gold-ruby glass cut and polished on the wheel, with engraved silver-gilt mounts

Height with cover, $8\frac{1}{2}$ in. (21·5 cm); without cover, $7\frac{3}{4}$ in. (18·8 cm)

Piriform jug with slender neck flaring slightly to the rim, which is tooled out in front to a pouring-lip. Plain, solid loop handle, slightly tooled at the lower sticking-part. The foot has been formed by the application of a flattened coil of glass: the pontil-mark has been ground out and part of the surrounding area of the foot smoothed off on the wheel. The body of the vessel has been decorated by a series of wheel-cut facets, circular, oval and pear-shaped. A pearl border of small circular 'printies' at the top of the neck is followed by a scale-border formed of truncated ovals: below these are three hexafoil motifs formed of small 'printies': below this frieze are four successive borders of plain oval and circular cuts: round the belly of the jug are six quatrefoil motifs, each heart-shaped petal formed of a double pear-shaped cut, single pear-shaped cuts radiating from between the petals, and small 'printies' filling the spaces between the lobes of the petals.

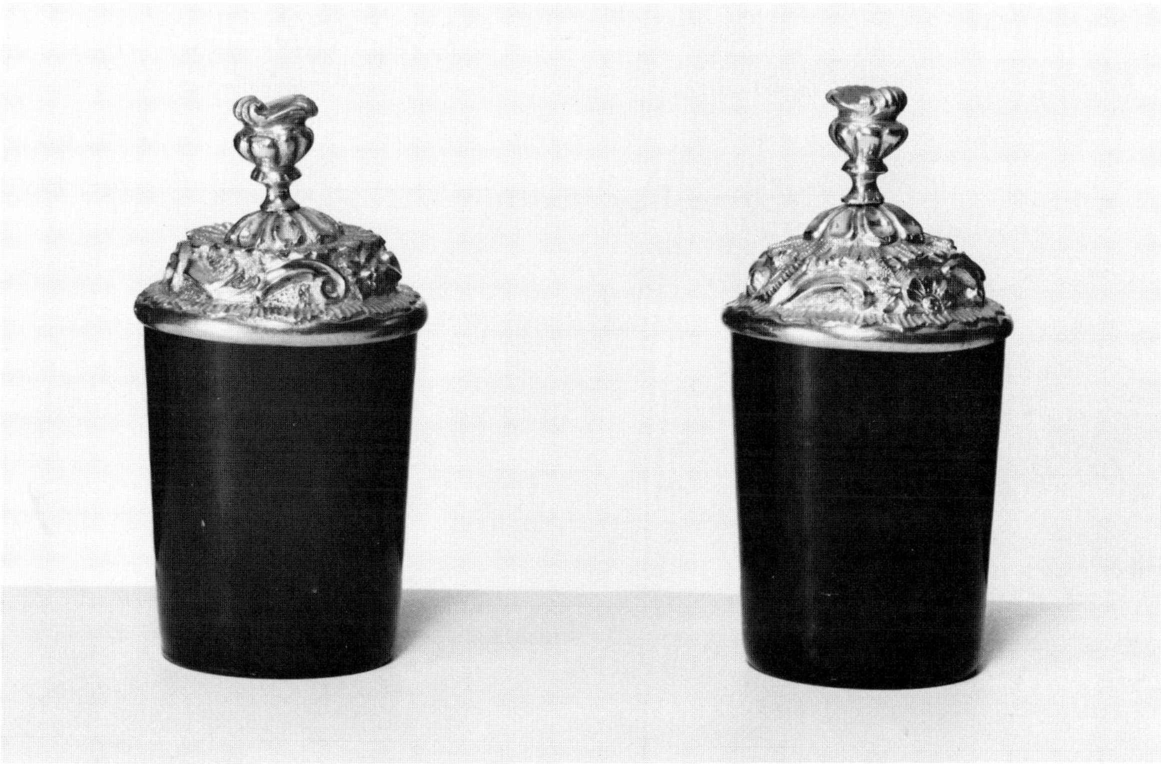

The jug is mounted in silver-gilt, a plain foot held in place by a row of acanthus-leaf lappets, and a flat cover fastened to a hinged plate, mounted on the handle by means of an upper and a lower curved plate pinned together at the sides. The rim of the cover has an ovolo border punched in firm zigzag lines, and a thumb-piece in the form of a cherub's head between scrolls.

The mounts are unmarked.

The decoration of this jug is distinctive, and unmatched on any other specimen so far traced. The area in which *par excellence* purely formal wheel-cut patterns were produced in the early years of the eighteenth century was Bohemia (see R. Schmidt, *Das Glas*, pp.261–2). It is difficult to match the quality of this work but similar grouping of cuts into rosette-like motifs may be found on a number of Bohemian goblets in colourless glass dating from 1700 and the period shortly thereafter (see, e.g., R. Schmidt, op. cit., Fig.146a, and d; J. R. Vávra, *5000 Years of Glass-making*, Prague, 1955, Figs.174 and 175; comparable work combined with wheel-engraving in Z. Pešatová, *Bohemian Engraved Glass*, Figs.24, 26, 29). A pair of crystal-glass jugs with gilt-metal mounts, in the Rosenborg Castle collection, Copenhagen, have a marked affinity in their wheel-cut decoration with the Waddesdon piece.

Comparable to the esteem in which ruby glass was held was that enjoyed by J. F. Böttger's red stoneware made at the Meissen porcelain factory, and the two substances were very similarly treated, being cut, engraved and polished on the wheel. Examples of Saxon cased ruby-glass exist with engraved decoration closely akin to that found on Meissen stoneware (W. Holzhausen 'Sächsisches Rubinglas und Steingefässe von J. F. Böttger', *Belvedere*, XII. Jahrgang, 1934/37, pp.16–22), and it is known that cutters and engravers were employed in Dresden for the decoration of the stoneware as well as glass (E. Zimmermann, *Die Erfindung und Frühzeit des Meissner Porzellans . . .*, Berlin, 1908, pp.131–5; Staatliche Kunstsammlungen, Dresden, *Böttgersteinzeug; Böttgerporzellan aus der Dresdener Porzellansammlung*, Dresden, 1969, pp.15 ff). Not infrequently the decoration on the stoneware accords stylistically with that found on Saxon engraved glass in general, with *intaglio* medallions surrounded by Baroque scrollwork, etc. (von Schlieben, 'Deutsche Glaskunst im 18. Jahrhundert', *Keramische Rundschau*, 1937, pp.105–7; 118–20; 142–3; 178–81; 190–3; 202–5). Often, however, the ornament on the stoneware consists of purely formal designs produced by facetting on the cutter's wheel, and it seems likely that a large part of this work was executed by Bohemian cutters, of whom at least ten are known to have been employed in this work (E. Zimmermann, op. cit., p.325). A group of Böttger red stoneware pieces '*gemuschelt*' in this way is illustrated in E. Zimmermann, op. cit., p.133, Fig.44. To the left of this group is a teapot decorated in a manner distinctly recalling the Waddesdon ruby-glass jug. Similarly, a bowl of presumably Bohemian ruby glass is known which in its allover facetting is very close to some of the red stoneware teapots (O. Drahotová, 'Bohemian Ruby Glass at the turn of the 17th and 18th centuries', *Glass Review*, *4*, 1973, Fig.4). The differences between Bohemian and Saxon glass-decoration, however, are often very slight, and it is not impossible that the Waddesdon jug is of Saxon origin. Ruby-glass is mentioned in the Neuostra glasshouse, near Dresden, as early as 1706 (von Schlieben, loc. cit., pp.204–5), and although J. M. Steinbrück boasted of the special qualities of Böttger's cased glass, this does not exclude the possibility that the solid ruby was also made (W. Holzhausen, loc. cit., p.17).

1922 Eythrope Inventory: 'Ruby jug, silver-gilt mount, German 17th century'.

Miss Alice de Rothschild — E1/14/2*

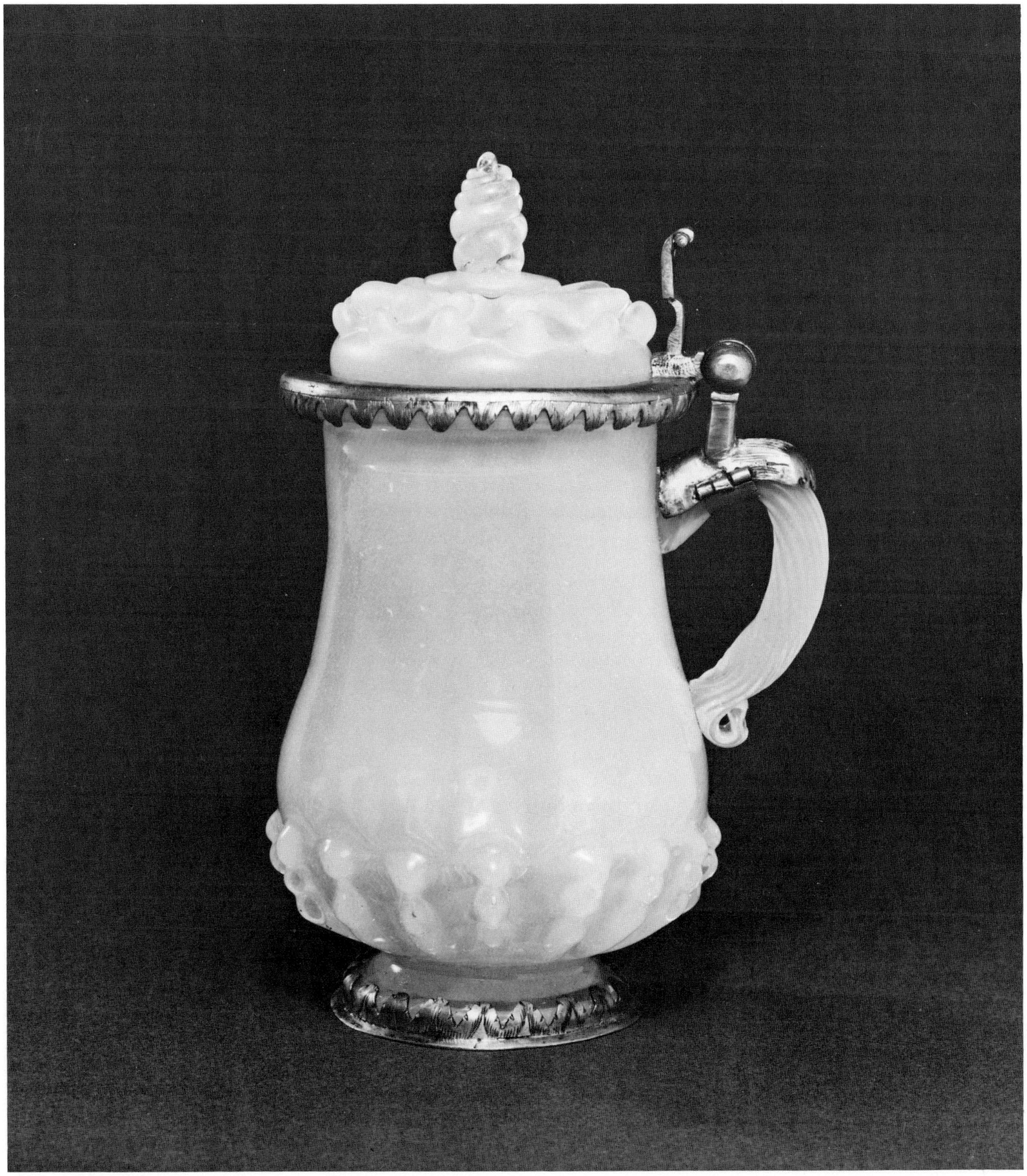

61. Goblet and cover, opalescent-white glass

South German or Bohemian; late 17th century

Goblet and cover of opalescent-white glass (*Milchglas*), the bowl, cover and sections of the stem decorated with vertical ribbing produced by blowing into a ribbed mould

Height, with cover, 12¼ in. (31 cm); without cover, 8⅝ in. (22 cm); diameter of rim, 3 in. (7·5 cm)

Round-funnel bowl with vertical ribbing, separated by a double merese from a smaller and a larger depressed knop, themselves separated by a double merese from a vertically ribbed stem-section composed of a depressed knop above a slender inverted baluster: this in turn is separated by a double merese from a straight section spreading into a disc which connects the whole stem with the slightly rising folded foot.

The domed cover is made in one piece and decorated overall with mould-blown ribbing, slightly 'wrythen' above the flange. A double merese separates it from a finial formed of a spire above a flattened knop.

Although this glass was referred to in the early inventories as Venetian, it is almost certainly of German or Bohemian origin. Although the Venetians certainly made such opal glasses, the form and decoration of this piece clearly suggest a northern origin. The compilation of its stem is unmistakably German, with the interpolation of pairs of mereses between the component knops. The mould-blown ribbing is also found repeatedly on shapes in *Milchglas* which are specifically German. These include a footless boat-shaped cup which is also found in characteristic 'crizzled' crystal metal as well as in coloured glasses; the shape is demonstrably German (examples in enamelled coloured glass with German inscription in the Victoria and Albert Museum (Mus. No.1837–1855): in crystal wheel-engraved with German inscription, perhaps of Nuremberg origin, in the Kunstmuseum, Düsseldorf – see E. Meyer-Heisig, *Der Nürnberger Glasschnitt des 17. Jahrhunderts*, Nuremberg, 1963, p.84, No.78, Fig.78). A second ribbed form in *Milchglas* is a sweetmeat with shell-bowl on a ribbed swelling stem and foot with radiating ribs, once in the Lanna Collection (*Sammlung des Freiherrn Adalbert von Lanna, Prag, Sale Catalogue*, Rudolph Lepke, Berlin, 1911, No.734, Pl.63, called Venetian). This piece has a handle on the long axis terminating in a scallop-shell motif, a feature found on two wheel-engraved crystal sweetmeats of similar form which from their engraving and stem-formation may reasonably be considered of Bohemian origin (W. B. Honey, *Glass . . .*, p.147 (355–1873), Pl.49, C, there attributed to Spain on grounds of its provenance). The third form is a ribbed *Milchglas Kuttrolf* with quadruple neck, mould-blown on a 'second gather', in an English private collection. This is a typical German form, quite unlike the analagous sprinkler-shapes made in Italy.

All these pieces are characterized by a finer ribbing than the parallels cited under *60*. They seem rather analogous to the ribbed specimens made in ruby-glass, for which a South German attribution has been suggested (see p.32–33). In the records of the Munich glasshouse between 1677 and 1681 mention is made of 'Kelchbecherl auf Füssen' ('goblets on feet') in 'Porzelän'

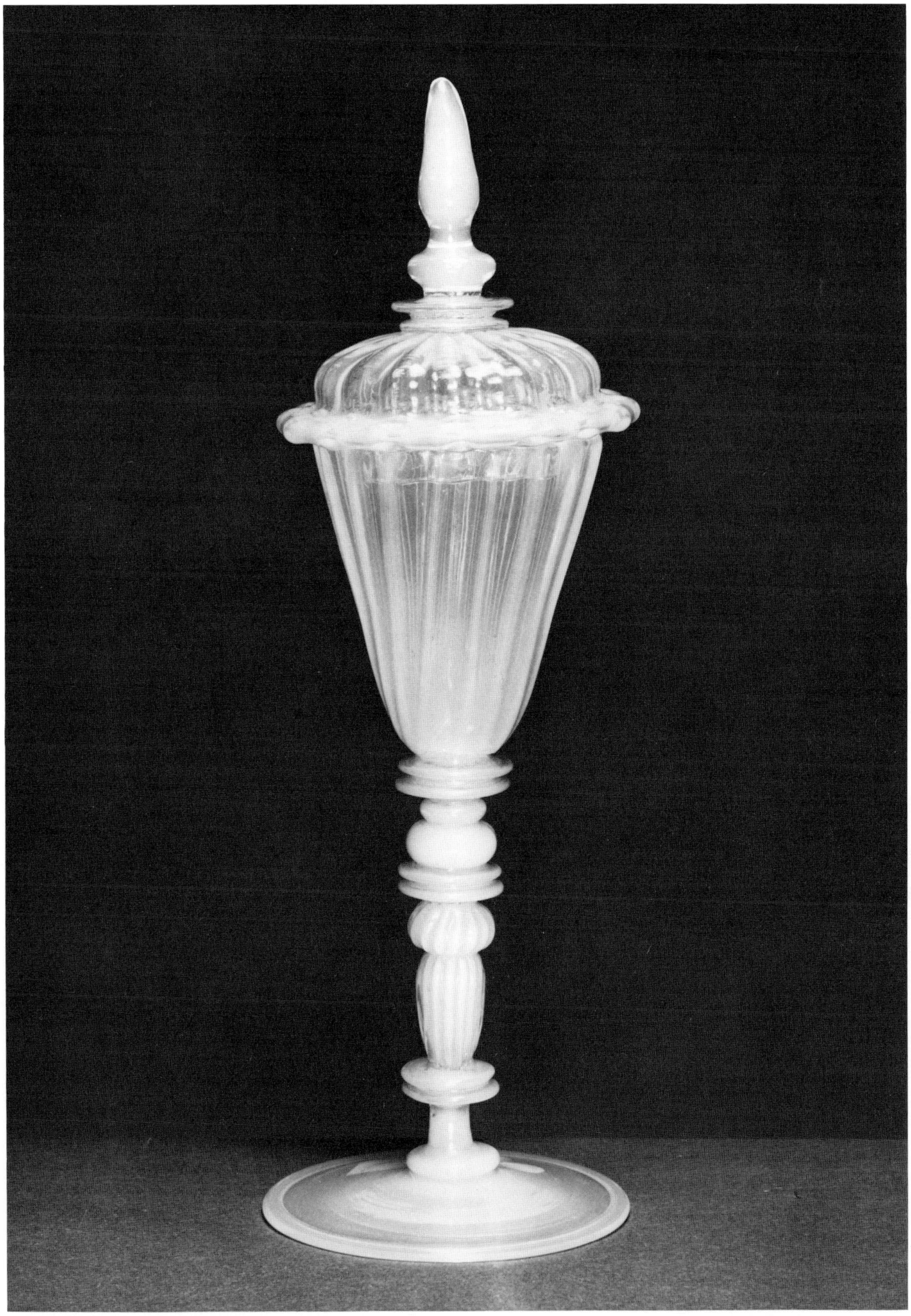

glass, a description which might reasonably be applied to a glass such as the Waddesdon piece (R. Berliner, loc. cit., p.120). In the same inventory a 'Muschel auf einem Fuss' (' a shell on a foot') and 'Längliche (Schalen)' ('oblong bowls') might well refer to objects of the two types discussed above (R. Berliner, loc. cit., p.121). The use of ribbed moulds in this glasshouse has been discussed in the section dealing with ruby-glass (p.33, above).

Probably from the Bernal Collection – see *Catalogue of the Celebrated Collection of . . . Ralph Bernal*, Christie and Manson, 25th Day's Sale (20 April, 1855), Lot 3331: 'A tall fluted glass and cover, opalised – 12½ in. high'.

1898 Waddesdon Inventory (Smoking Room): '1120. An opalised Venetian glass fluted vase with baluster stem and cover, 12½ in.'.
1922 Waddesdon Inventory (Smoking Room): 'Cup and cover on tall stand, opalesque, Venetian, 17th century'.

Baron Ferdinand de Rothschild W1/138/7*

62. Trick-glass

German or Bohemian; probably 17th century

Brownish-colourless bubbled glass with wrought decoration

Height, complete, $12\frac{1}{4}$ in. (31 cm); of goblet to rim, $8\frac{3}{8}$ in. (21·2 cm); diameter of foot, $4\frac{3}{8}$ in. (11 cm) (Illustrated overleaf)

The glass consists of two components – goblet and sucking-tube.

The goblet consists of a slightly flaring round-based bowl, decorated with 'wrythen' mould-blown ribbing and two circuits of self-coloured threading at the top and the bottom of the bowl. Rising from the centre of it inside is a tapering tube which leads down to a hollow knop, forming the stem of the goblet. This is connected to the bowl by three decorative vertical trails of glass, between which are three applied nipples, one of which is pierced through to the inside of the knop. The whole is supported on a folded pedestal foot, joined to the hollow stem by two discs.

The sucking-tube consists of the forequarters of a stag and a doe blown hollow and affixed on either side of a vertical tube with a rigaree band top and bottom. The stag's mouth is open and connects with this tube, while the doe's mouth is closed.

The trick consists of sucking the liquid in the bowl through the mouth of the stag, keeping one finger over the open nipple on the hollow knop, to avoid an intake of air. The liquid cannot readily be drunk direct from the bowl without the nose engaging on the central tube.

Trick-glasses based on the sucking-tube seem to have a long history in the Germanic countries. Some examples at least seem to go back to the sixteenth century (e.g. one with an antlered stag finial in the Kunstgewerbemuseum, Cologne (*Glas*, No.155, *ill.*)). The idea was adopted elsewhere in the seventeenth century, a glass with a plain tube, adorned with no more than some pincered frills, being included among the drawings sent by the London glass-seller John Greene about 1673 to his supplier in Venice, Allesio Morelli (A. Hartshorne, *Old English Glasses*, London, 1897, Pl.32). A Venetian or *façon de Venise* glass of this character is in the Museum of Decorative Arts, Prague (Uměleckoprůmyslové muzeum v Praze, *Benátske sklo* (*Venetian Glass*), *Catalogue*, 1973, No.142, Fig.35). In the main, however, this was a type of glass which won most favour in the German-Bohemian lands. Here the finial was usually in the form of a stag, and many examples survive – e.g., in the Victoria and Albert Museum, London (Museum No.C.413–1936; see W. Buckley, *The Art of Glass*, London, 1939, No.337, Pl.114, in green *Waldglas*); in the Österreichisches Museum für Angewandte Kunst, Vienna (Schlosser, *Das Alte Glas*, Fig.145); in the Musée Curtius, Liège (Baar Collection – see Musée Curtius, *Trois Millénaires*..., No.384, *ill.*); in the Peter H. Plesch Collection, Keele (*Journal of Glass Studies*, VII, 1965, p.81, Fig.9). The first of these, like the earlier example in Cologne, has no trick to it beyond the need to suck; the remaining three share with the Waddesdon example the finesse of the hollow bulb below and the open nipple which requires to be blocked before drinking can begin. This refinement presumably indicates a later stage in the development of this form.

Far less usual are the glasses with finials formed of two animals. The following examples may be cited – Corning Museum of Glass, New York (Mus.No.63.3.20; *Guide*, 1974, No.61);

private collection (K. Middlemas, *Continental Coloured Glass*, London, 1971, p.32, *ill.*). On both of these the second animal is a horse. A glass once in the Schlossmuseum, Berlin, has three stags grouped round the tube, their antlers meeting together at their tips (R. Schmidt, *Das Glas*, Abb.83).

From the Bernal Collection (*Catalogue of the celebrated Collection of . . . Ralph Bernal, Esq.*, Christie and Manson, 25th Day's Sale (20 April, 1855), Lot 3246: 'A curious drinking glass with tube in the centre, surmounted by two horses' heads' (*ill.*). It was bought by 'Rhodes' for £1.10.0.

1922 Waddesdon Inventory (Smoking Room): 'Puzzle cup and cover, two figures of animals. Venetian 17th century'.

Miss Alice de Rothschild W1/141/1*

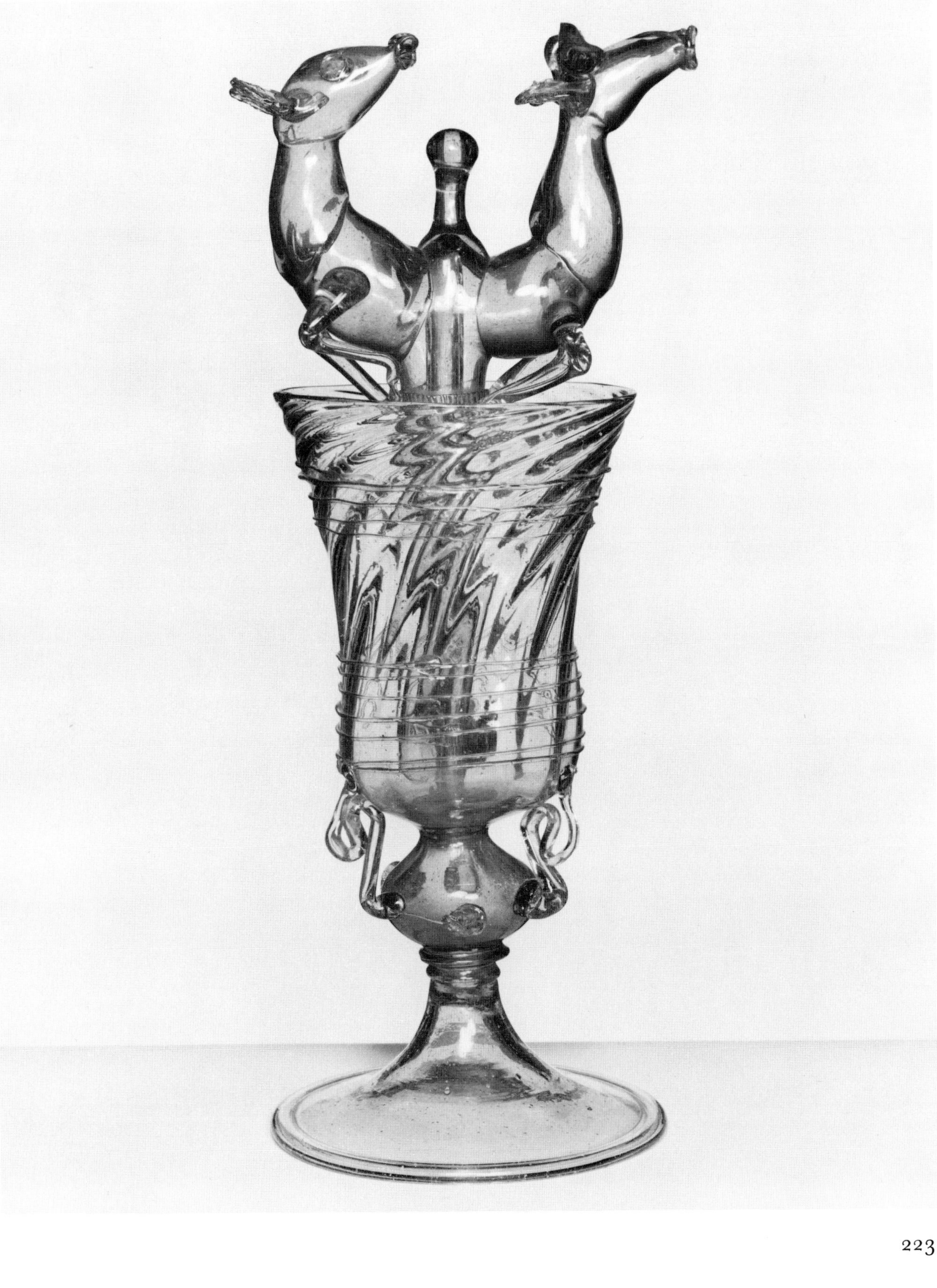

63. Goblet and cover, bubbled purple-toned glass

Bohemian or German; late 17th century

Goblet and cover with moulded ribbing on a 'second gather' which has been elaborately tooled

Height, with cover, 9 in. (23 cm); without cover, $6\frac{1}{8}$ in. (15·5 cm); diameter of rim, $2\frac{5}{8}$ in. (6·7 cm)

Goblet with barrel-shaped round-based bowl, tall stem composed of tripartite hollow-blown central component tooled into roughly quatrefoil cross-section and separated from bowl and foot respectively by a series of discs and mereses. Slightly rising foot with folded rim, and flanged domed cover surmounted by a finial corresponding in form to the central component of the stem, but solid.

Both bowl and cover have been decorated by forcing a second gather into a ribbed mould, the ribbing being then tooled into a form of 'spiky gadrooning'.

The purple tone of the glass is probably due to an overdose of manganese used as a decolorant, rather than as a deliberate colouring-agent.

In the treatment of the 'spiky gadrooning' of bowl and cover, this goblet resembles a number of glasses which on account of their engraving and stem-formation may with a reasonable degree of certainty be attributed to Bohemia (e.g. R. Schmidt, *Das Glas*, Pls. 84,d (already discussed under *60* here) and 144,b; Frankfurt-am-Main, Museum für Kunsthandwerk, *Europäisches and Aussereuropäisches Glas*, No.372). The pronounced purple tone and the slack workmanship of stem and finial, however, are unusual in Bohemian glass of this period, and the present glass may represent an essay in the Bohemian style made elsewhere in the Germanic lands.

1922 Waddesdon Inventory (Smoking Room): 'Cup and cover, ruby [sic], raised ornamentation. Venetian, 17th century'.

Miss Alice de Rothschild W1/139/3**

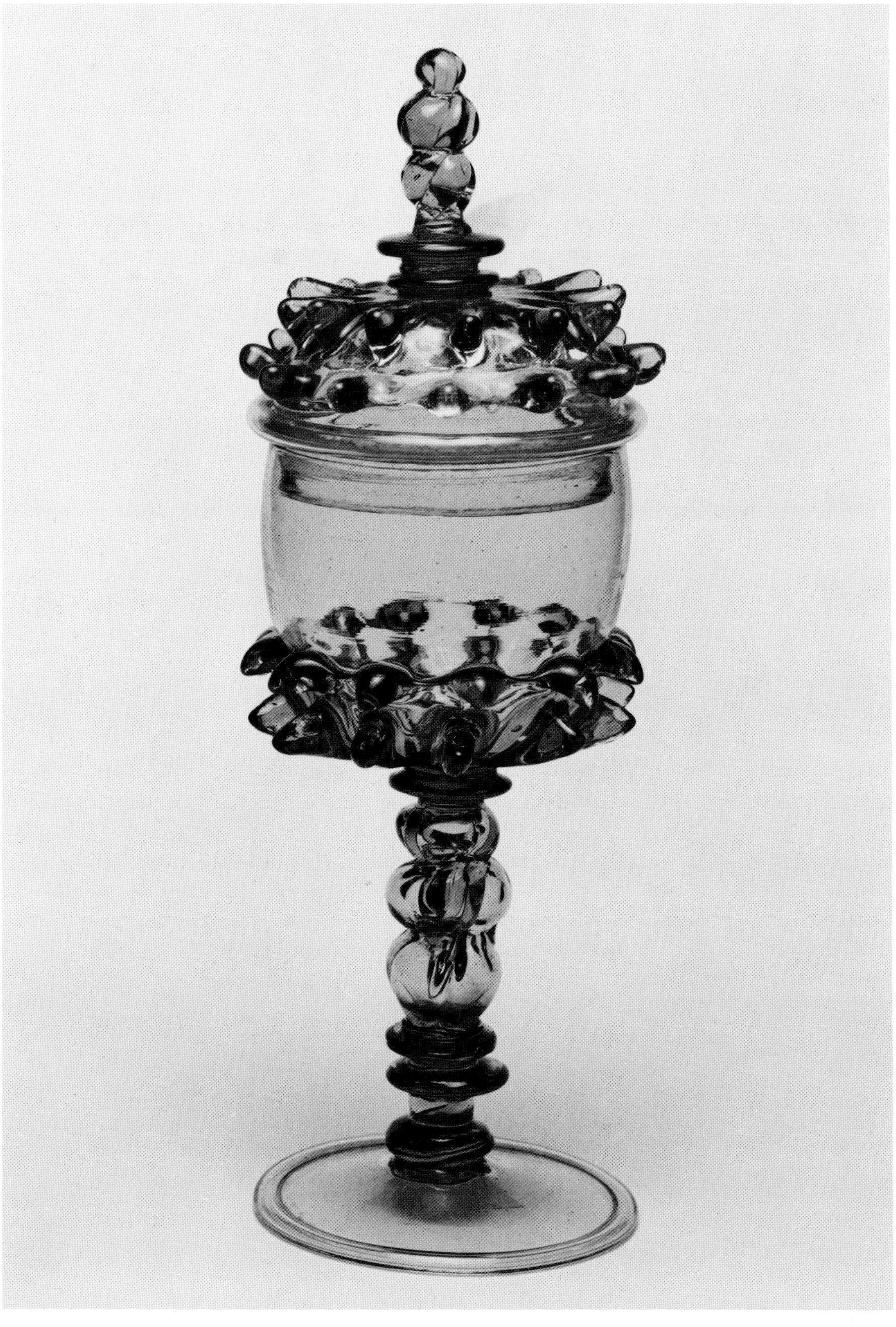

64. Goblet and cover, engraved with emblems

Bohemian; late 17th century

Clear colourless ('crystal') glass with wrought decoration and wheel-engraving

Height with cover, 16¼ in. (41·5 cm); without cover, 11 in. (28 cm); diameter of rim 4¼ in. (11 cm)

Goblet with round-funnel bowl, the base of which is decorated with 'spiky' gadrooning, produced by forcing a 'second gather' into a ribbed mould and then pincering it. The stem is composed of a solid double quatrefoil knop, separated by a triple merese from a solid four-lobed inverted baluster, double mereses separating the stem from the bowl and the plain folded foot. The domed cover is surmounted by an open work 'crown' topped by a cottage-loaf finial.

The bowl of the goblet is wheel-engraved with three oval panels enclosing the following emblems: two doves facing each other and billing, inscribed 'Zwey liebe sich kissen' ('Two love-[birds] kiss each other'); a table with three vessels on it, inscribed 'Auff gutte Gesundheit' ('To your good health!'); three hearts, inscribed 'Aus dreÿen Hertzen' ('From three hearts').

The decorative scheme of a series of medallions, round or oval, may be found on a number of Bohemian glasses of this period – e.g., on a double-goblet in the Krug Collection, Mülheim (Klesse, *Krug*, I, Munich, 1965, No.185, with doves billing above a heart); a goblet in the Museum für Kunsthandwerk, Frankfurt-am-Main (*Europäisches und Aussereuropäisches Glas*, No.370); a goblet in the Museum of Applied Arts, Prague (A. Vávra, *5000 Years of Glass-Making*, Prague, 1955, Pl.73, Fig.178). A covered goblet with doves billing above a heart is in the Kunstgewerbemuseum, Berlin (Günter Schade, *Deutsches Glas*, Leipzig, n.d., (1969?) Pl.43c).

The cover of this goblet, although a good fit, probably does not belong to it. It is normal on a goblet of this sort for the cover to match in decoration the base of the bowl, and for the finial to echo the formation of the stem. A parallel for this cover-finial may be found on a bowl once in the Schlossmuseum, Berlin (R. Schmidt, *Das Glas*, Fig.84e).

1922 Waddesdon Inventory (Smoking Room): 'Cup on tall stem, with cover with openwork top, the bowl engraved with three hearts, birds and a table with utensils. German, late 17th century'.

Miss Alice de Rothschild — W1/139/7*

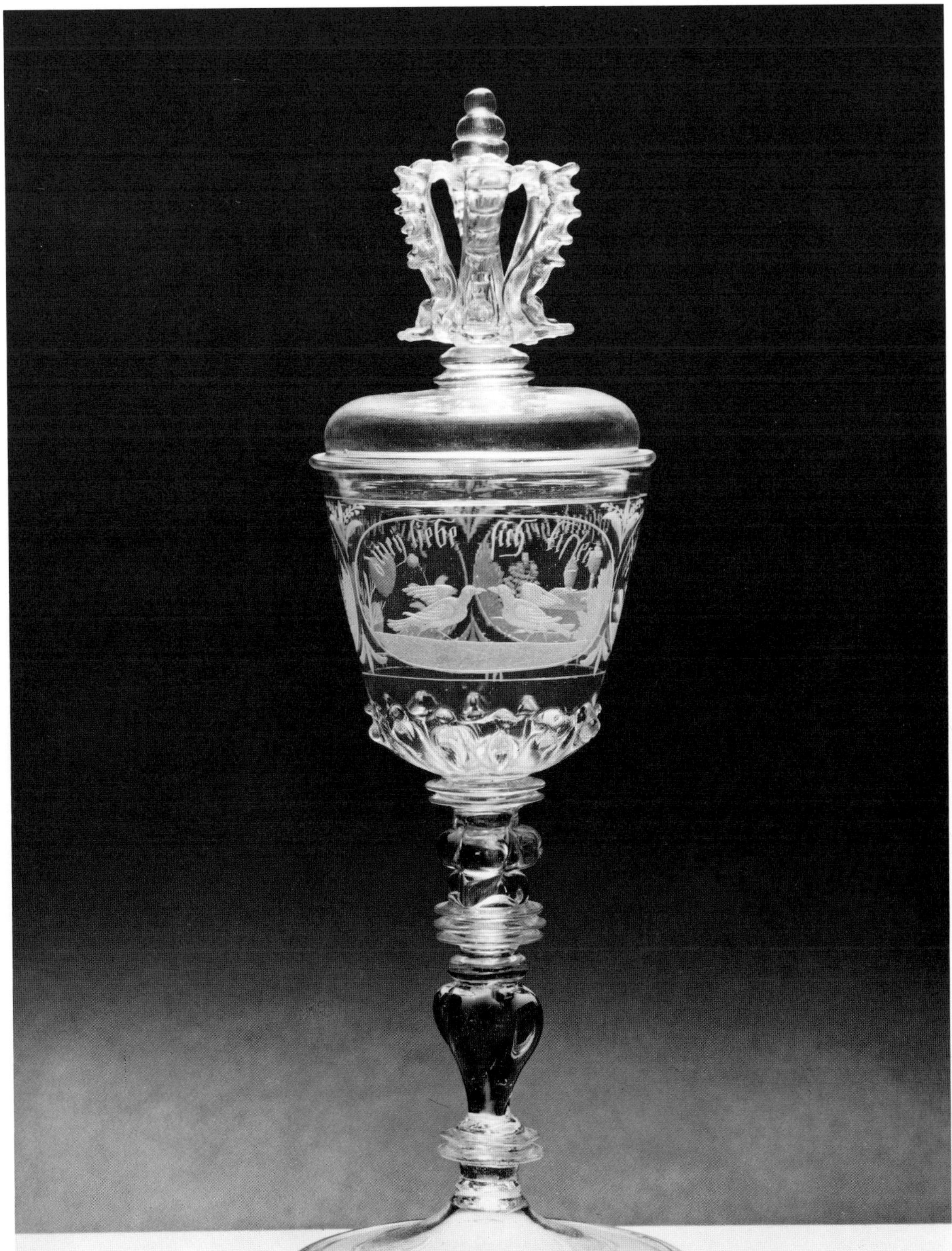

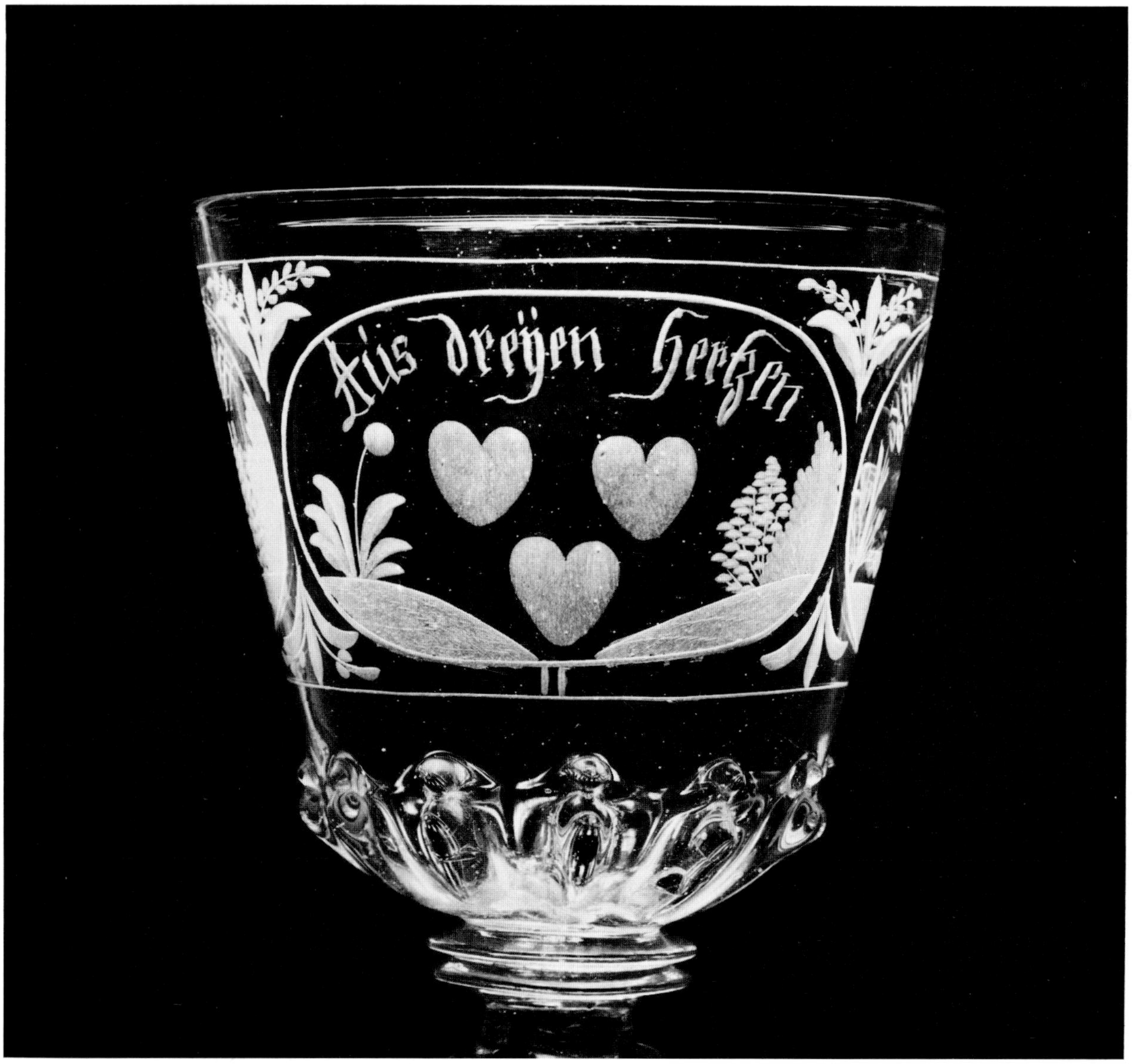

Detail of the bowl of 64

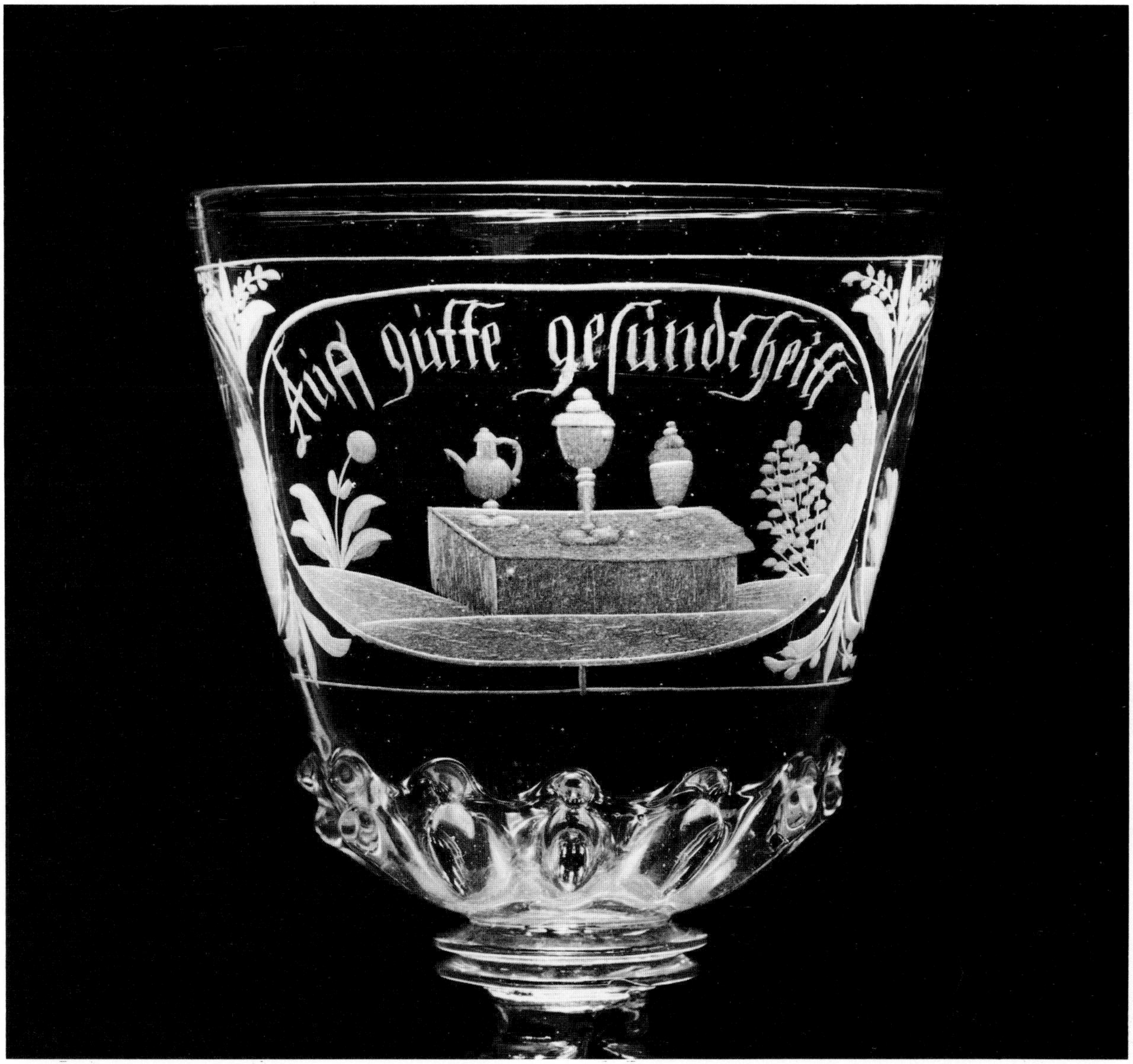

Detail of the bowl of 64

65. Goblet, engraved with flowers

Bohemian; late 17th century

Clear colourless ('crystal') glass with wrought decoration and wheel-engraving

Height, $11\frac{1}{2}$ in. (29·2 cm); diameter of rim, $3\frac{7}{8}$ in. (9·8 cm)

Goblet with round-funnel bowl, the base of which is decorated with 'spiky' gadrooning, produced by forcing a 'second gather' into a ribbed mould and then pincering it. The stem is composed of a central spherical hollow knop with 'wrythen' mould-blown ribbing, between elaborate sections of stem above and below, composed of various depressed knops and mereses. The slightly domed foot has a narrow fold.

The bowl of the goblet is wheel-engraved with a wide frieze enclosing three swags of formalized fruits and flowers between upper and lower borders of repeating paired leaves.

This goblet is somewhat unusual in the formation of its stem with its central slightly 'wrythen' ribbed knop. This feature recurs, however, on a tall goblet with multiknopped stem, elaborately gadrooned bowl and typical Bohemian engraving of a bird among plants, in the National Gallery, Prague (*Bulletin des 'Journées Internationales du Verre'*, *4*, Liège, 1965–66, p.57, Fig.48); a goblet with a very similar stem, but with three ribbed knops, is in the Sammlung Peill und Putzler, Düren (Düsseldorf Kunstmuseum, *Meisterwerke*, No.195); a double-goblet once in the Schlossmuseum, Berlin, had ribbed knops at the top and bottom of its stem, and displayed every variety of gadrooning on its upper and lower components (R. Schmidt, *Das Glas*, p.258, Fig.144,b). A goblet with a similarly composed stem, but with a vertically ribbed central knop, is in the Kunstgewerbemuseum, Berlin (Günther Schade, *Deutsches Glas*, Leipzig, n.d. (1969?) Pl.43c).

The engraving on this goblet betrays its Bohemian origins, if by nothing else, by the double border of paired leaves on either side of a central stem, a feature which occurs repeatedly on Bohemian goblets of this kind (e.g. a covered goblet in the Museum of Applied Arts, Prague, illustrated K. Hetteš, *The Bohemian Glass*, Fig.12, with a stylized hunting-scene; a double-goblet in the same Museum, illustrated Z. Pešatová, *Bohemian Engraved Glass*, Fig.12, with birds among plants, etc.). Similar floral engraving may be seen on a goblet once in the Schlossmuseum, Berlin (R. Schmidt, *Das Glas*, Fig.144a); and on another of squatter shape in the Museum für Kunsthandwerk, Frankfurt-am-Main (*Europäisches und Aussereuropäisches Glas*, No.372).

1922 Waddesdon Inventory (Smoking Room): 'Cup on tall stem, the bowl engraved with a floral ornament. German, late 17th century'.

Miss Alice de Rothschild — w1/139/6*

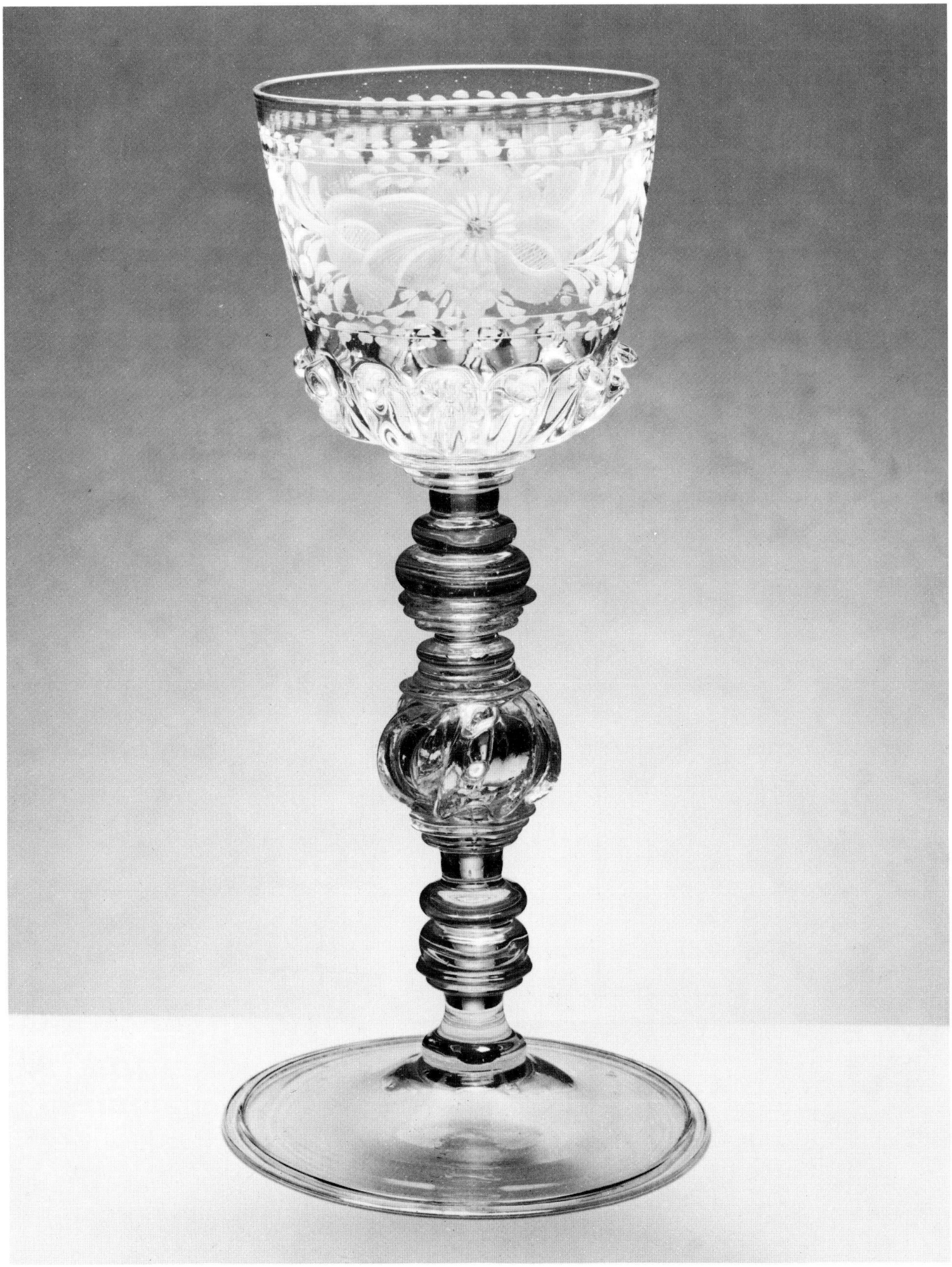

66. Beaker with gold mounts in a *piqué* stand

Probably Bohemian; about 1720–30

Height, without cover, $4\frac{1}{8}$ in. (10·5 cm); diameter of rim, $3\frac{1}{8}$ in. (8 cm)

Clear, colourless ('crystal') glass, blown in an octagonal mould with slight ribs at the angles and subsequently cut and engraved on the wheel, with plain reeded gold mounts at the rim and the foot, in a contemporary stand of tortoiseshell *piqué d'or*.

Both mounts have the import mark in use in Vienna, 1872–1901 (Rosenberg, *Der Goldschmiede Merkzeichen*, IV, Berlin 1922–28, No.7893.).

The beaker is octagonal on plan, with gently outward-sloping sides. The foot is made of a second gather which has been backed on and subsequently ground on the wheel to form a single *meniscus* on the underside. The eight sides are decorated with two types of design vertically arranged and alternating – one consisting of a quatrefoil between two circular 'printies', between two vertical rows of four smaller 'printies', all polished and contrasting with a matted ground; the other consisting of a basket of fruits and flowers above a double interlace design forming a central vertical lattice punctuated by small 'printies'. Below these motifs runs an arcade design of polished 'printies' on matt semi-circular 'arches'. The moulded ribs are notched from above this arcading upwards.

Wheel-engraving of this comparatively unambitious kind is frequently found on Bohemian glass of the late seventeenth or early eighteenth century. The vertical panels with mainly circular cuts singly or in groups, the work of the *Kugler* (or printie-cutter), may be matched in a number of glasses – e.g., a series of covered goblets once in the Schlossmuseum, Berlin (R. Schmidt, *Das Glas*, Fig.146); a slender covered goblet in the Museum of Arts and Crafts, Prague (London, Victoria and Albert Museum, *Bohemian Glass*, London, 1965, Pl.10); a goblet in the Museum Kulturen, Lund, Sweden (Sven T. Kjellberg, 'Europeiskt Glas', *Kulturen*, 1953, Fig.43); a covered goblet from the North Bohemian Museum in Libereč (Z. Pešatová, *Bohemian Engraved Glass*, Prague, 1968, Pl.26). These simple circular cuts are often combined with primitive wheel-engraved curves simulating natural forms such as leaves, stems, etc., as on most of the examples cited. Interlace designs like those on the Waddesdon glass may be seen on a beaker in the Kunstgewerbemuseum, Cologne (*Glas*, No.322), and a covered goblet once in Berlin (R. Schmidt, *Das Glas*, Fig.153, left). A goblet in the Musées Royaux d'Art et d'Histoire in Brussels, has a moulded body with vertical angles which have been notched exactly as on the Waddesdon glass (Liège, Musée Curtius, *Trois Millénaires . . .*, No.635).

A Venetian origin has been argued for some of the glasses of this general kind, and a beaker with very similar moulding and engraved decoration is in the Museo Vetrario, Murano (G. Mariacher, *Vetri Italiani del seicento e del settecento*, Milan, 1965, p.101). The presence of such glasses in Italy, however, seems more likely to be due to the importation of Bohemian glasses which heralded the ultimate dominance of the Bohemian style over Venetian glass-making in the eighteenth century.

See Geoffrey de Bellaigue, *Furniture, Clocks and Gilt Bronzes*, II, Fribourg 1974, p.827.

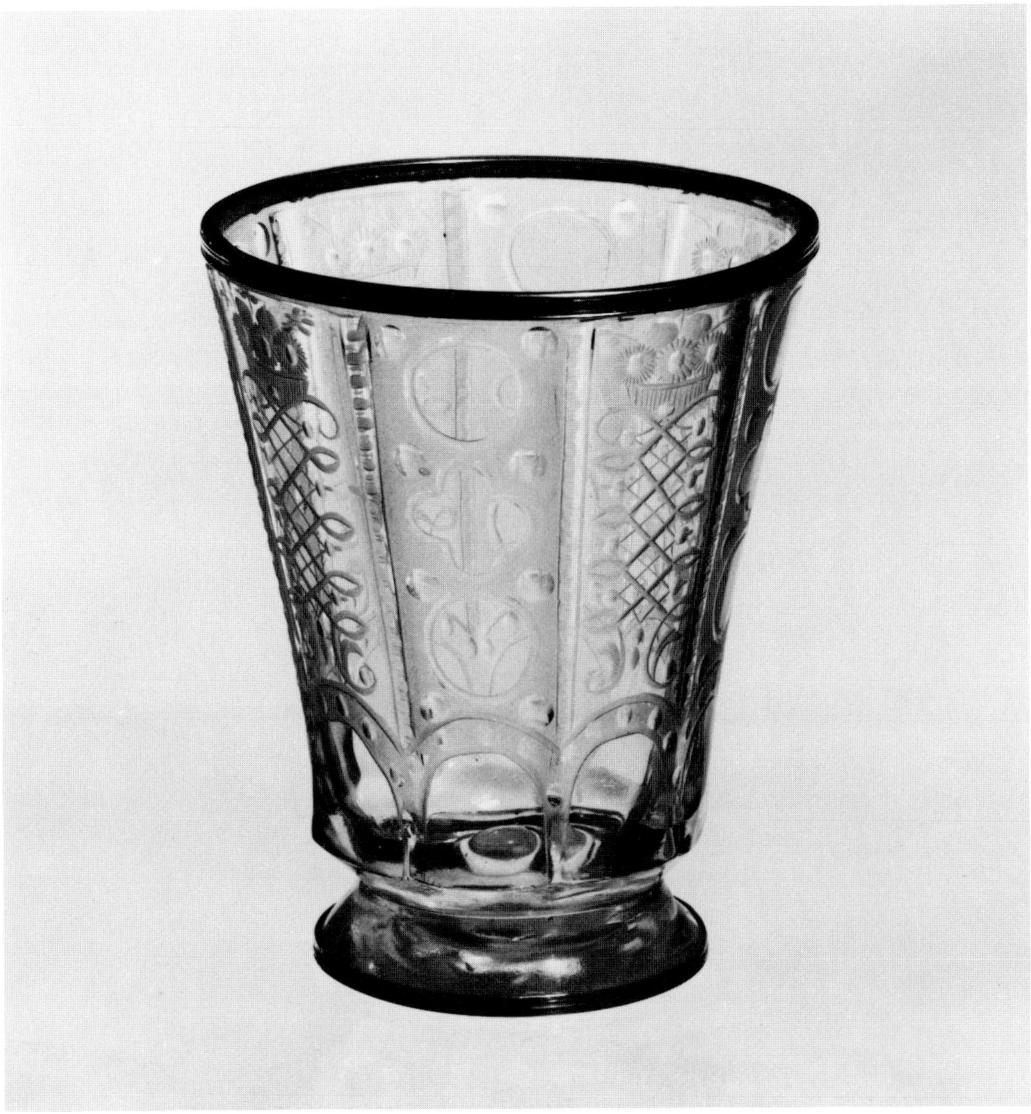

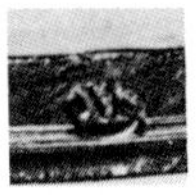

Import mark enlarged to twice actual size

'From the Hamilton Palace Collection' (Miss Alice de Rothschild).[1]

Baron Ferdinand de Rothschild WI/8/I

[1] Presumably acquired privately as it cannot be traced in Christie's sale catalogue of the Hamilton Palace sale of 1882.

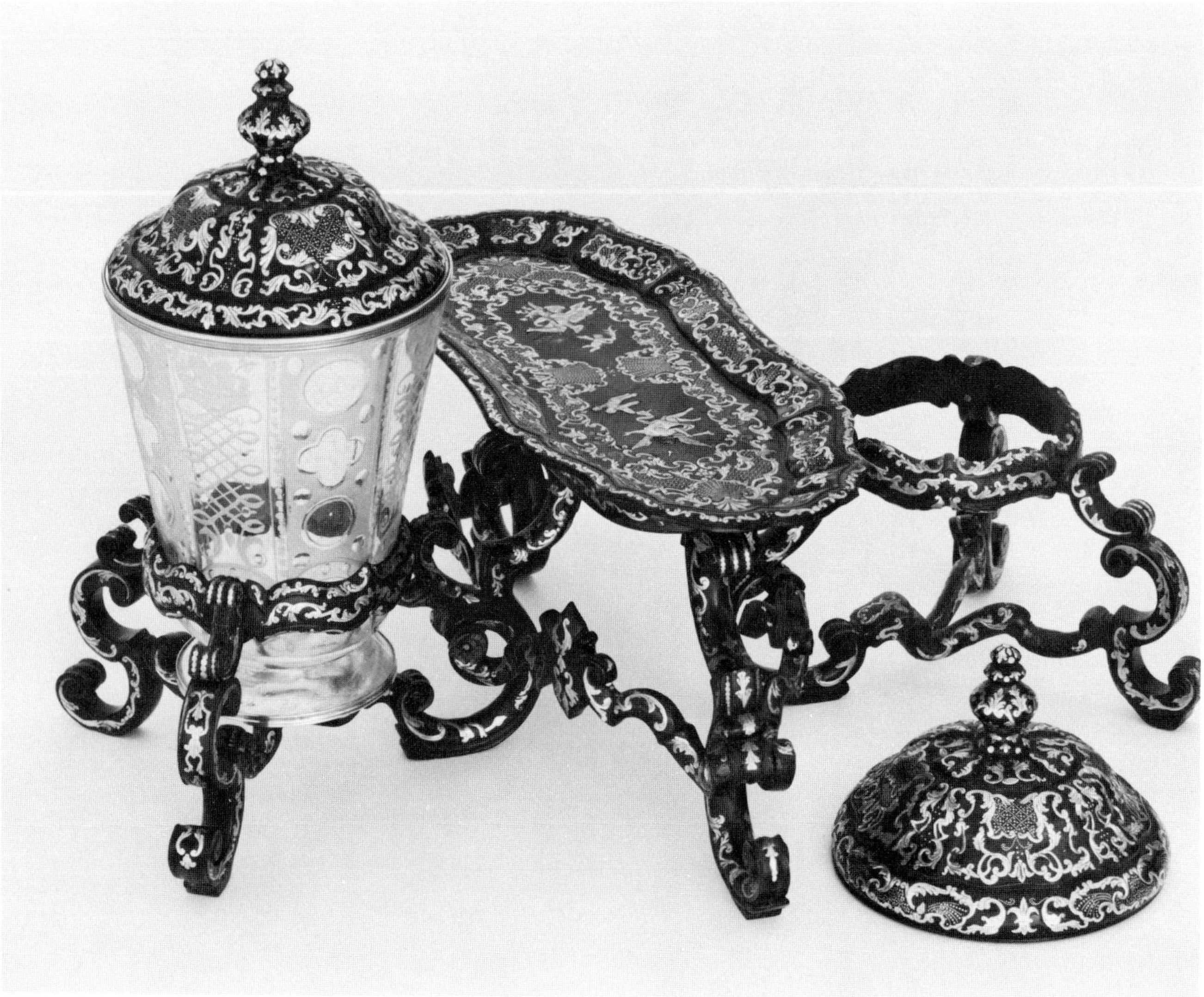

66 in its stand

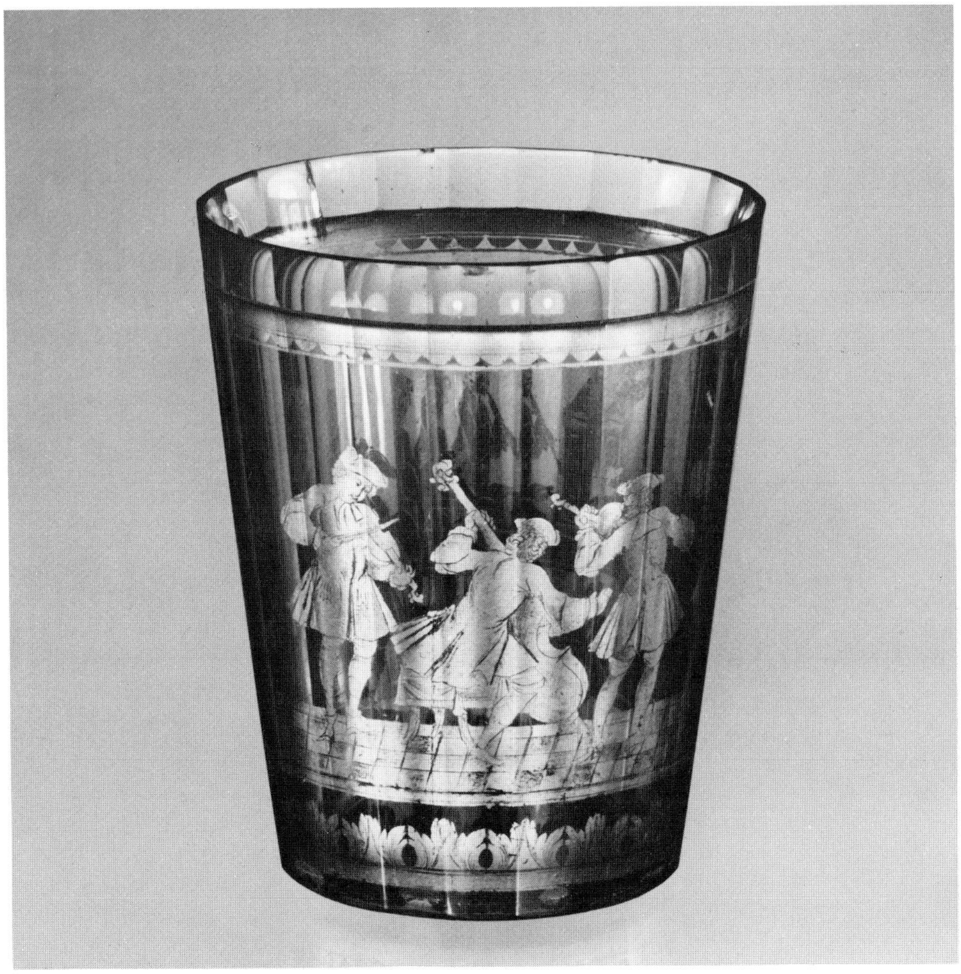

67. Beaker, with gold decoration

Bohemian; about 1730–40

Clear colourless glass facetted on the wheel, with double wall enclosing etched gold leaf (*Zwischengoldglas*)

Height, $3\frac{5}{8}$ in. (9·2 cm); diameter at rim, $2\frac{7}{8}$ in. (7·2 cm)

Straight-sided, slightly tapering, tumbler-form with flat vertical wheel-ground facets overall, the top of the inner glass projecting over the top of the outer glass and lying in the same plane. Below the foot of the inner glass a disc of glass coloured pink fits into the aperture framed by the outer glass, thus protecting a layer of etched gold leaf.

Round the wall of the beaker runs a music-making scene, with a man in a peruke seated at a chamber-organ, behind which two men in small tricorn hats sit playing recorders: behind them

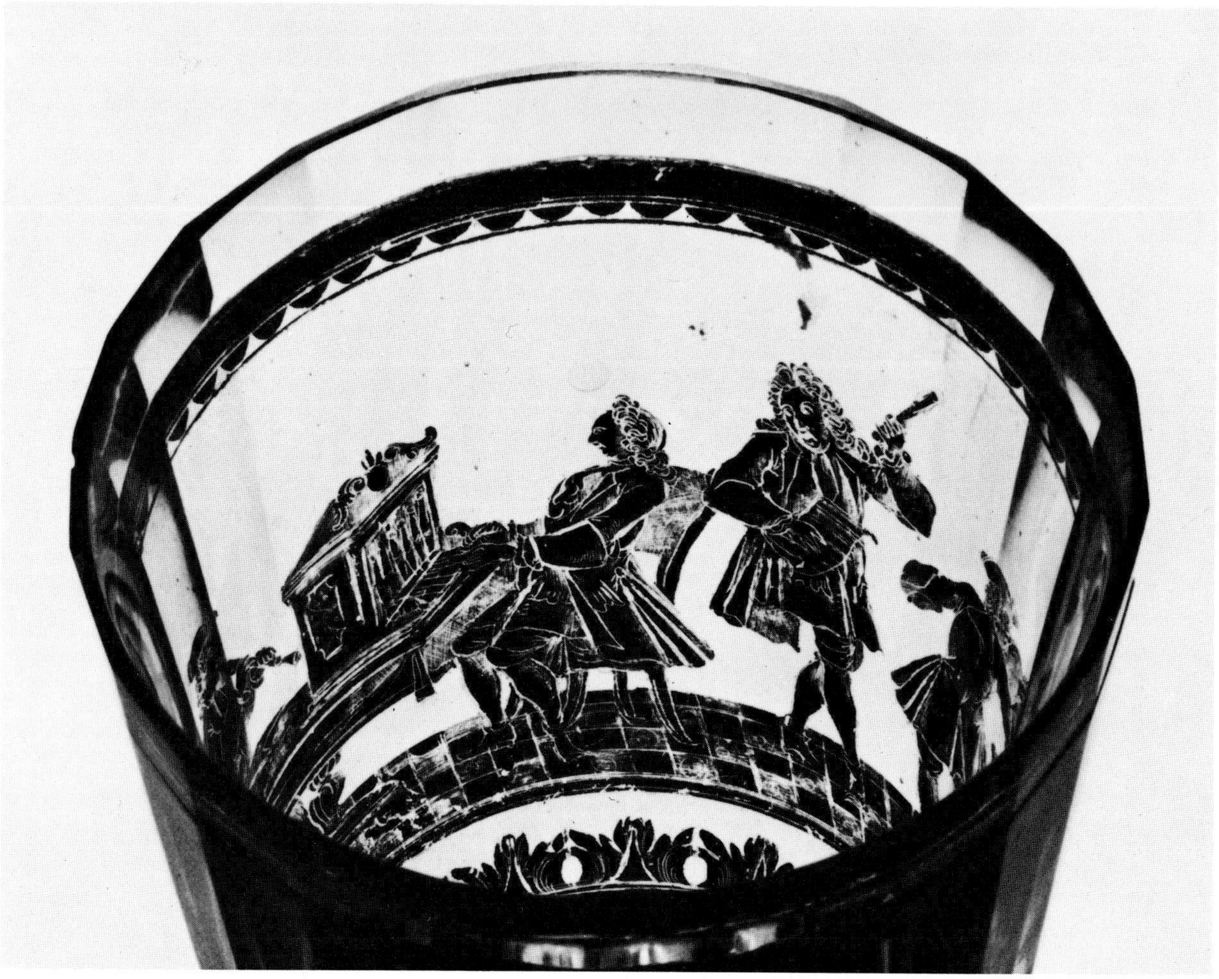

Interior view of 67, enlarged to twice actual size

but with their backs turned to them, stand two horn-players forming an ensemble with two violinists and a cellist with his back to the spectator; between them and the chair of the organist stands the conductor, a thick baton in his right hand, a sheet of music in his left. Round the foot of the beaker runs a border of upright acanthus-leaves, whilst at the point of junction of the walls of the beaker runs a dentil border. The design of the foot, visible to one looking into the beaker, and showing gold on pink, is of a horseman discharging his pistol behind him.

An exactly similar beaker was in the Reichenberg Museum Collection (Pazaurek, *Reichenberg*, Pl.35a). Comparable musical scenes are to be seen on beakers in the Mühsam Collection, now in the Metropolitan Museum of Art, New York. (Schmidt, *Mühsam*, I, No.266, Pl.33; Vávra, *5000 Years of Glass-Making*, Prague, 1955, Pl.238H), and the Krug Collection, Mülheim (Klesse, *Krug*, I, p.313, No.358).

1922 Waddesdon Inventory (Smoking Room): 'Pair of ditto (sc. tumblers), musicians. German 17th century' (see *68*).

Miss Alice de Rothschild — W1/140/4*

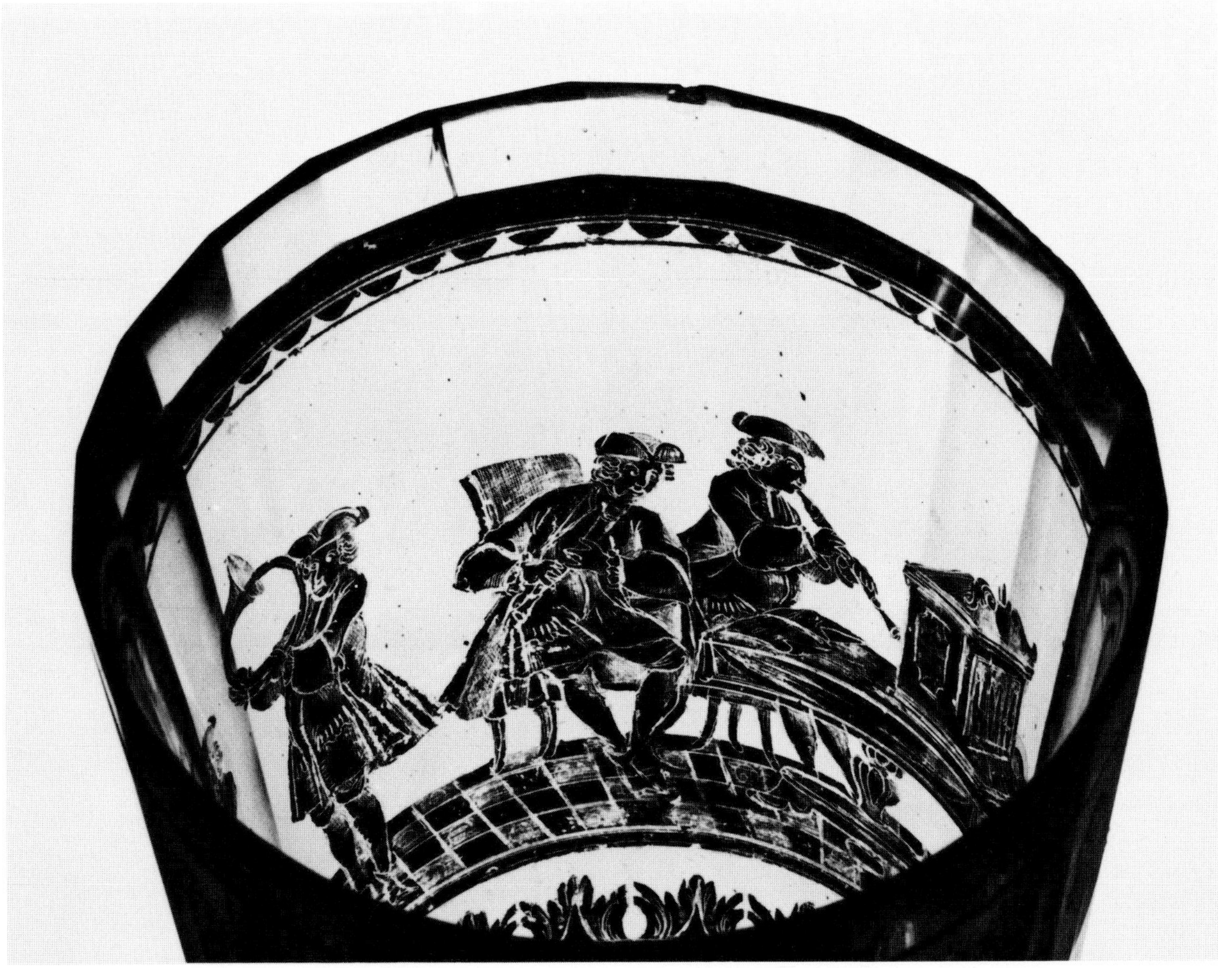

Interior view of 67, enlarged to twice actual size

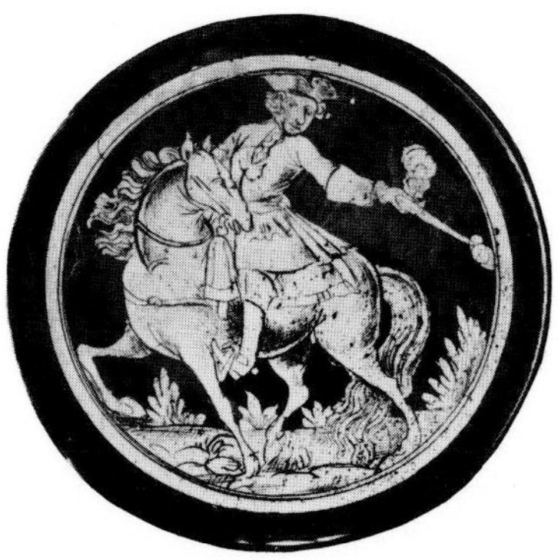

The base of 67, actual size

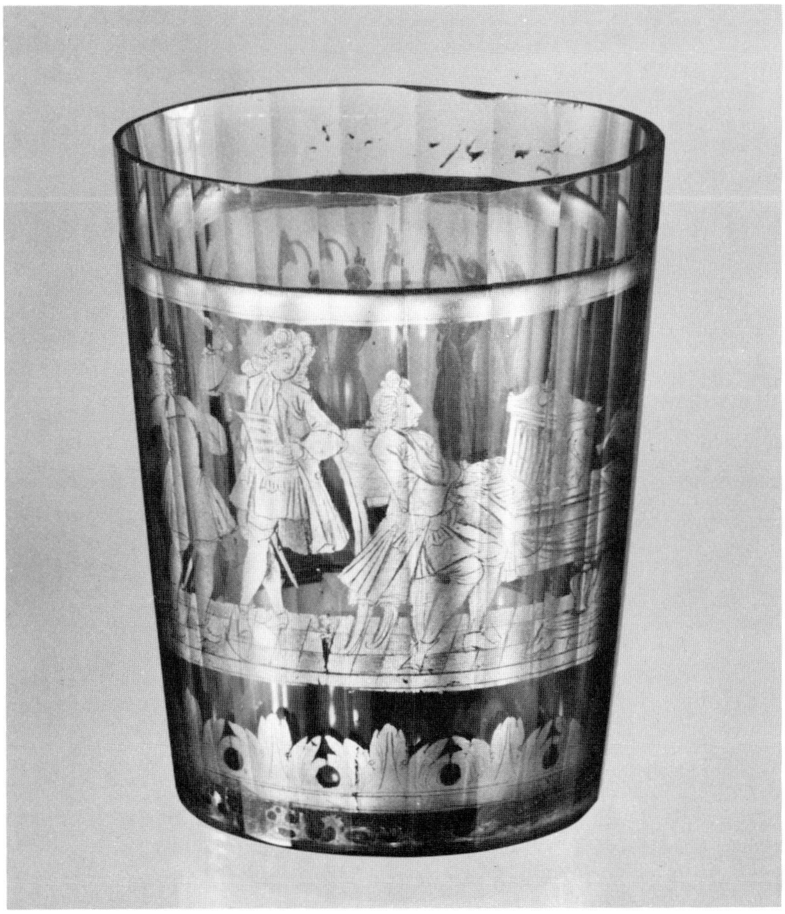

68. Beaker, with gold decoration

Bohemian; probably 19th or early 20th century

Clear colourless glass, facetted on the wheel, with double wall enclosing etched gold leaf (*Zwischengoldglas*)

Height, $3\frac{1}{2}$ in. (9 cm); diameter at rim, $2\frac{3}{4}$ in. (7 cm)

Straight-sided, slightly tapering, tumbler-form with flat vertical wheel-ground facets overall, the top of the inner glass projecting over the top of the outer glass and lying in the same plane. Below the foot of the inner glass a disc of glass coloured pink fits into the aperture framed by the outer glass, thus protecting a layer of etched gold leaf.

Round the wall of the beaker runs a music-making scene, with a man in a peruke seated at a chamber-organ, behind which two men in tricorn hats sit playing recorders: behind them but

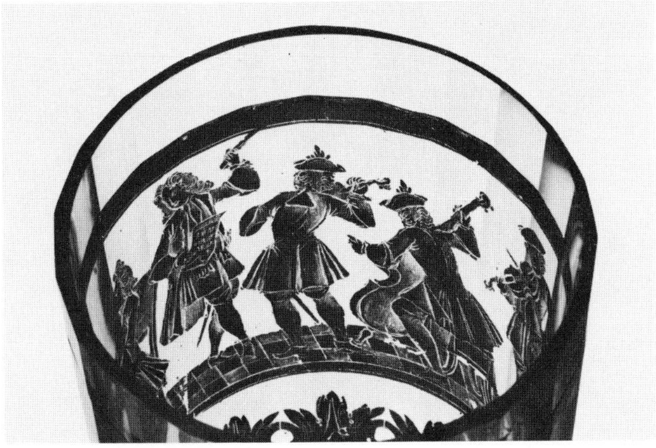

Interior view of 68, actual size

with their backs turned to them, stand two horn-players forming an ensemble with two violinists and a cellist with his back to the spectator; between them and the chair of the organist stands the conductor, a thick baton in his right hand, a sheet of music in his left. Round the foot of the beaker runs a border of upright acanthus-leaves, whilst at the point of junction of the walls of the beaker there is a plain band. The design of the foot, visible to one looking into the beaker and showing gold on pink, is of a hound pouncing on a hare in a landscape with trees and bushes (see also *70* here).

This beaker, which is condemned by the coarseness of the gold-leaf decoration and its spacing between the borders, was no doubt copied either from *67* here, or from a beaker exactly like it (e.g. that at Reichenberg – see observations on *67*). It is known that copies of these glasses have been made in relatively recent times (see Axel von Saldern, 'Originals, Reproductions, Fakes', *Annales du 5e. Congrès de l'Association Internationale pour l'Histoire du Verre*, Liège, 1972, p.316).

1922 Waddesdon Inventory (Smoking Room): 'Pair of ditto (sc. tumblers), musicians. German 17th century' (see *67*).

Miss Alice de Rothschild — W1/140/4*

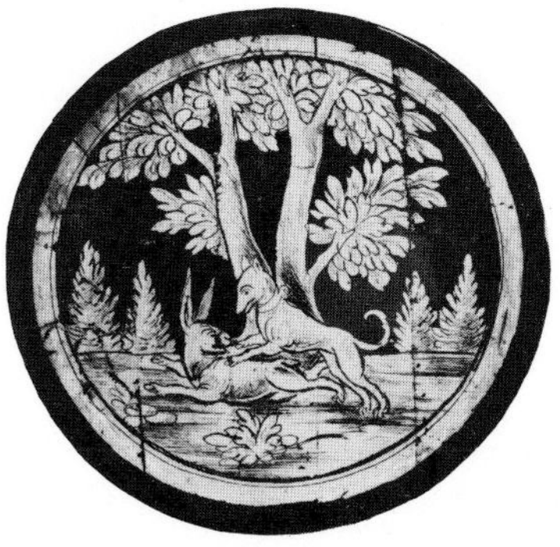

The base of 68, actual size

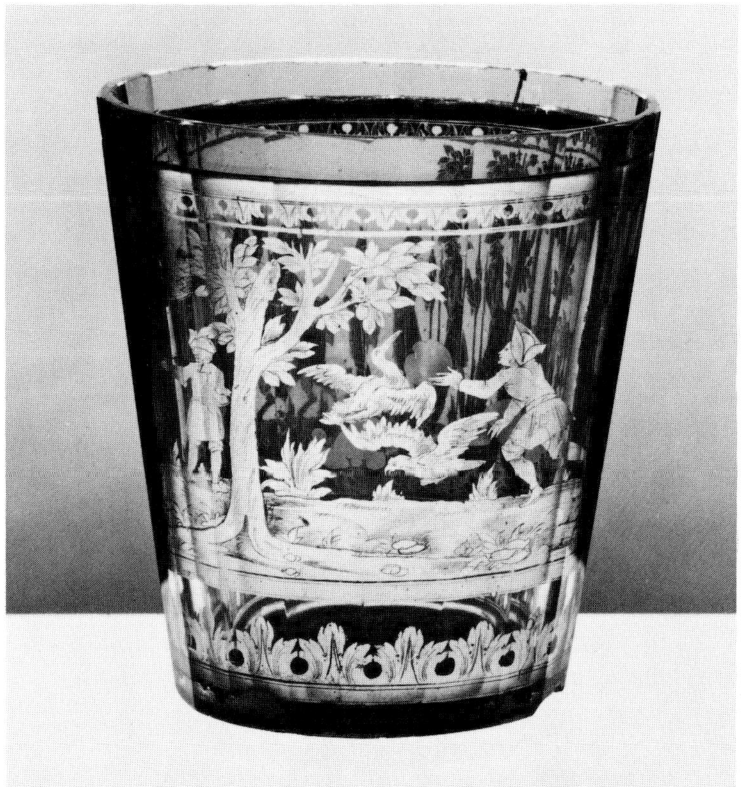

69. Beaker, with gold decoration

Bohemian; about 1730–40

Clear colourless glass, facetted on the wheel, with double wall enclosing etched gold leaf (*Zwischen-goldglas*).

Height, $3\frac{1}{4}$ in. (8·2 cm); diameter at rim, $2\frac{7}{8}$ in. (7·3 cm)

Straight-sided, slightly tapering, tumbler-form with flat vertical wheel-ground facets overall, the top of the inner glass projecting over the top of the outer glass and lying in the same plane. Below the foot of the inner glass a disc of glass coloured pink fits into the aperture framed by the outer glass, thus protecting a layer of etched gold leaf.

Round the wall of the beaker runs a fowling-scene, with three horsemen of whom one carries a falcon on his right wrist, another points forward with his switch, and a third urges his horse forward. Two huntsmen stand with their fowling-pieces, one leaning against a tree, while a second pair watch a falcon striking at a (?) heron; a fifth huntsman runs forward to where a

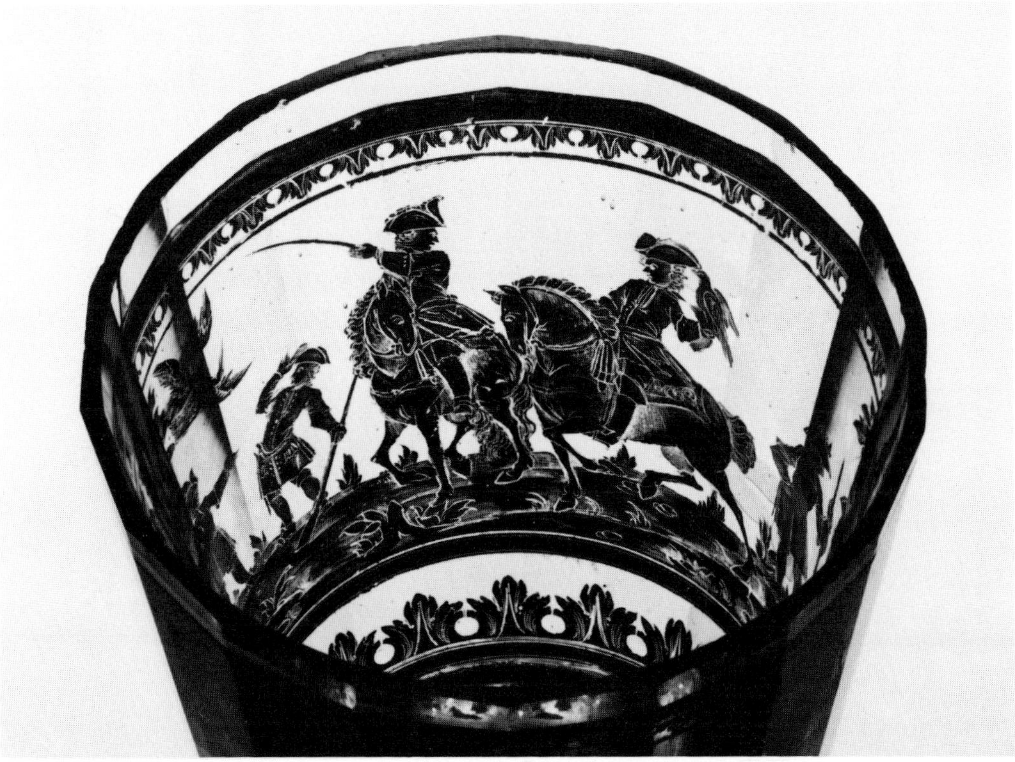

Two interior views of 69, enlarged one and a half times

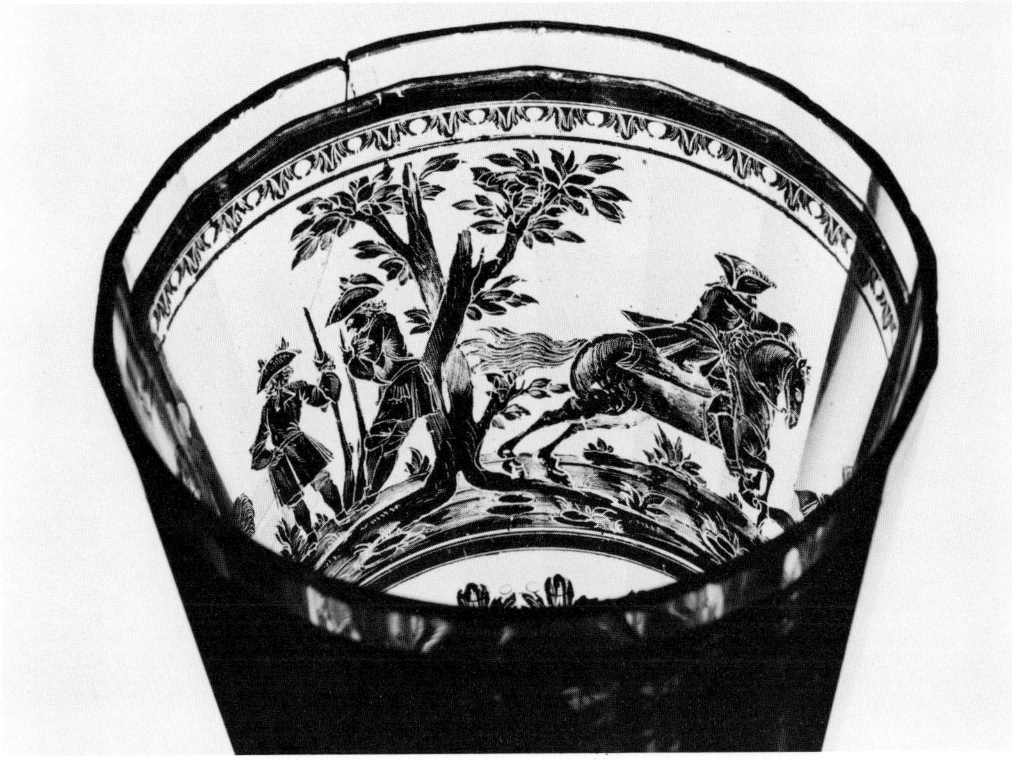

second falcon is engaged with another heron. Round the foot of the glass runs a border of large upright acanthus-leaves, whilst at the point of junction of the walls of the beaker runs another border of smaller pendent acanthus-leaves.

The design of the foot, visible to one looking into the beaker, and showing gold on pink, is of a huntsman with two hounds pursuing a stag, a tree in the background.

For a goblet with a closely comparable rendering of the galloping horseman, see R. Schmidt, *Das Glas*, Fig.221b.

1922 Waddesdon Inventory (Smoking Room): 'Five tumblers, sporting subjects in gold. German, 17th century' (see *70–73*).

Miss Alice de Rothschild W1/140/3*

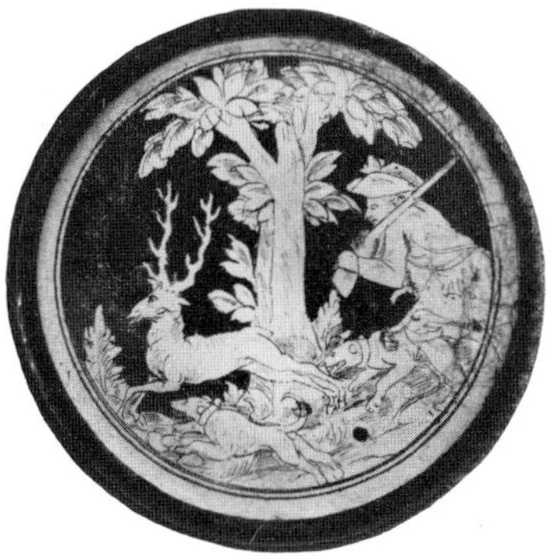

The base of 69, actual size

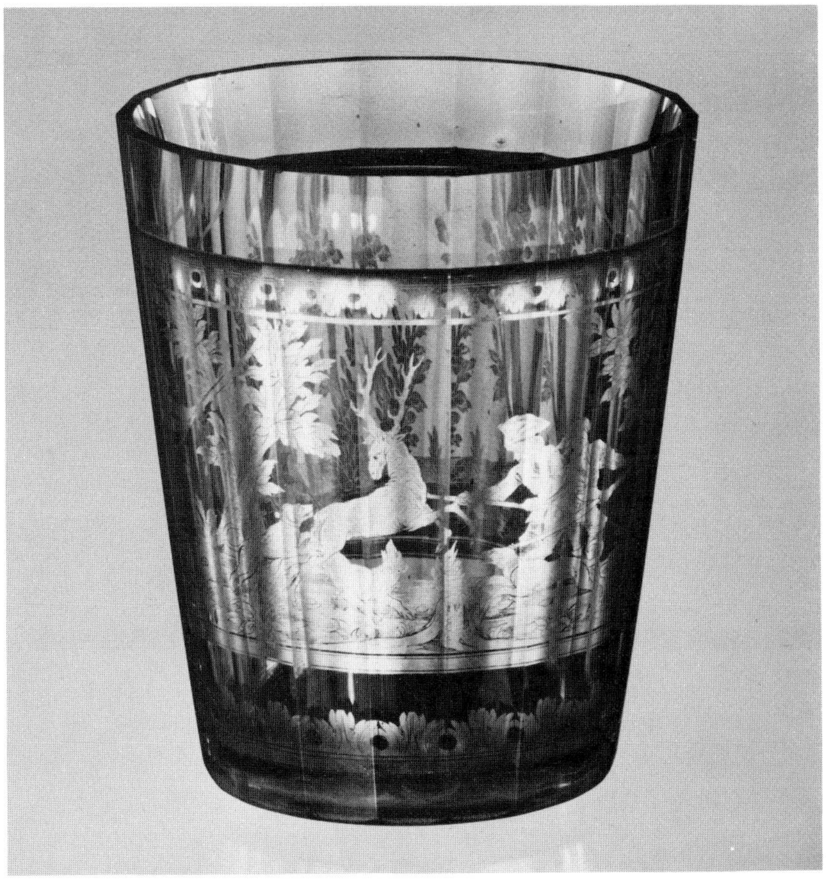

70. Beaker, with gold decoration

Bohemian; about 1730–40

Clear colourless glass, facetted on the wheel, with double wall enclosing etched gold leaf (*Zwischen-goldglas*)

Height, $3\frac{1}{2}$ in. (9 cm); diameter at rim, $2\frac{7}{8}$ in. (7·5 cm)

Straight-sided, slightly tapering, tumbler-form with flat vertical wheel-ground facets overall, the top of the inner glass projecting over the top of the outer glass and lying in the same plane. Below the foot of the inner glass a disc of glass coloured pink fits into the aperture framed by the outer glass, thus protecting a layer of etched gold leaf.

Round the wall of the beaker runs a hunting-scene, with two huntsmen and their hounds attacking a stag at bay, while a third huntsman levels his gun at it, two rustics holding staves in the background. Behind him another huntsman eviscerates a second stag, while a boy brings a bundle of twigs. On the other side of the scene, a huntsman stands with his gun grounded, two hounds beside him. Round the foot of the beaker runs a border of upright acanthus-leaves whilst at the point of junction of the walls of the beaker runs another border of smaller pendant acanthus-leaves.

The design of the foot, visible to one looking into the beaker, and showing gold on pink, is of a hound setting on a hare, two crossed trees in the background.

Glasses with very similar hunting-themes are to be found in the Mühsam Collection, one now in the Metropolitan Museum of Art, New York (Schmidt *Mühsam*, I, No.258, Pl.33: Vávra, *5000 Years of Glass-Making*, Prague, 1955, Pl.238c) and a second in the Art Institute of Chicago (Schmidt, *Mühsam*, I, No.255, Pl.31). Another is in the Victoria and Albert Museum, London (C.128–1934, *ill.* Honey, *Glass*, Pl.48a). The theme of the base-disc is repeated in *68* here.

1922 Waddesdon Inventory (Smoking Room): 'Five tumblers, sporting subjects in gold. German, 17th century' (see *69*, *71–3*).

Miss Alice de Rothschild — w1/140/3*

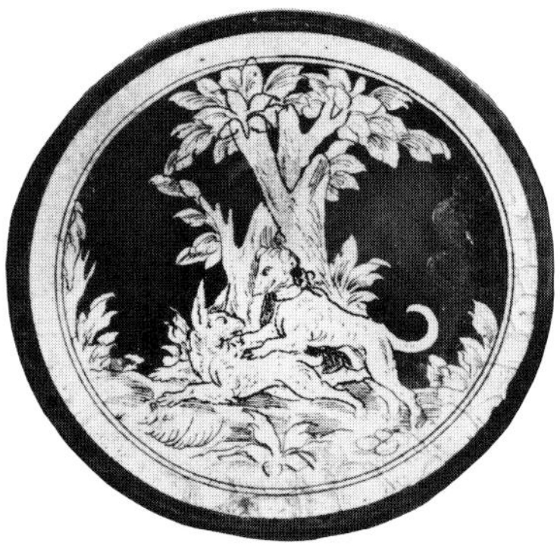

The base of 70 actual size

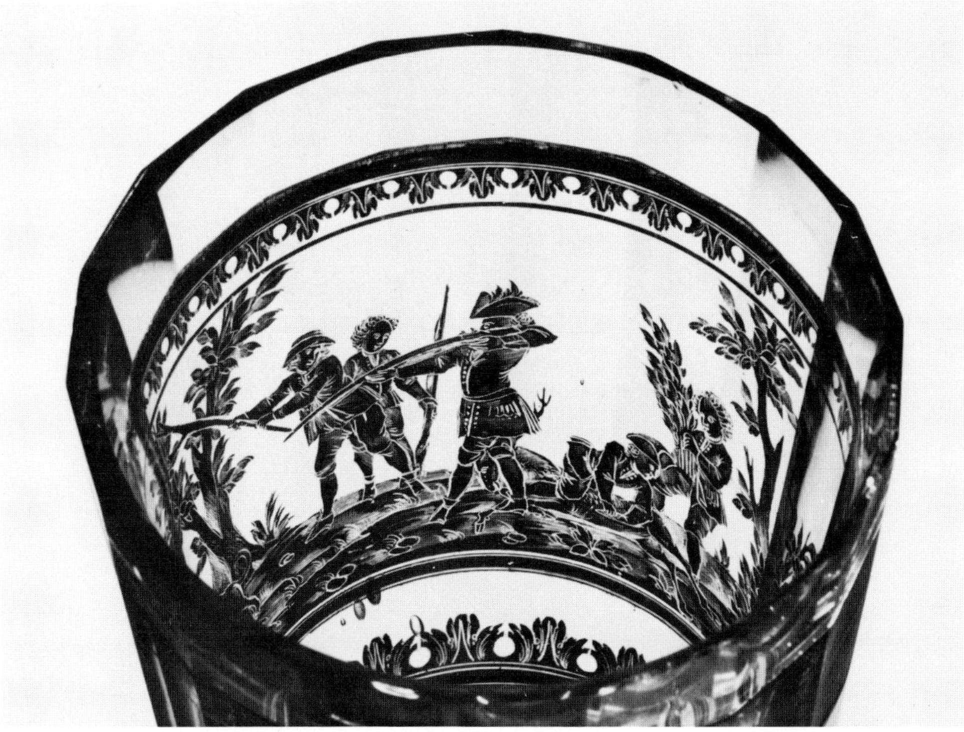

Interior views of 70, enlarged one and a half times

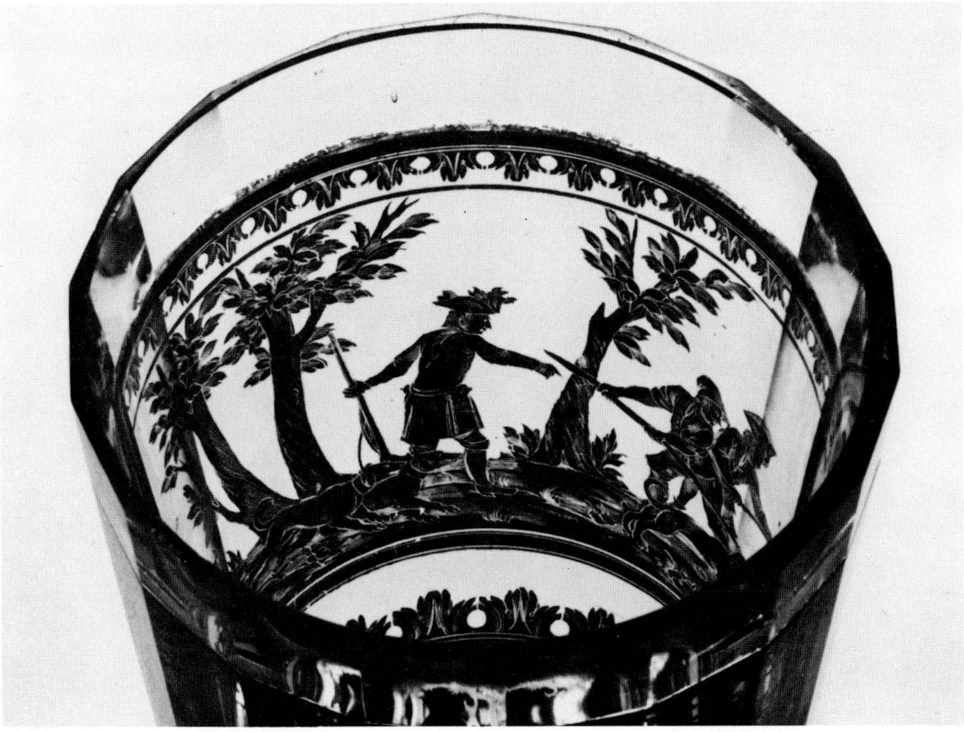

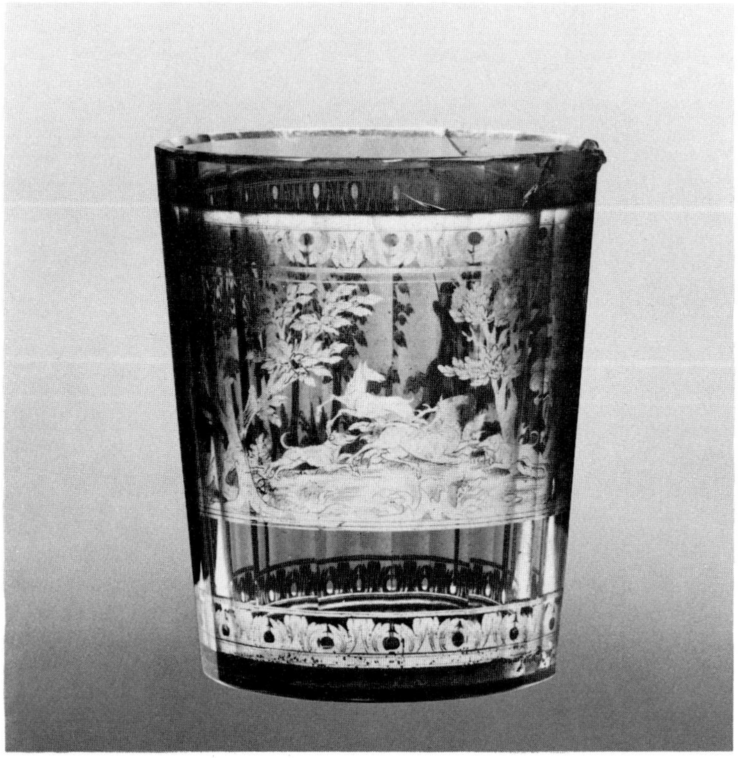

71. Beaker, with gold decoration

Bohemian; about 1730–40

Clear colourless glass, facetted on the wheel, with double wall enclosing etched gold leaf (*Zwischen-goldglas*)

Height, $2\frac{7}{8}$ in. (7·3 cm); diameter at rim, $2\frac{3}{8}$ in. (6 cm)

Straight-sided, almost cylindrical, tumbler-form with flat vertical wheel-ground facets overall, the top of the inner glass projecting over the top of the outer glass and lying in the same plane. Below the foot of the inner glass a disc of glass coloured pink fits into the aperture framed by the outer glass, thus protecting a layer of etched gold leaf.

Round the wall of the beaker runs a hunting-scene, with a stag and hind beset by two hounds, whilst to the right a huntsman fires at them with a gun. Behind him and separated from him by a tree is a mounted huntsman accompanied by a hound. At the other extremity of the scene is a huntsman on a prancing horse. Round the foot of the beaker runs a border of upright acanthus-leaves, whilst at the point of junction of the walls of the beaker runs another border of pendent acanthus-leaves.

The design of the foot, visible to one looking into the beaker, and showing gold on pink, is of a hound pursuing a stag, with two trees in the background.

1922 Waddesdon Inventory (Smoking Room): 'Five tumblers, sporting subjects in gold. German, 17th century'. (see *69–70*, *72–3*).

Miss Alice de Rothschild WI/140/6*

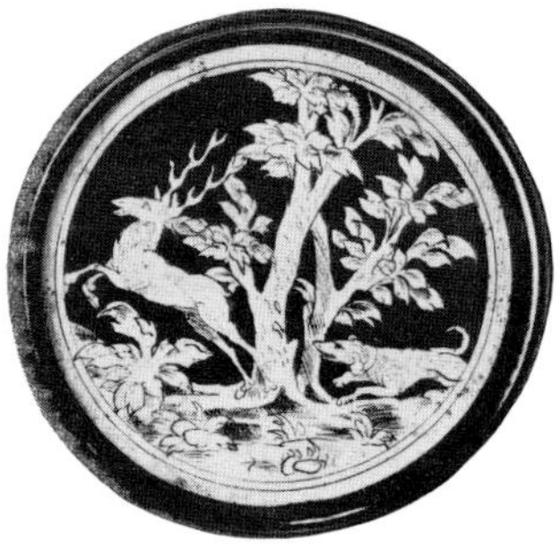

The base of 71, actual size

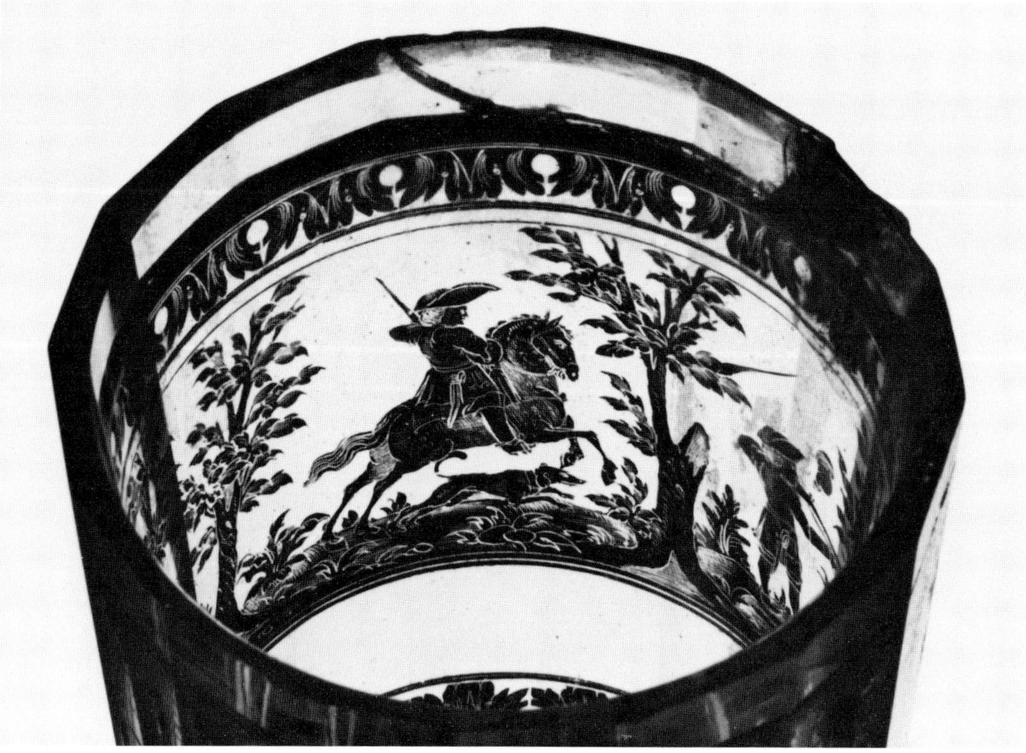

Two interior views of 71, enlarged to twice actual size

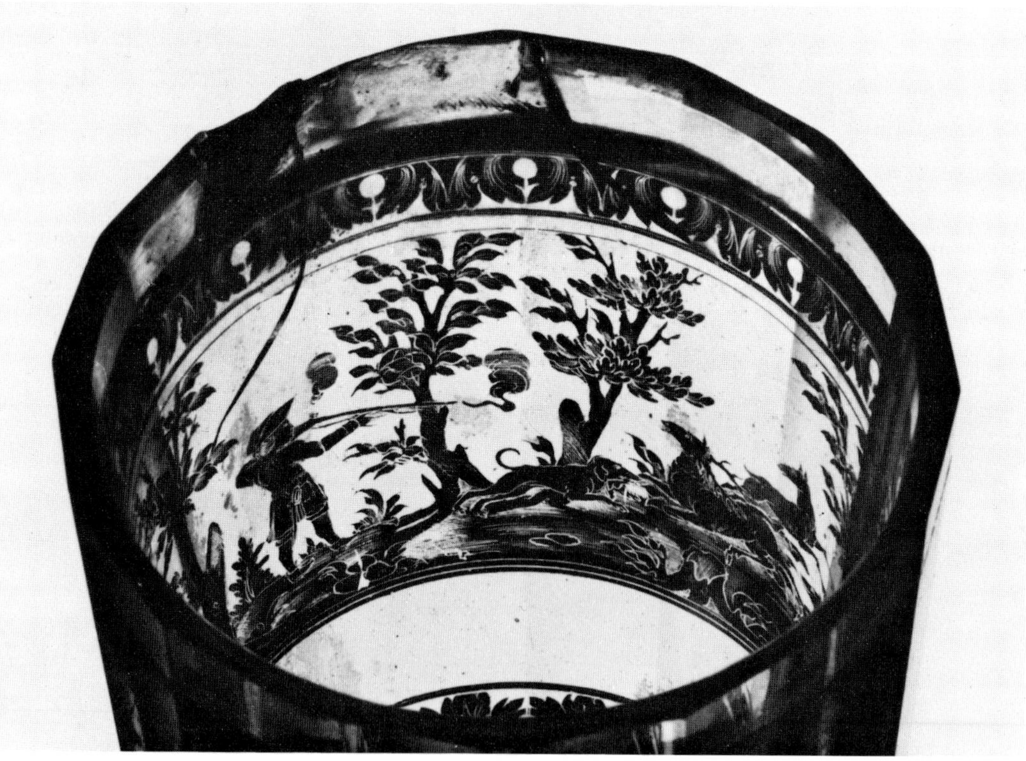

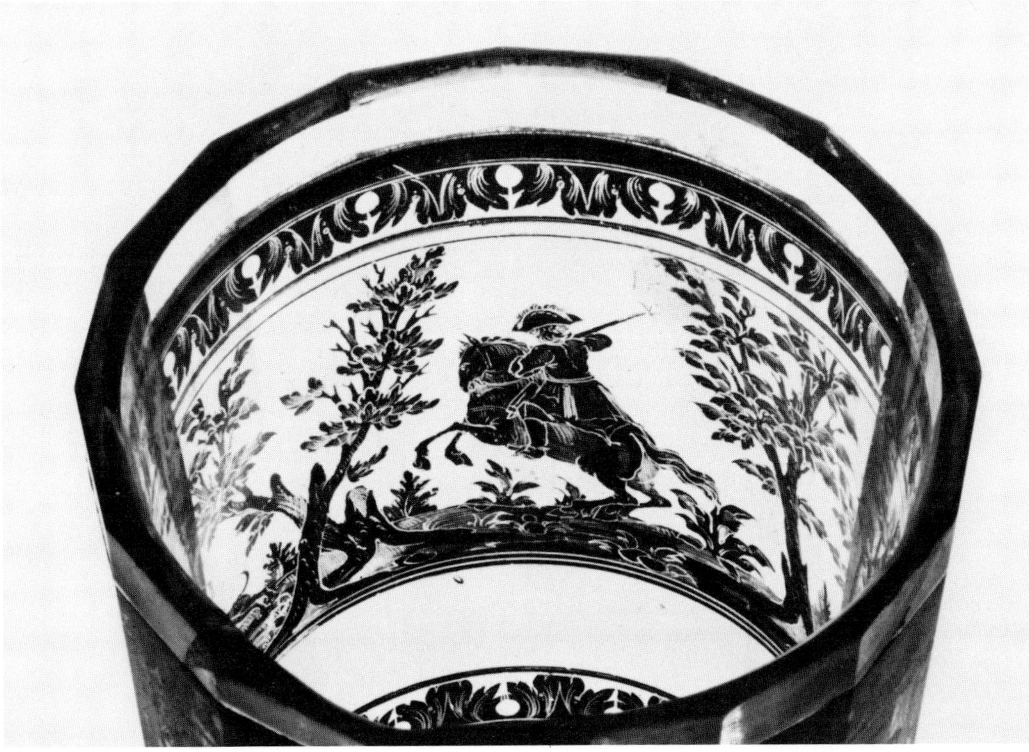

Another interior view of 71, enlarged to twice actual size

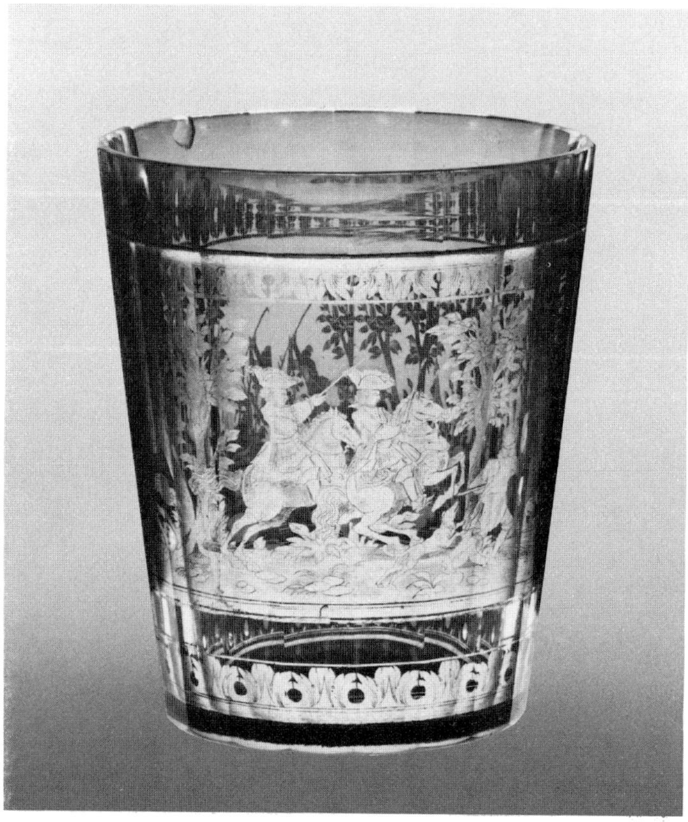

72. Beaker, with gold decoration

Bohemian; about 1730–40

Clear colourless glass, facetted on the wheel, with double wall enclosing etched gold leaf (*Zwischen-goldglas*)

Height, $3\frac{1}{4}$ in. (8·2 cm); diameter at rim, $2\frac{5}{8}$ in. (6·5 cm)

Straight-sided, slightly tapering, tumbler-form with flat vertical wheel-ground facets overall, the top of the inner glass projecting over the top of the outer glass and lying in the same plane. Below the foot of the inner glass a disc of glass coloured pink fits into the aperture framed by the outer glass, thus protecting a layer of etched gold leaf.

Round the wall of the beaker runs a hunting-scene, with a stag beset by two hounds and two huntsmen with spears, while to the left two horsemen bestride prancing horses, a hound beside them. Beyond them is a second scene with a stag brought down by two hounds, while to the left two rustics, one with a stave, and to the right a horseman, converge to attack it. Round the foot of the beaker runs a border of upright acanthus-leaves, whilst at the point of junction of the walls of the beaker runs another border of smaller pendent acanthus-leaves. The design of the foot, visible to one looking into the beaker, and showing gold on pink, is of the sacred monogram IHS, the cross above and the three nails of the Crucifixion below, within a flaming circle; outside this is a ribbon bearing the legend SIT NOMEN DOMINI BENEDICTUM ('The Lord's name be praised').

The base-disc of this glass rather surprisingly introduces a sacred note into an otherwise secular theme. The inscription, and the parallel inscription 'Gelobet sei Jesus Christus' ('Jesus Christ be praised'), occur on a number of glasses (R. Schmidt, *Das Glas*, p.366). A beaker once in the Jantzen Collection, Bremen, and now in the Kunstmuseum Düsseldorf, has the inscription 'Sit Nomen Domini Benedictum' and, more appropriately, a representation of St Ignatius Loyola on the wall of the vessel: a companion glass shows St Francis (J. Jantzen, *Deutsches Glas aus fünf Jahrhunderten . . .*, Düsseldorf, 1960, Nos.116–17, Pl.58). These religious themes lend weight to the supposition that these glasses may have been executed in a monastery or monasteries (see p.41 above).

1922 Waddesdon Inventory (Smoking Room): 'Five tumblers, sporting subjects in gold. German, 17th century' (see *69–71, 73*).

Miss Alice de Rothschild — W1/140/3*

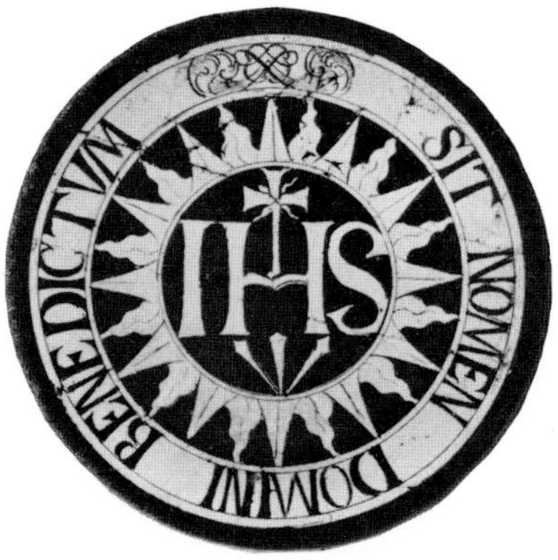

The base of 72, actual size

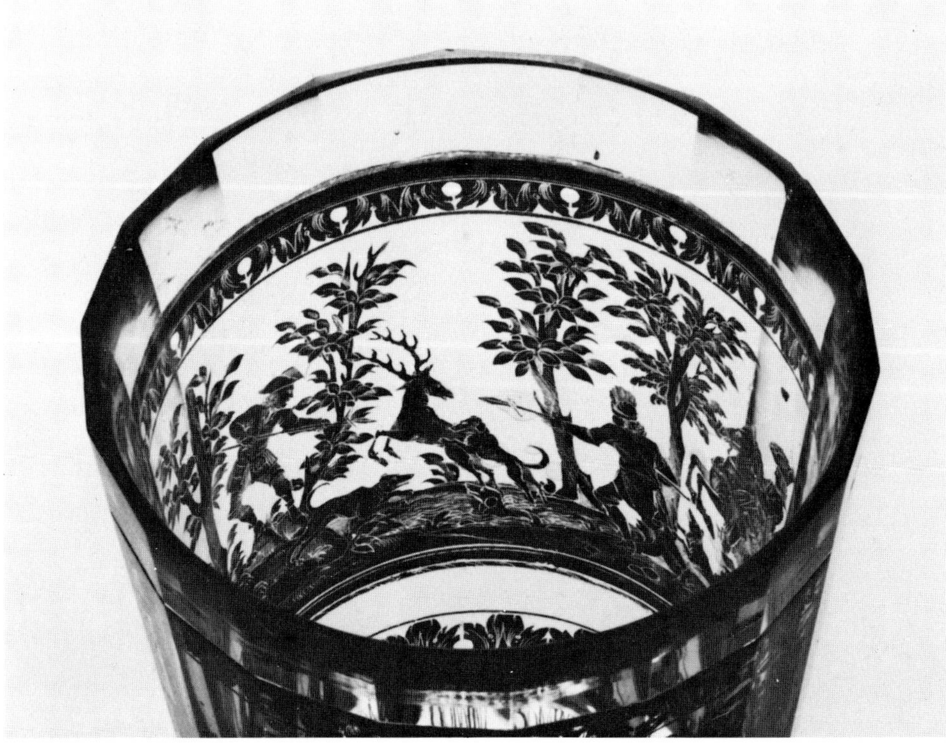

Two interior views of 72, enlarged one and a half times

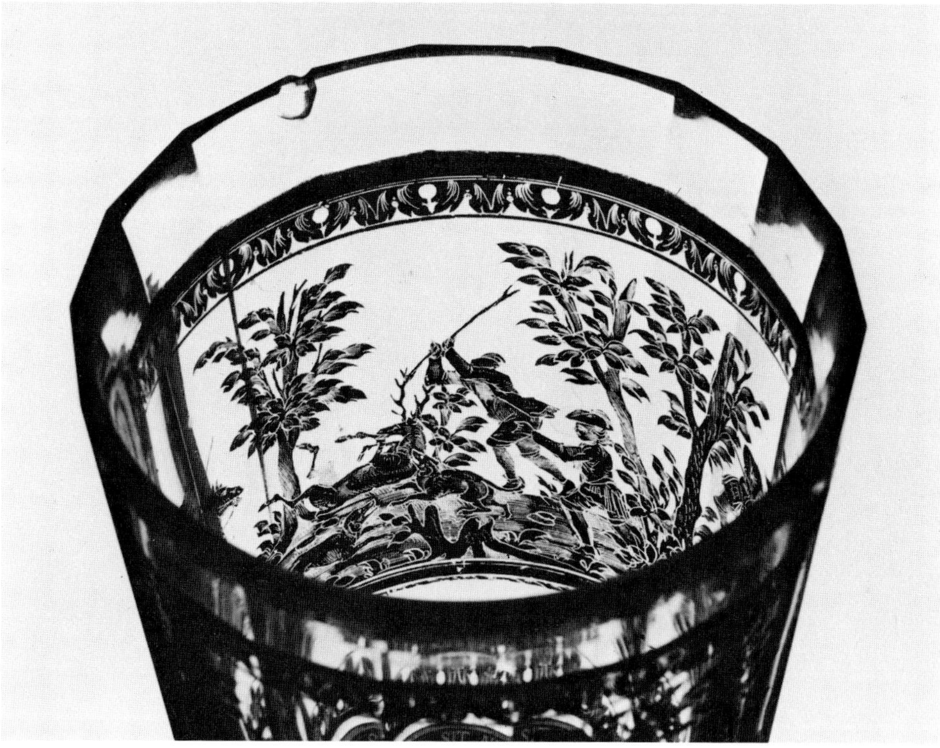

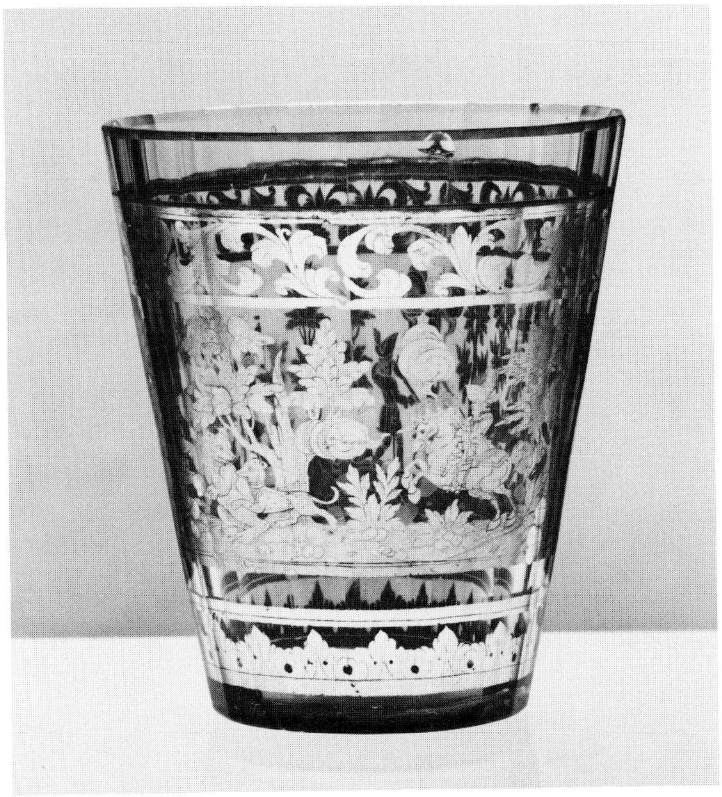

73. Beaker, with gold and silver decoration

Bohemian; about 1730–40

Clear colourless glass, facetted on the wheel, with double wall enclosing etched gold and silver leaf (*Zwischengoldglas*)

Height, $3\frac{1}{4}$ in. (8·2 cm); diameter at rim, $2\frac{3}{4}$ in. (7 cm)

Straight-sided, slightly tapering, tumbler-form with flat vertical wheel-ground facets overall, the top of the inner glass projecting over the top of the outer glass and lying in the same plane. Below the foot of the inner glass a disc of glass coloured pink fits into the aperture framed by the outer glass, thus protecting a layer of etched gold leaf.

Round the wall of the beaker runs a hunting-scene in gold, with a bear beset by two hounds, amidst trees, while a horseman discharges his pistol at the bear: to the left another huntsman stands between two trees with his gun grounded. In a second scene a huntsman, accompanied by a hound, discharges his gun at a hare. Round the foot of the beaker runs a border of upright acanthus-leaves in silver, whilst at the point of junction of the walls of the beaker runs a border of scrolling leaves, also in silver.

The design of the foot, visible to one looking into the beaker, and showing gold on pink, is of two flaming hearts on an altar, surmounted by an open coronet and flanked by scrolling leafage.

The relatively unusual theme of a bear-hunt is repeated in two closely similar beakers, one in the Kestner Museum, Hanover (*Glas-Sammlung*, No.257, Fig.70; also Savage, *Glass*, Fig.70), the other from the F. Neuburg Collection, Leitmaritz (Lempertz-Auktion, *Katalog, 447*, 15 May, 1957, Lot 631, Pl.26).

1922 Waddesdon Inventory (Smoking Room): 'Five tumblers, sporting subjects in gold. German, 17th century' (see *69–72*).

Miss Alice de Rothschild　　　　W1/140/3*

The base of 73, enlarged one and a half times

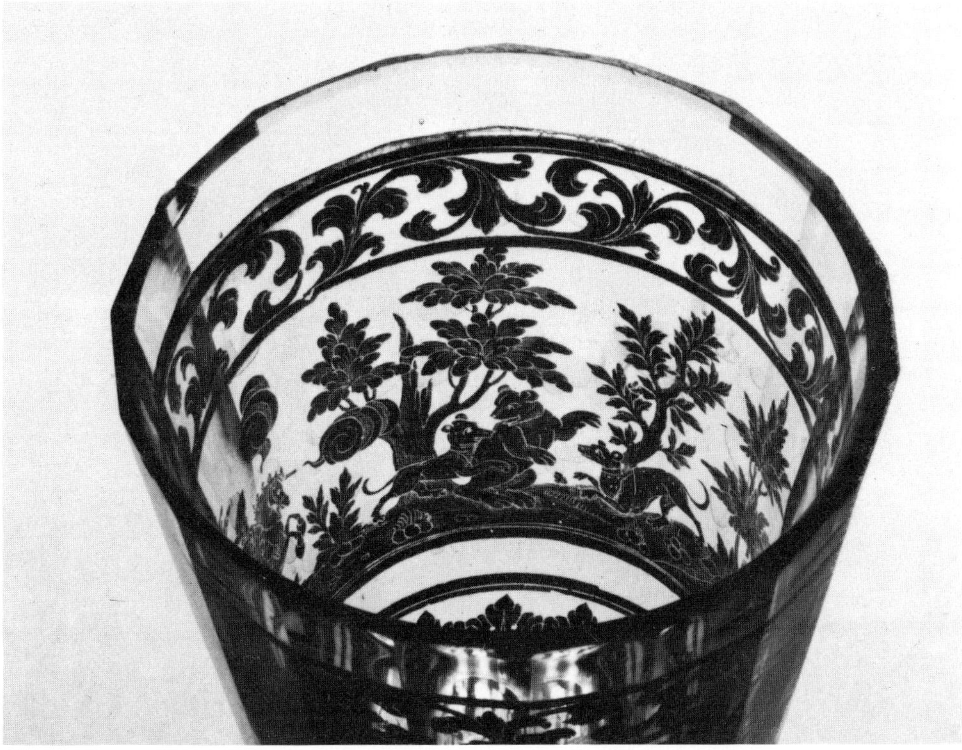

Two interior views of 73, enlarged one and a half times

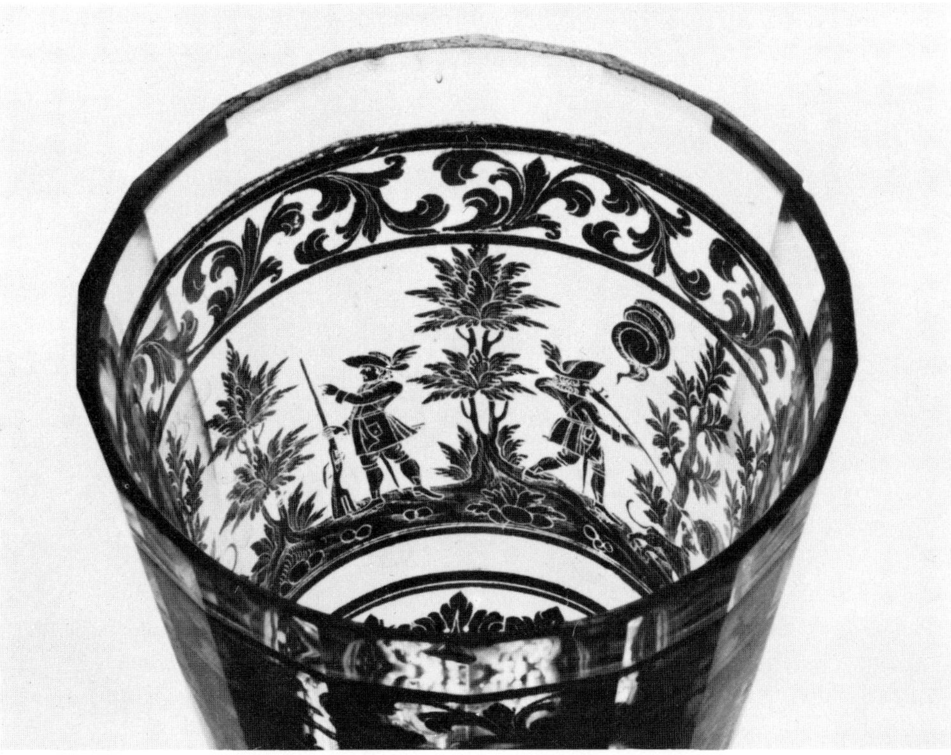

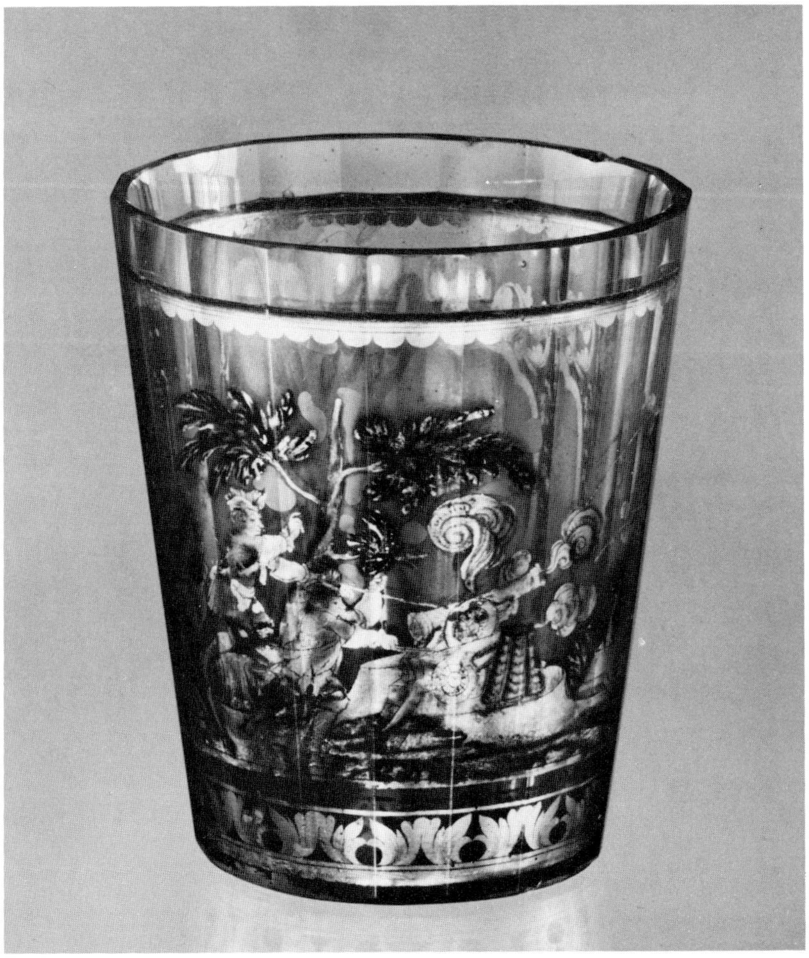

74. Beaker, with decoration in gold, silver and colours

Bohemian; about 1730–40

Clear colourless glass, facetted on the wheel, embodying etched gold and silver leaf, supplemented with colours (*Zwischengoldglas*)

Height, $3\frac{1}{2}$ in. (9 cm); diameter at rim, $2\frac{7}{8}$ in. (7·3 cm)

Straight-sided, slightly tapering, tumbler-form with flat vertical wheel-ground facets overall, the top of the inner glass projecting over the top of the outer glass and lying in the same plane. Below the foot of the inner glass a disc of glass coloured pink fits into the aperture framed by the outer glass, thus protecting a layer of etched gold leaf.

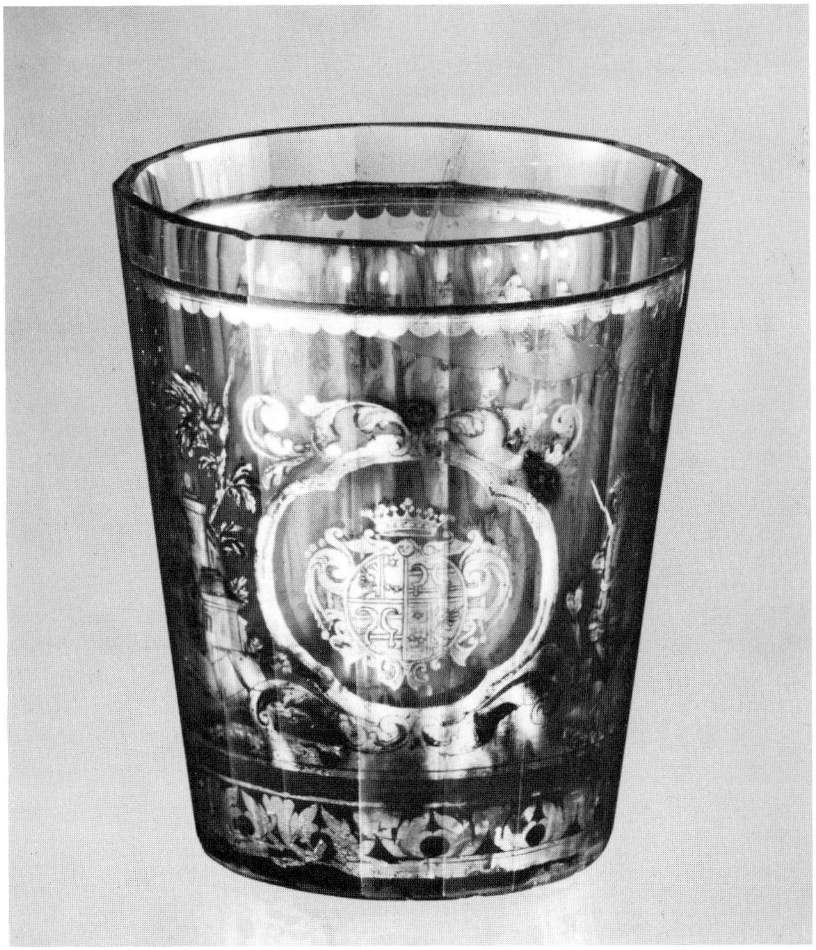

Round the wall of the beaker runs a battle-scene, with two soldiers touching off cannon firing into a fortified town, while behind them are two horsemen, one carrying a baton, the other a flag. This scene is rendered in silver-leaf with use of mainly red, green and brown colours overlying it. At the back of the glass is a coat-of-arms in gold, surmounted by an open coronet, within a scrolled panel. Arms quarterly: 1, per pale or a demi-eagle per pale displayed sable issuant from the sinister side, azure three mullets argent 1 and 2: 2 and 3, gules two bars of the fourth and over all two sickles addorsed proper: 4, as 1 reversed. The arms are those of the Von Bosch family (J. B. Rietstap, *Armorial Général*, I, Paris, 1903, Pl. CCLXXIII).

Round the foot of the beaker, a border of upright acanthus-leaves in gold, whilst at the point of junction of the walls of the beaker runs a dentil border, also in gold. The design of the foot, visible to one looking into the beaker, and showing gold on pink, is of a trophy of arms in a roundel.

A somewhat similar beaker was in the Mühsam Collection, now in the Art Institute of Chicago (Schmidt, *Mühsam* I, No.261, Pl.32).

1922 Waddesdon Inventory (Smoking Room): 'One ditto (sc. tumbler), battle scene and coats of arms, slightly coloured'.

Miss Alice de Rothschild WI/140/5

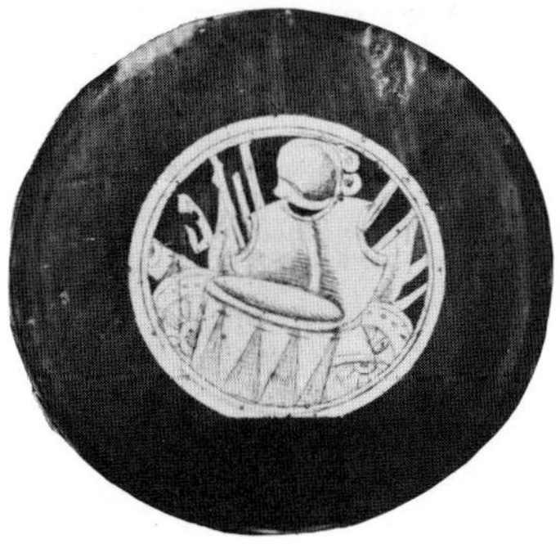

The base of 74, actual size

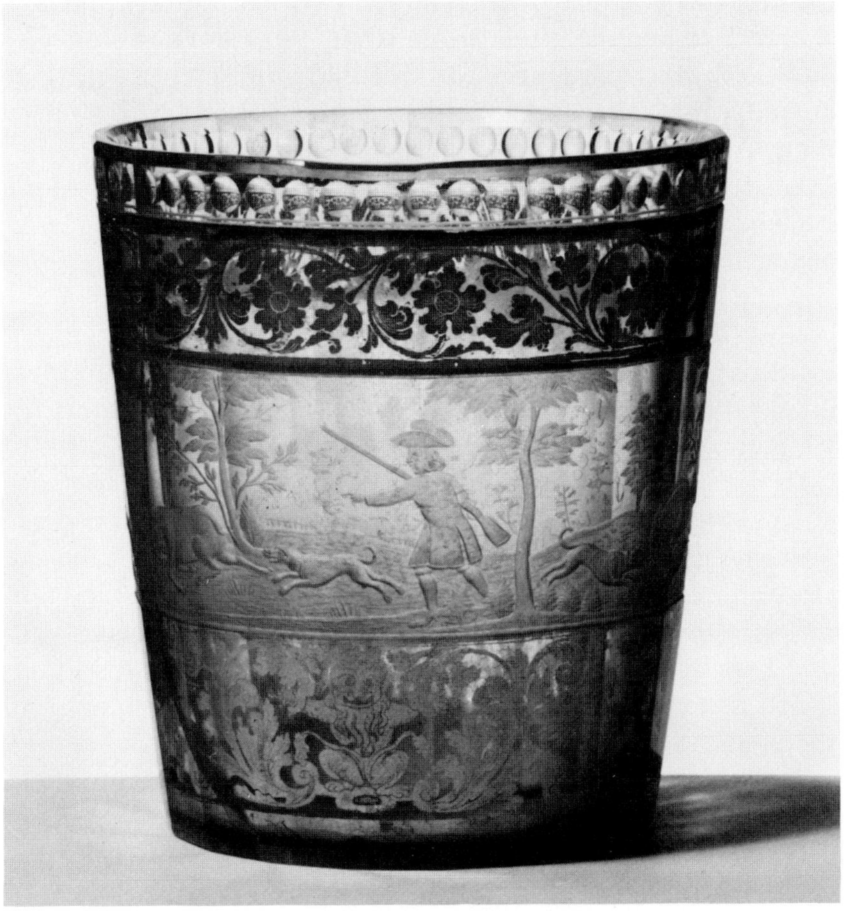

75. Beaker, wheel-engraved and with gold and silver decoration

Bohemian; about 1730

Clear colourless glass facetted and engraved on the wheel, with double wall enclosing etched gold and silver leaf (*Zwischengoldglas*)

Height, $3\frac{3}{4}$ in. (9·5 cm); diameter at rim, $3\frac{1}{4}$ in. (8·3 cm)

Straight-sided, slightly tapering, tumbler-form with flat vertical wheel-ground facets overall, the top of the inner glass projecting over the top of the outer glass and lying in the same plane. Below the foot of the inner glass a disc of glass partially stained pink fits into the aperture framed by the outer glass, thus protecting a layer of etched gold leaf.

Round the wall of the beaker runs a wheel-engraved hunting-scene of a bear beset by two hounds, while a huntsman to the right takes aim at it with his gun. Behind him two hounds, urged on by a huntsman with a gun, attack a fallen stag; behind the huntsman again is a third scene of a boar set on by two hounds, while a huntsman takes aim at him.

Round the foot of the beaker runs a broad border in gold, etched with a design of a frontal bird with outstretched wings alternating with a lion displayed with outstretched forepaws, these creatures being separated from each other by leafy scrolls. At the point of junction of the walls of the beaker runs another border in gold of a running scroll of leaves and flowers.

The design of the foot, visible to one looking into the beaker, and showing gold on a colourless ground with a border of pink, is of three hounds attacking a fallen stag, rendered in a lobed panel. From the outside may be seen a laurel-wreath in etched silver-leaf. Immediately below the rim of the beaker is a line of polished wheel-cut 'printies'.

The combination of wheel-engraving with the *Zwischengold* technique is rare and argues for the collaboration of a professional engraver with the presumed 'amateur' who etched the gold- and silver-foil. The Kunstgewerbemuseum in Cologne possesses a beaker of which the body is wheel-engraved with floral and scrollwork designs in typical Bohemian style, while the foot is decorated in the *Zwischengold* technique (Klesse, *Glas*, No.376). The base-disc has a stag in gold on a pink ground as a reminder of the more normal thematic range of the *Zwischengold* glasses. A beaker in the Brauser Collection, Munich, was decorated with medallions reserved in a ground of scrollwork in the *Zwischengold* technique, one medallion being wheel-engraved with an allegory of cupid knocking down a boundary-stone, with the motto 'Point de borne' ('No limit'); the other with a coat of-arms and initials. The base-disc was appropriately decorated with a symbolic theme of flowers growing from a crown, in the *Zwischengold* technique (Bernt, *Altes Glas*, p.62 and Pl.76a). A thick facetted beaker recorded by Robert Schmidt (*Das Glas*, pp.366–7) was decorated with wheel-engraved leafy scrollwork and swags of fruit, with a *Zwischengold* disc in the base etched with the Sacred monogram IHS and the inscription 'Benedictum sit Nomen Domini' (cf. *72* here). As an additional complexity, the lower part of the inner glass was cut conically so that a kaleidoscopic view was provided of the base-disc – a refinement perhaps comparable to the treatment of the disc of the Waddesdon glass, whereby it can be equally satisfactorily viewed from either side. A further glass combining wheel-engraving with the *Zwischengold* technique was in the Reichenberg Collection (Pazaurek, *Reichenberg*, p.24 and Pl.16d).

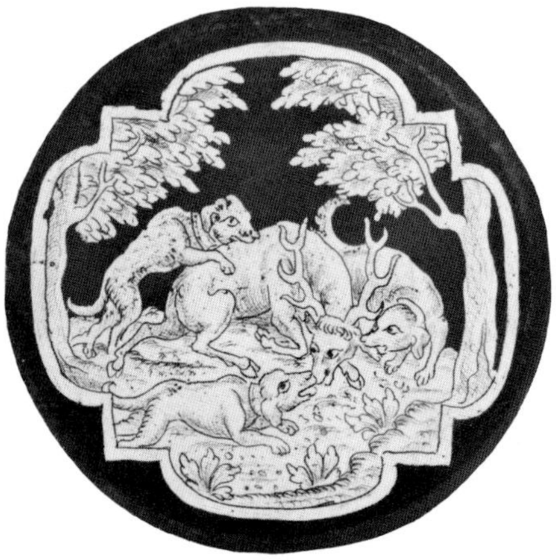

The base of 75, actual size

1922 Waddesdon Inventory (Smoking Room): 'One ditto (sc. tumbler), engraved sporting subjects'.

Miss Alice de Rothschild W1/140/6*

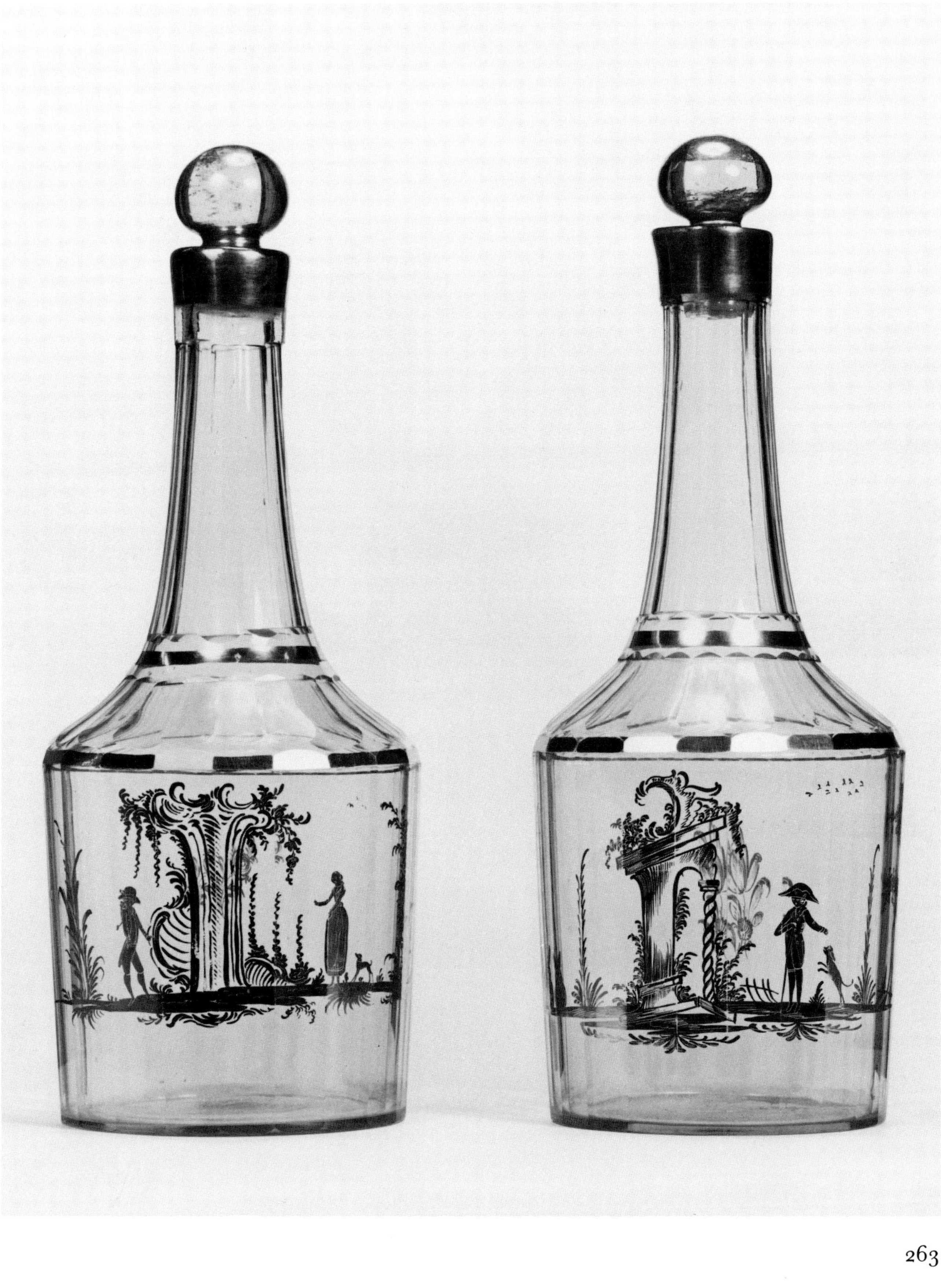

A decanter closely similar in shape, but with the neck cut in two zones of triangular facets separated by a plain gilt band, is in the Helfried Krug Collection, Mülheim (Klesse, *Krug*, I, p.239, No.253; also Lipp, *Bemalte Gläser*, Fig.182). The gilding on this decanter approximates more closely to that on the glasses 19–23, 28.

3 and 4. PAIR OF DECANTERS AND STOPPERS

Height with stopper, 10$\frac{7}{8}$ in. (27·6 cm) and 10$\frac{3}{4}$ in. (27·3 cm)
Height without stopper, 9$\frac{3}{4}$ in. (24·8 cm)

The decanters are straight-sided, almost cylindrical, with sloping shoulders and tall almost cylindrical necks, cut throughout with vertical facets, only the rim of the neck and the junctions of shoulder and neck and shoulder with body being left plain except for simple gilt bands. Both have gilt decoration of a man with a dog in a 'landscape' dominated by tall rococo architectural elements. The slightly depressed-globular stoppers are left uncut, but are painted over with solid gilding: they bear the numbers '12' and '33' inscribed underneath in gold, whereas the corresponding decanters have '44' and '20' inscribed inside the neck-rim.

See also observations on 1 and 2.

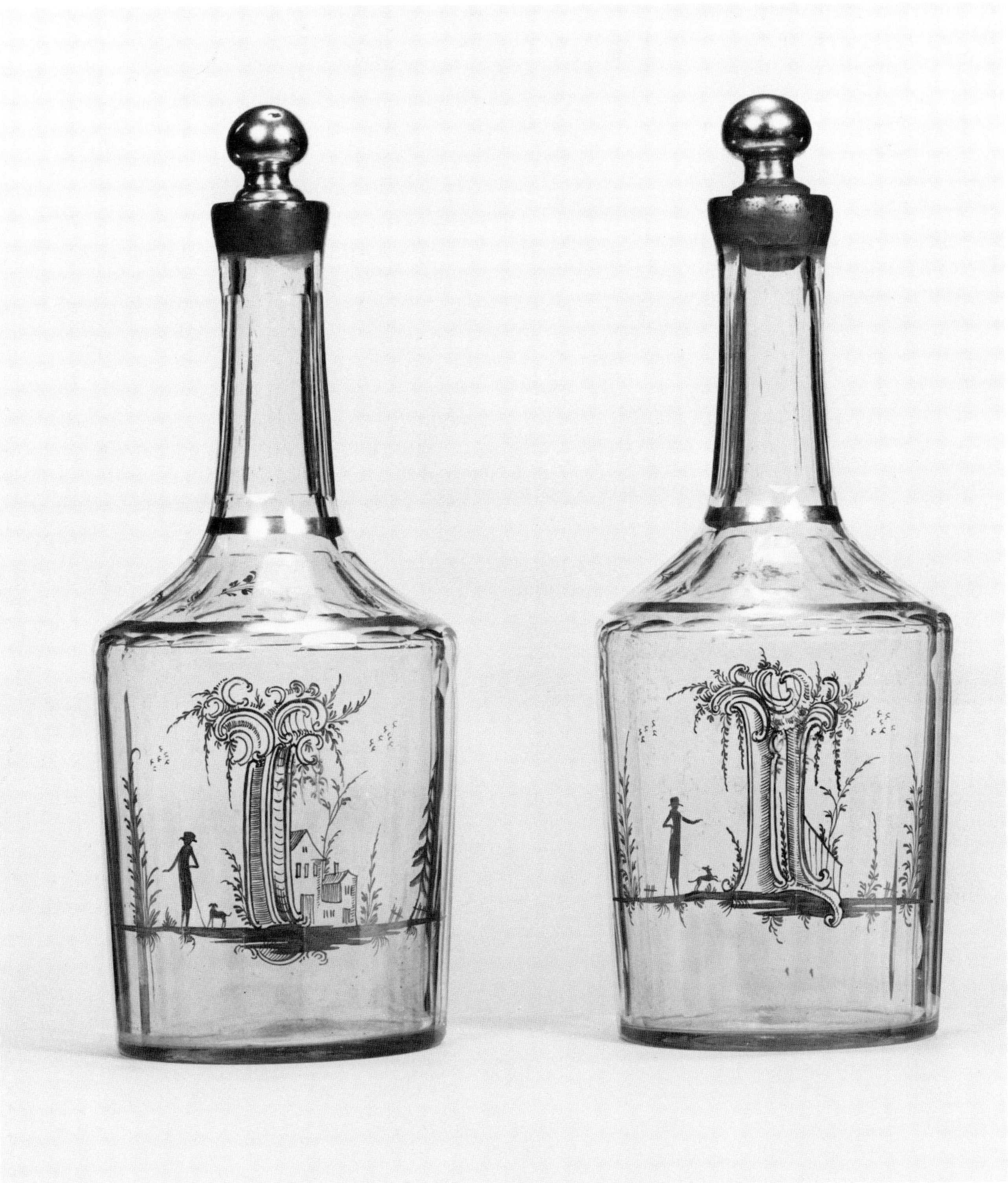

5 and 6. PAIR OF MINIATURE DECANTERS AND STOPPERS

Height with stopper, $9\frac{1}{2}$ in. (24 cm) and $9\frac{5}{8}$ in. (24·5 cm)
Height without stopper, $7\frac{5}{8}$ in. (19·5 cm) and $7\frac{3}{4}$ in. (19·7 cm)

The decanters are straight-sided, almost cylindrical, with sloping shoulders and straight slightly tapering necks, cut throughout with vertical facets, only the rim of the neck and the junction of neck and shoulder being left plain except for a simple gilt band. Both have gilt decoration of a human figure with a dog in a 'landscape' dominated by a rococo structure of a quasi-architectural character. The 'spire' stoppers are facetted and gilt with vertical ray motifs: one bears the number '20' inscribed underneath in gold, whereas the coresponding decanter bears the number '33' inside the neck-rim: the other (which clearly does not belong to its bottle) is un-numbered, whereas the decanter bears the number (?)'4'.

On the shapes of these decanters, probably intended for spirits or cordials, see the observations on 1–2. The spire-stopper seems to be a relatively late introduction in the Bohemian repertoire (see observations on 7–8), and the fact that the stoppers do not belong to these decanters suggest that they may have been 'married' at a later date. The tricorn hat of the man and the generally more rococo character of the decoration suggest a perhaps slightly earlier date than that of 1–4. Some of the characteristics of this decoration are shared by designs in the Harrachov pattern-book (Nos.182, 303, 322 – see below). The sheets on which these occur all fall before the pages which bear the dates 1787–8, and may well be earlier, as the shapes of the glasses themselves (Nos.303, 322) also suggest.

For a similar decanter with spire-stopper, decorated with a gilt shooting-scene, see Gasparetto, Pl.131, there by inference attributed to Venice (on which see p.43 here).

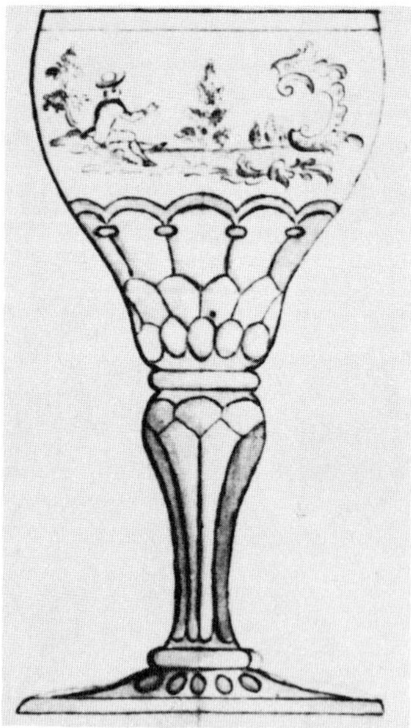

Page from the oldest extant Harrachov pattern-book: about 1785 (Pattern no. 303) (Muzeum skla a bižutérie, Jablonec nad Nisou). See p.42

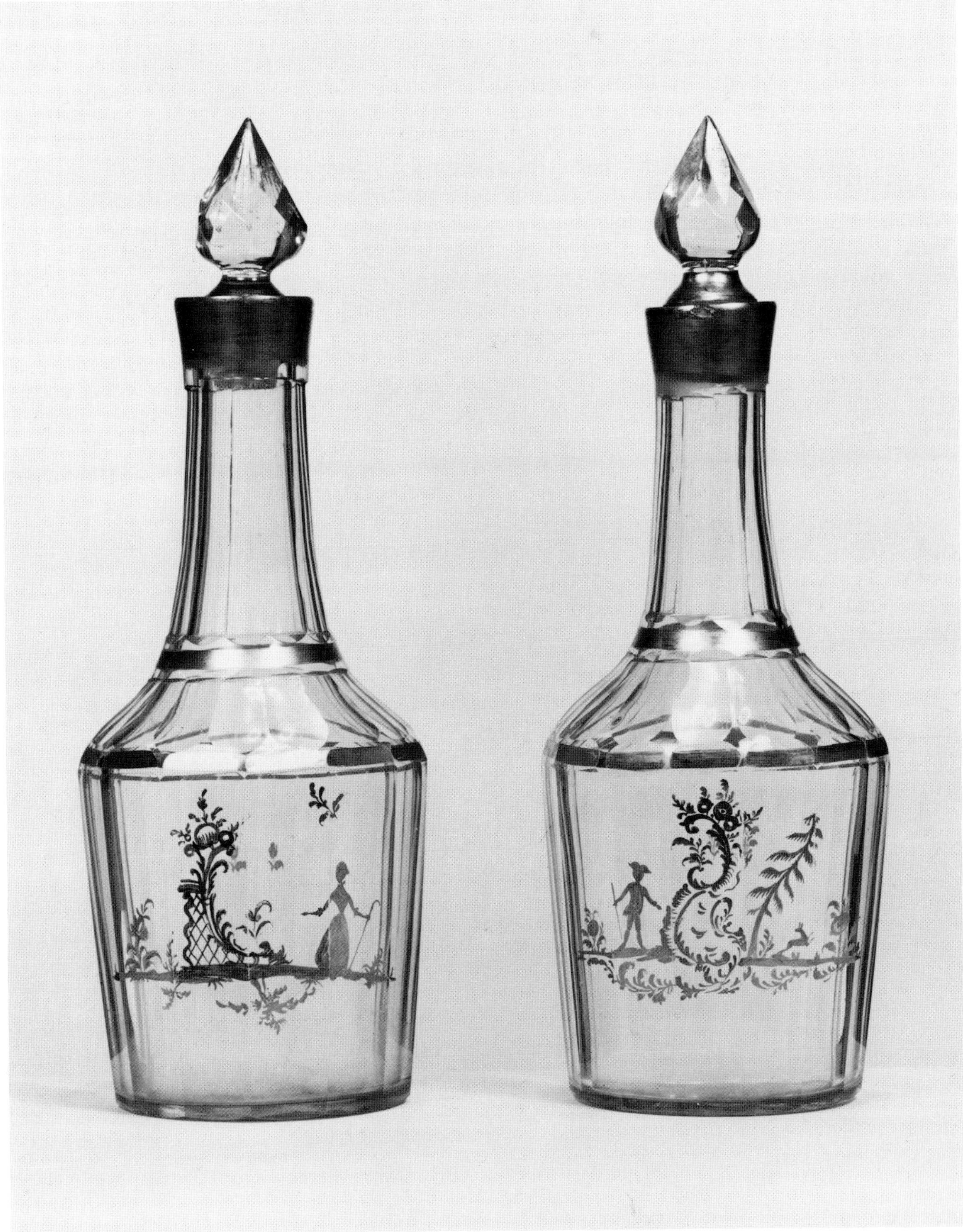

7 and 8. PAIR OF HANDLED MINIATURE DECANTERS, OR CRUET-BOTTLES, WITH STOPPERS

Height with stopper, $8\frac{7}{8}$ in. (20 cm) and $8\frac{3}{4}$ in. (22·3 cm)
Height without stopper, 7 in. (18 cm)

The bottles are straight-sided, almost cylindrical, with sloping shoulders and slightly tapering necks having small pouring-lips, cut throughout with vertical facets (except for the inaccessible areas under the handles), only the rim of the neck, and the junction of neck and shoulder being left plain except for a simple gilt band. Both have gilt decoration of swags of flowers and leaves hanging down from a horizontal chain and tied at the points of junction with bows terminating in pendent tasselled ribbons. The 'spire' stoppers, which clearly do not belong to the bottles, are facetted and their points gilt; they bear the numbers (?) '1' and (?)'8', but the bottles are unnumbered.

These bottles are probably to be regarded as cruets, since they have handles. A pair of such bottles (presumably for oil and vinegar respectively) are shown as part of a condiment-set in the

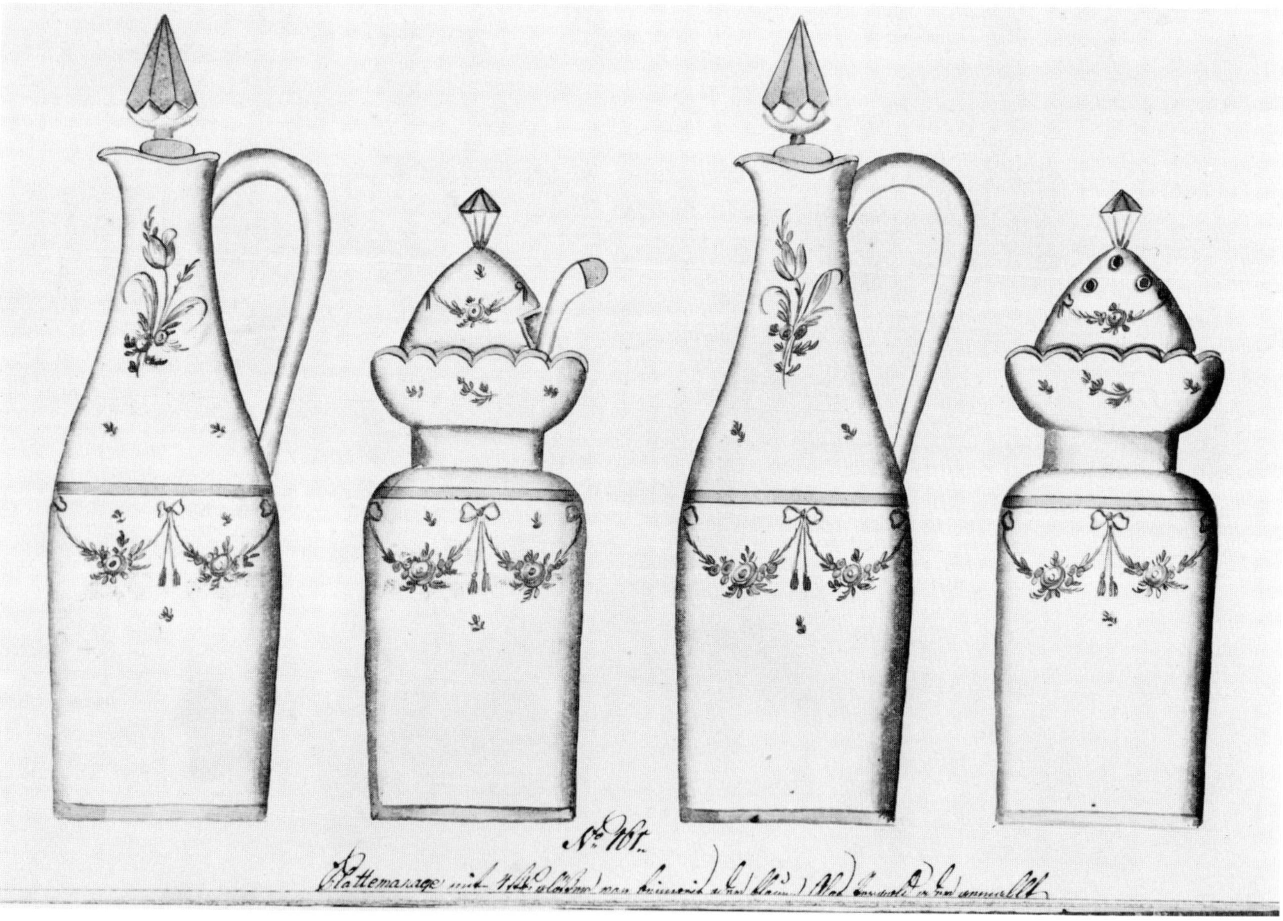

Page with four drawings for cruets numbered 161 from the Gardiner Glassware Catalogue. Bohemian, for the American Market; early 19th century. Inscribed: 'Plattemanage mit 4 Stk. glässer von Beinweiss oder Blau Glas vergold oder gemahlt' (*'Cruet consisting of four pieces, of bone-white or blue glass, gilt or painted . . .'*) (*Henry Francis du Pont Winterthur Museum Libraries*). *See p.42*

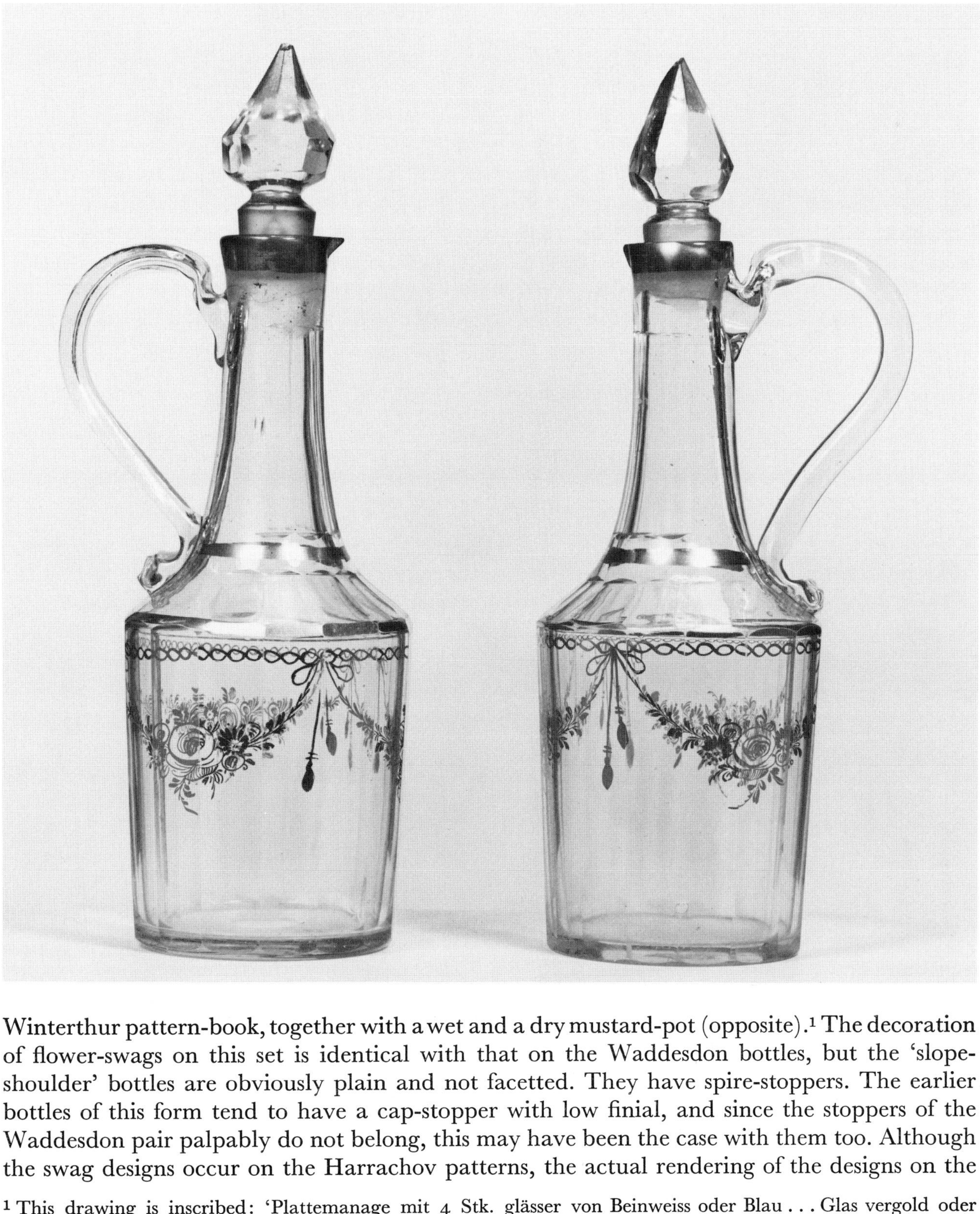

Winterthur pattern-book, together with a wet and a dry mustard-pot (opposite).[1] The decoration of flower-swags on this set is identical with that on the Waddesdon bottles, but the 'slope-shoulder' bottles are obviously plain and not facetted. They have spire-stoppers. The earlier bottles of this form tend to have a cap-stopper with low finial, and since the stoppers of the Waddesdon pair palpably do not belong, this may have been the case with them too. Although the swag designs occur on the Harrachov patterns, the actual rendering of the designs on the

[1] This drawing is inscribed: 'Plattemanage mit 4 Stk. glässer von Beinweiss oder Blau . . . Glas vergold oder gemahlt in Holtz gestelln' (Cruet consisting of four pieces, of bone-white or blue glass, gilt or painted, in a wooden stand').

Waddesdon bottles is very close to that recurring again and again in the Winterthur book, where scenic and figural designs do not appear at all. These general considerations suggest that this pair of bottles form the latest component in the Waddesdon 'service', and may well not belong to it at all.

Pairs of cut-glass handled cruet-bottles occur not infrequently in France and the Netherlands, in silver stands bearing marks dating from the last forty years of the eighteenth century. One such set is clearly shown in a still-life painting dated 1787 by Anne Vallayer-Coster (Royal Academy Exhibition, *France in the 18th Century* (Winter, 1968), Cat. No.690).

A 'slope-shouldered' uncut handled cruet, with closely similar gilt swags depending from a chain is in the Victoria and Albert Museum (C.31–1941). A pair of facetted cruets of the same general shape as the Waddesdon pair (and with spire-stoppers) is also in the Victoria and Albert Museum (C.156 and A–1934). They are decorated in gold with small figures in landscapes, more in the style of 5 and 6 and 9–23 here.

9 to 23. FIFTEEN WINE-GLASSES WITH GILT DECORATION

Normal height, $5\frac{1}{2}$ in. (13·5 cm)

Straight-sided polygonal funnel-bowls with concave-profiled bases, separated by an uncut 'merese' from a facetted inverted-baluster stem which swells slightly towards the bottom, at the junction with the gently rising foot.

The bowls of these glasses vary between having nine facets (9, 12) and fourteen (16). Nos.10, 11, 14 and 15 have decagonal bowls; 13, 18, 23 eleven-sided; 17, 19, 20 and 22 twelve-sided; and 21 thirteen-sided. These variations seem to have no special significance. There seem, however, to be two different styles of decoration represented. The glasses of the first group (9–18) have gilt decoration which is characterized by a tall C-shaped *rocaille* at the left and a lush spray of foliage at the right, between these being sometimes a human figure, sometimes an animal, sometimes merely another plant-form (10 and 11). The glasses of the second group (19 to 23) show a human figure (sometimes with a dog), and in one instance a riderless horse, in a 'landscape' characterized by an architectural 'ruin' or a feathery tree.

These glasses are probably to be identified generically as 'wine-glasses', being smaller than the 'Champagnes' of the Harrachov pattern-book (design No.121, p.273 here) and larger than the 'Liquer' glasses of the later Winterthur book (p.273 here), and the remaining glasses in the Waddesdon 'service' (*24–29*). Whether they are of the larger 'Burgundy' or the smaller 'Tokay' size, it is impossible to say, but perhaps the latter is more likely. The ratio of size to the cordial-glasses seems to bring them more into line with the smaller wine-glasses of the Winterthur pattern-book (p.273, No.70) than with the larger (No.71), which presumably represents the 'Burgundy' size.

Some of the gilt ornaments on these glasses find echoes among the designs of the Harrachov pattern-book (Nos.303, 320 and 322 for the glasses of the first group, and No.325 for the second – see p.266 here). For the dating of these, see note on 5 and 6 above. None of the designs on these wine-glasses, although having superficial affinities with them, seem necessarily connected with those on the decanters (1–4).

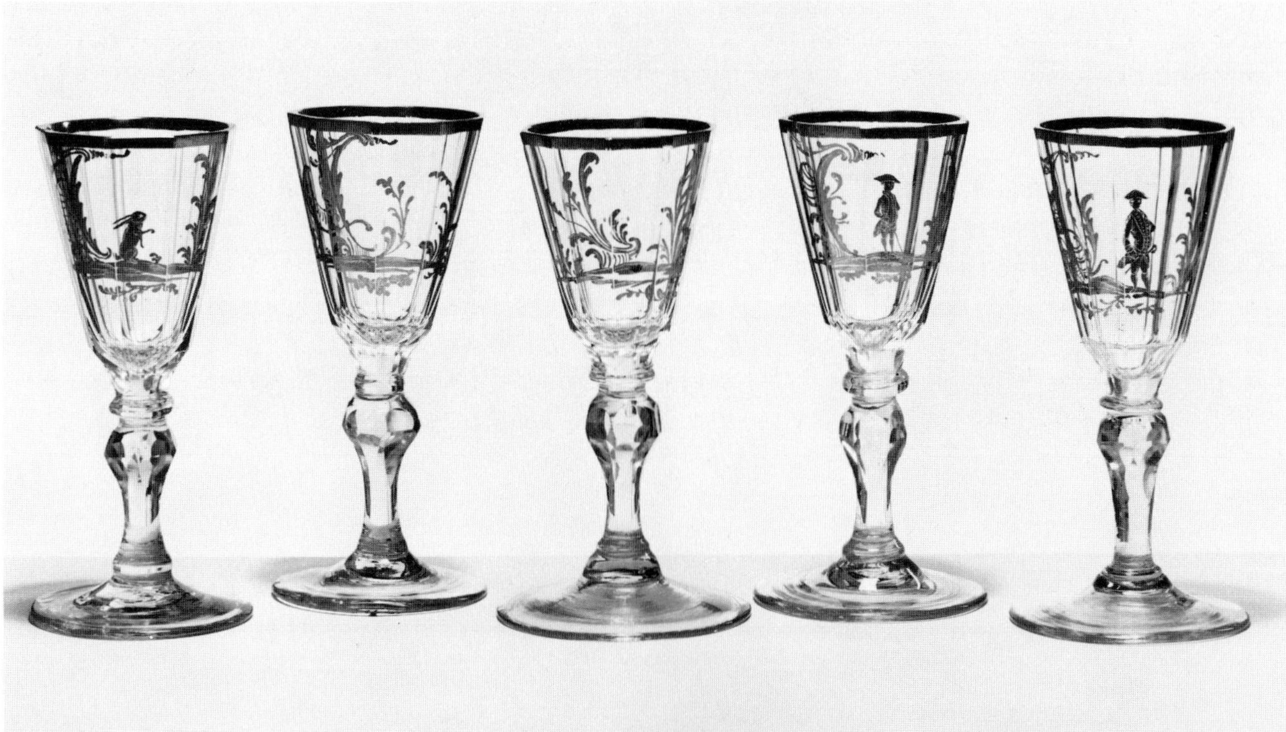

9 10 11 12 13

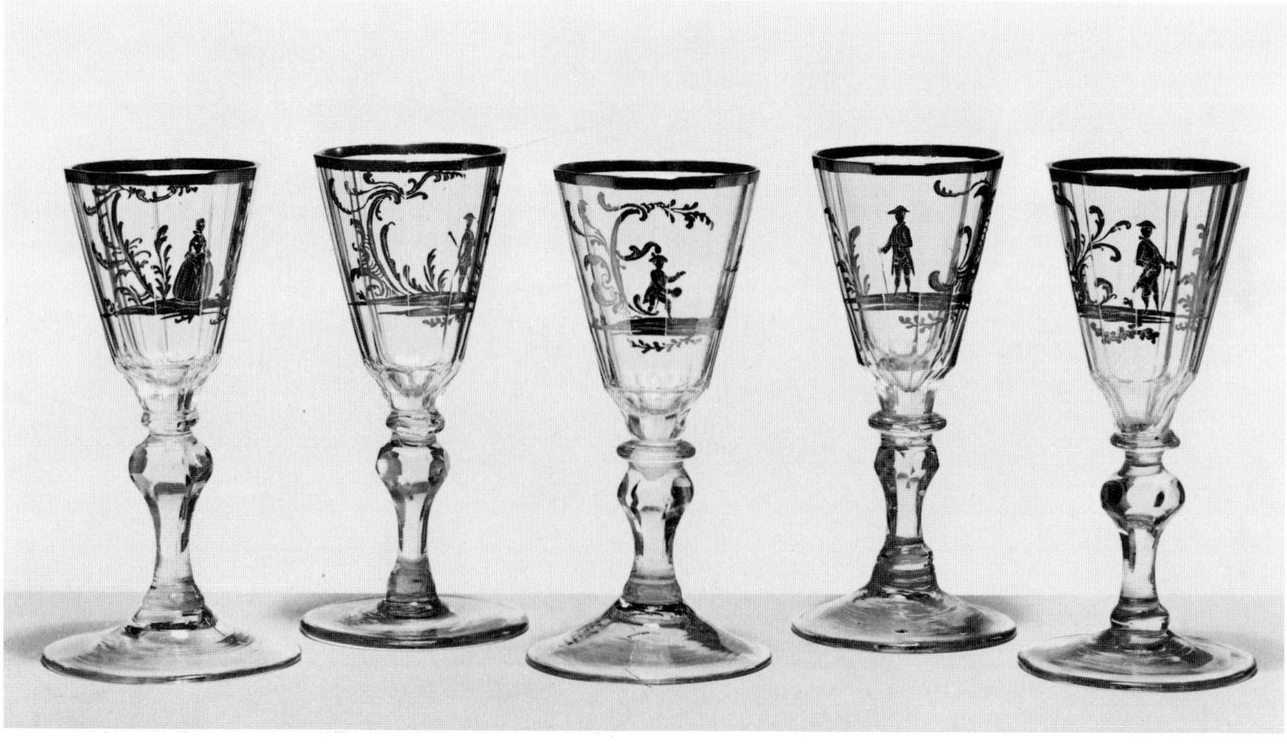

14 15 16 17 18

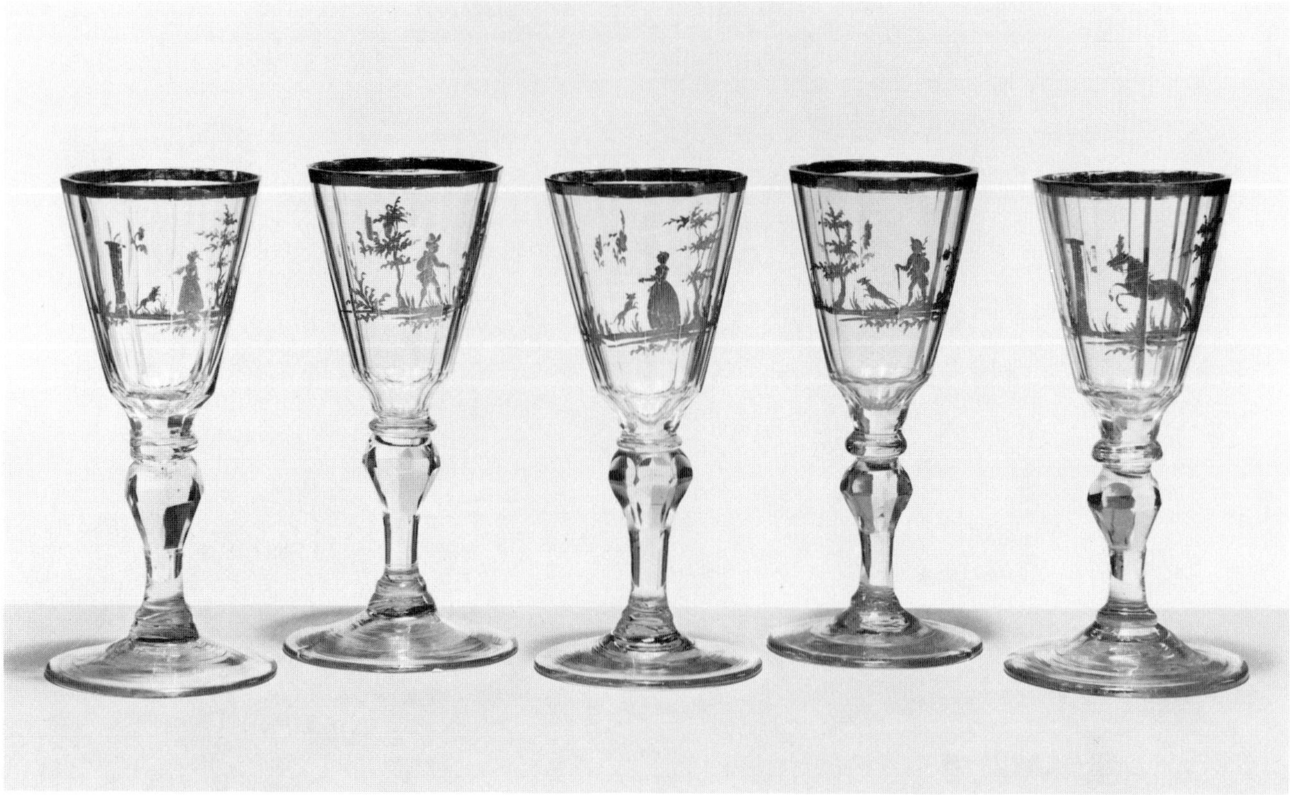

19 20 21 22 23

Opposite above: From the oldest extant Harrachov pattern-book, about 1785 (Pattern nos. 121 and 122) (Muzeum skla a bižutérie, Jablonec nad Nisou). *See p.270.*

Opposite below: Detail from the Gardiner's Island Glass Catalogue. Bohemian, for the American Market: early 19th century. Inscribed: 'No.69. 30m Liquer glässer geschliffene. No.70. 30m Weinglässer derglch. No.71. 24m grössere derglch.' *(Henry Francis du Pont Winterthur Museum Libraries).* *See p.42*

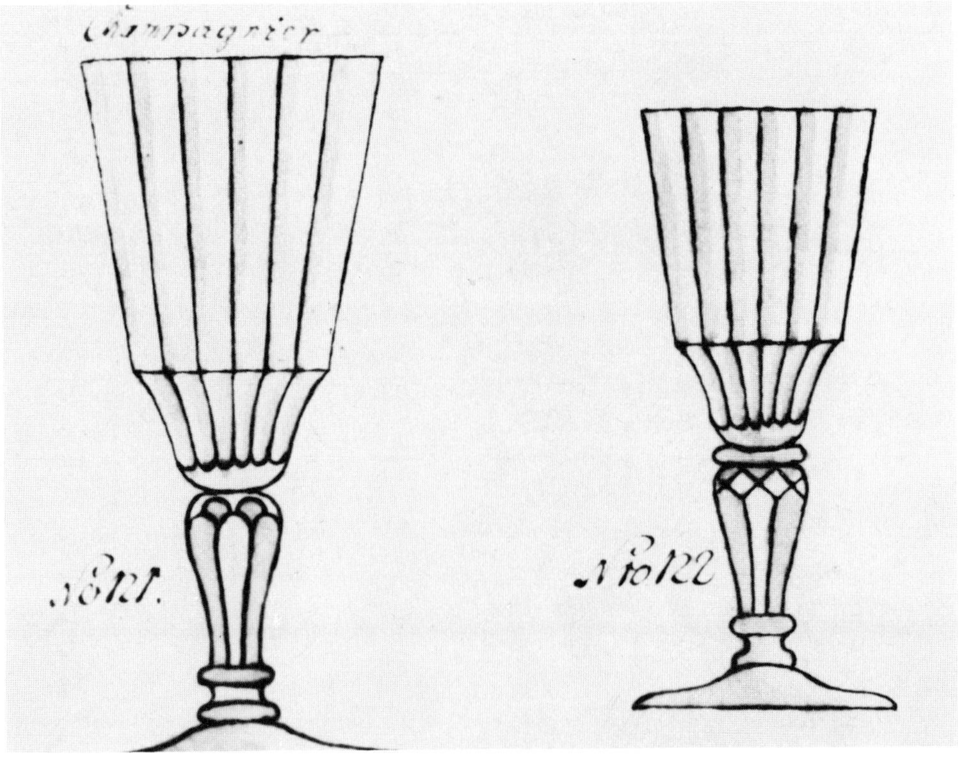
No 121.
No 122

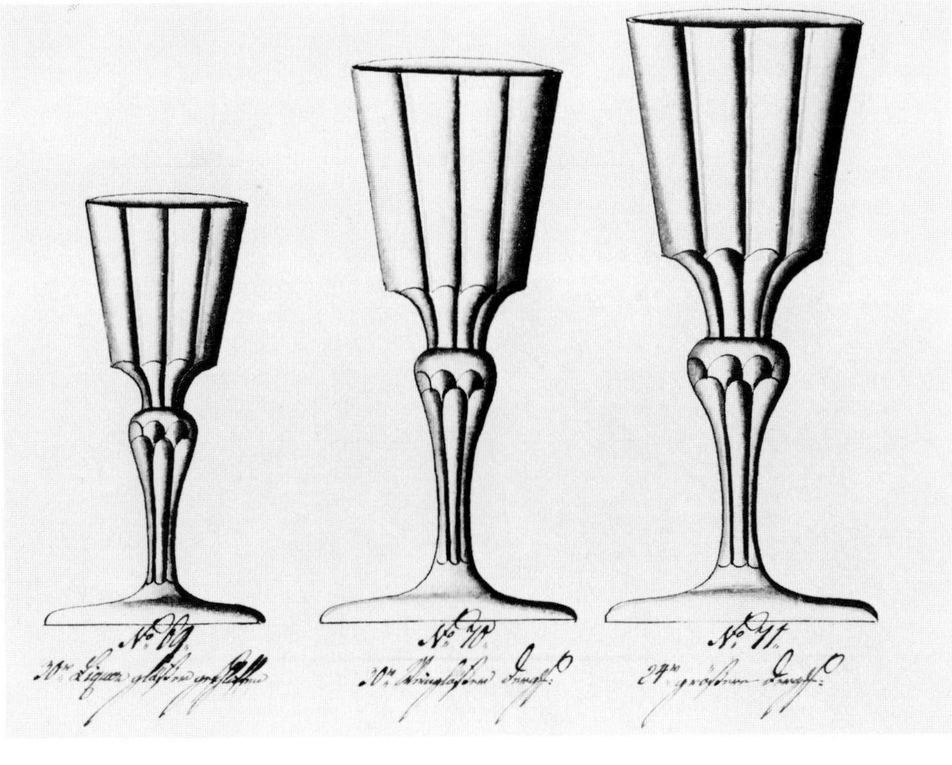
No 69.
No 70.
No 71.

24 to 29. SIX CORDIAL-GLASSES

Normal height, $4\frac{1}{4}$ in. (11 cm)
No.28, height, $4\frac{3}{8}$ in. (11·2 cm)

Straight-sided polygonal funnel-bowls with concave-profiled bases, separated by an uncut 'merese' from a facetted inverted-baluster stem which swells slightly towards the bottom, at the junction with the gently rising foot.

Nos.24, 25 and 27 are eleven-sided, but 26, which on grounds of decoration appears to be *en suite* with them, is dodecagonal. Of the two glasses which do not appear to be *en suite*, 29 is also dodecagonal, while 28 is decagonal.

The glasses of the first group bear gilt decoration of a building (usually a ruin) in a 'landscape', normally with a human or animal figure. The decoration of 29 is so rubbed as to be indistinguishable, but appears to include a C-shaped *rocaille* and a human figure; while 28 shows a hare in a 'landscape' with feathery trees. Both these glasses are distinguished either in shape or height (or both) from those of the first group.

For a commentary on the shape and size of these glasses, see the notes on 9–23.

The gilt decoration of the first group finds an echo in some of the designs found in the Harrachov pattern-book (e.g., Nos.185, 314[1] and 329). The glass 28 may be compared with design No.182 (below).

Miss Alice de Rothschild E2/22/7**

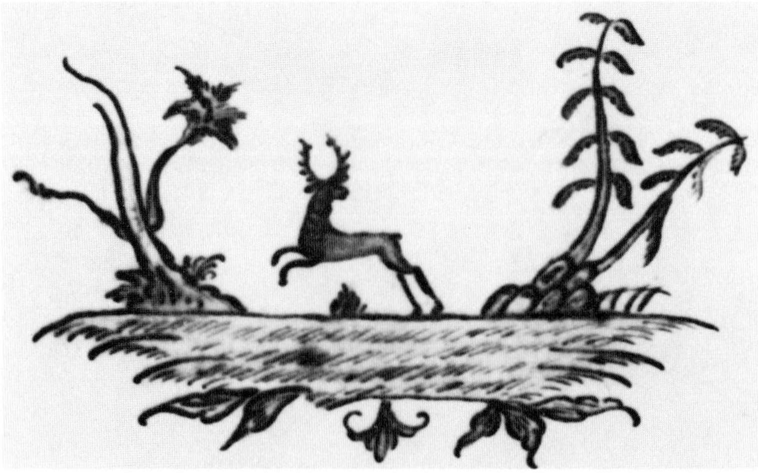

Decoration on a design for a case-bottle in the oldest extant Harrachov pattern-book; about 1785 (Muzeum skla a bižutérie, Jablonec nad Nisou). See p.42

[1] Illustrated by S. Urban, loc. cit. on p.42, n.160, Fig.23.

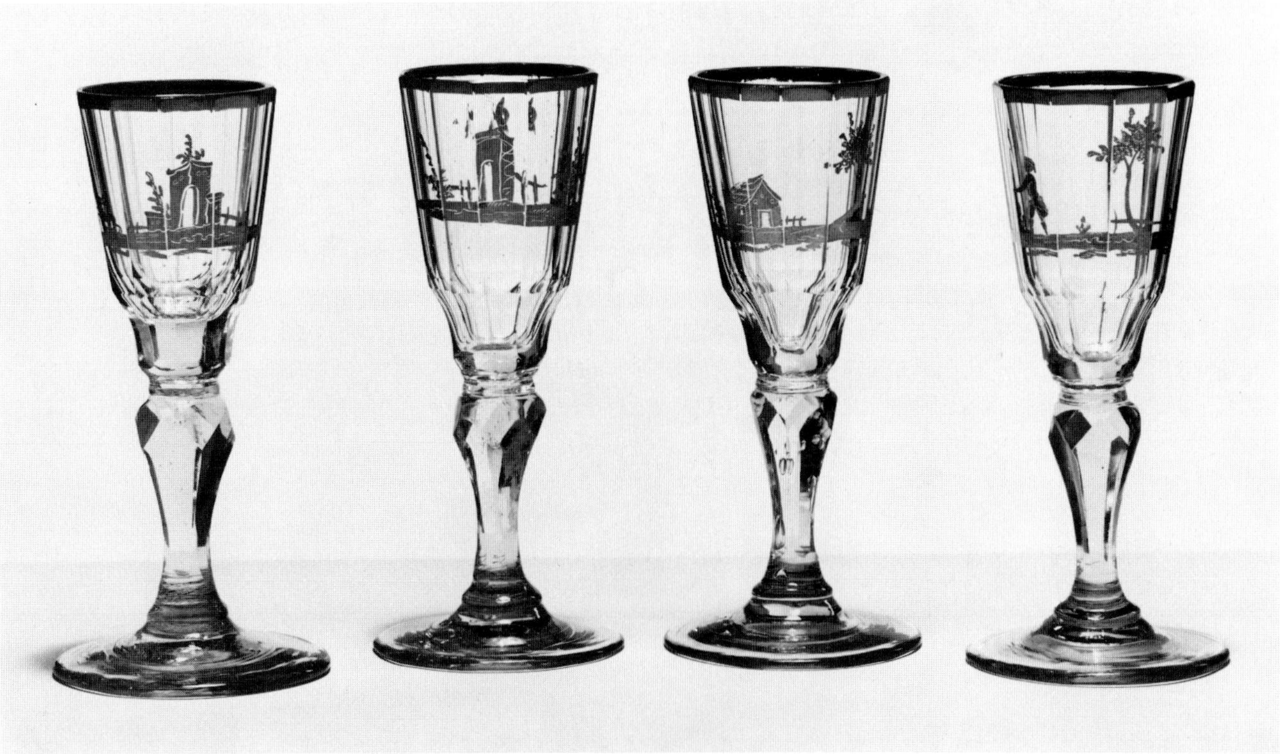

24 25 26 27

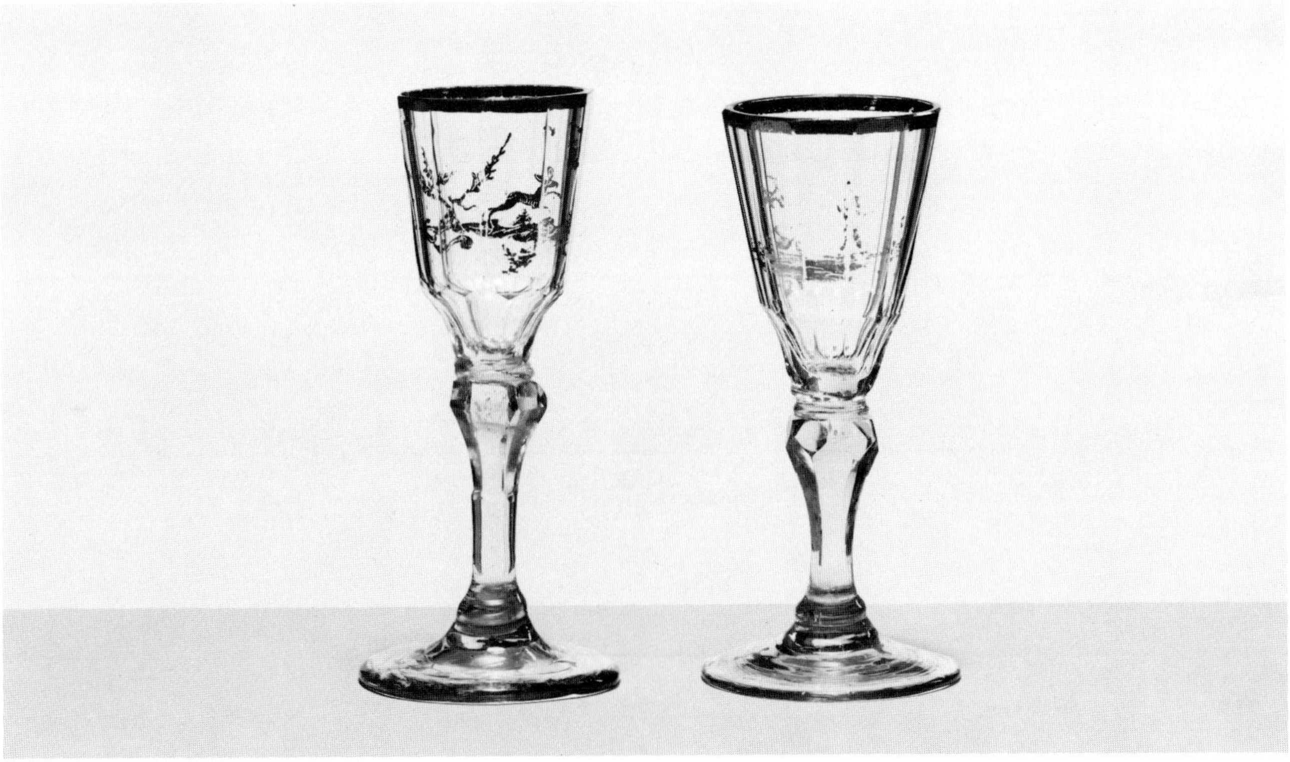

28 29

Profile and Sectional Drawings of the Roman Glass

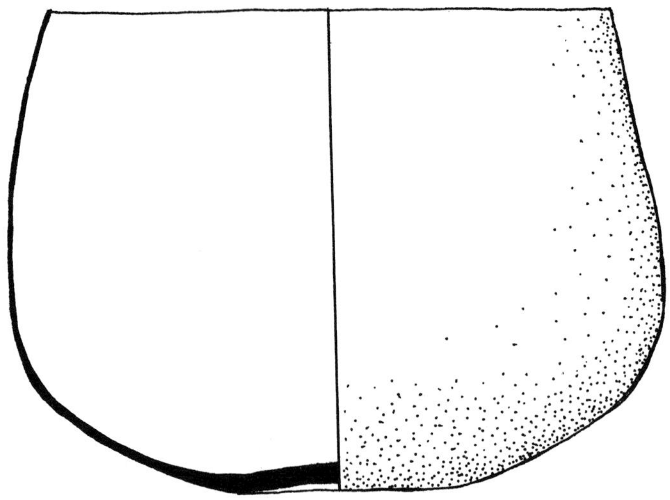

No.1 Page 52

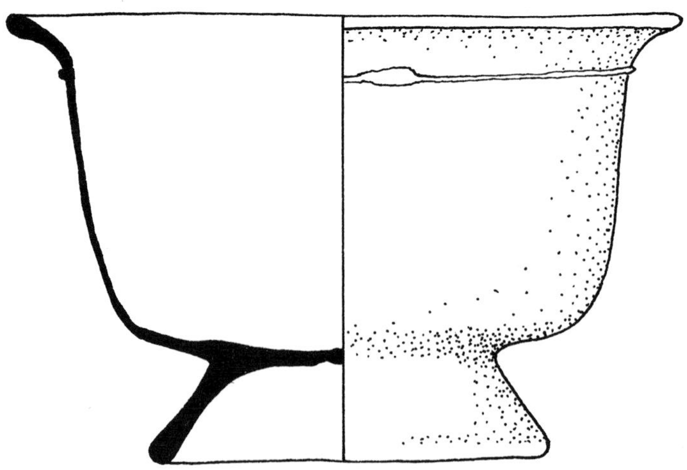

No.2 Page 54

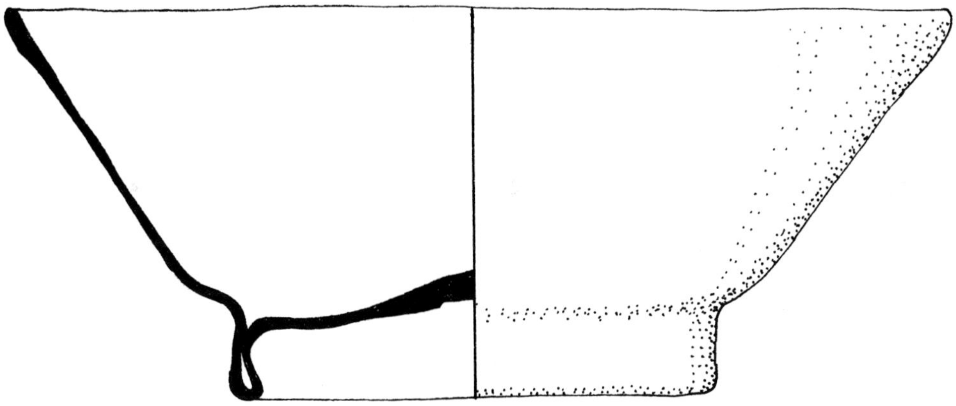

No.3 Page 56

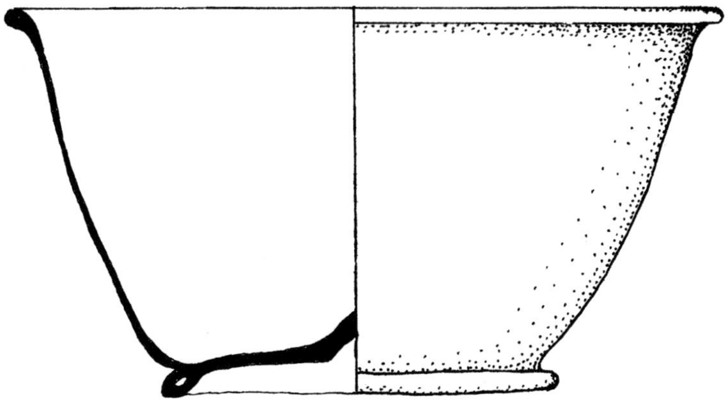

No.4 Page 58

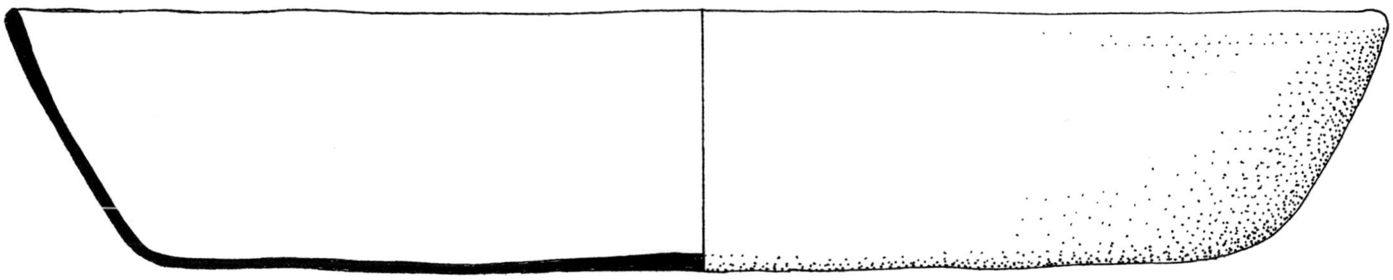

No.5 Page 60

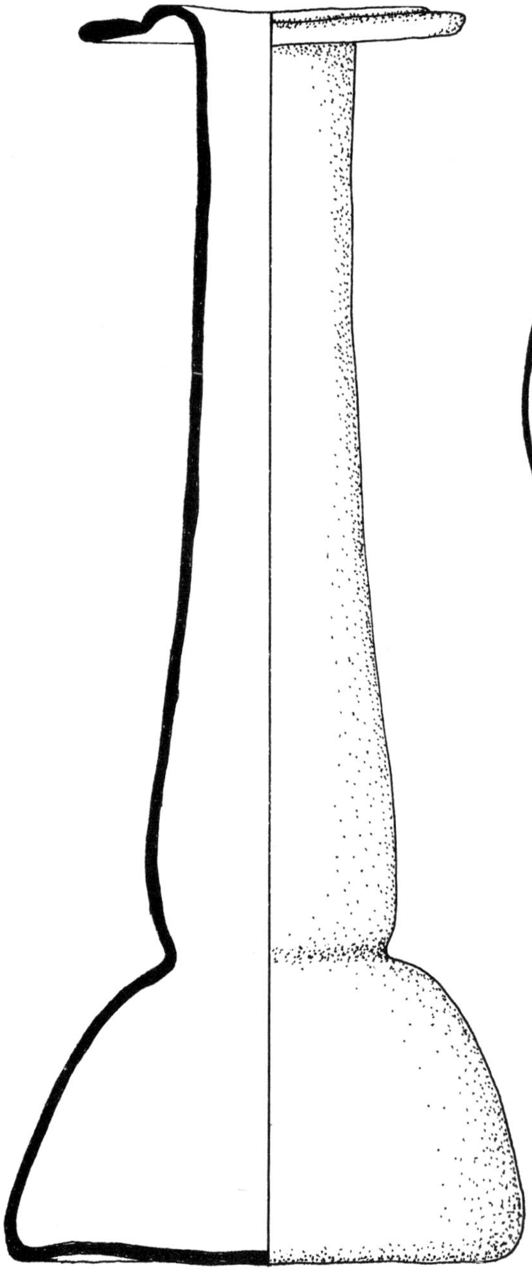

No.6 Page 62

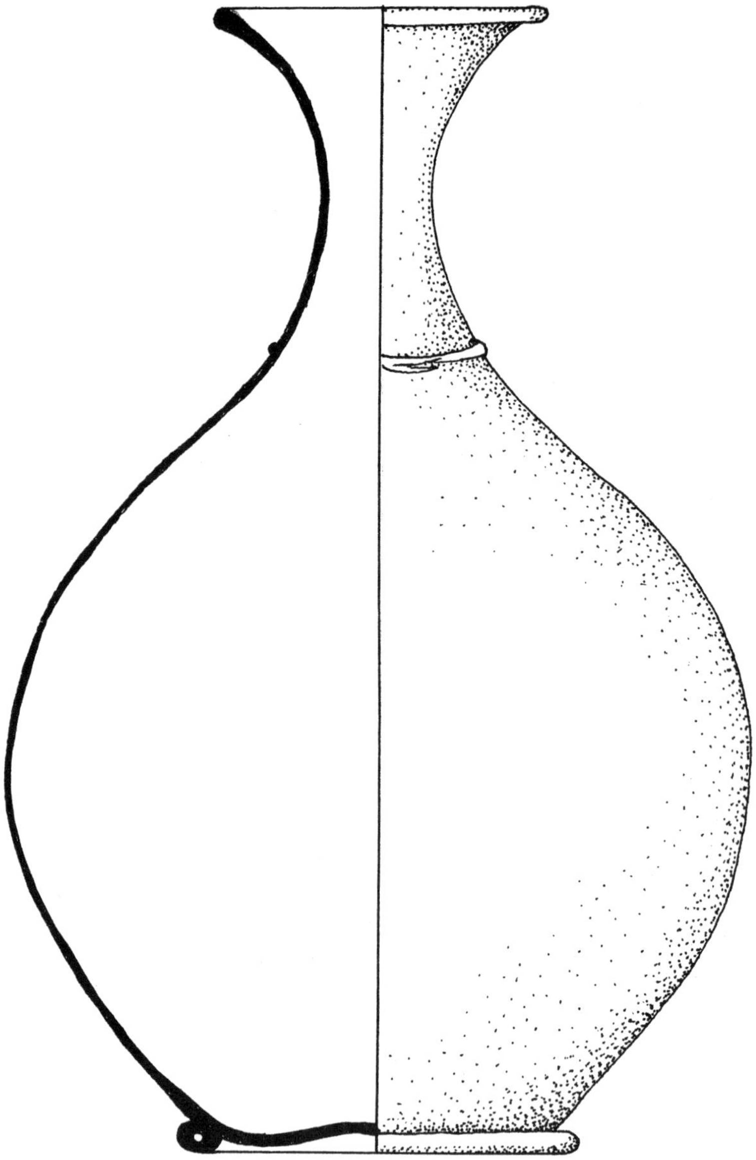

No.7 Page 64

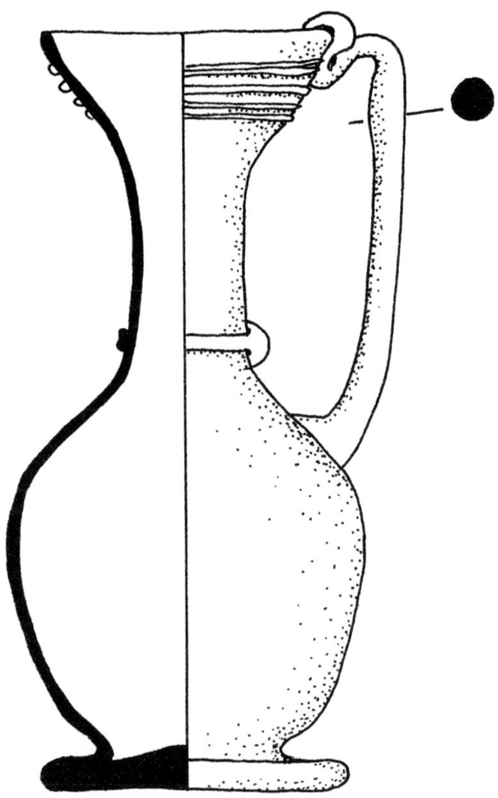

No.8 Page 66

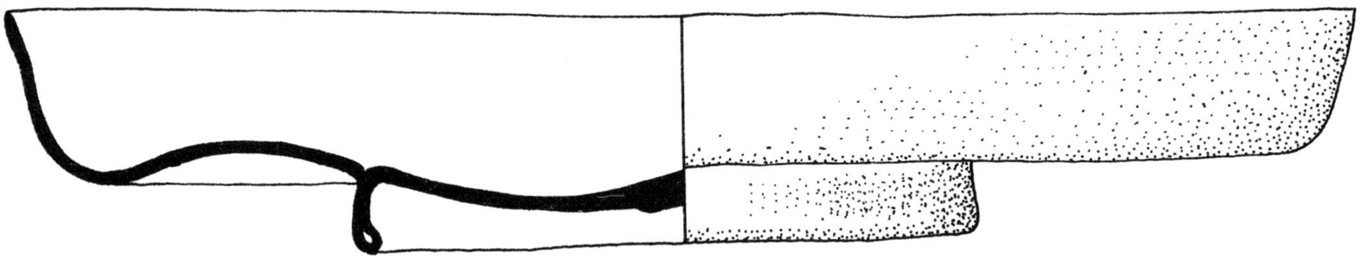

No.9 Page 68

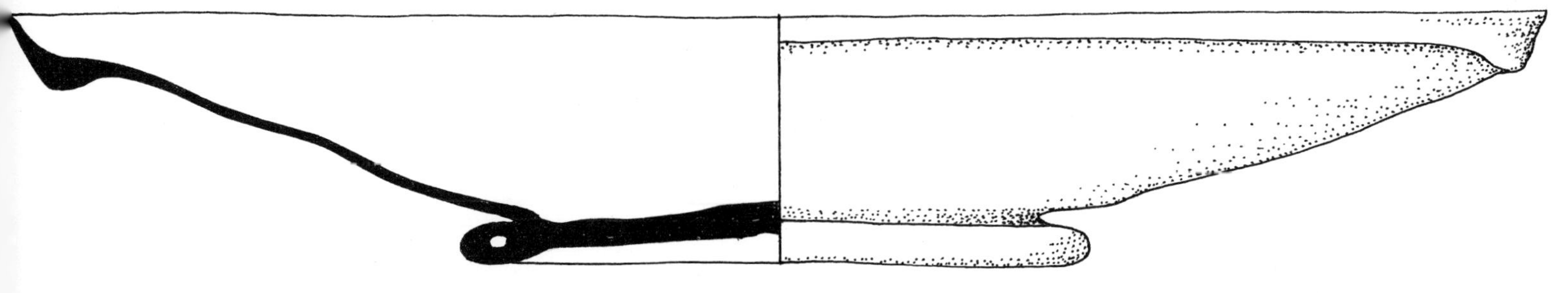

No.10 Page 70

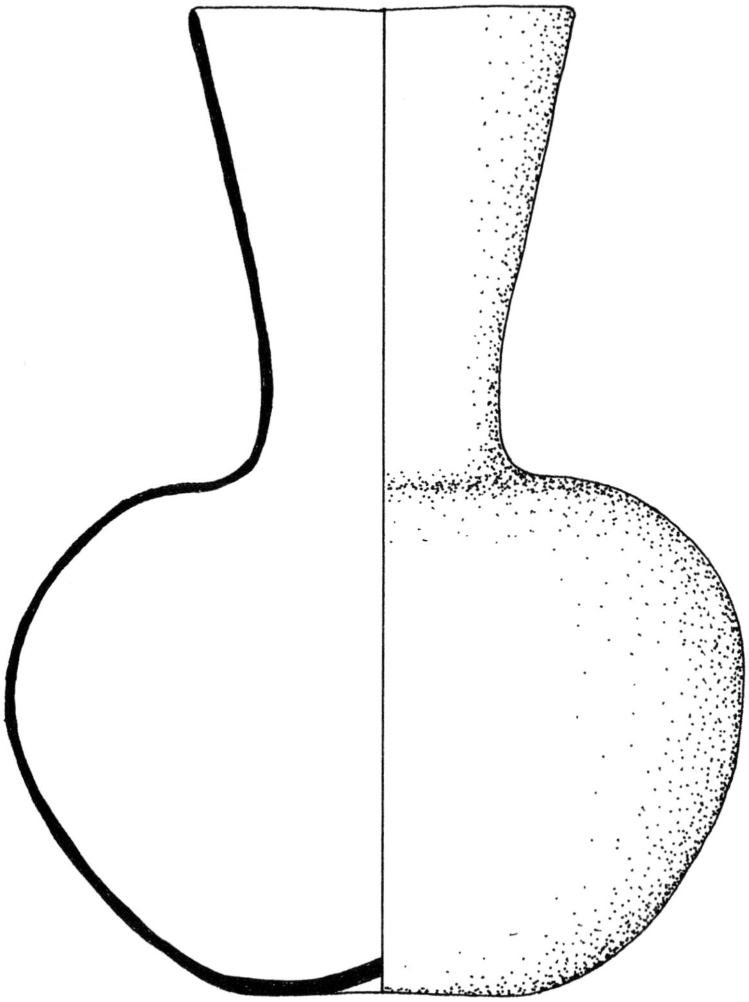

No.11 Page 72

No.12 Page 74

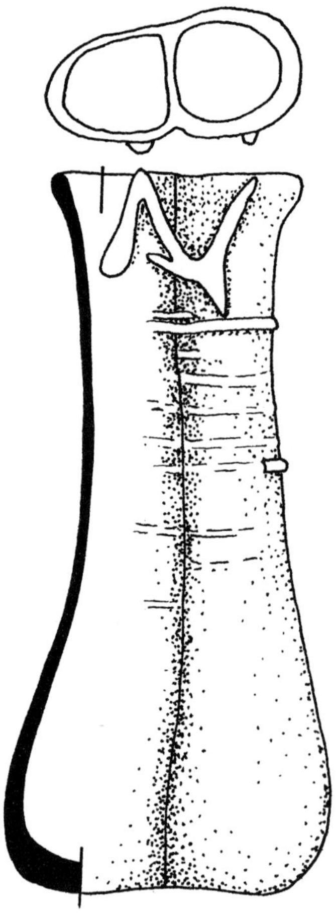

STAINED GLASS

Michael Archer

Introduction

The stained glass in the Waddesdon collections consists of three unrelated groups. There are two medieval English roundels, seven small German roundels painted with coat-of-arms, and six Swiss armorial panels. The only common characteristic is that all are secular in subject matter. Nothing is known of their earlier history, other than that they were bought by Miss Alice de Rothschild. So disparate are the three groups that it seems likely that they were acquired in order to form part of a decorative effect rather than with the intention of making a historically coherent collection.

The earliest pieces are the two English roundels, one of which can be dated to about 1400, the other to the late fifteenth or early sixteenth century. They are painted with black enamels and silver stain. The former is the conventional pigment used from early times for line and shading work and which is still in use today for these purposes. The latter is silver nitrate, which when fired gives a yellow colour ranging from a pale lemon to a deep orange. The technique of using silver stain was first discovered early in the fourteenth century and was soon widely practised since it allowed yellow and white to appear together on one piece of glass, where before a separate piece was needed for each colour. The coloured glass used throughout the Middle Ages is known as 'pot-metal' glass. This was made by mixing the glass when liquid in a crucible, or 'pot', with various metallic oxides so that the colour penetrated right through the material. Colours such as red and blue were so dark in their natural condition that they could only be used by melting a thin film of coloured glass on to a sheet of clear white glass. Sometimes parts of this coloured film of 'flashing' would be abraded in order to expose the clear glass. This was particularly useful for rendering heraldic devices and appears on some of the German roundels in the collection.

In the sixteenth century, glass-painters increasingly used enamel colours. These were translucent pigments of a type employed on vessel glass from a much earlier date. Their great advantage to the glazier was that they were suitable for rendering detailed work such as heraldry, in a variety of colours. Although a reddish flesh-tint appears at least as early as 1507–13 at Auch in France,[1] blue enamel has not so far been found in use earlier than 1538, the date of the *Last Supper* window at St Jacques, Antwerp, where it is used in the landscape.[2] The earliest dated French window in which blue enamel appears is that showing the *Childhood of Christ* in the Church at Montfort-L'Amaury dated 1543, but it is also found in the *Triumph of the Virgin* at St Patrice in Rouen, which was probably executed a little earlier in about 1540.[3] A number of enamel pigments are used on the later Swiss panels in the Waddesdon Collection and are also found on the German heraldic roundels.

[1] J. A. Knowles, 'The Transition from the Mosaic to the Enamel Method of Painting on Glass', *The Antiquaries Journal*, Vol. VI, No. 1, January 1926, pp.26–35.

[2] Illustrated as Plate LXXV, Fig. 168 in J. Helbig, *De Glasschilderkunst in Belgie*, Vol. I, Antwerp, 1943.

[3] J. Lafond, *Essai Historique sur le jaune d'argent*, Rouen, 1943, p.99, n.135.

The two English roundels are widely separated in date and have nothing in common other than that both have secular subject matter. The boar playing the bagpipes is a lively example of the medieval taste for amusing grotesque subjects of the type found in the margins of manuscripts or carved on misericords. The coat-of-arms is of very high workmanship and, unlike the boar, may have come from an ecclesiastical setting. No closely comparable pieces have come to light elsewhere.

All the coat-of-arms which have been identified on the German roundels (*78–83*) belonged to families prominent in Nuremberg in the sixteenth century. Stylistic considerations suggest that they were executed in the general period 1560–80 and they appear to belong to a single series. It is not possible to make any suggestion for whom they were made, or where they came from, other than that they are likely to have been in a domestic setting, perhaps the home of one of the families recorded. There is a comparable but slightly larger roundel which is painted with the arms of Gugel (*c.*1600), and although in different style it is also attributed to Nuremberg and is illustrated as Plate 231 in S. Beeh-Lustenberger (ed.), *Glasmalarei um 800 – 1900 im Hessichen Landesmuseum in Darmstadt*, Frankfort, two Vols. 1967 and 1974. Other small roundels which may be attributable to Nuremberg are in the Victoria and Albert Museum (C.100 and C.101–1934 and 9064, and 9066–1863) and the parish church of Wisbech St Mary, Cambridgeshire.

The Swiss glass at Waddesdon can be dated between about 1500 and the mid-seventeenth century. Towards the end of the fifteenth century it became customary in Switzerland for individuals or groups of people to make gifts of small panels of stained glass, generally painted with the arms and name of the donor. These were given as personal presents to friends or were intended for installation in monasteries, churches, town-halls, inns or other public buildings. They commemorate a variety of occasions, including weddings, appointment to a post or the completion of a building. The designs of these panels soon became stereotyped and compositions were frequently repeated. The earliest in the collection (*85*) seems likely to have come from the workshop of Lukas Zeiner of Zurich, the greatest Swiss artist in stained glass of the end of the late Gothic period. In this panel the basic convention is established of a coat-of-arms, a figure standing beside it in an architectural setting, and the name of the donor. Catalogue numbers *87–90* illustrate a later development in which the centre of the panel is filled with figures or coats-of-arms with subsidiary scenes and inscriptions above and below. The latest panel in the collection (*90*) is dated 1667 and demonstrates how the architectual setting could dominate the design.

NOTE. In the catalogue entries which follow, the measurements given include the leading on the outer edge.

Bibliography

ENGLAND

Herbert Read, *English Stained Glass*, London and New York, 1926.

E. A. Greening Lambourn, *The Armorial Glass of the Oxford Diocese, 1250–1850*, London 1949.

C. Woodforde, *English Stained and Painted Glass*, Oxford, 1954.

L. M. C. Randall, *Images in the margins of Gothic manuscripts*, Los Angeles, 1966.

G. L. Remnant, *A Catalogue of Misericords in Great Britain*, Oxford, 1969.

GERMANY

Nürnberger Geschlechterbuch, MS 94402 in the library of the Germanisches Nationalmuseum, Nuremberg.

H. Schmitz, *Die Glasgemälde des Königlichen Kunstgewerbemuseums in Berlin*, Berlin, 1913.

J. L. Fischer, *Handbuch der Glasmalerei*, Leipzig, 1937.

S. Beeh-Lustenberger (ed.), *Glasmalerei um 800 – 1900 im Hessischen Landemuseum in Darmstadt*, Frankfurt, two vols., 1967 and 1974.

SWITZERLAND

H. Lehmann, *Zur Geschichte der Glasmalerei in der Schweiz*, Leipzig, 1925.

H. Lehmann, 'Lukas Zeiner und die spätgotische Glasmalerei in Zürich', *Mitteilungen der antiquarischen Gesellschaft in Zürich*, Band XXX, Heft 2, Zurich, 1926.

B. Rackham, 'The Glass-Painter Lukas Zeiner of Zurich and his School', *Old Furniture*, Vol. VIII, October 1929, pp.54–60.

J. Schneider, 'Die Standescheiben von Lukas Zeiner im Tagsatzungssaal zu Baden (Schweiz)', *Basler Studien zur Kunstgeschichte*, Band XII, Basel, 1954.

P. Boesch, *Die Schweizer Glasmalerei*, Basel, 1955.

J. Schneider, *Glasgemälde: Katalog der Sammlung des Schweizerischen Landesmuseums*, Zurich, 1971 [Schneider].

In the discussion of armorial bearings, the work cited as *Rietstap* is J. B. Rietstap, *Armorial Général*, Berlin, 1934.

CATALOGUE OF STAINED GLASS

77. Roundel, painted with a boar playing the bagpipes

English; *c.*1375–1425

Colourless glass painted in brown enamel and yellow stain

Diameter, 4¾ in. (12 cm)

The edging of the cape, the bagpipes, the spray of leaves and the narrow area outside the circular border fillet are painted in yellow, the other parts in dark brown. Some of the lines look as if they may have been strengthened a little at a later date. The colours used are enamel with silver stain for the yellow.

This roundel is illustrated as Fig.8 and mentioned on page 116 in Herbert Read, *English Stained Glass*, London, 1926. As this piece and *78* are the only two pieces of English stained glass at Waddesdon, it seems possible that they may have been acquired together and come ultimately from the same source. The subject matter suggests that *77* originally came from a domestic setting, though decorated domestic glass was not common in England until the latter part of the fourteenth century. A man pushing a wheelbarrow on which is seated a boar playing bagpipes appears in the border of the illumination showing the Visitation in the *Très Riches Heures du Duc de Berry* (Musée Condé, Chantilly) which can be dated to the second decade of the fifteenth century.

Miss Alice de Rothschild E2/24/5a*

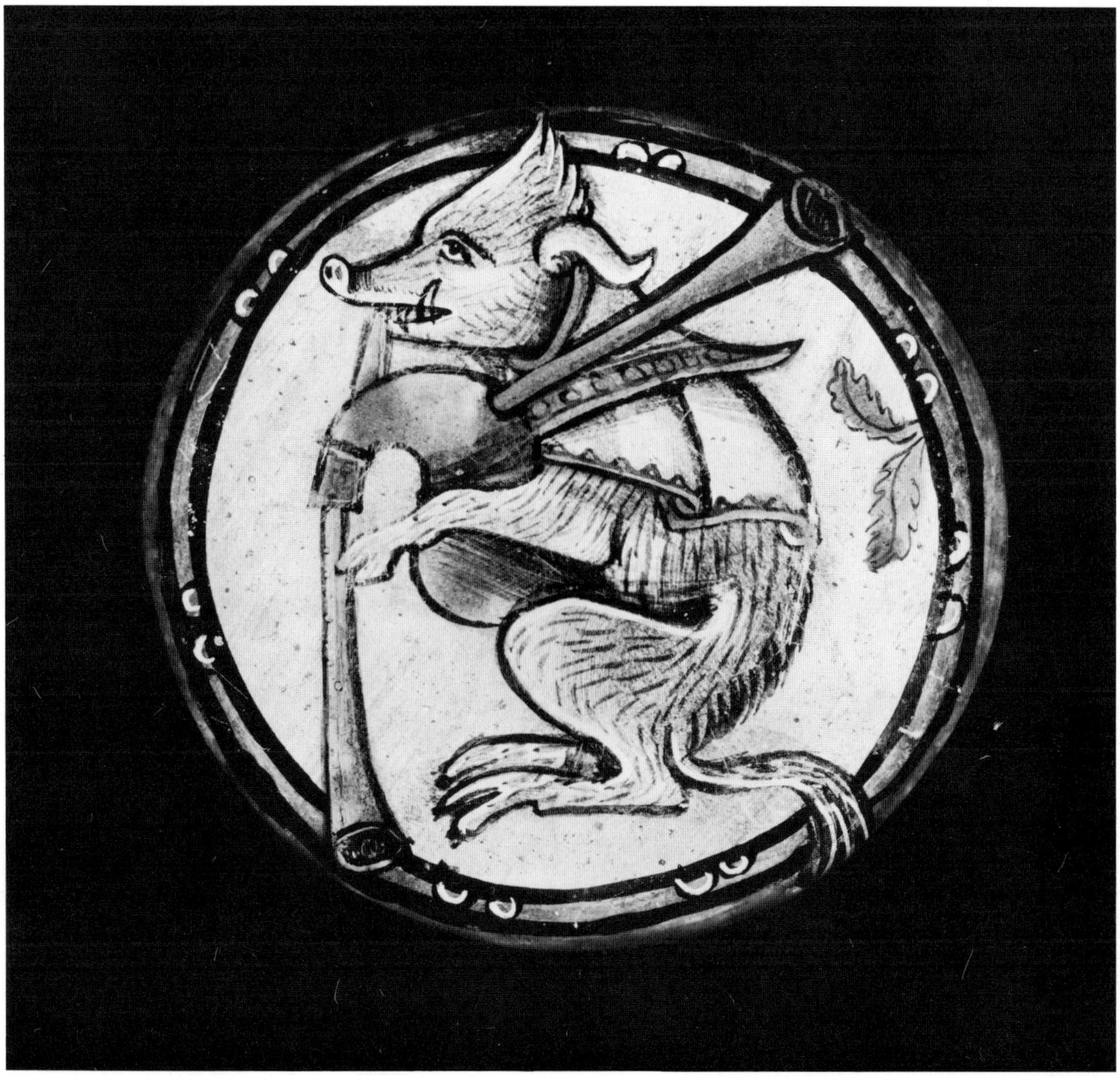

78. Roundel, painted with a coat-of-arms

English; late 15th or early 16th century

Colourless glass painted in black and brown enamels and yellow stain

Diameter, $4\frac{5}{8}$ in. (11·7 cm)

The coat-of-arms is in black and yellow on a background of dark-brown stippling with three cusps and a thin line in yellow. The colours used are enamels with silver stain for the yellow. The decoration is painted, with scratched detail. The arms are sable, a chevron or between three boars' heads couped argent facing sinister.

This roundel is illustrated as Fig.7 and mentioned on page 104 in Herbert Read, op. cit. It is also referred to on page 91 in E. A. Greening Lambourn, *The Armorial Glass of the Oxford Diocese, 1250–1850*, London, 1949, where it is compared with a shield at Nether Winchendon, Buckinghamshire which has the boars' heads and the chevron, with the difference that the chevron is argent and the boars' heads are tusked or. Other similar arms are those of Thirlwall, given by Rietstap as sable, a chevron argent between three boars' heads couped argent. The arms in the Waddesdon panel differ from these in the tincture of the chevrons and in having the heads facing sinister rather than the usual dexter. Greening Lambourn suggests that it may have come from Nether Winchendon but gives no evidence to support this supposition. Although it is not possible to identify finally the arms on the Waddesdon panel, it seems likely that they belong to an English family. On grounds of style the panel can be dated to the late fifteenth or early sixteenth century.

Miss Alice de Rothschild E2/24/5b*

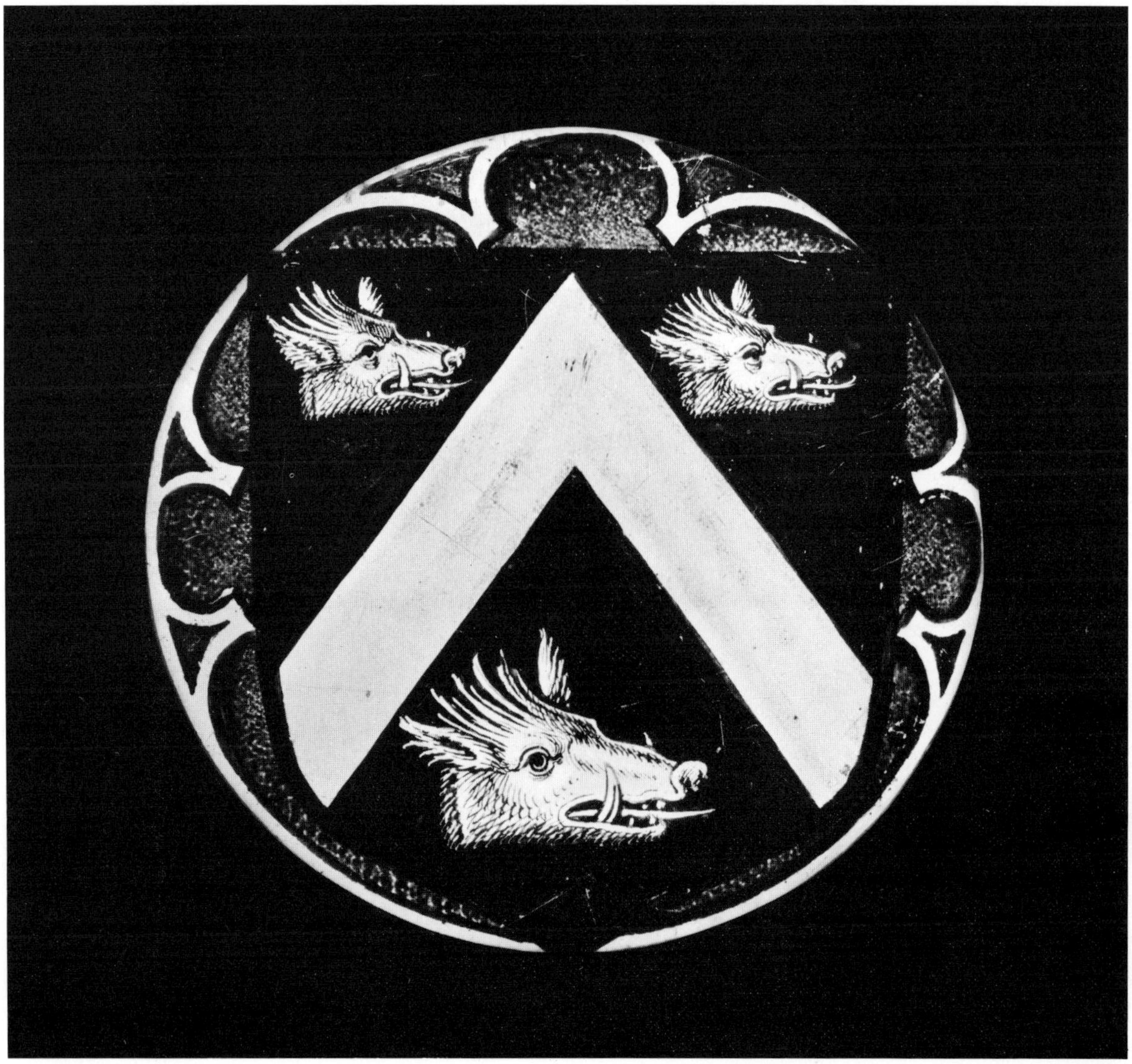

79. Roundel, painted with a coat-of-arms

German (Nuremberg); *c.*1560–80

Red and colourless glass painted in brown enamel and yellow stain

Diameter, $4\frac{5}{8}$ in. (11·7 cm)

The coat-of-arms is painted in red, brown and yellow on a background of foliate scrolls in yellow. The colours used are pot-metal for the red, silver stain for the yellow, and brown enamel. The decoration is painted with scratched detail. The flashed red has been abraded through to colourless glass in places. The arms are gules, a demi-lion rampant argent crowned or. They are those of Grundherr, an old Nuremberg family who still use the lion as part of their arms.

Miss Alice de Rothschild E2/24/5e*

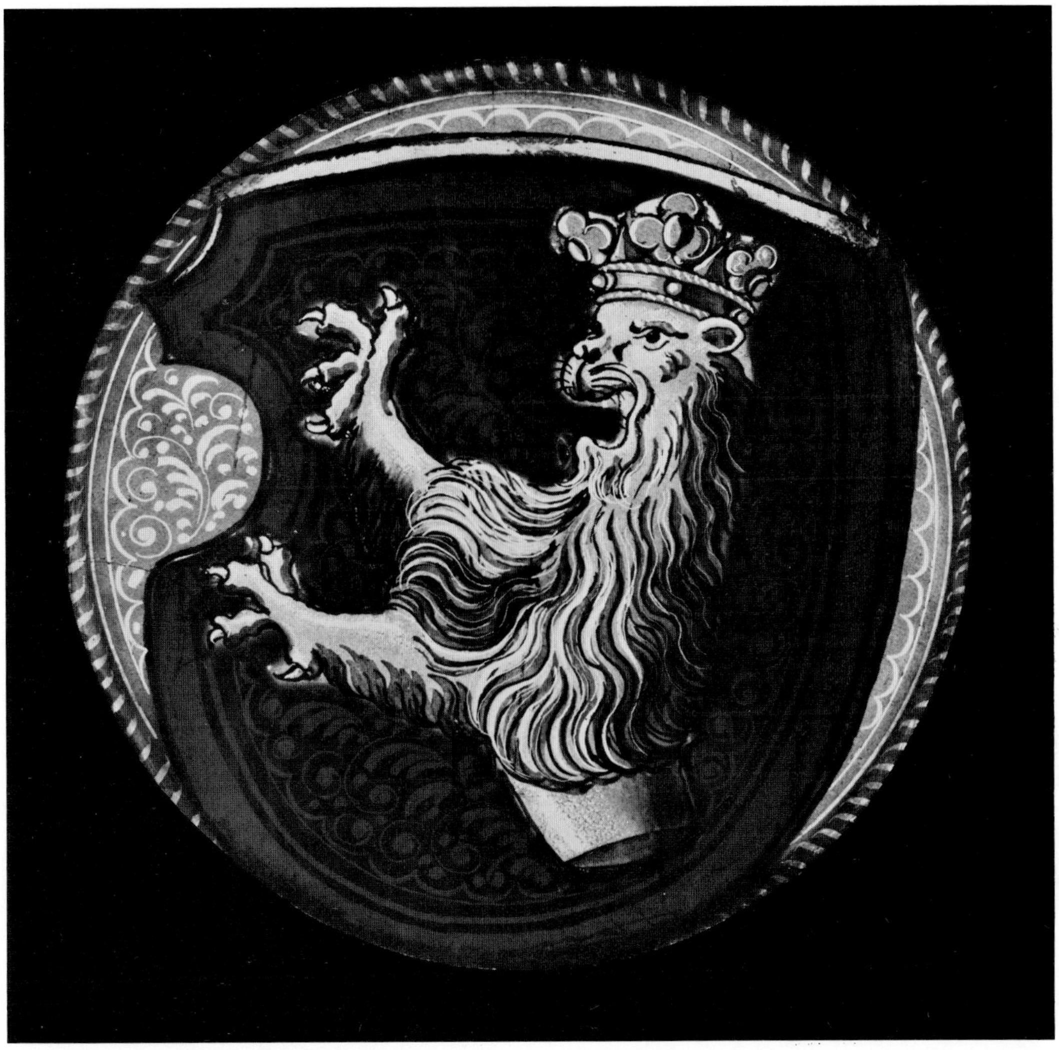

80. Roundel, painted with a coat-of-arms

German (Nuremberg); *c.*1560–80

Colourless glass painted in black and blue enamels and yellow stain

Diameter, $4\frac{1}{2}$ in. (11·4 cm)

The coat-of-arms is painted in black and yellow on a background of rays in blue. The colours used are enamel pigments with silver stain for the yellow. The decoration is painted, with scratched detail. The arms are parti per pale, dexter a lion rampant sable; sable, three fleurs-de-lis in pale or. It has not been possible to find the family to which these arms belong.

Colour illustration on page 315

Miss Alice de Rothschild E2/24/5d*

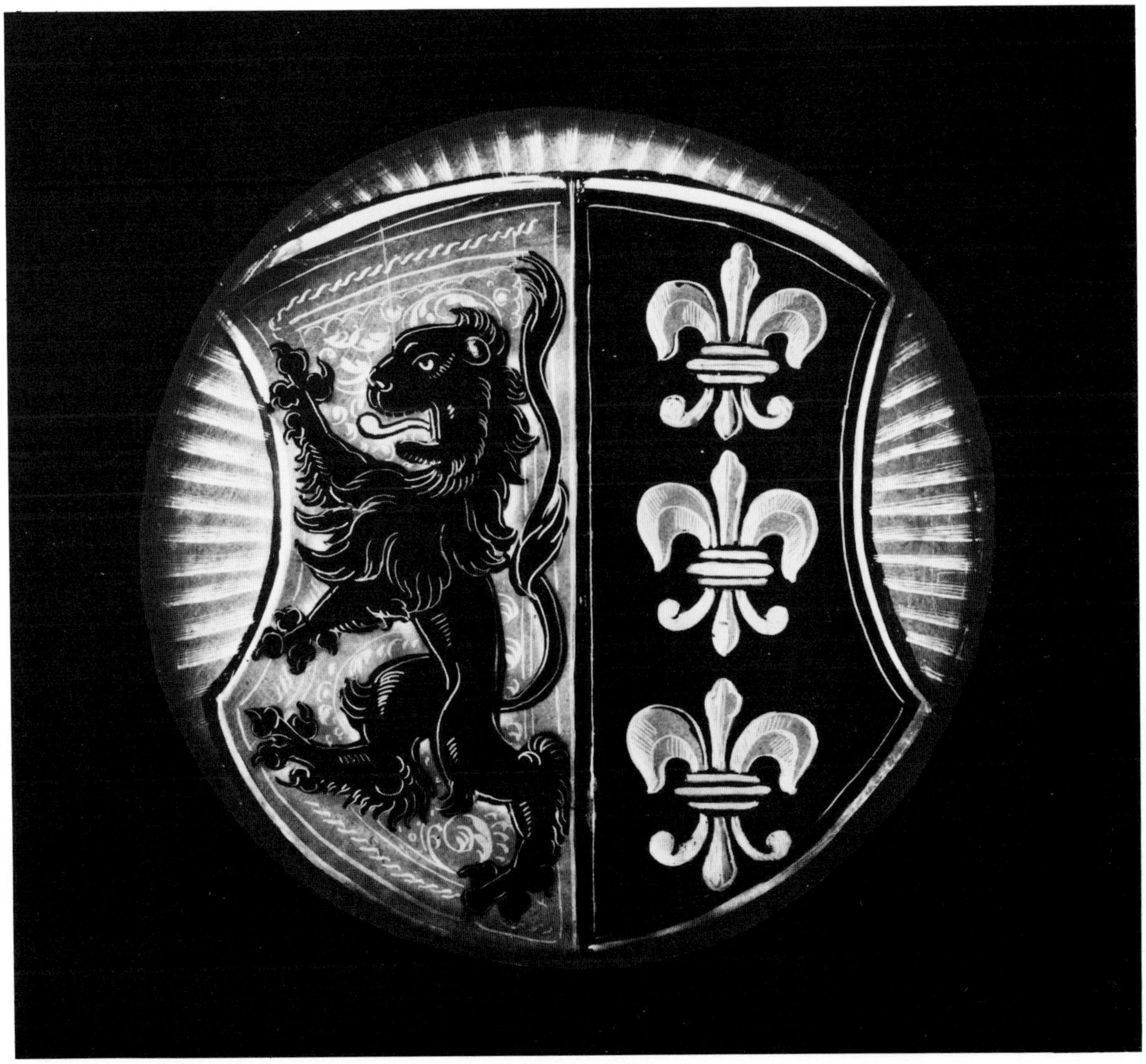

81. Roundel, painted with a coat-of-arms

German (Nuremberg); *c.*1560–80

Red and colourless glass painted in brown enamel and yellow stain

Diameter, $4\frac{1}{2}$ in. (11·4 cm)

The coat-of-arms is painted in red, brown and yellow on a background of rays in yellow. The colours used are pot-metal for the red, silver stain for the yellow, and brown enamel. The decoration is painted, with scratched detail. The arms are gules, a demi-hind rampant pierced with an arrow proper. They are those of Ayrer, a patrician family of Nuremberg.

The identical arms appear on a panel dated 1646 which was item 96 in the sale catalogue of various private collections (Helbing, Munich, 21 November 1912). The painting of the background has points in common with the Waddesdon roundel. The date on it appears to have been added later, since it is roughly done without any consideration for the design.

Colour illustration on page 315

Miss Alice de Rothschild E2/24/5e*

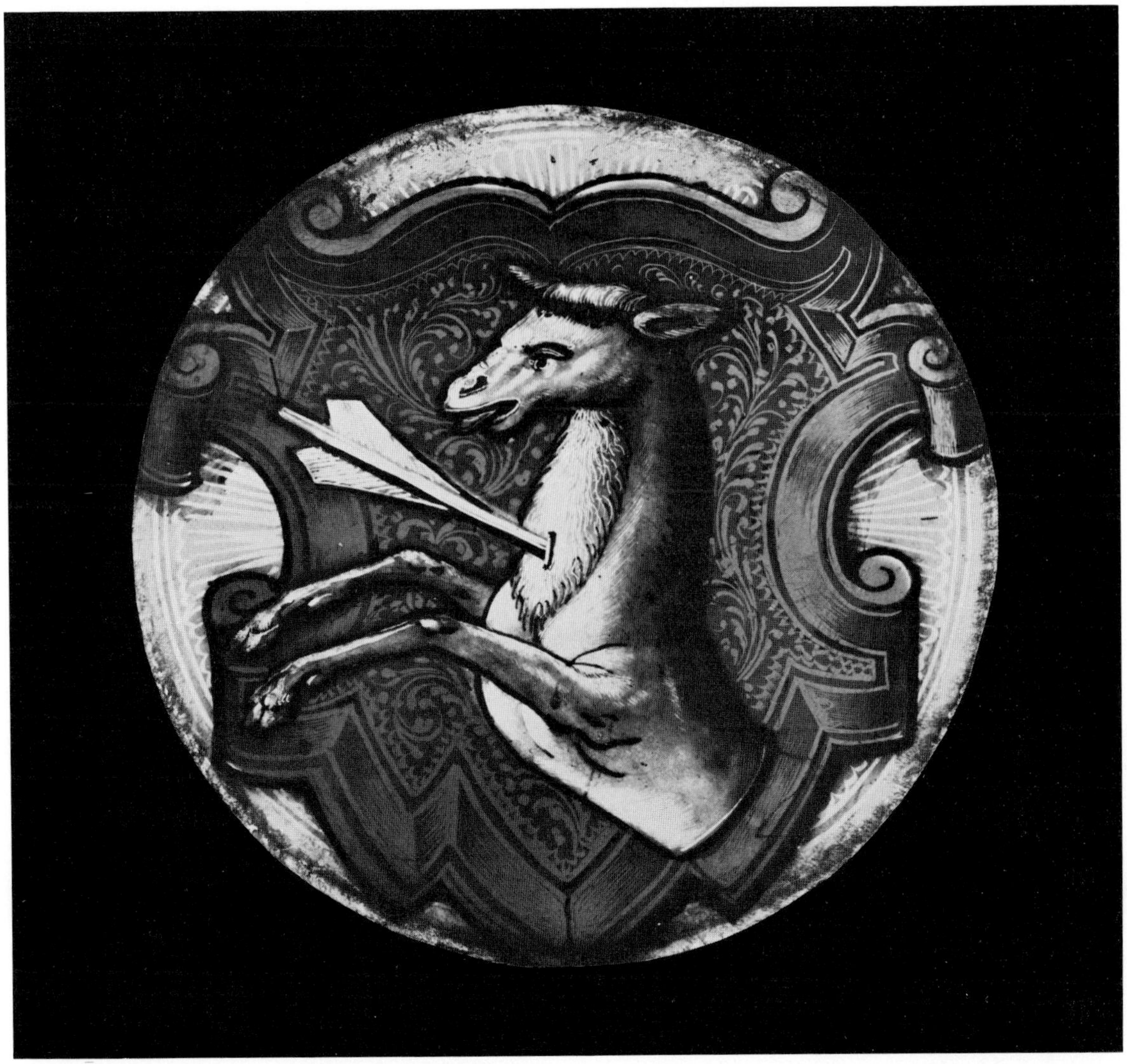

82. Roundel, painted with a coat-of-arms

German (Nuremberg); *c*.1560–80

Colourless glass painted in black, brown and blue enamels and yellow stain

Diameter, $4\frac{5}{8}$ in. (11·7 cm)

The coat-of-arms is painted in black, brown and yellow on a background of foliate decoration in blue. The colours used are enamel pigments with silver stain for the yellow. The decoration is painted, with scratched detail. The arms are quarterly, 1, sable, a demi-hind pierced with an arrow proper; 2, or, a falcon sable gorged with a coronet on a mound or; 3, argent, a falcon sable gorged with a coronet on a mound or; 4, argent, a demi-hind pierced with an arrow proper. It has not been possible to identify the individual to whom these arms belong but he seems likely to have been related to the Ayrer family, as the first and fourth quarterings are identical with the arms shown on the preceding roundel (*81*) but with the field differing in tincture. The anonymous author of the 'Nürnberger Geschlechterbuch' (the Nuremberg genealogical register, MS 94402 in the library of the Germanisches Nationalmuseum, Nuremberg) states that quarterings two and three were supposed to have been awarded to the Ayrer family in 1610 by Rudolf II. The author gives no evidence for this statement and indeed appears himself to have doubts about its truth.

Miss Alice de Rothschild E2/24/5f*

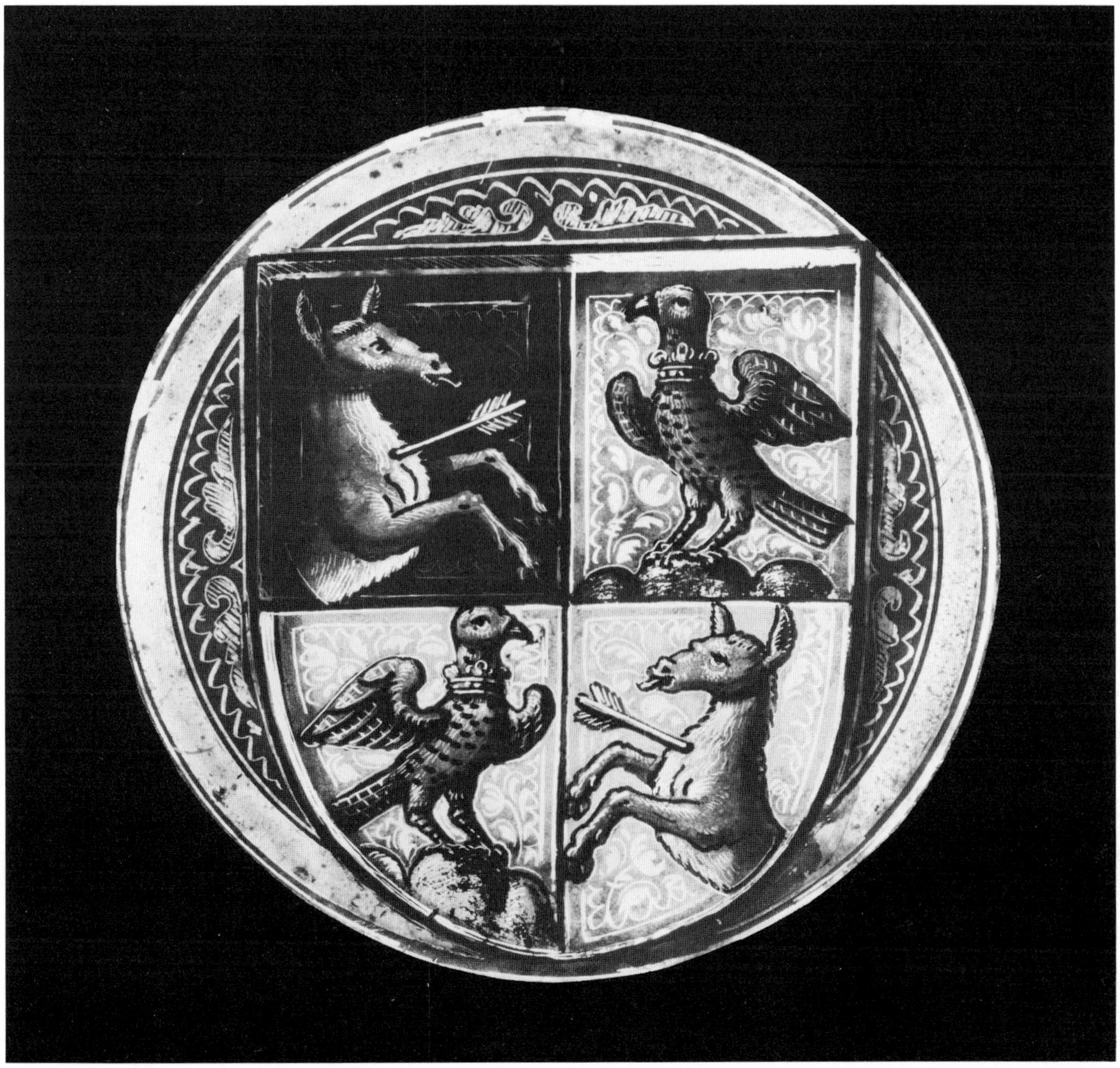

83. Roundel, painted with a coat-of-arms

German (Nuremberg); *c.*1560–80

Colourless glass painted in black and light brown enamels and yellow stain

Diameter, $4\frac{5}{8}$ in. (11·7 cm)

The coat-of-arms is painted in black and yellow on a background of foliate decoration in light brown. The colours used are enamel pigments with silver stain for the yellow. The decoration is painted, with scratched detail. The arms are per fesse or and sable, a fleur-de-lis countercharged. They are those of the Nuremberg family of Bamberger.

Miss Alice de Rothschild E2/24/5g*

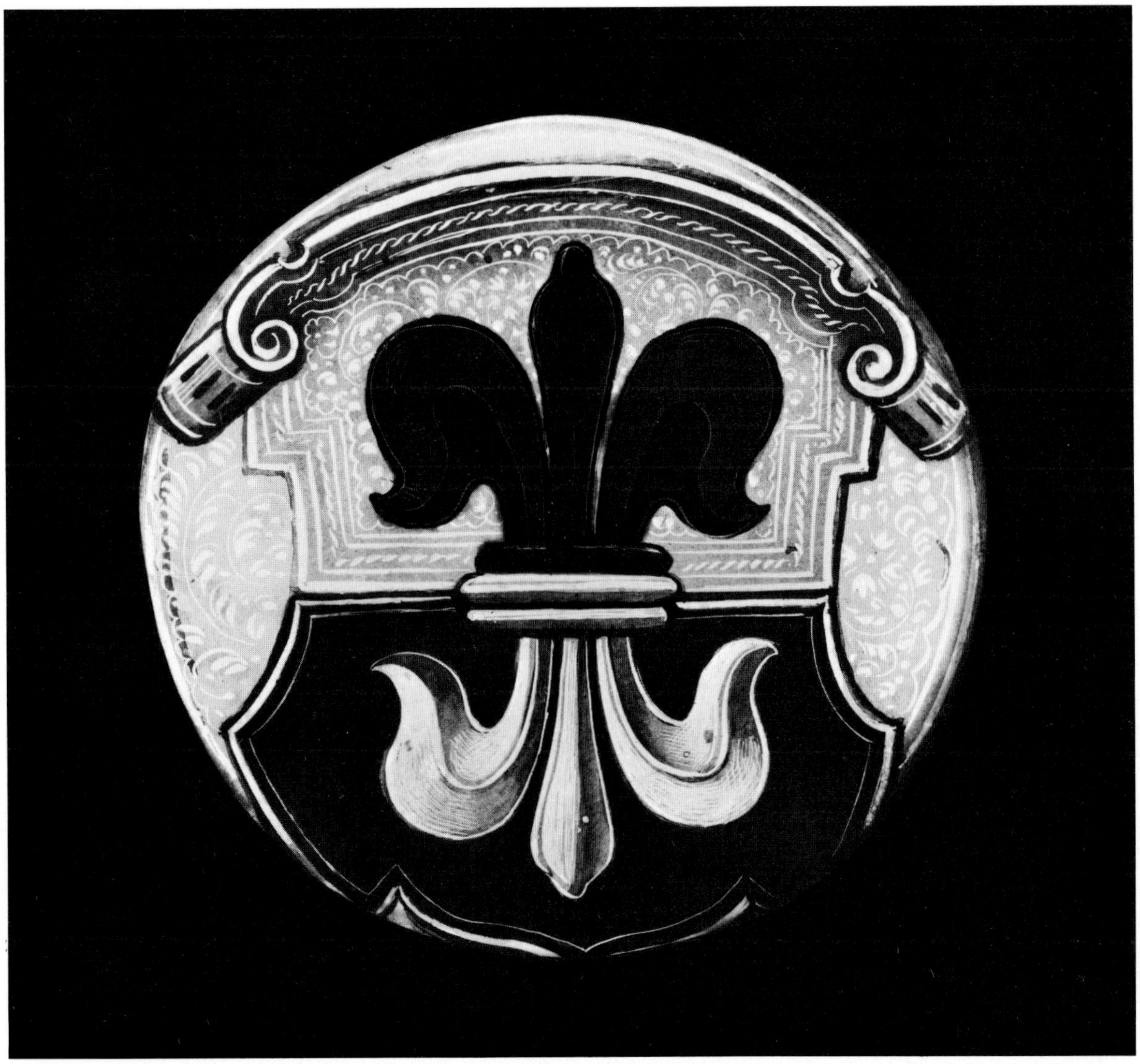

84. Roundel, painted with a coat-of-arms

German (Nuremberg); *c.*1560–80

Colourless glass painted in black and blue enamels and yellow stain

Diameter, $4\frac{3}{4}$ in. (12 cm)

The coat-of-arms is painted in black and yellow on a background of rays in blue. The colours used are enamel pigments with silver stain for the yellow. The decoration is painted, with scratched detail. The arms are parti per pale, dexter, per pale sable and argent, three mullets of six points in pale or; per fesse or and sable, a fleur-de-lis countercharged. It has not been possible to identify the individual to whom these arms belong but he is likely to have been related to the Bamberger family, as the arms shown on the sinister half of this roundel are identical with those shown on the preceding roundel (*83*). The arms shown on the dexter half are those of Dillherr, a Nuremberg family.

Miss Alice de Rothschild E2/24/5h*

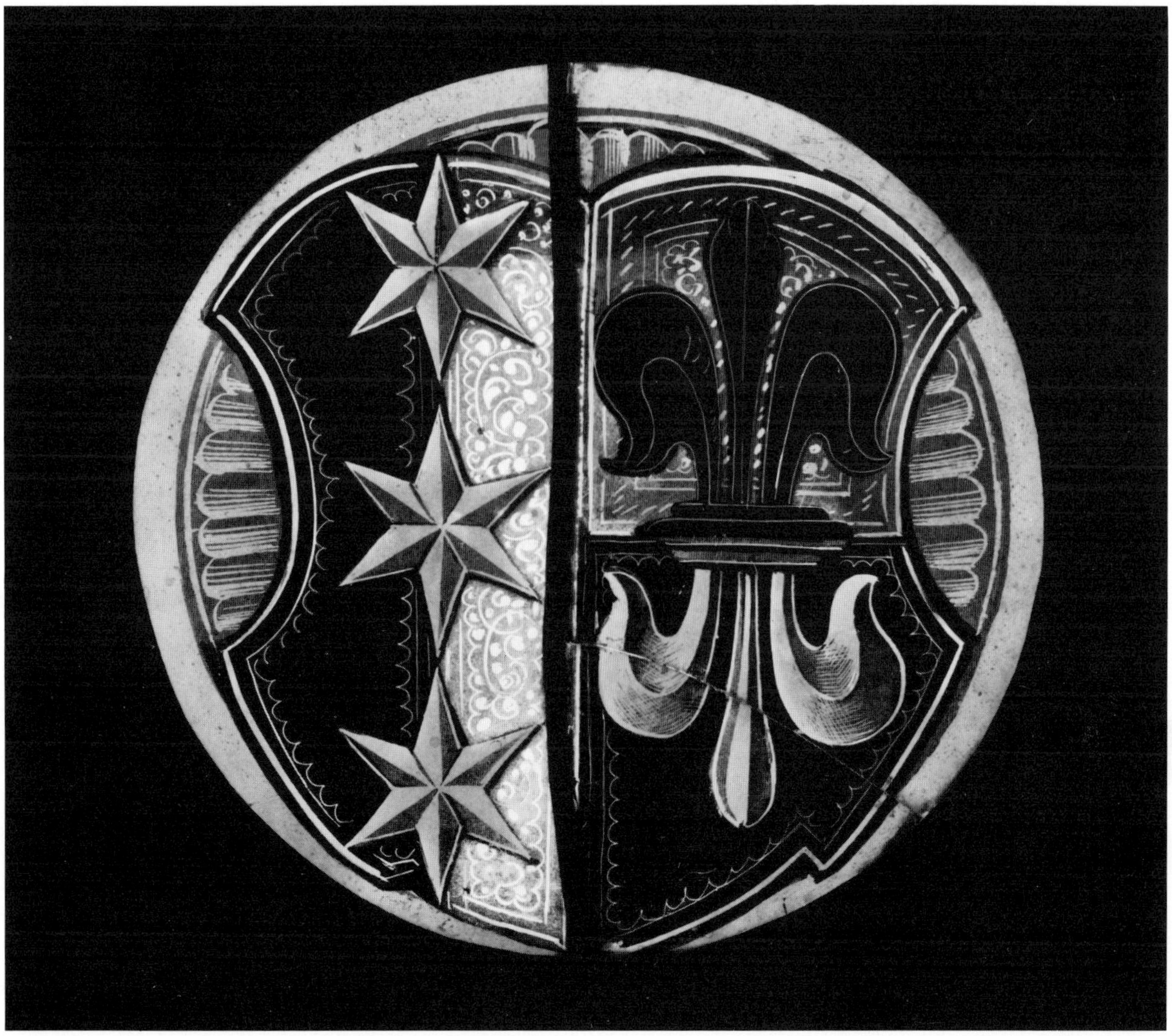

85. Panel, painted with a standing figure and a coat-of-arms

Swiss (Zurich), by Lukas Zeiner or his workshop; *c.*1500

Blue, red, murry and colourless glass painted in black enamel and yellow stain

Height, 14⅞ in. (37·8 cm); width, 10½ in. (26·7 cm) *Colour illustration on page 315*

Panel decorated with a female figure in a robe of dull blue, green and light murry standing beside a coat-of-arms with black and yellow mantling, against a background of red foliate scrolls above and green grass and leaves below, within an architectural frame painted in shades of grey with touches of yellow. The two figures standing in the spandrels are playing musical instruments. Above the coat-of-arms is a pilgrim's staff and scrip.

The arms are those of Stapfer (Basel) and are given by Rietstap as 'd'or, à cinq palissades gabionnées de sable, rangées en fasce, chargé d'un pont volant d'or'. As shown here, the arms differ in having six palissades gabionnées, and they and the pont volant are on a sable field.

Inscribed on the arch at the top, '1550', and in a rectangular panel at the base, 'Hans Růdolff Stapffer/Būrger zūo Zürich' (Hans Rudolff Stapffer Burgher of Zurich).

The colours used are pot-metal with silver stain for the yellow. The green on the robe is produced by painting silver stain on the back of flashed blue. Flashed glass is abraded through to white in some places. The decoration is painted, with scratched detail.

The architectural frame and the inscription are probably modern. The mouldings on the columns, the method of stippling and the blind arcading in the spandrels are all quite unlike the work of Zeiner or his workshop to which the panel is ascribed. The date 1550 is some forty years too late for this type of panel and inscriptions on stained glass by Zeiner and his followers are usually done in scrolls of irregular outline and in a heavy, somewhat clumsy Gothic script. Furthermore, the yellow stain used on the architectural frame differs noticeably in colour from that used on the centre part.

This panel is illustrated as Fig.7 in Bernard Rackham, 'Glass Painting as an Art for Today. Part 1, Antecedents', *Artwork*, Vol.2, No.5, October–December, 1925, pp.5–12, and there attributed to the Zurich School. However, in his article 'The Glass-Painter Lukas Zeiner of Zurich and his School', *Old Furniture*, Vol. VIII, October, 1929, pp.54–60, Rackham attributes this panel 'to the hand of Lukas Zeiner'. He also states that the border is restored. An attribution to Zeiner or his workshop is highly likely. The composition places this panel in the same general group as Schneider 34, 35, 43, 53, 54, 54A, 65, 71 and 73. The treatment of the hair and headdress is particularly close to Schneider 43, 54, 54A and 73, while the placing of the figure's right hand is similar to 43, 53, 65 and 71. The grass and leaves below the shield can be compared with 70 and 71 and the foliate scrolls above the shield are similar to those found on 67, 68 and 73. The delicate modelling of the face, particularly the mouth and chin, has much in common with that of the right-hand standard bearer of 69. Schneider attributes 65 and 67–70 to Lukas Zeiner himself and 43, 53, 54 and 71 to him or his workshop.

Miss Alice de Rothschild E2/24/51*

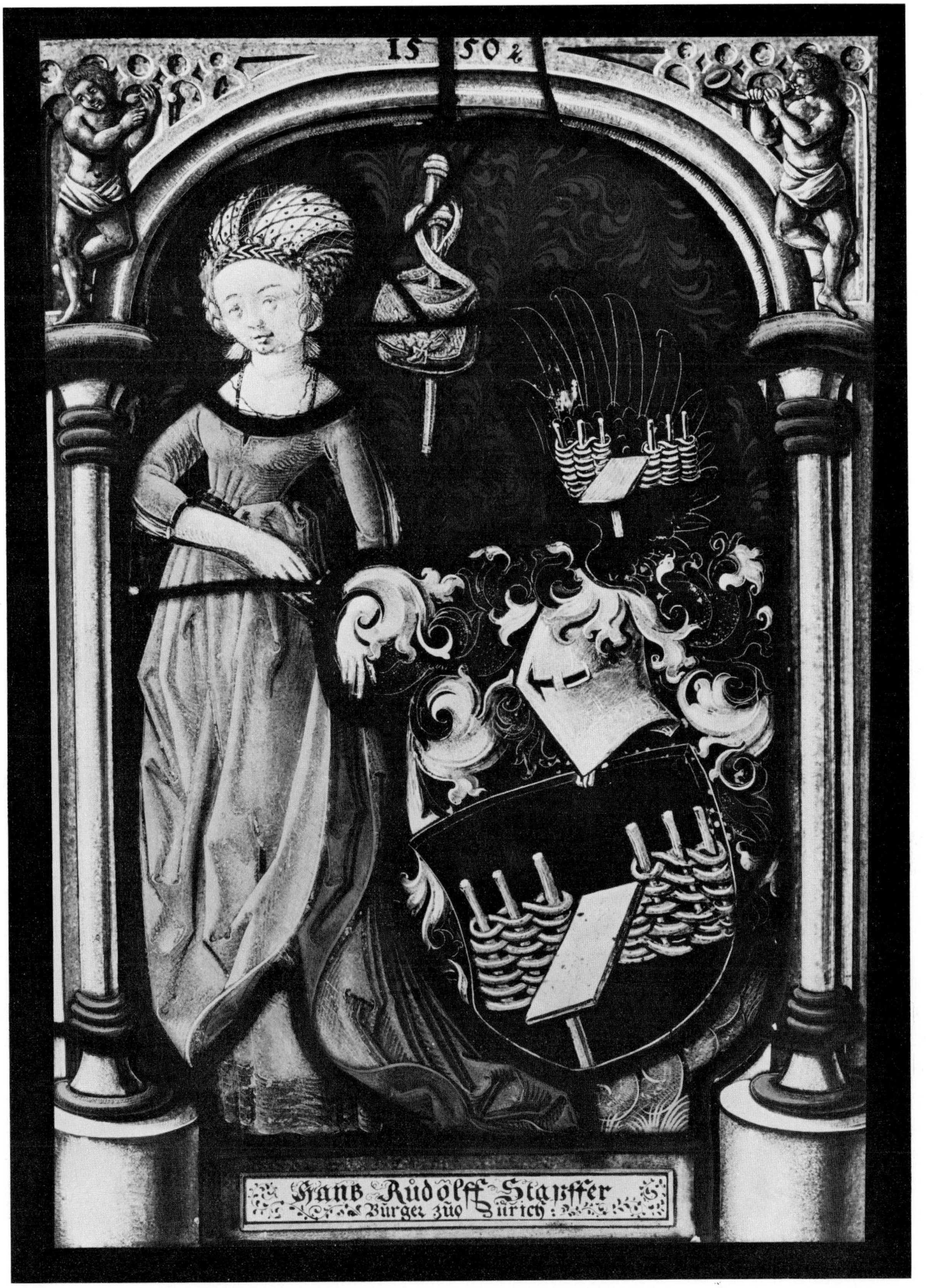

15 50
Hans Růdolff Stauffer
Bürger zuo Zürich

86. Panel, painted with a coat-of-arms

Swiss; first quarter of the 16th century

Blue, green, red and colourless glass painted in black enamel and yellow stain

Height, $15\frac{1}{8}$ in. (38·3 cm); width, $10\frac{1}{2}$ in. (26·7 cm)

Panel decorated with a coat-of-arms with red and white mantling standing within an architectural frame painted in shades of grey and yellow, against a dull blue ground above and dull green grass below. The arms are those of Arsent (Fribourg and Neuchâtel), parti per pale argent and gules, two roses countercharged slipped and leaved. The lower edge of the panel consists of a scroll, inscribed 'ANNO DOMINI MDXXIII'. The colours used are pot-metal with silver stain for the yellow. Flashed glass is abraded through to clear glass in some places. The decoration is painted, with scratched detail. The painting on the whole panel, particularly the shield, is badly rubbed. The back is extensively scratched. Part of the upper border is restored and the inscription is modern. The glass on which it is painted differs from most of the rest of the panel. The right hand column acting as a frame and possibly some of the foliage in the spandrels above the arms seem likely to be additions, as such details are normally symmetrical.

A different version of the arms of Arsent, in which the roses are without stalks or leaves, can be seen in Schneider 182 which is dated 1529. Rietstap also omits the stalks and leaves and gives the dragon on the crest as or rather than vert. However, L. and M. Jéquier, *Armorial Neuchâtelois* Neuchâtel, 1939, Vol. 1, p. 81, shows the arms precisely as here.

The scratched decoration flanking the crest is of a type not frequently found before 1500 and which became more common in the first quarter of the sixteenth century (cf. Schneider 79, 159, and 168 dated *c.*1505, *c.*1520 and *c.*1525 respectively). If this dating is accepted, the date of 1523 on the scroll may conceivably be correct. The regular capital letters, roman numerals and indeed the whole nature of the scroll are inadmissable for an early sixteenth-century date. The inscription may however repeat an earlier one containing the same date. If this is so, the panel may well have been made for Peter Arsent, a citizen of Fribourg in Switzerland, prominent between 1519 and 1525, who died in 1533 (see Schneider 182).

Colour illustration on page 315

Miss Alice de Rothschild — E2/24/5j*

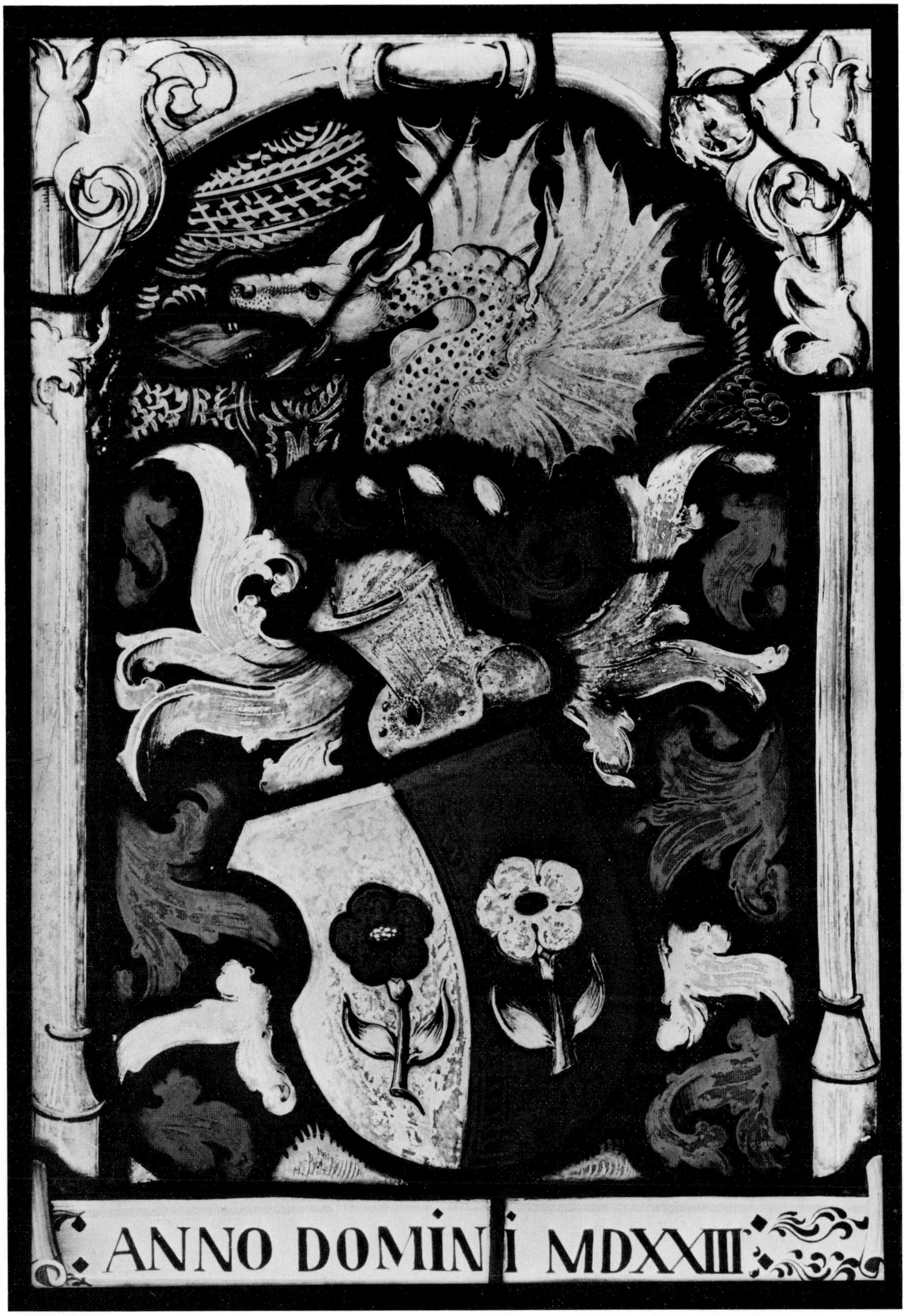
ANNO DOMINI MDXXIII

87. Panel, painted with a mounted knight, an attendant and a dog

Swiss; dated 1610

Red, green, purple and colourless glass painted in black, purple, blue and brown enamels and yellow stain

Height, $13\frac{1}{4}$ in. (33·7 cm); width, $8\frac{3}{4}$ in. (22·2 cm)

Panel decorated with an architectural frame of red, green and dull purple within which a man in armour, accompanied by a page painted in purple and dull blue, and a black dog, rides on a brown horse through a landscape painted in brown and yellow. Above is a scene of two men driving cattle, painted in shades of black, brown and yellow. Below is a coat-of-arms and an inscription within a yellow strapwork frame. The arms are gules, a ram (?) rampant to sinister argent between two mullets standing on a mound or.

Inscribed across the upper part of the main scene:

> 'Ob ich mich schon von Viehzucht năhr
> So bin ich doch ein Mann zur Wehr'
> (Although I live by the breeding of cattle
> I am none-the-less a knight)

Inscribed across the base:

> 'Samsson Mülli
> Stein'
>
> 'Und Caspar von
> Müllistein Bedt
> Brüderen 1610'
>
> (Samson von Müllistein and Caspar von Müllistein, both brothers, 1610)

All the colours are enamel pigments with the exception of the red, green and purple of the architectural frame, which are pot-metal. The flashed red is abraded through to white in some places and the yellow is silver stain. The decoration is painted, with scratched detail.

The lower inscription does not go with the arms and the two parts of the inscription may also not belong together. The strapwork on the two parts differs in certain details, as does the tone of the silver stain. The small piece of glass along the base in the lower left corner of the panel is evidently a restoration. Parts of the columns of the architectural frame may also be later as they do not match each other. Generally speaking, such architectural details in Swiss panels are more or less symmetrical. In this case the base of the left-hand column seems to be original but that on the right, which is different in shape and colour, seems to be an insertion. The arms are evidently an insertion and are close to those shown in the dexter coat-of-arms of a pair painted on a panel, dated 1605, and inscribed 'Hans Petter Kilchsperger und Ursula Zwick', which formed Lot 48 in the sale catalogue of Lord Sudeley's Collection (Helbing, Munich, 4 October, 1911). They differ in that the Waddesdon arms have a mullet missing which can be explained by later restoration.

A stained glass panel is known of a Hans Müllistein and family, dated 1611, from Ennetbühl, in the canton of St Gall.

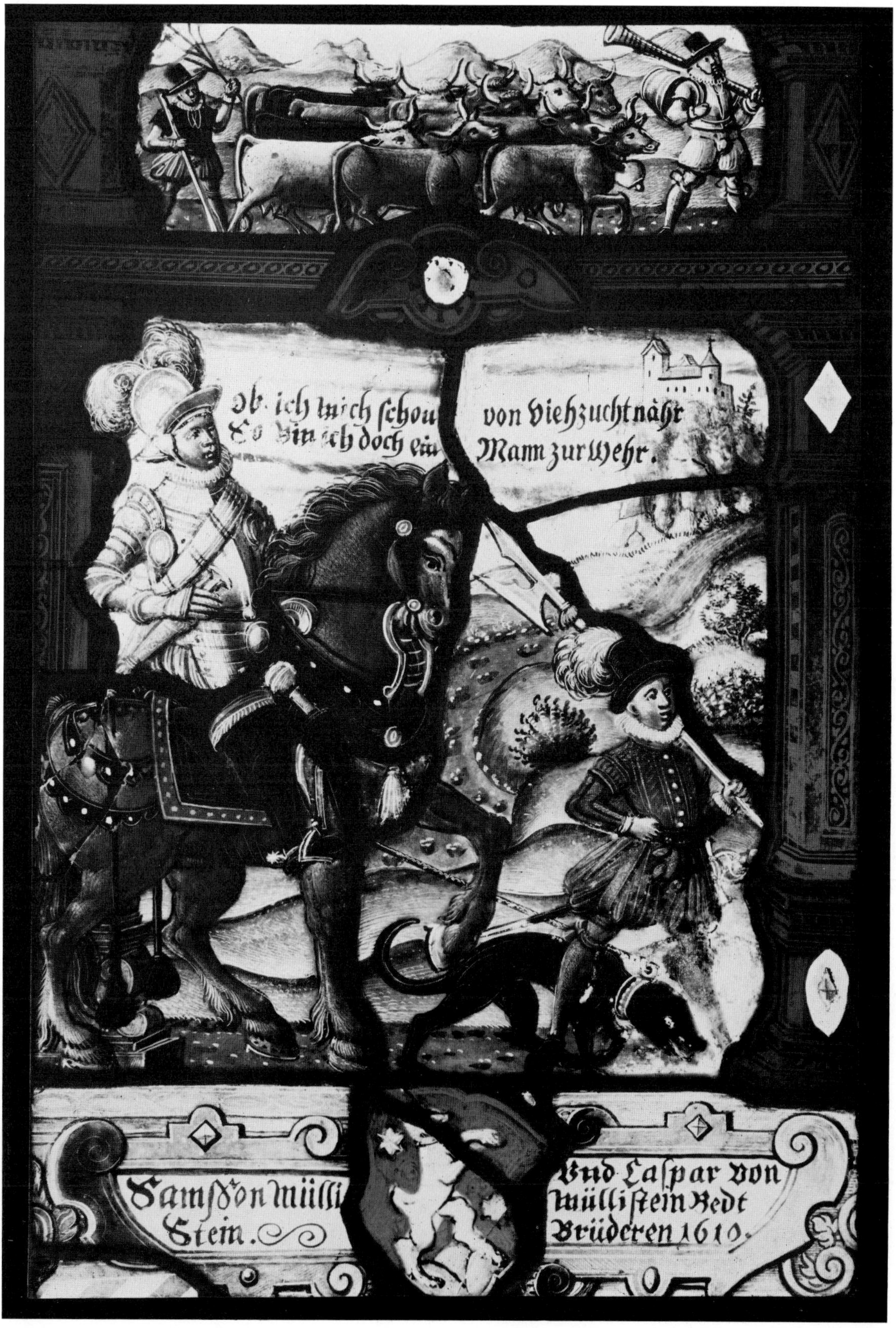
Ob ich mich schon von Viehzucht nähr
Mann zur Wehr.
Samson Mülli
Stein.
Und Caspar von
Müllistein Redt
Brüderen 1619.

The theme of a mounted figure can be found on Schneider 547 dated 1632, where a similar scene showing oxen also appears at the top. Comparable scenes are also painted on Schneider 432 and 495, dated 1599 and 1613 respectively.

Miss Alice de Rothschild E2/24/5*

88. Panel, painted with two coats-of-arms

Swiss; probably first half of the 17th century

Red, dull purple, murry and colourless glass painted in black, blue and green enamels and yellow stain.

Height, $13\frac{1}{8}$ in. (33·3 cm); width, $8\frac{3}{4}$ in. (22·2 cm) *Colour illustration on page 315*

Panel decorated with two coats-of-arms, with black, blue and yellow mantling, standing in an elaborate architectural frame in red, blue, dull purple, murry and yellow. Below are two figures painted in yellow with blue, dull purple and green robes inscribed 'ERALTIN (?)' and '[T?] ERRA'. Above are two scenes, painted in green, blue, purple and yellow; on the left, two men are trapping birds driven into a net by a dog; on the right, an owl on a perch is attacked by birds.

The arms are, on the left, Hochreutiner (St Gall), partie per pale azure and argent, dexter, three branches of oak bendwise in pale proper; sinister, on a tree-trunk two burning logs in saltire all proper, and on the right Schlappritzi (St Gall), or, bulls' horns reversed sable, between the horns a cross patté and in base two roundels argent.

Inscribed within a tablet at the top:

'Gelt unnd gŭt,
Ist als dann gût,
Wens hat den Man
Ders Recht bruche kan'
(Money and possessions are good if they are with the man who uses them in the right way)

and within a tablet at the base:

'Jacob Schlapritze, burger zu St
Gallen, und Frauw Ana Schlapritzin gebor
ne Blankin von Zürich sein Ehegem
achel. Anno 1594'
(Jacob Schlappritze, burgher of St Gall and Ana
Schlapprizte, née Blank of Zurich, his wife.
Year 1594).

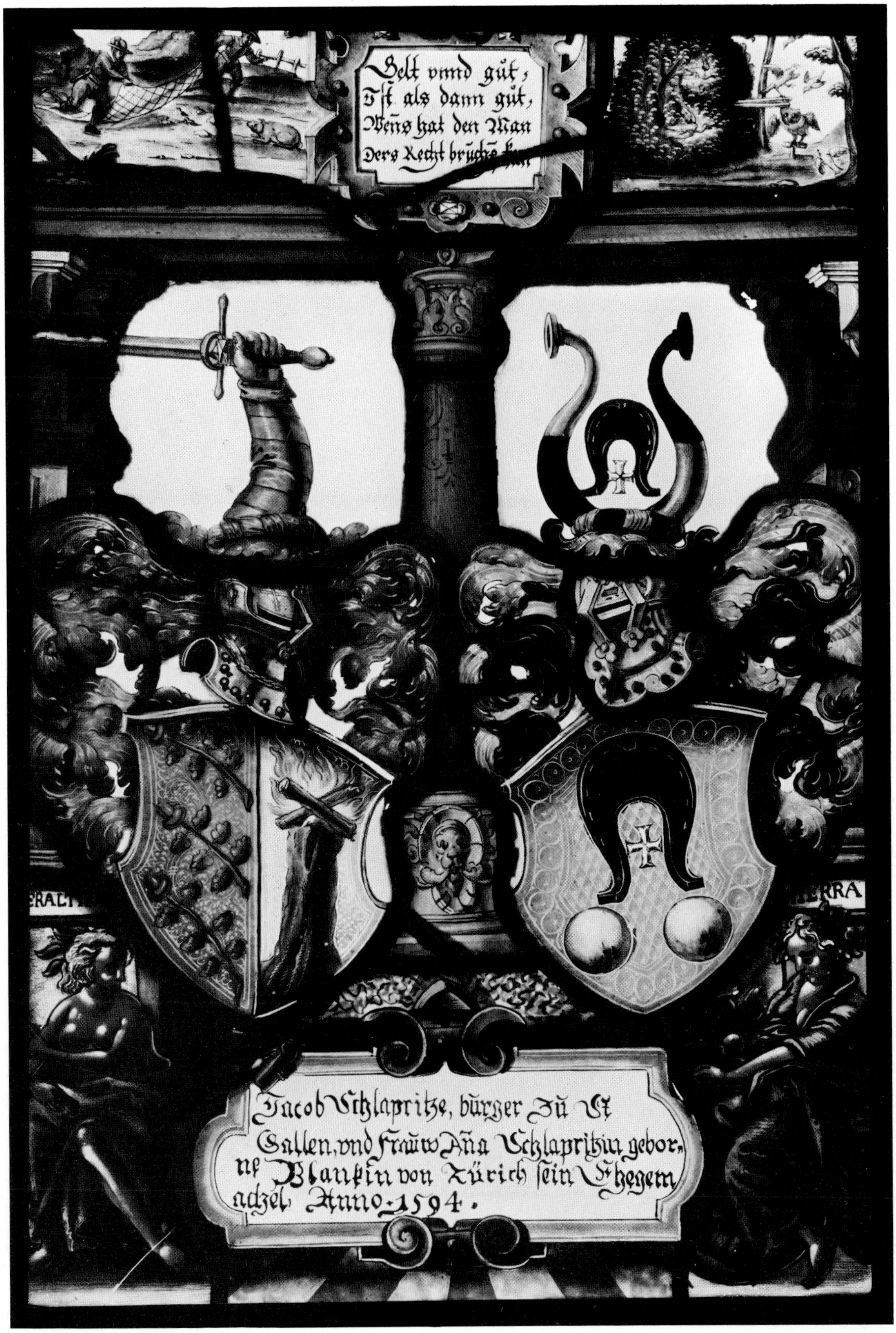
Gelt vnnd gůt,
Ist als dann gůt,
Wēns hat den Man
Ders Recht brüchē
Jacob Vchlapritze, burger zů St
Gallen, vnd Fraüw Aña Vchlaprizin gebor-
ne Blankin von Zürich sein Ehegem
achel, Anno. 1594.

The colours used are pot-metal for some of the murry, red and purple. The yellow is silver stain. All the other colours are enamel pigments. The decoration is painted, with scratched detail.

The lower inscription and the figures flanking it are probably later additions. The name Blankin is improbable and does not correspond with the arms. Since the husband's arms always appear on the left, this panel should commemorate a male member of the Hochreutiner family and a female of the Schlappritzi family, making nonsense of the inscription. The script differs noticeably from the upper inscription, which is probably genuine as it forms part of the fine and convincing scene at the upper left. The figures at the base are obscurely titled and painted in a heavy dark manner unlike the similar allegorical figures to be found on the panels that can be associated with this one, which is evidently by the same hand as *88*. Other than this, the most closely comparable panel is Schneider 495 dated 1613, which shows the same arms and was commissioned by Dominikus Hochreutiner of St Gall (1551–1635), who married Magdalena Schlappritzi (1566–1647) in 1584. This is very similar to the Waddesdon example, if not actually by the same hand, and it is therefore likely that the latter was also commissioned by Dominikus Hochreutiner. The Schlappritzi were one of the patrician families of St Gall. A number of panels showing the Schlappritzi arms are described and illustrated in Dr J. Egli, *Die Glasgemälde des historischen Museums in St Gallen*, Part 1, St Gallen, 1925.

Miss Alice de Rothschild — E2/24/5*

89. Panel, painted with two coats-of-arms

Swiss; probably first half of the 17th century

Purple, murry, green, red, blue and colourless glass painted in black and blue enamels and yellow stain

Height, 13⅛ in. (33·3 cm); width, 8¾ in. (22·2 cm)

Panel decorated with two coats-of-arms with black and yellow mantling, standing in an elaborate architectural frame in red, blue, green, dull purple, murry and yellow. Below are two figures painted in yellow with green, blue and purple robes inscribed '(R)HETORICA' and 'MUSICA'. Above are two fishing scenes in green, blue and yellow.

The arms are those of Schlappritzi (St Gall), or, bulls' horns reversed sable, between the horns a cross patté and in base two roundels argent. The other arms, or, three hunting horns in pale stringed sable and garnished argent, in sinister chief a mullet of six points also sable, are those of Schlumpf (St Gall), but with the addition of a mullet.

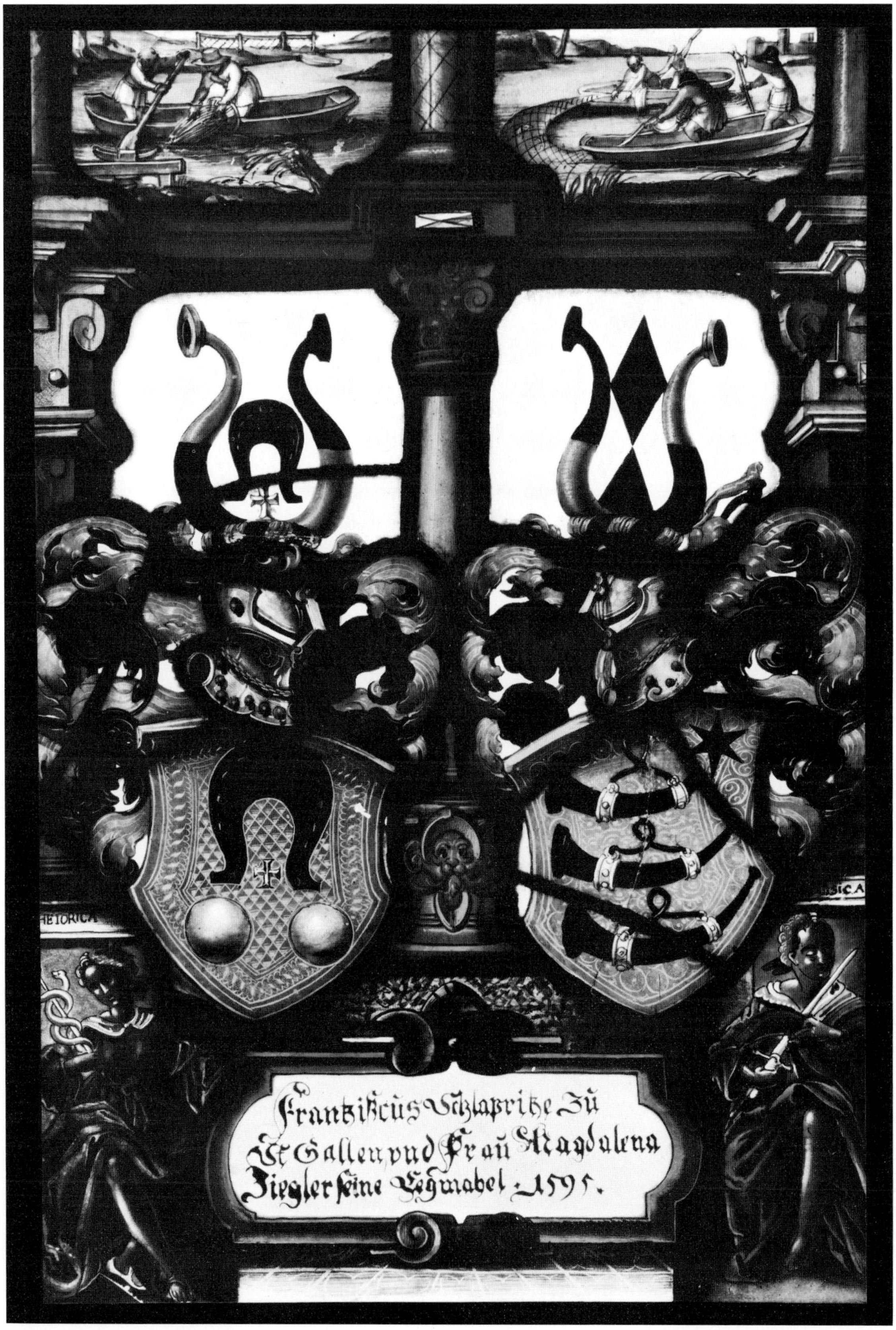
Frantziscus Schlapritze Zu
St Gallen und Frau Magdalena
Ziegler seine Eegmahel 1595.
HISTORICA
MUSICA

Inscribed within a tablet at the base:

'Frantzißcus Schlapritze zŭ
St Gallen und Fraß Magdalena
Ziegler seine Eegmahel. 1595'
(Frantzißcus Schlappritzi of St Gall and Magdalena
Ziegler his wife. 1595).

The purple, murry, green, red and some of the blue are pot-metal colours. The other colours are enamel pigments and the yellow is silver stain. The decoration is painted, with scratched detail.

The panel is evidently by the same hand as *88*. Here again the inscription and the figures flanking it appear to be later. The mullet on the Schlumpf arms is also intrusive but probably contemporary, since the scratched background and border designs are finely executed but differ from those on the rest of this shield. The inscription is in the same hand as that responsible for *88*, and here also the name of the wife, in this case Ziegler, does not correspond with the relevant arms.

The combination of the Schlappritzi and Schlumpf arms in the same positions occurs on a panel illustrated as item 42 in J. Egli, op. cit. It is dated 1648 and was commissioned by Christof Schlappritzi and Margrit Schlumpf. Above the arms are fishing scenes similar to those on the Waddesdon panel. In terms of style, the Waddesdon panel has details in common with both this panel of 1648 and the Schneider group 492–495, one of which is dated 1612 and the rest 1613. This suggests that the Waddesdon panel should be dated somewhere between 1612 and 1648 and that it was commissioned by Christof Schlappritzi and Margrit Schlumpf.

Miss Alice de Rothschild E2/24/5*

90. Panel, painted with a group of male figures in long robes

Swiss (Zurich); dated 1667

Red and colourless glass painted in black, blue, red, brown, green and purple enamel colours and yellow stain

Height, 13 in. (33 cm); width, $8\frac{7}{8}$ in. (22·5 cm)

Panel decorated with a group of men in long robes accompanying a bearded man whose mitre-like hat may show that he is a priest, painted in blue, brown, grey, green, purple and yellow. They stand beside a portico in an architectural setting of red, blue, green, purple and yellow. Above them is a harbour scene in shades of grey, green, blue and yellow. Below, an oval wreath

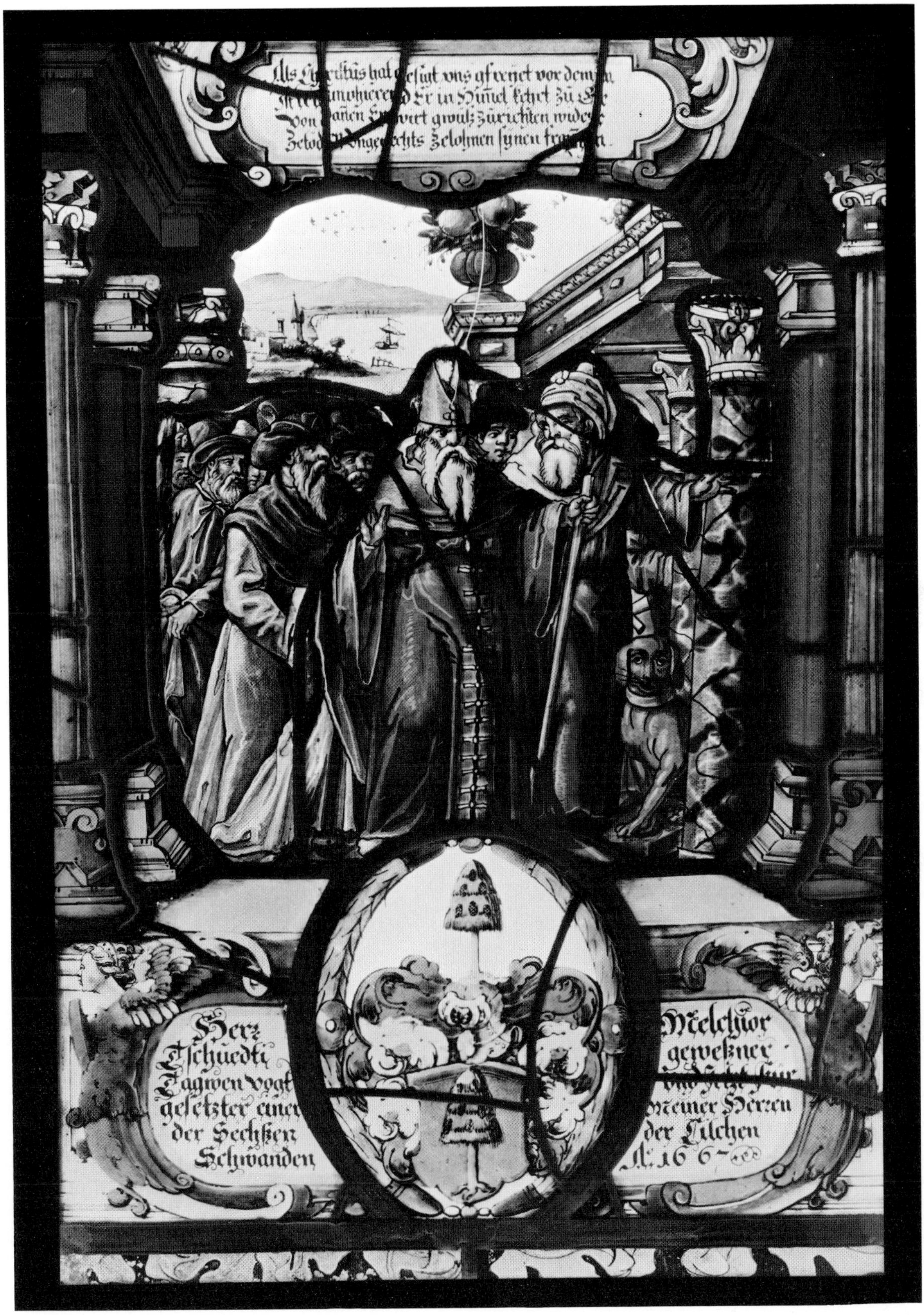
Herr
Tschuedti
Tagwen vogt
gesetzter einer
der Sechßen
Schwanden
Melchior
geweßner
meiner Herren
der Kilchen
A: 1667

in green, yellow, blue and purple containing a coat-of-arms with yellow and blue mantling. The arms are those of Tschudi (Glarus), azure, on a fir tree vert with trunk gules, six fir cones, three, two, one, gules, though Rietstap gives the field as or rather than azure. Along the base there is a narrow panel of pale blue leaf decoration. All the colours are enamel pigments with the exception of some of the red, which is pot-metal. The yellow is silver stain. The decoration is painted with scratched detail.

Inscribed within the tablet at the top:

'Als Christus hat gesigt uns gfreyet vor dem (?a) (demonen)
Ist thriumphierend Er in Himel kehrt zu Ehr
vo dañen Er wirt gwüss zürichten wide (? ker)
Ze töden Ungerechts zelohnen synen fromen'
(When Christ was victorious and freed us from evil (?)
He returned triumphantly to honours in heaven
Whence He will certainly return to judge,
To kill the unjust and to reward His faithful).

Inscribed on either side of the coat-of-arms at the base:

'Herr	Melchior
Tschuedti	geweßner
Tagwen vogt	und jetzt Für
gesetzter einer	meiner Herren
der Sechßen	der Cilchen
Schwanden	A°. 1667'

(Herr Melchior Tschudi formerly steward and now Principal, one of the sixteen elders of the church of Schwanden. Year 1667.[1])

The only restoration is the right-hand end of the upper cartouche, including the final letters of the inscription it contains. The narrow panel of leaf decoration at the base is evidently an unrelated addition.

The subject matter is obscure but is likely to be biblical as scenes from both the Old and New Testaments appear comparatively frequently on mid-seventeenth century panels. Schneider 573 shows a figure identifiable as a High Priest wearing a hat somewhat like that worn by the central figure in this panel. It is possible that he and the figure to the right are meant to represent Moses and Aaron. Schneider 657 has an Old Testament scene with an architectural setting, an upper cartouche and a wreath round the coat-of-arms which are closely similar to those on the Waddesdon panel. These details, the large scale of the subject, taking up all the central space and the numerous columns, are all typical of Zurich panels.

Miss Alice de Rothschild E2/24/5*

The panels of stained glass illustrated opposite are as follows: top left, 85; top right, 86; below left, 80; below right, 81; centre 88

[1] Schwanden is a village in the canton of Glarus.

LIMOGES ENAMELS

Madeleine Marcheix

Introduction

The greater part of the collection of Limoges enamels at Waddesdon Manor was assembled by Miss Alice de Rothschild (1847–1922) in order to furnish the two rooms, particularly the Smoking Room, left empty after the death of her brother Baron Ferdinand in 1898. Baron Ferdinand had bequeathed the enamels, jewellery, arms and armour, glass and maiolica which filled these rooms to the British Museum, where it formed the Waddesdon Bequest. Other enamels now at Waddesdon were formerly in the collection of Baroness James de Rothschild of Paris. They were inherited by her son, Baron Edmond (1845–1934), who in turn bequeathed them to his son, Mr James A. de Rothschild.

These pieces cover the entire period of the Renaissance, from the enamels of the Aeneid Master right at the beginning, up to a mirror-back by Léonard II Limosin. Of course not all the Limoges artists are represented, and this catalogue will only contain notes on the enamellers whose works decorate the walls of the rooms at Waddesdon Manor. Entries on these artists are arranged as far as possible in chronological order. It would have been interesting to put together objects presenting the same subject matter, but this would have necessitated dividing up the works of a single artist. These comparisons are however pointed out by cross-references in the catalogue since it is fairly certain that Miss Alice did not bring these objects together by pure chance.

The word enamel inevitably calls up the name Limoges. For this one need only skim through the catalogue of one of those great nineteenth-century exhibitions of *objets d'art* – such as that held at South Kensington in 1862 – and compare the number of enamels from Limoges with those from elsewhere.

The production of enamels at Limoges was, however, divided into two great periods, one covering the later Middle Ages, till the middle of the fourteenth century, the other beginning in the later fifteenth century, leaving a gap of more than a century between the two phases, when production was interrupted by the events of the Hundred Years War, which included the Sack of Limoges by the Black Prince in 1370. This led to the final closing down of the workshops in Limoges, but there is evidence to show that the craft had already been in a decline for at least a generation.

When enamel began again to be made in Limoges it was in a technique different from that used in the city during the earlier phase. Medieval enamels were made in two different techniques, both using opaque enamel: *cloisonné* (on gold), and *champlevé* (on copper). In the former, which was never practised at Limoges, the powdered enamel was placed in areas separated by partitions (*cloisons*) built up on the surface of the gold; in the latter it was put into areas hollowed out of the plate with a burin. In both cases the plate was then fired at a fairly high temperature. A third technique, that of translucent enamel, in which the enamel allowed the silver support to shine through it, appeared at the end of the thirteenth century in Italy and in the fourteenth century in France. This technique was not used at Limoges, perhaps because silver was too expensive.

The revival of the production of enamels, and in particular, painted enamels in the late fifteenth century was accompanied by developments in other arts; easel-painting was introduced and stained-glass painters were able now to paint on white glass and fire it as opposed to the earlier practice of joining together pieces coloured all through. Enamelling and glass-painting have more than one similarity at this date and an artisan such as Pierre Pénicaud was, at times, involved in both fields.

The oldest examples of French painted enamel which have survived do not, curiously enough, originate from Limoges. These are two round plaques, enamelled on both sides and attributed to the painter Jean Fouquet, dating from about 1450. One (in the Louvre) is a self-portrait; the other, formerly in the Schlossmuseum in Berlin, but now destroyed, represented the *Descent of the Holy Spirit*. The black ground was applied to the plaques and covered with a layer of gold through which the subject was drawn on to the black enamel. A little later come two very small round medallions by the Milanese artist Filarete, in the Dresden Museum, dated 1465, which have gold decoration on a blue ground.

The first painted enamels made at Limoges were on plaques of various shapes, the subject matter appearing on the principal side only. At first these copper plaques were thick and flat, in order to resist the distortion which might be caused by the temperature at which the enamels were fired. Later it was discovered that the same result was obtained when a thinner plaque was used, provided it was made slightly curved.

At first the reverse was covered with a 'counter-enamel' made of the residue of enamels used elsewhere and which after firing was likely to be streaky, irregular and dark. After about 1530 enamellers began to cover the back with a flux which was really just colourless enamel, close to glass in composition, composed of silica, to which were added soda and potassium chlorate and even lime, which enabled fusion to take place at a lower temperature.

After this stage, work began on the principal face of the plaque. Here the enamel powders were applied straight on to the plaque without dividing partitions, as in *cloisonné* and *champlevé*.

The plaque was first covered with a ground or preparation (*apprêt*). In the earliest work this *apprêt* consisted of a black or very dark layer overlaid with a white one, through which the subject was drawn with a sharp point, thus revealing the black or dark blue enamel. At this stage, the piece resembled an engraving. After it had been fired, it could be retouched and details in black added with the brush on the white layer. In one case, however, the Aeneid series (see *1*), the subject was drawn in bistre on the copper plaque which had been covered not with one or more grounds of *apprêt* but with a layer of silver.[1] Next, coloured enamels were applied together containing a flux and coloured by metallic oxides. The new technique allowed the use of transparent enamel which retained its lustre if it was applied on a white ground, made opaque by the addition of tin oxide, or took on depth from a dark ground, but a colourless enamel was sometimes applied discreetly on the copper to produce a beautiful bronze shade.

During the whole of the first period manganese was used instead of red, because the latter became distorted in the firing, but later this weakness was rectified by applying the red after the other colours and subjecting it to less intense heat. Gold highlights were also applied on the surface, but these have often disappeared, leaving only slightly matt traces on the brilliance of the enamel.

[1] P. Verdier, *Walters Art Gallery, Catalogue of the Painted Enamels of the Renaissance*, Baltimore, 1967, p.76.

During the first phase of painted enamels, represented by the artist called Monvaerni and by the Master of the Orléans Triptych (called after the enamelled triptych of the Annunciation in the Orléans Museum), a black ground was often used, but this was abandoned by the next generation and not taken up again until the introduction of *grisaille* enamel in the 1530s. Instead, the white ground was applied directly on to the copper and the subject was drawn on it in black or bistre with a brush.

Sometimes the enameller applied to the copper small squares of gold or silver in leaves, known as *paillons* or foils, before he added the coloured enamel, as a base for the decorative details. Several of the earlier enamellers – Nardon Pénicaud, the Master of the Orléans Triptych and the Master of the Louis XII Triptych – also decorated the borders of dresses, hangings, clasps and flowers with little drops of enamel on foils to simulate precious stones.

In the last third of the sixteenth century, after the period from 1540 to 1560 when grisaille was the fashion, the taste for coloured enamel returned and the use of foil spread increasingly not only for small decorative details but also for large surfaces, particularly clothes, ornaments, birds and flowers.

All the works which date from before about 1530 were either single plaques or groups assembled in polyptychs, and all were in coloured enamel. With very few exceptions – the *Sultan* in the Brunswick Museum, attributed to the Master of the Orléans Triptych, and the Aeneid series – these enamels represented religious subjects, nearly all of which were taken from the New Testament and the lives of the saints.

The enamellers of the first period must have found their sources in engravings from the Low Countries or from Burgundy, although few have so far been identified. Some details of engravings by Schongauer and Israel van Meckenem turn up here and there in enamel plaques; but it is rare that one can say with certainty that they served as models. A rare exeption is Monvaerni's *St Christopher* in the Louvre which is an exact copy of an engraving of 1460 by the Master E.S.

The sources used by Jean I Pénicaud can be identified more exactly. He frequently bases his plaques on Dürer's *Small Passion*. However, the copies made at this time were not slavish: each enameller stamped his own personality on the figures and even on the composition, and his powers of invention had free rein in the translation into colour of the black and white engravings.

From about 1530 onwards one can in many cases trace the sources of Limoges enamels. Collections of engraved prototypes continued to circulate in France and they were followed by French illustrated books, particularly bibles, courtly romances and mythological anthologies. The enamellers of Limoges owed much to these sources, not only in their painting of small enamel pictures in the manner of their predecessors, but also in the decoration of dishes and vessels.

The Aeneid Master had already shown the way by adapting, if not copying, the plates of Grüninger's Virgil in his enamels. After him the models still came mainly from outside France. The Passions by Dürer and Lucas van Leyden and the prints of Marcantonio Raimondi – sometimes known through copies – inspired the best masters from Léonard Limosin to Pierre Reymond, but the artists of Limoges were soon attracted by the freedom and novelty of inspiration introduced from Italy by the painters of the School of Fontainebleau. The wind of the Renaissance had reached Limoges. In 1532 Jean de Langeac became Bishop of Limoges. He was much travelled and during a period of fifteen years had occupied important diplomatic posts in Portugal, Poland, Switzerland, Scotland, Rome and Venice. He was deeply influenced by what he saw of Italian art and his tomb in Limoges cathedral is one of the finest monuments of the French Renaissance. It must have been through him that Léonard Limosin was introduced at Court.

In the second half of the century, after the death of François I, the so-called 'Little Masters', such as the engravers of book-illustrations, Etienne Delaune and Bernard Salomon replaced Primaticcio, Rosso and Niccolò dell'Abbate as the favourite models of the artists of Limoges. The change of taste was one of the features which marked the gradual decline of the art of enamel in Limoges, though the tradition, and even to some extent the style of the later sixteenth century was continued until well after 1600.

Bibliography

The abbreviation *B.S.A.H.L.* has been adopted for references to the *Bulletins de la Société archéologique et historique du Limousin.*

EXHIBITION CATALOGUES

Catalogue of the Special Exhibition of Works of Art of the Medieval, Renaissance and more recent Periods, on loan at the South Kensington Museum, June 1862, Ed. J. C. Robinson, London, 1863 (see Cat. Nos.*7* and *13*).

Union centrale des Beaux-Arts appliqués à l'industrie. Exposition de 1865. Musée rétrospectif, Paris, 1867 (see Cat. Nos.*6* and *32*).

Exposition universelle. Catalogue général . . . Histoire du travail et monuments historiques. Paris, 1867 (see Cat. Nos.*1*, *6*, *8*, *32*).

Catalogue of the Special Loan Exhibition of Enamels on Metal held at the South Kensington Museum in 1874, London, 1875 (see Cat. No.*12*).

Exposition en faveur des Alsaciens-Lorrains, Paris, 1874 (see Cat. No.*1*).

Burlington Fine Arts Club, London, 1897 (see Cat. No.*7*).

L'Ecole de Fontainebleau, Paris, Grand Palais, 1972, and National Gallery of Canada, Ottawa, 1973.

SALE CATALOGUES

Strawberry Hill sale, 25 ff April, 1842 (see Cat. No.*6*).

Debruge-Duménil, Paris, 23 January–12 March 1850 (see Cat. Nos.*9* and *39*). (cf. Jules Labarte, *Description des objets d'art qui composent la collection Debruge-Duménil. . .*, Paris, 1847).

Ralph Bernal, London, 5 ff March 1855 (see Cat. Nos.*28*, *42*).

Charvet, Paris, 11–12 May 1883 (see Cat. No.*30*).

Andrew Fountaine, London, 16 ff June 1884 (see Cat. No.*28*).

Hollingworth Magniac of Colworth, London 2 ff July 1892 (see Cat. Nos.*7* and *13*)

Frédéric Spitzer, Paris, 17 ff April 1893 (see Cat. Nos.*14*, *15*, *33*). (cf. *La Collection Spitzer, antiquité, moyen âge, renaissance*, Ed. Emile Molinier; émaux peints par Claudius Popelin, Paris, 1891, Vol.II).

Martin Heckscher, of Vienna, London, 4–6 May 1898 (see Cat. Nos.*3*, *9* and *39*).

John Edward Taylor, London 1 ff July 1912 (see Cat. Nos.*7*, *14*, *15*).

GENERAL BIBLIOGRAPHY

Jean Adhémar, 'Les portraits dessinés du XVIe siècle au Cabinet des Estampes', *Gazette des Beaux-Arts*, September and December 1973.

Jean Adhémar, 'Une galerie de portraits Italiens à Amboise en 1500', *Gazette des Beaux-Arts*, October 1975.

Anon, 'Limoges enamels in the Wernher Collection, Luton Hoo', *Apollo*, May 1949.

Maurice Ardant, 'Notice historique sur les émaux, les émailleurs, leurs divers ouvrages, et les procédés de fabrication en usage à Limoges', *Bulletin de la Société royale d'agriculture* [de Limoges], No.2, Vol.20, 1842.

Maurice Ardant, 'Léonard Limosin émailleur; esquisse biographique', *B.S.A.H.L.*, Vol. VI, 1855.

Maurice Ardant, 'Les Pénicaud, émailleurs', *B.S.A.H.L.*, Vol. VIII, 1858.

Maurice Ardant, 'Les Limosins, émailleurs', *B.S.A.H.L.*, Vol. IX, 1859.

Maurice Ardant, 'Pierre, Jean et Martial Courteys', *B.S.A.H.L.*, Vol. X, 1860.

Maurice Ardant, 'Emailleurs limousins: Jehan Court dit Vigier, Susanne Court, Jean de Court', *B.S.A.H.L.*, Vol. XI, 1861.

Maurice Ardant, 'Emailleurs limousins: Pierre Ier et II, Jean, Joseph et Gabriel Reymond, *B.S.A.H.L.*, Vol. XII, 1862.

Isa Belli Barsali, *Lo smalto in Europa*, Milan, 1966.

J. Adam von Bartsch, *Le Peintre graveur*, Vienna, 1803–1821.

Louis Bourdery, *Les émaux peints à l'exposition rétrospective de Limoges en 1886*, Limoges, 1888.

Louis Bourdery, 'Les Jean Limosin, émailleurs, *B.S.A.H.L.*, Vol. XXXVI, 1887.

Louis Bourdery, 'Léonard Limosin émailleur', *B.S.A.H.L.*, Vol. XLIV, 1895.

Louis Bourdery and Emile Lachenaud, *Léonard Limosin peintre de portraits*, Paris, 1897.

Joseph Breck, 'The Aeneid Enamels', *Bulletin of the Metropolitan Museum of Art*, Vol. XX, 1925.

Wilfred Buckley, 'A Limoges enamel plaque of 1434', *Burlington Magazine*, September 1931.

Georges-Emmanuel Clancier, 'L'œuvre de Limoges', *Cahiers de la Céramique et des Arts du Feu*, No.54, 1974.

H. Cunynghame, *European Enamels*, London, 1906.

Alfred Darcel, 'Exposition de Manchester. Les arts industriels du moyen âge et de la renaissance', *Revue française*, 1858.

Alfred Darcel, *Paris, Musée national du Louvre. Notice des émaux et de l'orfèvrerie*, Paris, 1867.

André Demartial, 'Chronique de l'orfèvrerie et de l'émaillerie de Limoges en 1910, 1911, 1912, 1913', *B.S.A.H.L.*, Vols. LX, LXI, LXII, LXIII.

André Demartial, 'Léonard Limosin, émailleur et graveur', *Revue de l'art chrétien*, Vol. LXII, 1912.

O. Dobroklonskaya, *Painted Enamels of Limoges XV and XVI Centuries. The State Ermitage Collection*, Moscow, 1969.

Antoinette Faÿ, *Un chandelier de Pierre Reymond*, Berne, 1971.

Edouard Garnier, *Histoire de la verrerie et de l'émaillerie*, Tours, 1886.

Marie-Madeleine Gauthier and Madeleine Marcheix, *Emaux de Limoges*, Prague, 1962.

Louis Guibert, *Catalogue des artistes limousins*, Limoges, 1909.

A. Ilg., 'Die Limousiner grisaillen in den Kaiserlichen Haus Sammlungen', *Jahrbuch der kunsthistorischen Sammlungen des allerhöchsten Kaiserhauses*, Vol. II, Vienna, 1884.

Bertrand Jestaz, *L'influence flamande en France à la fin du XVIe siècle*, Paris, 1975.

Jules Labarte, *Histoire des arts industriels au moyen âge et à l'époque de la Renaissance*, Paris, 1864–1866.

Léon de Laborde, *Paris, Musée national du Louvre. Notice des émaux, bijoux et objets d'art divers exposés dans les galeries du Musée du Louvre*, Paris, 1853.

F. de Lasteyrie, *Histoire de l'orfèvrerie depuis les temps les plus reculés jusqu' à nos jours*, Paris, 1875.

Pierre Lavedan, *Léonard Limosin et les émailleurs français*, Paris, 1914.

Charles de Linas, *Emaillerie limousine*, Liège, 1887.

London University, Courtauld Institute Galleries, *The Gambier-Parry Collection, Provisional Catalogue*, 1967.

Luigi Malle, *Smalti . . . del Museo d'arte antica. Catalogo*, Turin, 1969.

Madeleine Marcheix, 'Les émaux de Limoges à l'exposition de l'école de Fontainebleau', *B.S.A.H.L.*, Vol. C, 1973.

Madeleine Marcheix, 'Plaque émaillée de Léonard Limosin, Limoges, musée municipal', *B.S.A.H.L.*, Vol. XCVIII, 1971.

J-J. Marquet de Vasselot, 'Une suite d'Emaux limousins à sujets tirés de l'Enéide', *Bulletin de la Société d'histoire de l'art français* (*B.S.H.A.F.*), fasc. 1, 1912.

J-J. Marquet de Vasselot, 'L'orfèvrerie et l'émaillerie au XVIe siècle', in *Histoire de l'art*, ed. A. Michel, Vol. V,1, Paris, 1912.

J-J. Marquet de Vasselot, *Paris, Musée national du Louvre. Catalogue sommaire de l'orfèvrerie, de l'émaillerie et des gemmes du moyen âge au XVIIe siècle*, Paris [1914].

J-J. Marquet de Vasselot, *Les émaux limousins de la fin du XVe et de la première partie du XVIe siècle; étude sur Nardon Pénicaud et ses contemporains*, Paris, 1921.

J-J. Marquet de Vasselot, 'Emaux de la suite de l'Enéide du legs Salomon de Rothschild au Musée du Louvre', *Bulletin des Musées de France*, 1923.

J-J. Marquet de Vasselot, *Bibliographie de l'orfèvrerie et de l'émaillerie françaises*, Paris, 1926.

Alfred Meyer, *L'art de l'émail de Limoges, ancien et moderne. Traité pratique et scientifique*, Paris, 1895.

Peter E. Michaels, 'Technical observations on early Painted Enamels of Limoges', *The Journal of the Walters Art Gallery*, Vols. XXVII-XXVIII, 1964–1965 (1968).

Emile Molinier, *L'émaillerie*, Paris, 1891.

Emile Molinier, *Dictionnaire des émailleurs depuis le moyen âge jusqu'à la fin du XVIIIè siècle*, Paris, 1885.

Roger Pinkham, 'Attributions to the Aeneid Master', *Apollo*, May, 1972.

Roger Pinkham, *Limoges Painted Enamels*, London, Victoria and Albert Museum, 1974.

Roger Pinkham, 'The Art of the Limoges Enameller', *Country Life*, 5 June, 1975.

Claudius Popelin, *L'émail des peintres*, Paris, 1866.

Claudius Popelin, *Les vieux Arts du feu*, Paris, 1878.

Bernard Rackham, 'Limoges Enamels of the Aeneid at Alnwick Castle', *The Burlington Magazine*, Vol. XXXVIII, 1921.

Olga Raggio, 'Decorative Portraits by Léonard Limousin', New York, Metropolitan Museum of Art, 1951.

Ris-Paquot, *Etude sur les émaux anciens*, Paris, 1881.

A. Robert-Dumesnil, *Le peintre-graveur français, ou catalogue raisonné des estampes gravées par les peintres et les dessinateurs de l'école française* . . ., Paris, 1835–1871.

Marvin C. Ross, 'Léonard Limousin, Enameller and Engraver', *Print Collectors' Quarterly*, October 1938.

Marvin C. Ross, 'Notes on Enamels by Pierre Reymond', *The Journal of the Walters Art Gallery*, Vol. II, 1949.

Marvin C. Ross, 'A Method for Detecting False Limoges Painted Enamels', *The Journal of the Walters Art Gallery*, Vol. XII, 1949.

Edmond du Sommerard, Paris, *Musée des Thermes et de l'Hôtel de Cluny. Catalogue et description des objets d'art de l'antiquité, du moyen âge et de la Renaissance exposés au musée*, Paris, 1883.

W. F. Stohlman, *Gli smalti del Museo sacro Vaticano*, Vatican City, 1939.

Abbé Texier, *Essai historique et descriptif sur les argentiers et les émailleurs de Limoges*, Poitiers, 1843.

Philippe Verdier, 'French Renaissance Enamels', *Apollo*, May, 1971.

Philippe Verdier, *Enamels by Jean III Pénicaud and Pierre Reymond*, City Art Museum of St Louis, Museum Monographs, II 1971.

Philippe Verdier, 'Les Sibylles dans l'émaillerie limousine', *B.S.A.H.L.*, Vol. LXXXIX, 1962.

Philippe Verdier, *The Walters Art Gallery. Catalogue of the Painted Enamels of the Renaissance*, Baltimore, 1967.

Henri Zerner, *L'école de Fontainebleau*, Paris, 1969.

CATALOGUE OF LIMOGES ENAMELS

THE AENEID MASTER

In 1502 there appeared at Strasbourg the complete works of Virgil (*Publii Virgilii Maronis Opera*), edited by Sebastian Brandt, printed by Johannes Grüninger and illustrated with wood engravings by an anonymous artist. At least two further editions were published at Lyons in 1517 and 1529 when the same blocks were used, though their state had deteriorated. An enamel worker – whose name is also not known – made, presumably on commission, a single set of reproductions in enamel of most and possibly all of the engravings which illustrated the first nine books of the Aeneid, that is to say 104 out of the 144 plates in the book. Marquet de Vasselot[1] identified 64 enamels and this number has been increased to 74 by M. P. Verdier[2].

It is difficult to attribute these enamels to a precise date, or to establish whether the artist used the first edition or a later one. The counter-enamel of these plaques is of flux which leads one to think of a date after 1525–1530. The enamel worker more or less standardized the format of the compositions which varied in the printed version.

[1] In his first article 'Une suite d'émaux limousins à sujets tirés de l'Enéide', *Bulletin de la Société d'histoire de l'art français*, 1912, pp.6 ff. he describes 55 plaques and mentions eight further items in the collection of Baroness Salomon de Rothschild which he had been unable to see. Later in 'Emaux de la suite de l'Enéide du legs Salomon de Rothschild au Musée du Louvre', *Bulletin des Musées*, 1923, he described these eight plaques and added a ninth from the same collection.

[2] P. Verdier, *Walters Art Gallery, Catalogue of the Painted Enamels of the Renaissance*, Baltimore 1967, p.75.

1. The Aeneid Master

Aeneas founding the town of Acesta in Sicily

A rectangular plaque

$8\frac{3}{4} \times 7\frac{3}{4}$ in. ($22{\cdot}2 \times 19{\cdot}8$ cm)

The subject is described in Book V of the Aeneid. The Trojans have disembarked on the shore. Aeneas sees his father Anchises in a dream. On the right of the plaque Aeneas, Nautes and Acestes discuss the plan of the town. They are building the Temple of Venus on Eryx. The town occupies the whole of the background of the composition; the foreground shows the Trojans rejoicing in their new kingdom.

The grisaille ground is painted with blue, green and brown translucent enamels. The architecture is rendered in mauve, the details in rust-red with highlights in gold. The waves are finely drawn by *enlevage*, the needle revealing the underlying grisaille through the blue. The counter-enamel is translucent, colourless and thickened in places.

The enamel is copied from the plate on f. 248 of Sebastien Brandt's edition of Virgil (*see overleaf*).

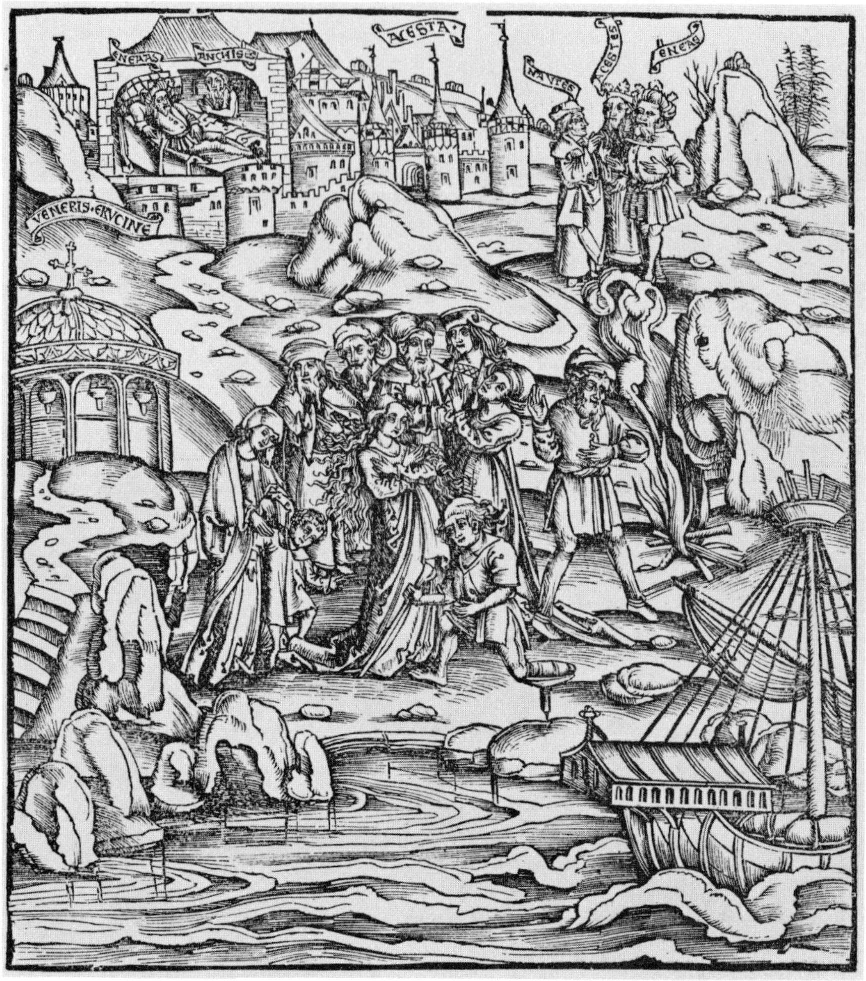

Illustration to Sebastien Brandt's edition of Virgil's Aeneid, *Strasbourg, 1502*
(*Paris, Bibliothèque Nationale*)

Exhibited: Exposition Universelle, Paris, 1867 (No.2858), lent by Baron James de Rothschild. *Exposition des Alsaciens-Lorrains*, Paris, 1874 (No.24), lent by Baroness James de Rothschild.

Bibliography: Marquet de Vasselot: 'Une suite d'émaux Limousins à sujets tirés de l'Enéide', *B.S.H.A.F.*, No.33.

Baroness James de Rothschild (Collection No.H.81); Baron Edmond de Rothschild P/1697

He was the first enameller to make extensive use of a white ground, for example in the plaques of the twelve Apostles or the two plaques showing goddesses in chariots at Waddesdon, and he achieved a very luminous effect, which was entirely personal to him, by mixing white with blue-green washes.

2. Léonard Limosin: The Chariot of Juno

An oval plaque in a nineteenth century gilt wood frame

$11\frac{1}{4} \times 8\frac{1}{16}$ in. (28·5 × 20·5 cm); with gilt wood frame, $18\frac{3}{8} \times 12\frac{1}{8}$ in. (46·7 × 30·8 cm)

Unsigned. Companion to *3*

Cupid flies in a cloud which emits shafts of light. Beneath, Juno sits in a chariot harnessed to three peacocks.

The abundance of white gives an opaque effect to the whole composition, of which the components are outlined in black. The colours are light and there are numerous gold highlights in the upper part of the piece.

The counter-enamel is translucent.

Miss Alice de Rothschild WI/77/I

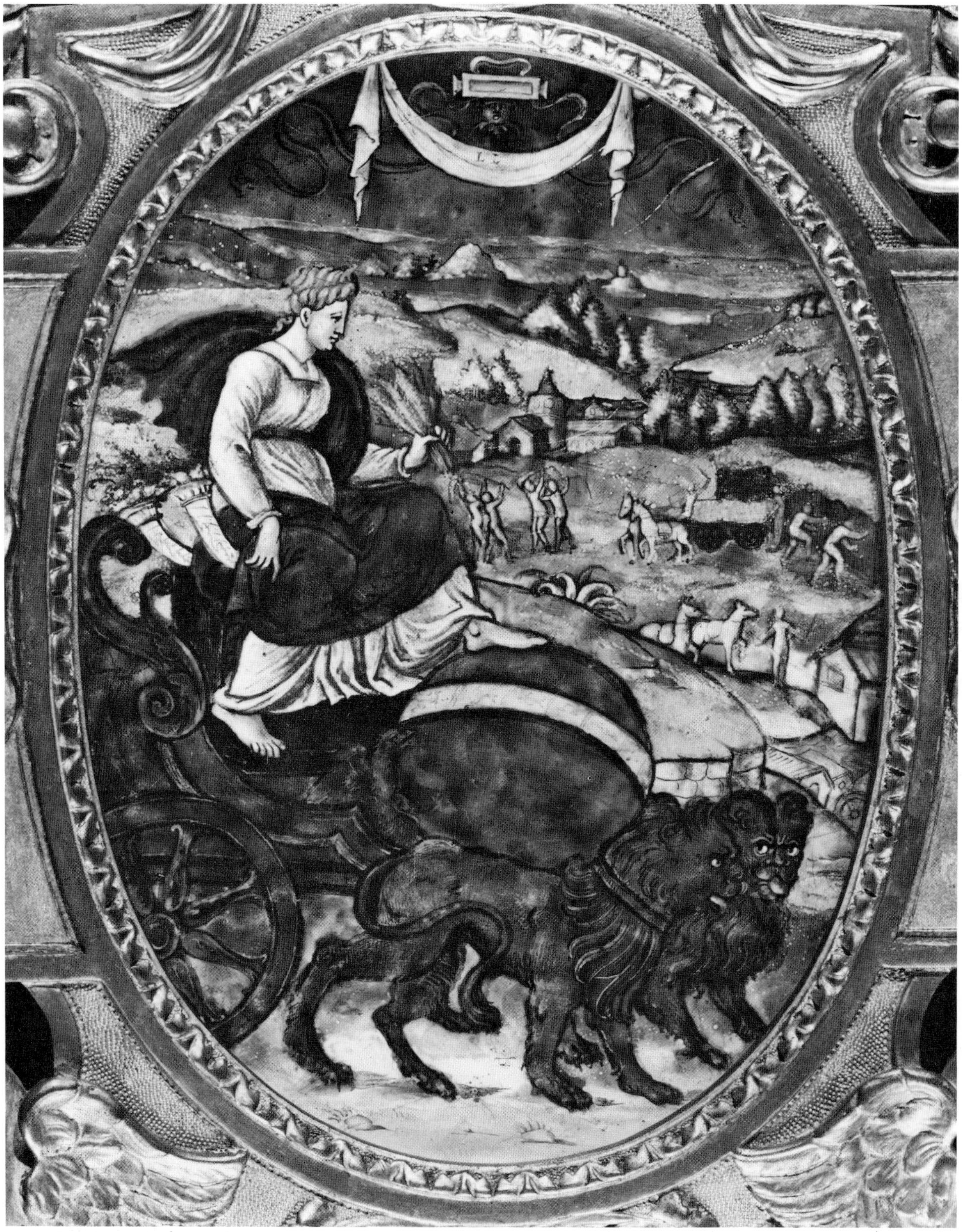

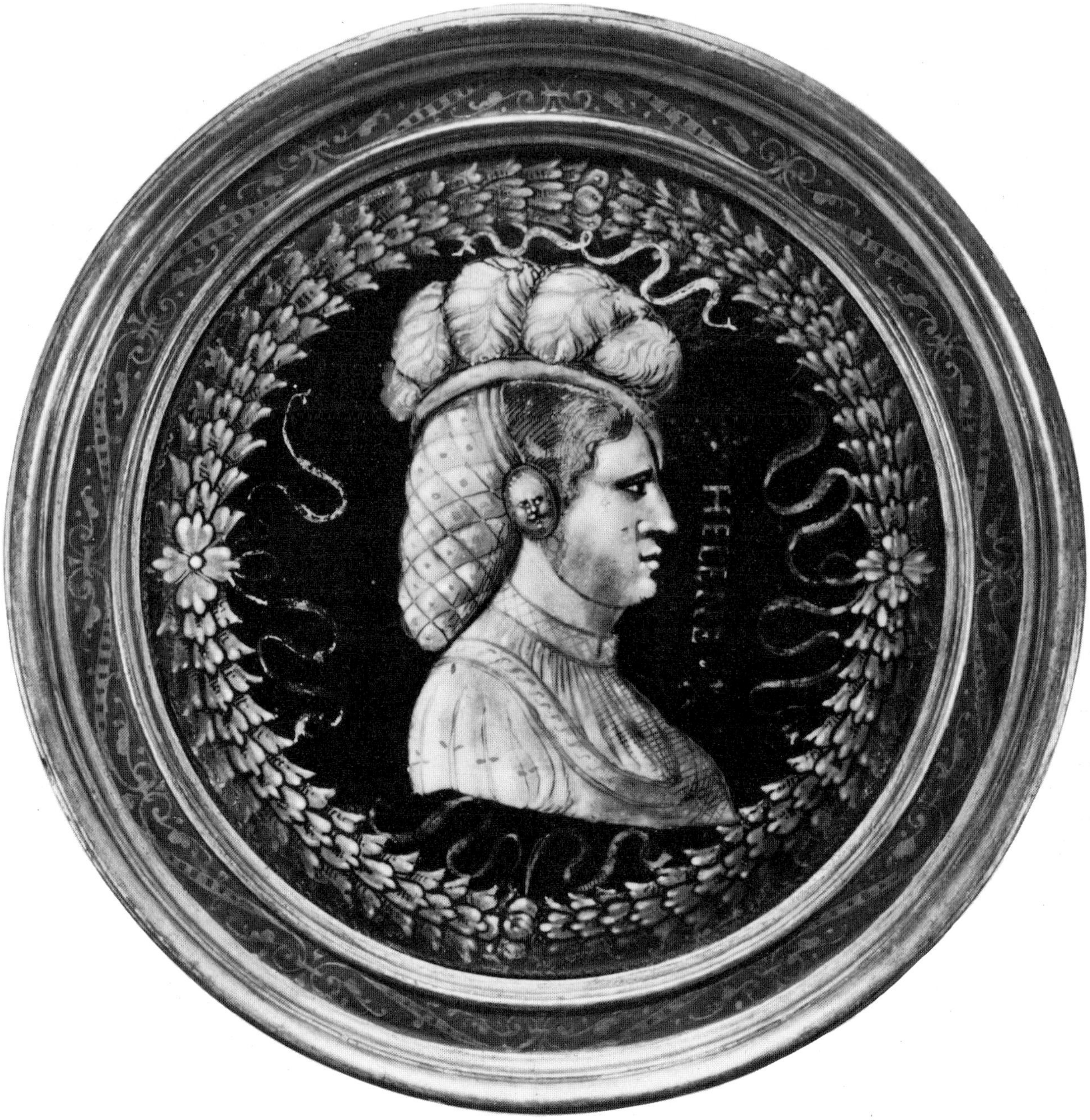

4. Attributed to Léonard Limosin or his studio: Helen and Paris

A pair of oval plaques in nineteenth century gilt wood frames

About $5 \times 4\frac{11}{16}$ in. ($12{\cdot}7 \times 11{\cdot}9$ cm) (the plaques are partly covered by the frames)

The profile busts identified as Helen and Paris by inscriptions are shown facing each other on black grounds. The borders of the plaques consist of wreaths of overlapping leaves of blue verging on turquoise, intersected horizontally by small flowers or spherical motifs from which

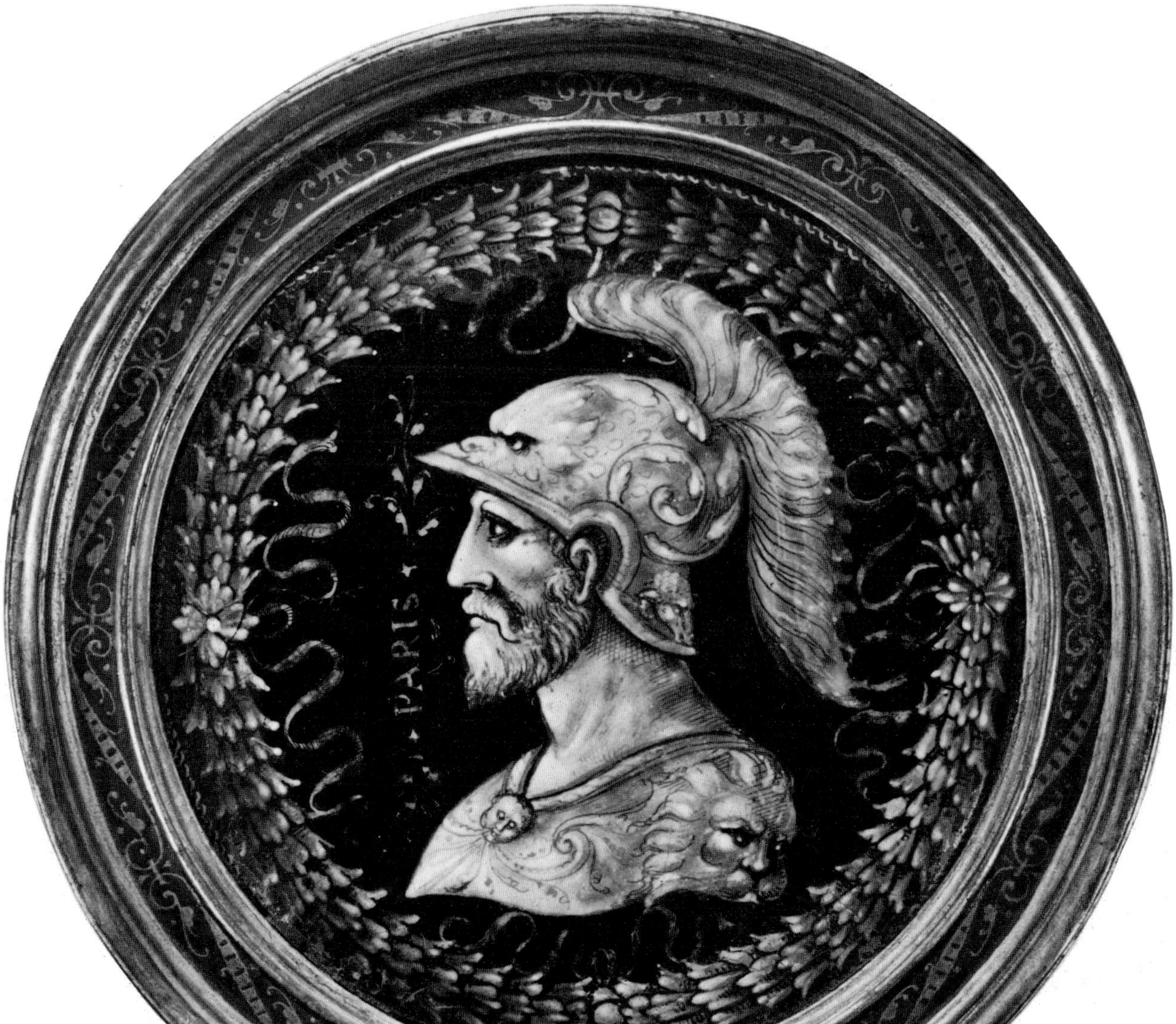

flow cascades of gold ribbons engraved with the needle.

Almost identical wreaths, but without the small flowers, are to be found in some of Léonard Limosin's square plaques of Sibyls and in his lozenge-shaped plaques such as that in the Museum at Limoges (Inv. 11) representing a bust of Helen.

The faces are painted in grisaille without additional colours and modelled by impasto, the shadows being indicated by hatching. Helen's hair is drawn with parallel hatching and covered with a layer of gold (partly rubbed), the waves of the hair being shaped by delicate *enlevage*.

Miss Alice de Rothschild — WI/77/2

5. Limoges, mid-16th century: 'Rolant filius Samson'

An oval plaque in a rectangular frame with enamel corner-pieces

Plaque, $7\frac{5}{8} \times 6\frac{3}{8}$ in. ($19{\cdot}4 \times 16{\cdot}2$ cm); overall dimensions, $8\frac{1}{8} \times 6\frac{13}{16}$ in. ($20{\cdot}7 \times 17{\cdot}4$ cm)

A haughty blue-bearded figure, mounted on a grey caparisoned horse, rides towards the right on a bank covered with delicate blue flowers. He wears a sort of violet toque and a blue coat with a violet lining and wide sleeves over a short light-brown tunic, and holds the reins in his left hand.

A white banderole with an inscription in black: ROLANT FILIVS SAMSON stands out from a black ground scattered with gold dots. The subject evokes the numerous equestrian figures and bust-portraits of Charlemagne's *Neuf Preux*, often rendered by the enamel workers of Limoges, particularly Couly Noailher and M. D. Pape.

The borders are restored. FILIV is original and is painted on the white enamel base, but the other words and the final S of FILIVS have been repainted in oils with a brush. It is possible that they do not conform to the original inscription because, though one of the *Preux* was called Samson, he is not usually identified as the father of Roland.

The whites are thick; gold is used for ornamental highlights and for the check pattern on the lining of the coat which is turned back to form a sort of cape. The corner-pieces are decorated with golden cascades on a black ground.

Miss Alice de Rothschild — WI/77/3

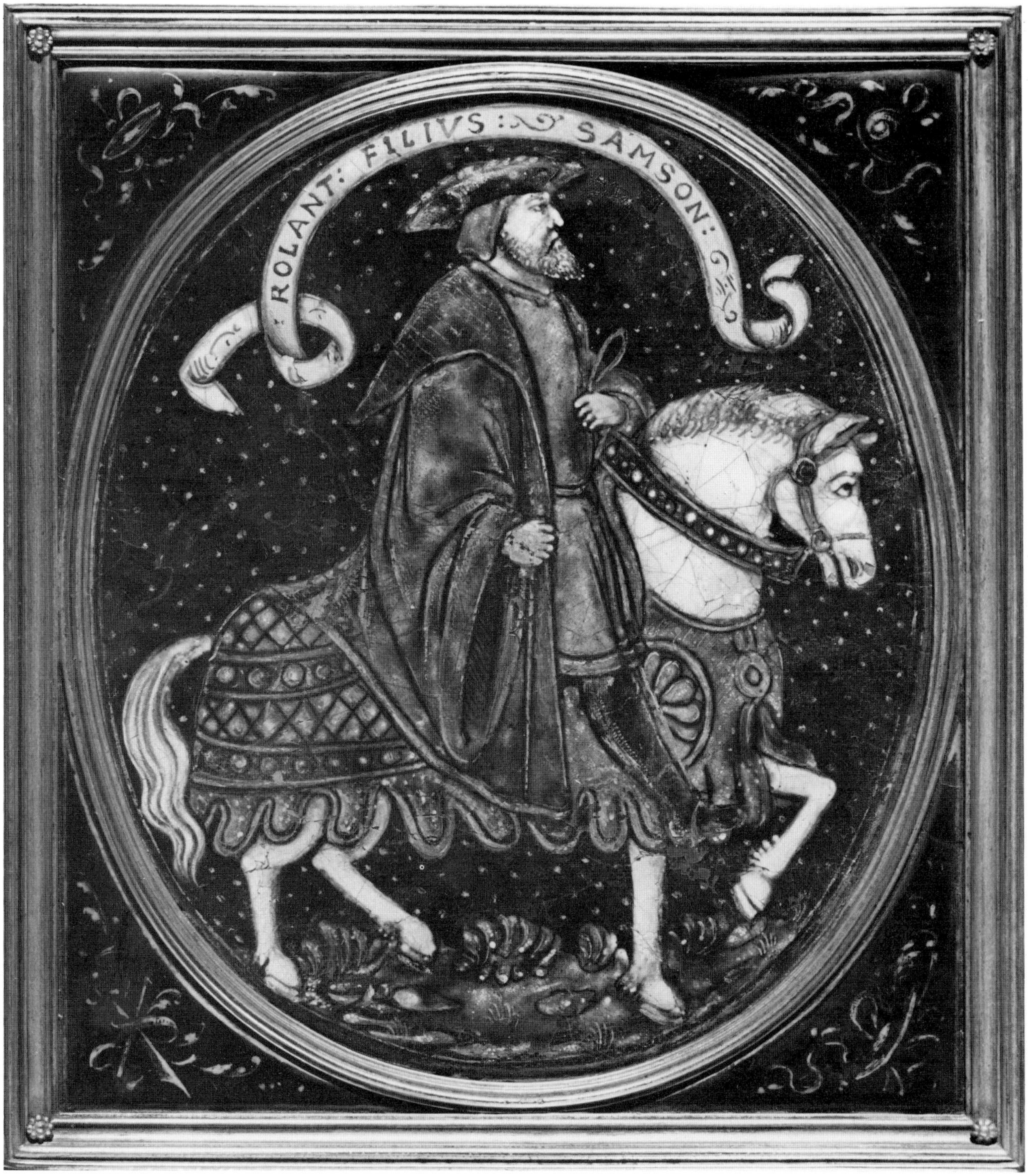
ROLANT: FILIVS: SAMSON:

PIERRE PENICAUD

The lineage of the Pénicaud family is complicated and the fact that three enamel workers of this name had the Christian name Jean does not simplify the problem, but from the dates of their works and taking into account the little information we have on their lives, it seems that Jean II and Pierre were the first and third sons of Nardon.

Pierre worked in enamel and stained glass. In 1556 he had just completed a window representing the *Last Supper* (destroyed in 1770). His name is last mentioned in 1590, at which date he must have been very old.

He does not seem to have continued working in enamel after about 1565. The works which bear his signature – the monogram P.P. – include cups with or without lids, a dish and some plaques. His medium was almost exclusively grisaille, sometimes slightly tinted, on a black or blue ground. The sources of his compositions were contemporary engravings, and his figures are characterized by their exaggerated length, in the manner of Primaticcio.

It is possible that he shared a studio with Jean III who may have been his nephew, and this would explain the great similarity of their techniques.

6. Attributed to Pierre Pénicaud, 1557
The Israelites gathering the Manna

A dish, nearly circular

$17\frac{3}{8} \times 17\frac{1}{16}$ in. ($44 \times 43{\cdot}3$ cm)

Unsigned but dated 1557 on the border

The dish, which shows a closely packed group of men, women and children collecting the manna, is painted in grisaille with pink flesh tints applied in patches, with slight detailing in gold in the falling manna. It is taken from an engraving by Agostino Veneziano (Bartsch, XIV, p.10, No.8) said to be after Raphael but more probably after Luca Penni. The engraving is rectangular and the composition has been slightly compressed in order to fit it into the round shape of the dish. The tents shown in the background on the dish do not appear in the engraving and must derive from another model.

In the centre of the reverse is an oval medallion of a double superimposed profile showing a woman dressed in light material knotted on her shoulder and a bearded man with a very expressive face, surrounded by a white frame in six sections filled with fruit, gold-winged putti sounding trumpets, and grotesque masks. On the rim are eight oval medallions containing the heads of bearded men seen in profile, executed in gold on black and reclining figures in bistre on white, separated by S-shaped foliated scrolls terminating in dolphins' heads.

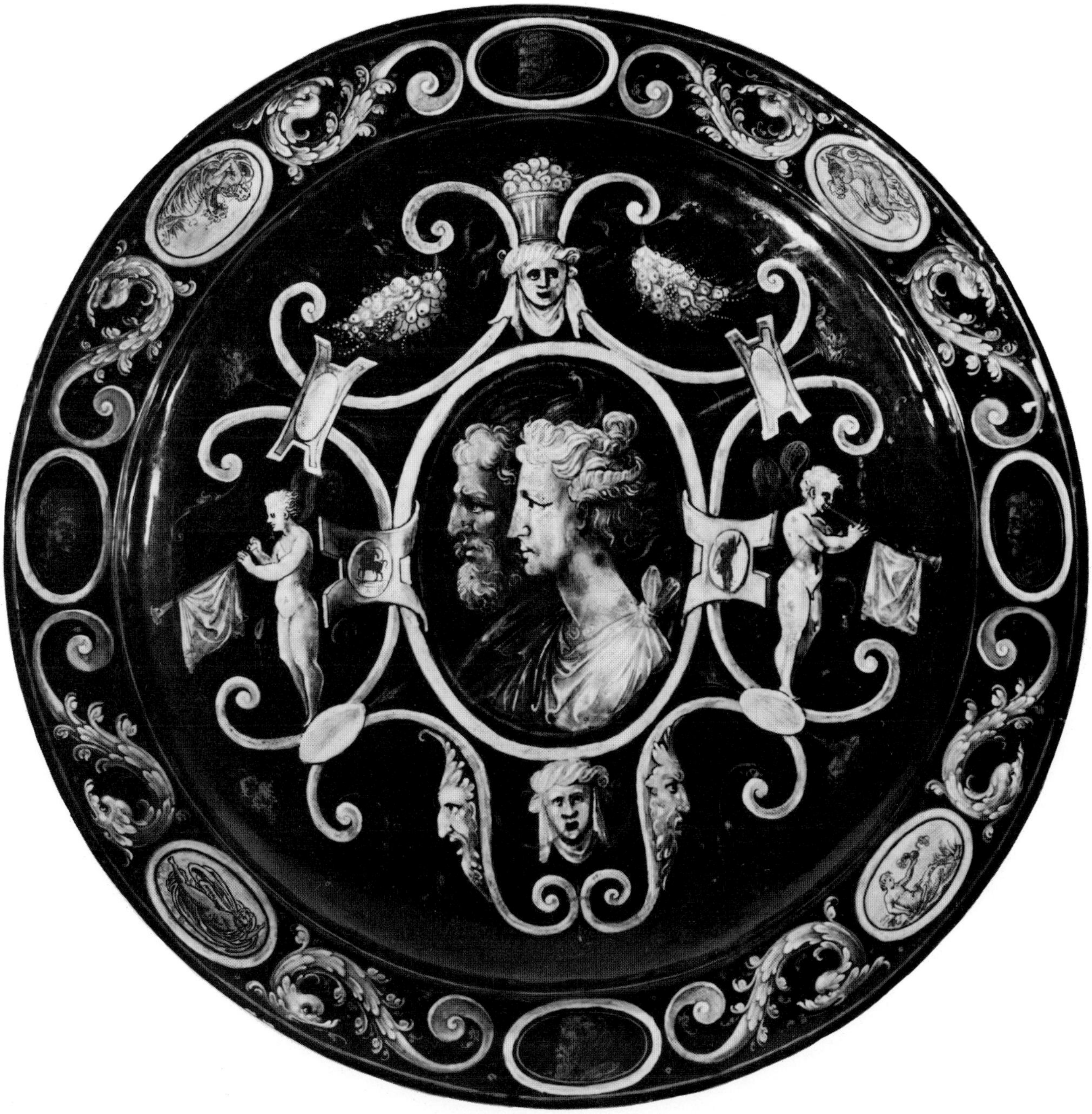

The reverse side of the dish

In the Musée de Cluny in Paris is a round dish about 43 cm in diameter, undated but signed with the monogram P.P., and with a coat-of-arms at the base of the composition. It represents in grisaille *Moses seated in Judgement beside Jethro* (wrongly identified as 'Moses explaining the tablets of the law'). Its border is in a very bad state, but the sections in good preservation are close to the present piece. Instead of putti it shows naked old men grouped round clusters of fruit; the cavetto is decorated with closely packed foliated scrolls, and the dish shows in the central scene the slender elongated silhouettes common to Pierre and Jean III, the airy lightness and transparency of the clothes, and the small touches of white on the clumps of trees which appear in the Waddesdon dish. A comparison with this dish supports the attribution of the Waddesdon piece to Pierre Pénicaud, already proposed when it was lent to the 1865 exhibition in Paris.

Exhibited: Musée Rétrospectif, Paris, 1865 (No.2443), and *Exposition Universelle*, Paris, 1867 (No.2892) lent by Baron James de Rothschild.

Provenance: In the Strawberry Hill sale beginning 25 April, 1842, the following item occurs: 'Lot 80. A magnificent large circular dish, a splendid specimen of the Limoges enamel, with the history of Moses, most elaborately displayed, and fine arabesque border, the back equally beautiful. It is one of the finest pieces of this enamel, of a very early period, 1557'. Apart from the fact that the precise subject is not specified, this description would fit the Waddesdon dish. The purchaser at the Strawberry Hill sale is recorded as Baron C. de Rothschild, who could only be Baron Carl de Rothschild of Frankfort and Naples. The Waddesdon dish is recorded by Baron Edmond as coming from the collection of his mother and was lent on two occasions

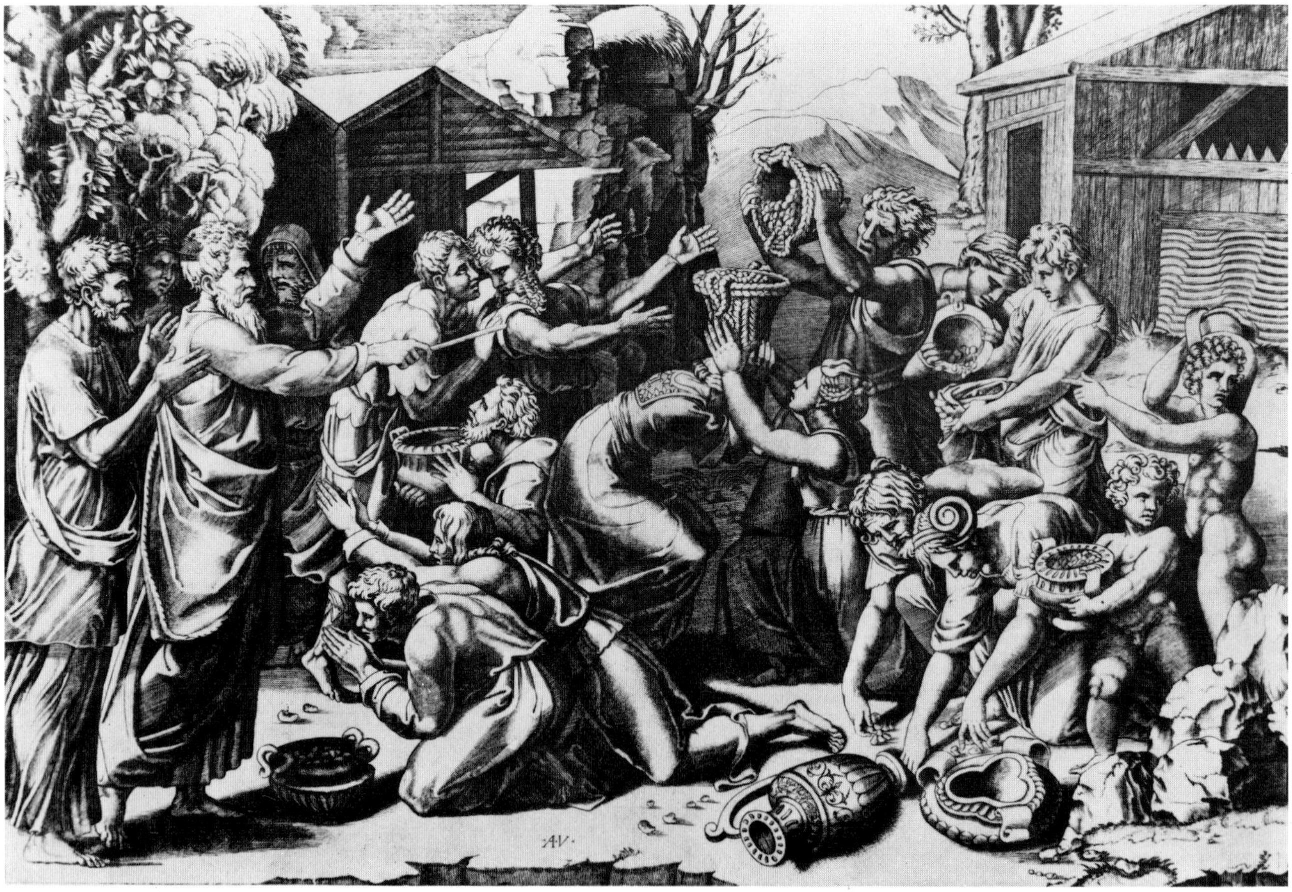

Engraving by Agostino Veneziano

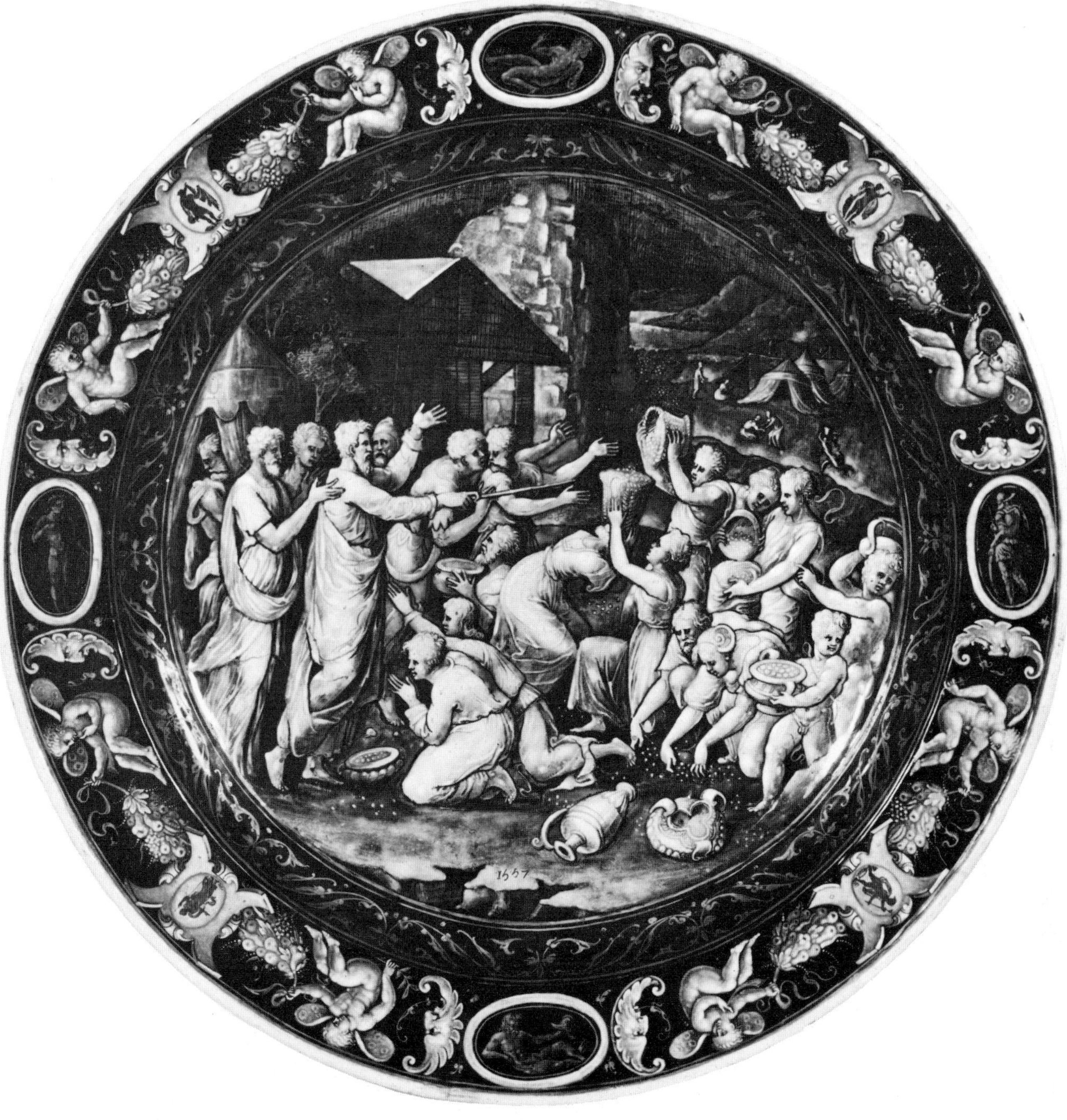

The face of the dish

as from the collection of his father. As both of these were closely related to Baron Carl, it is possible that the dish might have been given as a present and so passed from one branch of the family to another.

Probably Baron Carl de Rothschild; Baroness James de Rothschild (Collection No.H.51); Baron Edmond de Rothschild. P/1699

JEAN III PENICAUD

He was the last of the enamel workers in the Pénicaud family, and ceased production in the 1570s. Laborde[1] maintains that he was superior to all the other Limoges enamel workers because of two facts: he never copied anyone, and he never signed his works. Philippe Verdier on the other hand[2] wrote: 'Jean III . . . followed in the wake of Jean II, copying the models created in the workshop of the latter, and lacking the capacity for personal creation.'

In fact a number of Jean IIIs works are decorated with scenes taken from Agostino Veneziano, Marcantonio Raimondi, the Master of the Die and Bernard Salomon. Laborde's mistake originates from the fact that many works now recognised to be by Jean II, notably those in the Louvre, were attributed to Jean III in the nineteenth century.

Most of the works of Jean III were in grisaille, and his principal originality lies in the deep blue background of some of his grisailles. Like Pierre Pénicaud he filled the borders with small oval medallions in imitation of antique cameos, sometimes painted in gold on a coloured, black or white ground, or in bistre on a white ground. Further, his figures were very much elongated like those of Jean II. The reverses of his plaques are stamped with the crowned mark of the Pénicaud family.

7. Attributed to Jean III Pénicaud
Profile of a young woman

A circular plate

Diameter, $7\frac{11}{16}$ in. (19·6 cm)

This head of a young woman with blonde hair and with lips and forehead lightly tinted pink, stands out strongly from a royal blue ground. A kind of toga of brownish-red over foils is knotted on her right shoulder. The purplish-blue border is decorated with a fine gold geometric interlacing design, divided into four sections by oval medallions showing figures or scenes in gold on blue, green or mauve grounds painted on foils (two of the subjects are nearly effaced).

The cartouche on the reverse shows a female bust in profile painted in grisaille on a black ground with gold hatching. The border consists of fine symmetrical foliated scrollwork painted in gold with four cameo-type medallions showing figures or mythological scenes in gold on a white ground.

[1] Laborde, *Notice des émaux . . . du Musée du Louvre*, Paris, 1857, p.161.
[2] P. Verdier, *Walters Art Gallery Catalogue*, p.xx.

PIERRE REYMOND

Pierre Reymond lived from about 1513 to about 1584. He was a consul of Limoges in 1560 and 1567, and was active as an enamel worker for about forty years. Although he made original illustrations for the register of the brotherhood of the Holy Sacrament in the church of Saint-Pierre-du-Queyroix, he generally took his compositions from Marcantonio Raimondi – and sometimes through him Lucas van Leyden – and later from Bernard Salomon and Jacques Androuet du Cerceau the Elder.

He is at his best in decorative work, and although the grotesques, the cartouches, the fantastic animals which cascade round his plates, cups and ewers were derived from contemporary books of ornaments or earlier Italian sources,[1] he used them with great imagination and freedom and with a fine decorative sense.

His earliest works were fresh, gay plaques enamelled in colour, but later he specialized in table-ware for which he had many commissions. On one occasion he added enamel decoration to a bronze ewer made by the celebrated Nuremberg goldsmith, Wenzel Jamnitzer, and between 1558 and 1562 he made a service for the Tucher family of Nuremberg (now in the Tucher Haus there). In 1566 he made a set of plates decorated with the arms of Pierre Séguier, president of the Parlement of Paris, with scenes illustrating the Labours of the Months after Etienne Delaune.

Nearly all the large services made by Pierre Reymond during the middle of his career are in grisaille highlighted with gold, with subdued flesh tones and sometimes with delicate blue-green washes for trees or ground. The strongly outlined contours are softened by the languid grace of the figures. At the same time, however, he made a few plaques or triptychs in colour, and at the end of his life he used colour extensively, though not always with great success.

[1] They may even be indirectly inspired by Ariosto's description in the *Orlando Furioso* (canto 6): 'Some were of human stature from the neck down to the feet, but they all had the heads either of aged monkeys or of cats. Others had satyrs' heads and hooves. Others took the forms of slender centaurs always poised as if about to begin racing . . . Some of them . . . leaped on to the backs of the centaurs and sat there, others were carried by eagles, ostriches or cranes . . . One of these monsters was of almost human form, though he had the paws, belly, neck, ears and muzzle of a large mastiff . . .'

8. Pierre Reymond, 1560: The Creation of Eve

A circular plate

Diameter, $7\frac{5}{8}$ in. (19·4 cm)

Signed with monogram PR and dated 1560

This scene is executed after an engraving by Lucas van Leyden, dated 1529 (Bartsch, 1)[1]. In the border is a frieze of symmetrical grotesques between oval cartouches on which the monogram PR and the date 1560 alternate with GENESE (above) and II (below) indicating the reference

[1] J. Adam von Bartsch, *Le Peintre graveur*, Vienna, 1803–21.

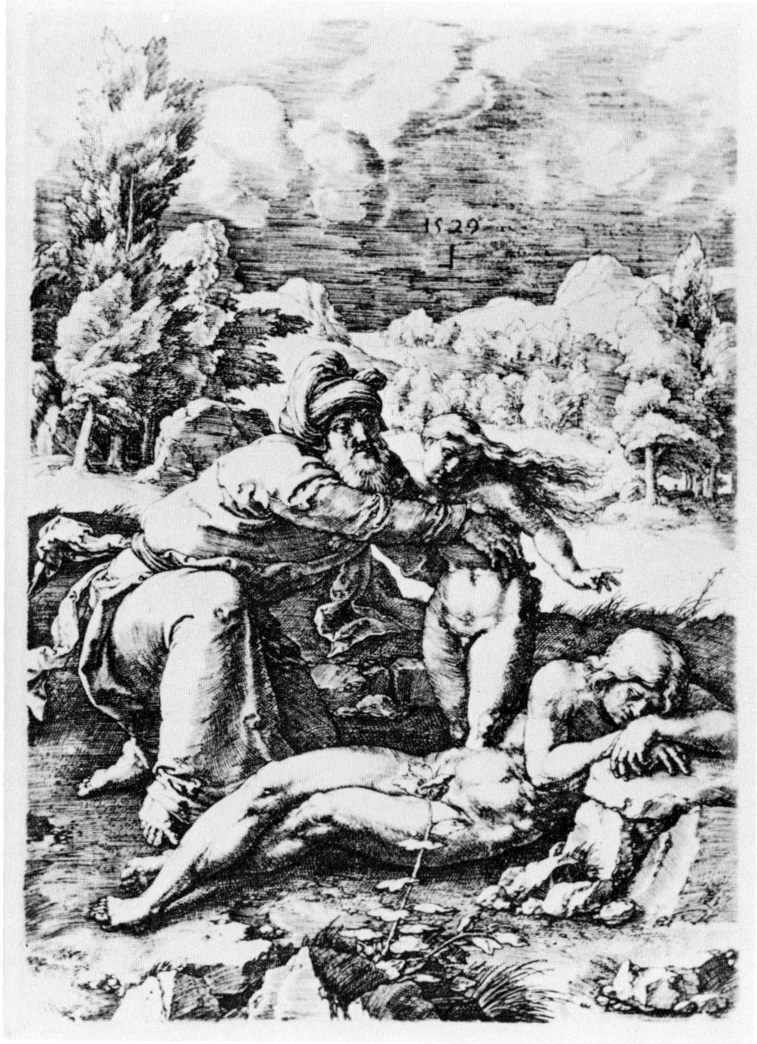

Engraving by Lucas van Leyden

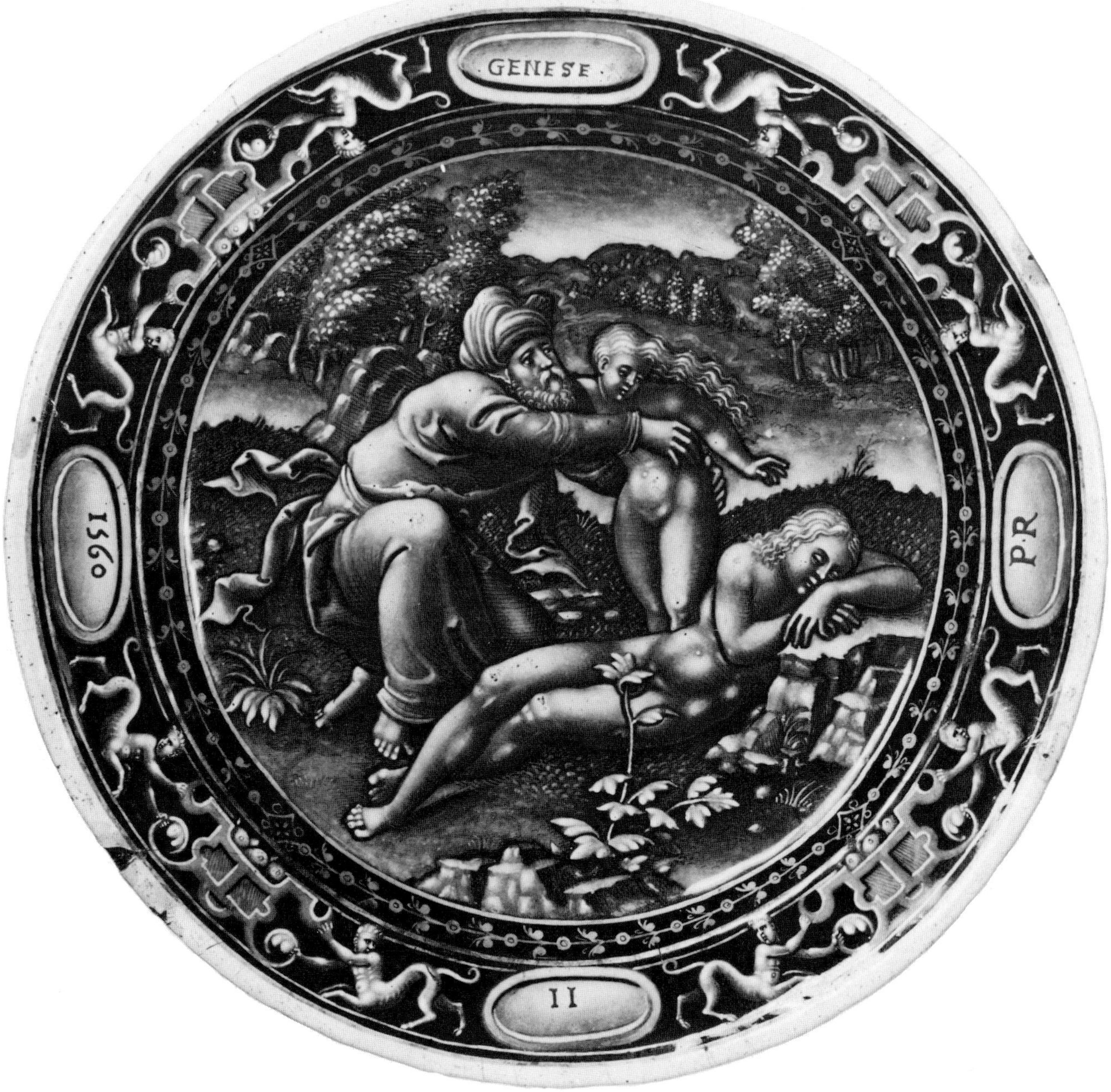

to the second chapter of Genesis. An oval dish also showing the Creation of Eve with the same inscription and signed PR is at Luton Hoo. Three of the plates belonging to the same set are known: two dated 1560 and representing Eve offering the apple to Adam and the Expulsion from Paradise are in the Louvre (Inv. Nos. 1590, 1591), and the third, inscribed GENESIS IX and showing the Drunkenness of Noah, is in the Victoria and Albert Museum. All three plates have exactly the same border as the Waddesdon plate.

On the reverse is a bust of a bearded man shown in profile against a black background scattered with gold dots and surrounded by a double border, strapwork and foliated scrolls.

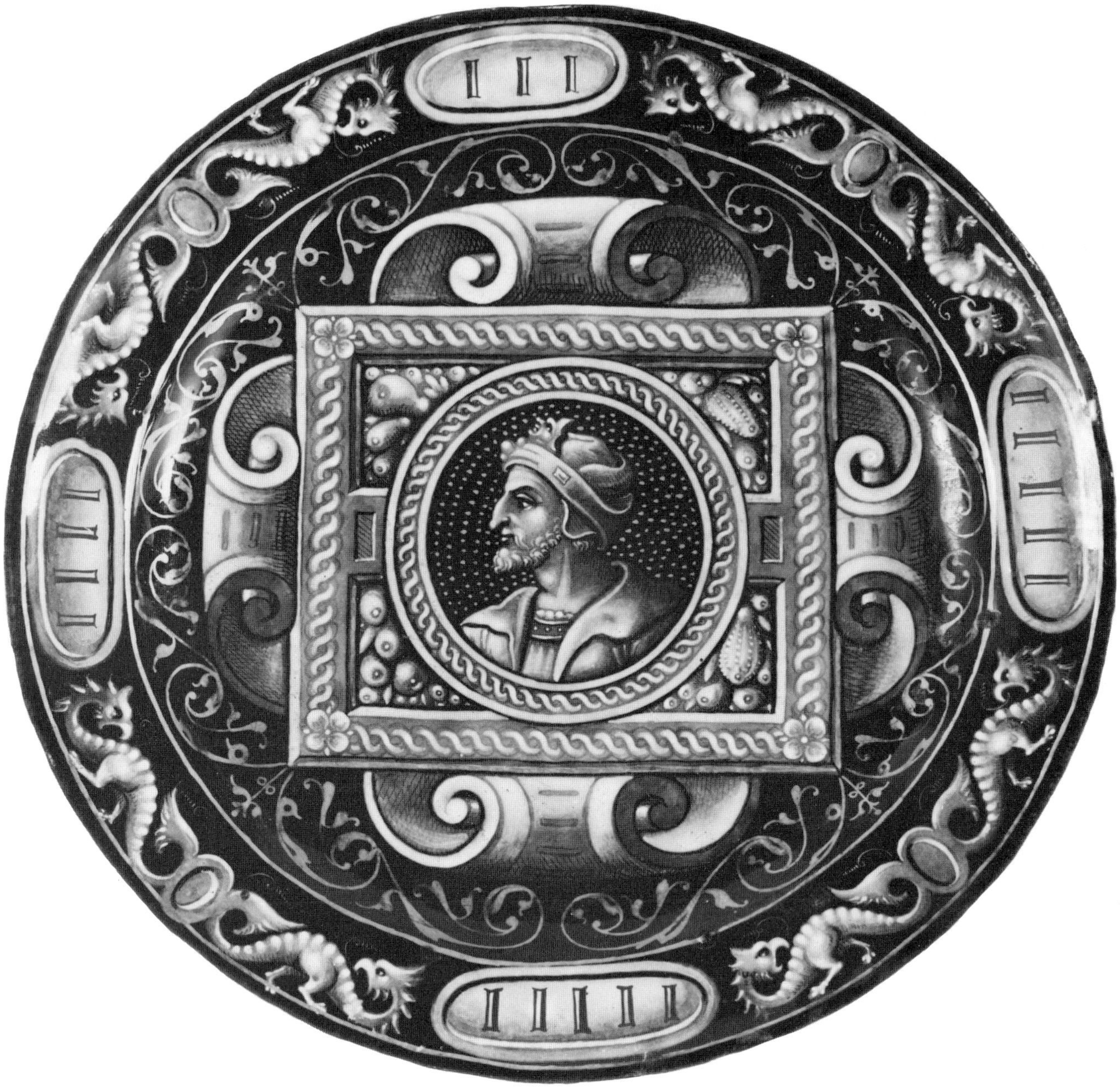

The reverse of 8

The plate is enamelled in grisaille. The flesh in the main subject is tinted slightly salmon pink, but is white in the border and on the reverse. The highlights are in gold.

Exhibited: Exposition Universelle, Paris 1867 (No.2974) lent by Baron James de Rothschild.

Baroness James de Rothschild (Collection No.H.30); Baron Edmond de Rothschild P/1709

JEAN DE COURT (The Master I.C.)

It now seems certain that the works signed with the monogram I.C. or I.D.C., and the portrait of Margaret of Navarre in the Wallace Collection inscribed IEHAN DE COURT MA FAICT 1555 must be attributed to one and the same enamel worker.

Jean de Court enjoyed a certain renown and, after the death of François Clouet in 1572, he was appointed painter to the king.

We know of a considerable number of enamels by him, but only one dated piece, the portrait mentioned above which is dated on the reverse – a very unusual occurrence in the sixteenth century. Most of the works by this enameller must belong to the second half of the century, because many of them, particularly pieces from dinner services, were inspired by the *Quadrins historiques de la Bible* illustrated by Bernard Salomon and first published by Jean de Tournes in Lyons in 1553. The Master I.C. was one of the first to enamel the backs of little mirrors. These remained fashionable for about fifty years, and the practice of enamelling them was confirmed by Suzanne Court and the later members of the Limosin family. On the other hand, neither Léonard Limosin nor Pierre Reymond nor the Pénicaud family produced any such mirrors, though Pierre Reymond was not above painting very small surfaces when occasion demanded. Moreover, though the regular use of foil became normal for ornamental objects, it did not come into general use before the 1560s.

The Master I.C. decorated plates and caskets with various scenes from the *Story of Joseph* after the *Quadrins* and the scene of the *Hebrews crossing the Red Sea* is to be found on many oval dishes, but the Old Testament was not his only source of inspiration. The collection at Waddesdon includes several little mirrors by this artist which illustrate subjects taken from classical mythology.

The Master I.C.'s works are characterised by rich and sometimes complicated decorative composition, and abundant use of foil for clothes, birds and flowers, much gold highlighting, profile heads in which the nose is drawn in a continuous line from the forehead – a feature also to be found in the work of Suzanne Court – and salmon-pink flesh tones. Sometimes, however, the flesh is painted white with touches of red on the cheeks, and sometimes completely white, particularly in the rare works – such as *15* – in which the subject is embossed in relief. In the *grisailles*, which are often very fine, dark tones predominate.

9. Jean de Court (The Master I.C.)
Minerva and the Muses on Mount Helicon

An oval enamelled plaque and a mirror in a gilt-metal frame

$4\frac{3}{16} \times 3\frac{1}{8}$ in. (10·6 × 7·9 cm); with frame, $4\frac{11}{16} \times 3\frac{1}{2}$ in. (11·8 × 8·9 cm)

Signed I.C. in gold

Minerva wears a short turquoise-blue tunic over a red dress and a plumed helmet and approaches the nine Muses who are playing various musical instruments. The landscape consists of a foreground with green turf and soft blue waves, a middle distance with trees in which are perched the Pierides transformed into magpies, and in the background violet-coloured mountains in which the artist's monogram can be read.

The clothes and the plumage of the magpies are enriched by the use of foils. The flesh is slightly tinted with salmon pink.

The subject is freely interpreted from the plate of Pallas and the Muses in *La Métamorphose d'Ovide figurée* which was illustrated by Bernard Salomon and printed by Jean de Tournes in 1557. The same plate was used for *28* below.

Provenance: Debruge-Duménil sale, 23 January – 9 February 1850 (lot 741); Heckscher sale (from Vienna) London, 4–6 May 1898 (lot 67).

Miss Alice de Rothschild — WI/77/81

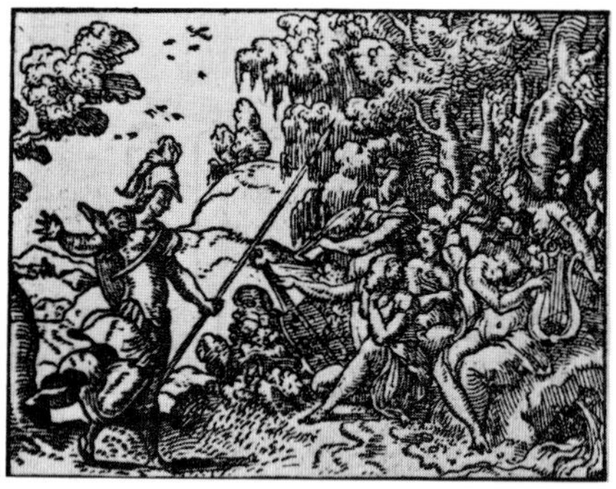

Illustration by Bernard Salomon to La Métamorphose d'Ovide, *1557*
(Paris, Bibliothèque Nationale)

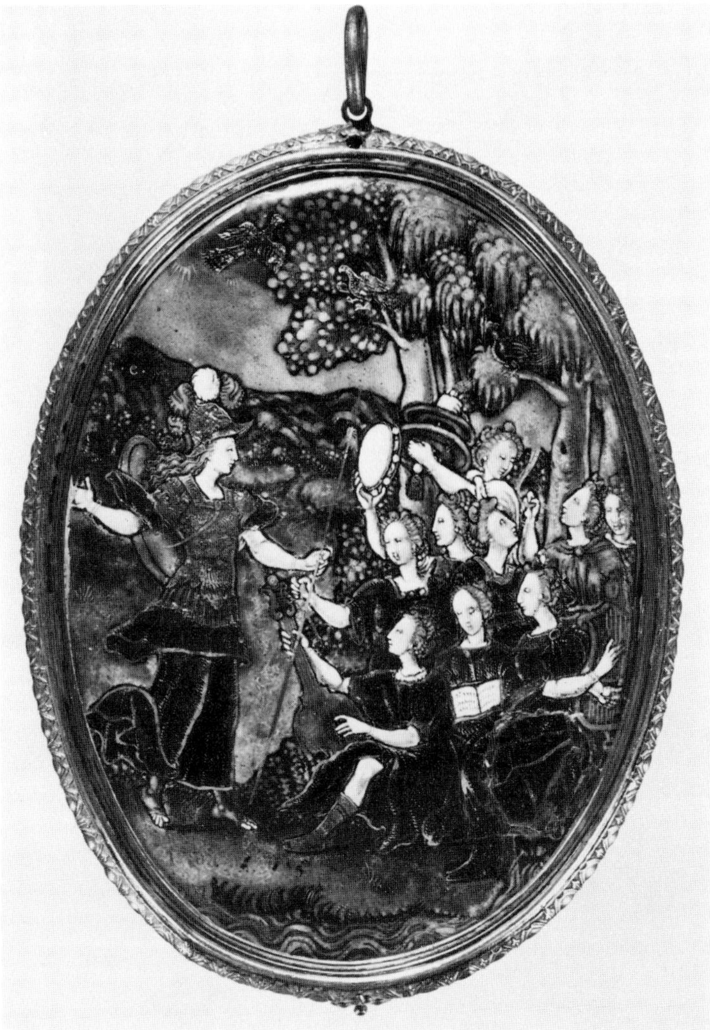

10. Jean de Court (The Master I.C.)

A man and a woman hunting

An oval enamelled plaque in a gilt-metal mirror frame; the glass is now missing

$3\frac{9}{16} \times 2\frac{11}{16}$ in. ($9 \times 6{\cdot}9$ cm); with frame and floriated loop, $4\frac{9}{16} \times 3$ in. ($11{\cdot}4 \times 7{\cdot}5$ cm)

Signed I.C. above the stump of the tree on which the woman rests her left foot

There is no means of deciding the exact subject as neither figure has any attribute. The protagonists might be Cephalus and Procris, or just a youth and a girl.

The man wears a tunic of green verging on turquoise with pink fringes and short blue sleeves; the woman has a blue dress with a pink scarf and boots. The sky is dark blue, the church slate grey. The trees are scored with black hatching on the shaded side and with gold on the side on which the light falls. The foreground rises in four ridges, painted alternately blue and green. The details of the ground are outlined in black and decorated with small details in gold. The use of foil is restricted to the clothes worn by the two figures and the collars of the dogs.

Miss Alice de Rothschild — WI/77/8j

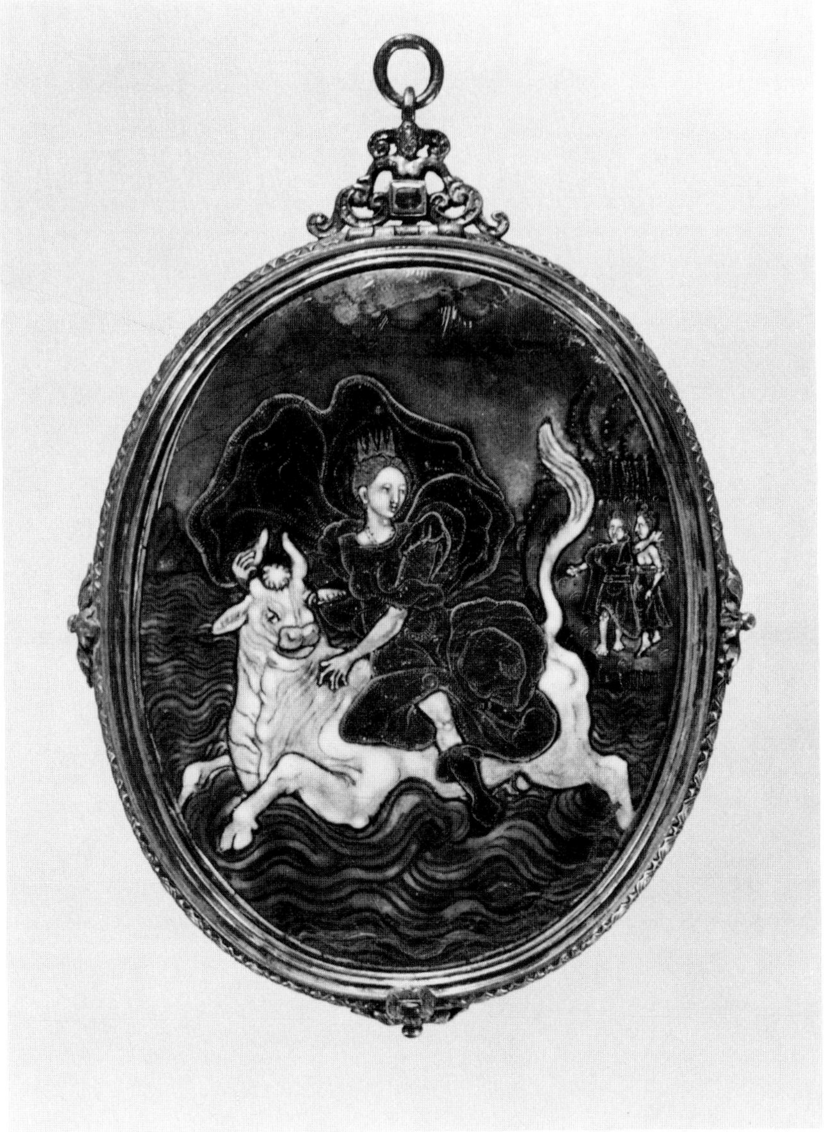

12. Jean de Court (The Master I.C.) The Rape of Orithyia and The Rape of Europa

An oval plaque decorated on both sides, in a fine perforated gilt-metal frame

$2\frac{5}{16} \times 1\frac{11}{16}$ in. (5·8 × 4·3 cm); with frame, $2\frac{13}{16} \times 2\frac{3}{16}$ in. (7·1 × 5·6 cm)

Signed I.D.C. at the top of the convex side

The Rape of Orithyia is shown on the convex side. The figures are set against an area of gold sky, surrounded by grey clouds tinged with pink. Orithyia wears a blue gown and green scarf enamelled over foils. Below, the sea is depicted with delicately drawn waves, and above it is a greyish-mauve sky with blue clouds.

The subject is taken from Bernard Salomon's engraving of the same subject in *La Métamorphose d'Ovide figurée* (cf. *9* above). The engraving was copied in reverse by Virgilius Solis in *Les Métamorphoses d'Ovide*, Frankfort, 1563.

On the reverse is shown the Rape of Europa (cf. *11* above). Europa wears a blue gown under a deep pink tunic with a green belt, a wide blue cloak and pink boots. The two girls on the shore are painted in gold, pink and blue; one of the cows in the distance is yellow-brown and the other white. The bull is powerfully modelled in white, with subtle touches of grisaille. The flesh tones are reddish. The details have emphatic dark outlines, and there is much use of *enlevage*.

Exhibited: South Kensington Museum, 1874 (No.654), lent by Henry Willett.

Miss Alice de Rothschild — W1/77/7

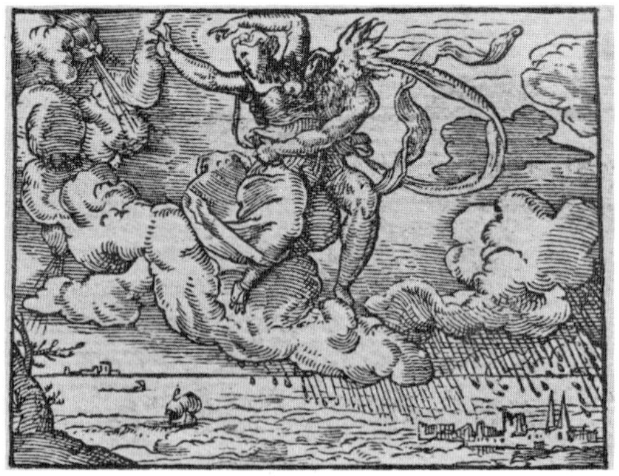

Illustration by Bernard Salomon to La Métamorphose d'Ovide, *1557*
(*Paris, Bibliothèque Nationale*)

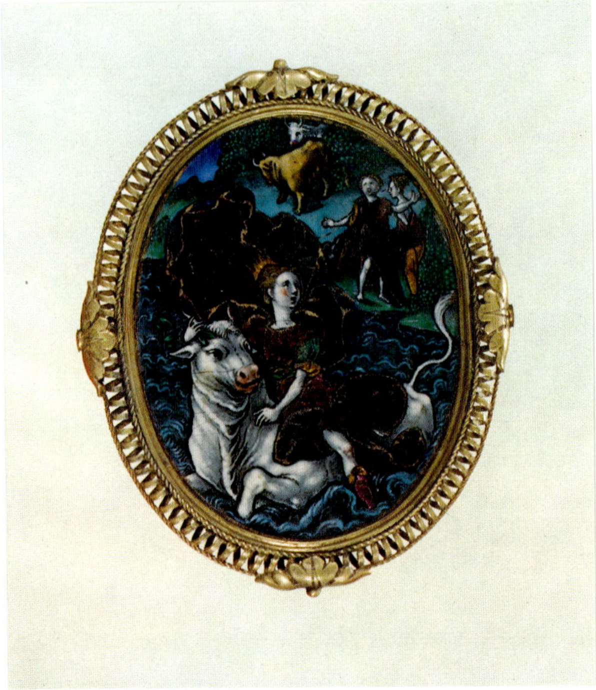

Above : the concave side

Below : the convex side

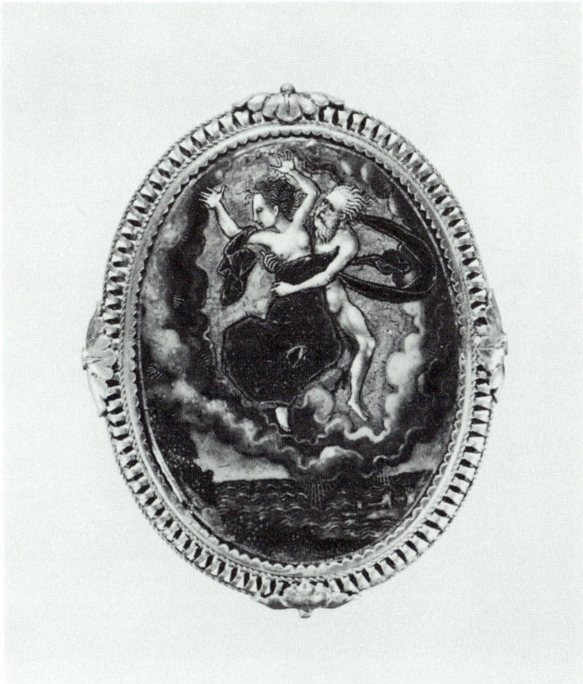

13. Jean de Court (The Master I.C.)
The departure of Diana or a nymph for the hunt

An oval enamelled plaque in a gilt-metal mirror frame, the glass now missing

$4\frac{3}{8} \times 3\frac{1}{4}$ in. (11·1 × 8·3 cm); with frame and floriated loop, $5\frac{3}{8} \times 3\frac{9}{16}$ in. (13·5 × 9 cm)

Signed I.D.C. at the top of the composition

The fact that the figure does not have a crescent moon in her hair suggests that she is probably a nymph rather than the goddess Diana. Her tunic is purple, the sash brown, the scarf blue. Her hair is tied with a flowing green ribbon. She wears blue boots with brown cuffs. The fabrics and the red collar of the dog on the left are all painted on foils.

Above the green undergrowth the middle distance is painted in shades of light turquoise above which are dark green trees. I.D.C. is painted in gold in the upper part of the china blue sky which darkens towards the top. The light touches of gold and little clouds are partially effaced.

The flesh is tinted deep salmon pink, the dog on the left is light grey, the two on the right dark grey. All three have red-tipped muzzles.

The counter-enamel is transparent and very thin, showing the red copper.

Exhibited: South Kensington Museum, 1862 (No. 1857), lent by Hollingworth Magniac.

Provenance: Magniac sale, Christie's, 2–8, 11–15 July 1892, lot 513, bought by Harding.

Miss Alice de Rothschild — W1/77/8f

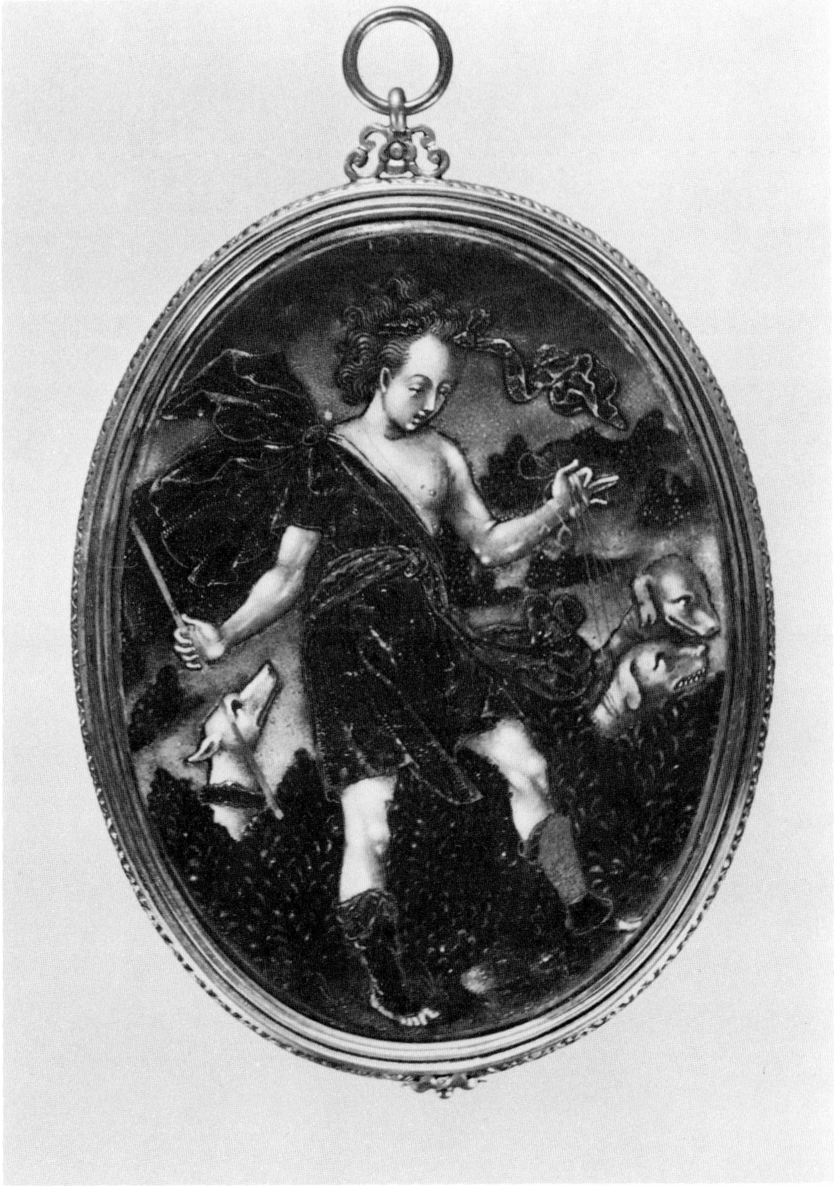

14. Attributed to Jean de Court (The Master I.C.)
Minerva

An oval enamelled plaque and a mirror in a gilt-metal frame

$4\frac{1}{16} \times 3\frac{3}{16}$ in. (10·4 × 8·2 cm); with frame, $4\frac{1}{2} \times 3\frac{11}{16}$ in. (11·5 × 9·5 cm)

Unsigned

The goddess, who stands between three vases filled with flowers and fruit, wears a blue gown beneath a light green tunic, a helmet of the same colour with plumes in grisaille, and turquoise gaiters with brown cuffs. The flesh tints are slightly salmon-pink. The treatment of the folds, the manner of modelling the muscles, the dark outline of the right arm, the delicacy of the painted face surmounted by waved hair, justify an attribution to the Master I.C.

Provenance: Spitzer Collection (*Catalogue* 1891, II, No.162, illus.). Spitzer sale, Paris, 17 April – 16 June, 1893, lot 577 (attributed to I.C.). J. E. Taylor sale, Christie's, 1–4, 9–10, July 1912, lot 152 (said to be in the 'manner of Jean Limosin').

Miss Alice de Rothschild — w1/77/8w

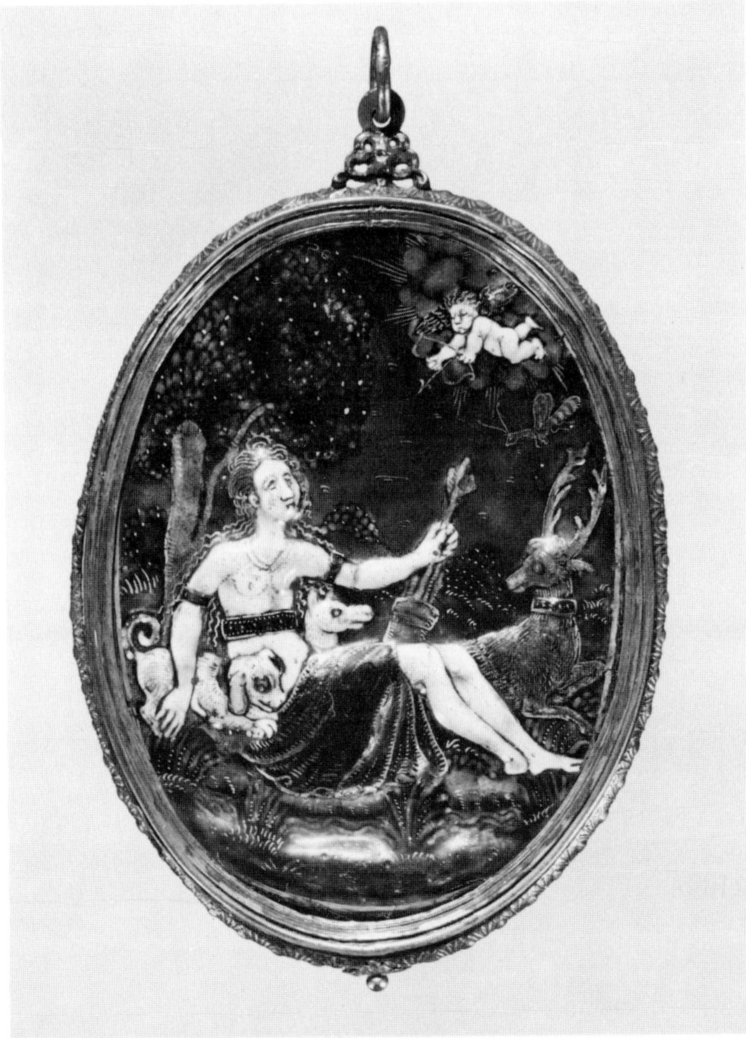

16. Attributed to Jean de Court (The Master I.C.) Diana departing for the hunt with her hounds

An oval enamelled plaque and a mirror in a hinged gilt-metal frame

$3\frac{5}{8} \times 2\frac{3}{4}$ in. (9·2 × 6·9 cm); with frame and floriated loop, $4\frac{1}{2} \times 3$ in. (11·4 × 7·6 cm)

Unsigned

The iconography is unusual in that Diana holds a crescent moon in her hand instead of wearing it in her hair.

The scene is oval, worked in relief and framed in a wide black border with floral and foliated scrolls and a bird at the top and bottom. The birds and the flowers on the border are on foils. The background of the main subject is green and blue. The flesh is painted white, heightened with pink in places. The muscles are indicated by shading. The waves of the hair are drawn in gold.

The Master I.C. seems to have made several enamels in which the designs are treated in relief embossed in the copper plaque. Two pieces are known, signed with the monogram: an oval plaque (probably the reverse of a mirror) representing the *Rape of Europa*, surrounded by flowers, birds and arabesques, from the A. von Parpart sale, Cologne, 20–25 October, 1884 (lot 551), and the reverse of an oval mirror, illustrating *The Continence of Scipio*, surrounded by flowers, foliage and birds, in the Spitzer sale, Paris, 17 April–16 June, 1893, lot 576.

Miss Alice de Rothschild — WI/77/8c

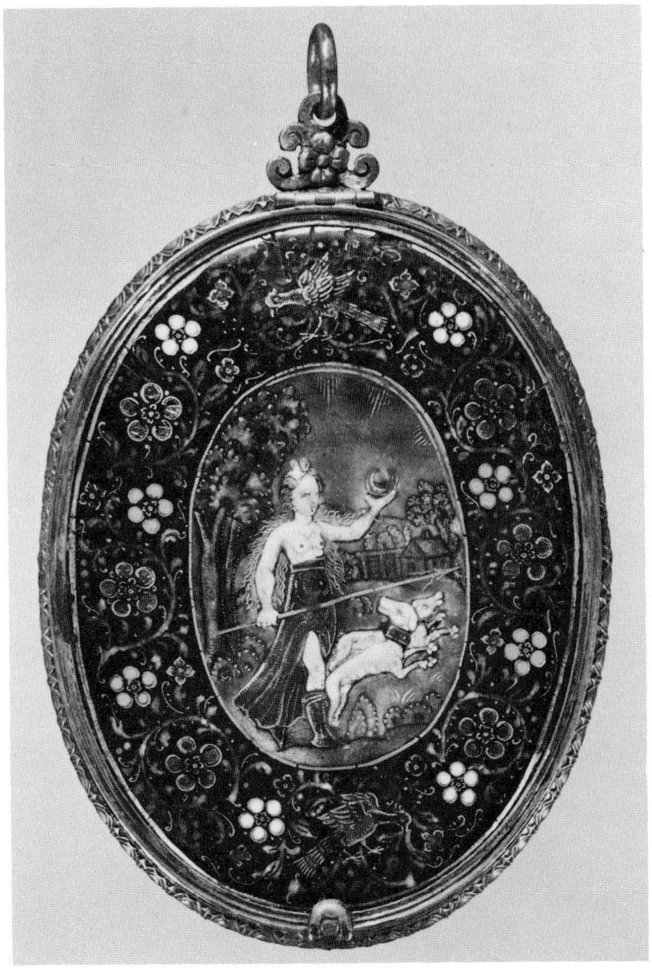

17. Attributed to Jean de Court (The Master I.C.)
Britomartis, pursued by Minos, throws herself into the sea

An oval enamelled plaque and a bevelled mirror in a gilt-metal frame

$4\frac{1}{8} \times 3\frac{1}{16}$ in. (10·5 × 7·7 cm); with mount and floriated loop, $5\frac{1}{8} \times 3\frac{7}{16}$ in. (13 × 8·7 cm)

Unsigned

The story is told by Callimachus (*Hymn to Diana*, 189ff). Britomartis, a Cretan girl, was pursued by Minos and escaped by throwing herself over a cliff. She was caught in a fisherman's net and escaped to Aegina, where she took refuge in a temple of Diana.

King Minos stands by the shore. By him are a cupid and two dogs, one of which rushes into the sea. The sea is depicted as calm in the distance, but Britomartis swims among waves, strongly defined with black lines, against a mauve sea. The boat, outlined in gold, is greyish with a white sail. The greens of the trees and ground verge on blue, the low cliff is pinkish-mauve and above is a dark blue starry sky.

Minos is dressed in a blue robe, turquoise tunic, deep purplish-mauve cloak and red cap and boots. Cupid's wings are green. Britomartis wears a green gown with a mauve belt and red scarf. One of the dogs is pale grey and the other salmon-pink. The flesh is tinted slightly salmon-pink. The king's beard and hair are nearly red. The clothes are painted on foils.

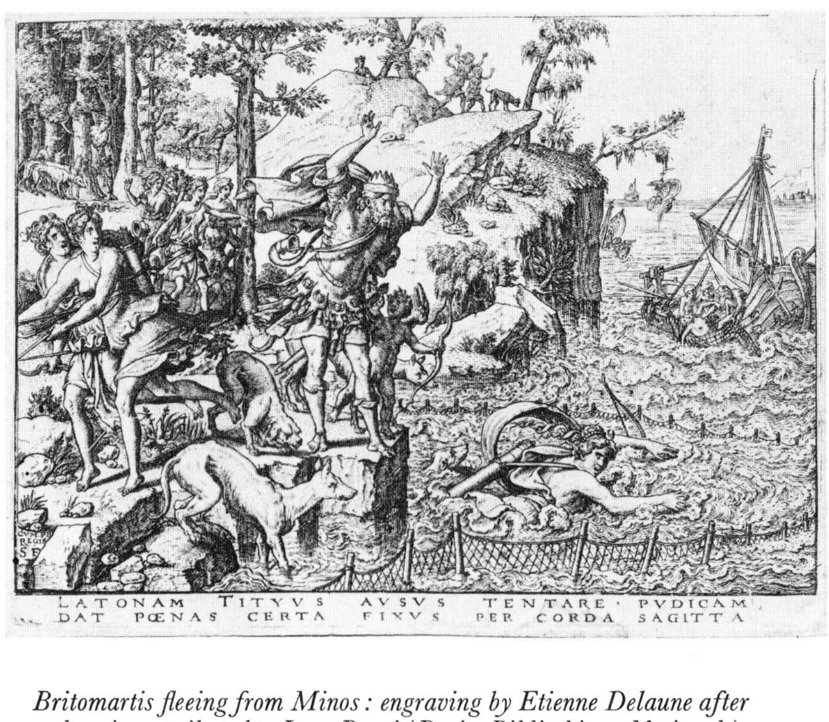

Britomartis fleeing from Minos: engraving by Etienne Delaune after a drawing attributed to Luca Penni (Paris, Bibliothèque Nationale)

18. Attributed to Jean de Court (The Master I.C.)
Orion pierced by the arrows of Apollo

An oval enamelled plaque and a mirror in a hinged gilt-metal mount

$4\frac{1}{4} \times 3\frac{1}{16}$ in. (10·7 × 7·7 cm); with frame and fleurette, $5\frac{3}{8} \times 3\frac{11}{16}$ in. (13·6 × 9·3 cm)

Unsigned

From the top of a precipitous cliff Apollo draws his bow at Orion who is swimming in the sea. In the foreground a nymph crowned with a laurel wreath stands at the foot of a tree. At her feet a dog raises his muzzle towards three other nymphs, each wearing a flower in her hair, who point to the standing nymph.

Translucent enamel is used for the clothing, the modelling and hatching being visible on the foil. The sea is midnight blue and marked with waves.

The pointed faces with small round chins and long straight noses are very characteristic of I.C. The muscles are indicated by areas of opaque white. The bodies and faces are outlined with thick black lines.

The design is taken from the fifth print of the Apollo and Diana series (cf. *17*) by Etienne Delaune after drawings attributed to Luca Penni (Robert-Dumesnil 137) inscribed: IMPROBVS ORION CASTAM VIOLARE DIANAM AVSVS APOLLINEI CADIT ICTVS ACVMINE TELI. In the original

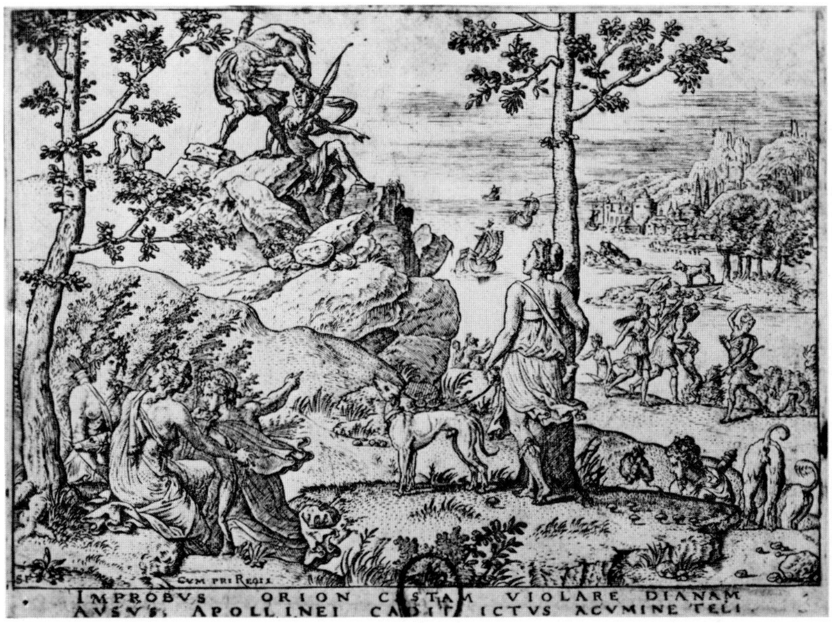

Orion pierced by the arrows of Apollo: engraving by Etienne Delaune after a drawing attributed to Luca Penni (Paris, Bibliothèque Nationale)

the standing nymph turns her back to the spectator. The subject was treated by Pierre Reymond on the right half of the front of a ridged casket which was in the Stein sale, 1866, lot 77 and appears again on the back of a mirror by François Limosin now in the Louvre (M.V. 740). The Delaune plate was also used for *43* below.

Miss Alice de Rothschild WI/77/8e

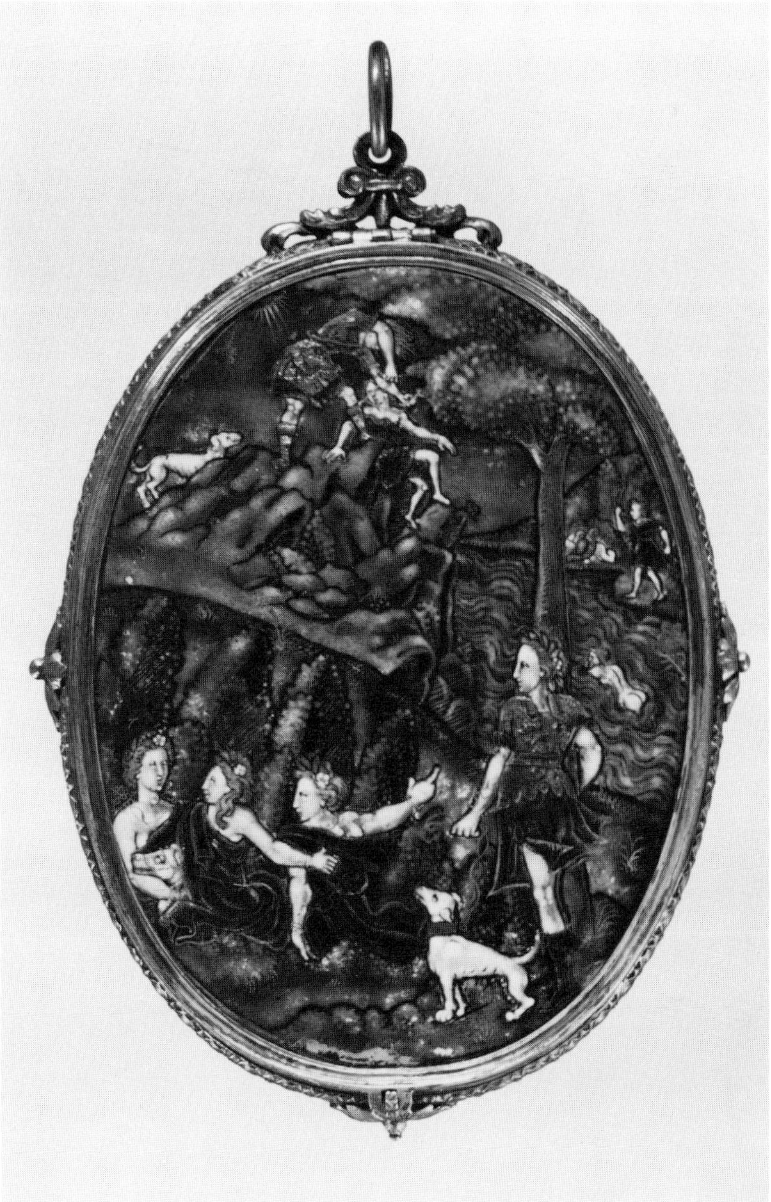

19. Attributed to Jean de Court (The Master I.C.) Pyramus and Thisbe

An oval enamelled plaque and a mirror in a hinged gilt-metal frame.

$4\frac{1}{16} \times 3\frac{1}{16}$ in. (10·3 × 7·8 cm); with frame and floriated loop, $5\frac{9}{16} \times 3\frac{13}{16}$ in. (14 × 9·6 cm)

Unsigned

Pyramus is shown lying on the ground, while Thisbe falls on the sword. In the background is depicted an earlier scene in the story, showing Pyramus hurrying forward on seeing the lion on the left carrying off Thisbe's scarf.

The turf is green, greenish-blue and blue, the water a dull blue with waves indicated in black and gold. The two towers on the right are chocolate-coloured, the town and clouds mauve. The contours of the hills are edged with dotted gold and at the top a small section of the sky is gold. The clothes and the scarf being carried away by the lion are enamelled on foils. Pyramus is dressed in a violet robe, turquoise tunic and brown boots, and Thisbe wears a blue gown, a purplish blue tunic and has a green jewel in her hair. The small figure in the middle distance is dressed in brown and blue. The lion is brownish-gold and the scarf bright green. The flesh is salmon-pink without hatching and with numerous gilt highlights.

The drawing of the folds, the wavy hair touched with gold, the softness of tone of the water, and the sky dappled with clouds suggest a tentative attribution to the Master I.C.

The counter-enamel is partly of greenish-blue and of a transparent flux which shows the pink copper.

The composition was inspired by the illustration by Virgilius Solis in *Les Métamorphoses d'Ovide* (Frankfort, 1563, Book IV, No. 11).

Miss Alice de Rothschild — WI/77/8d

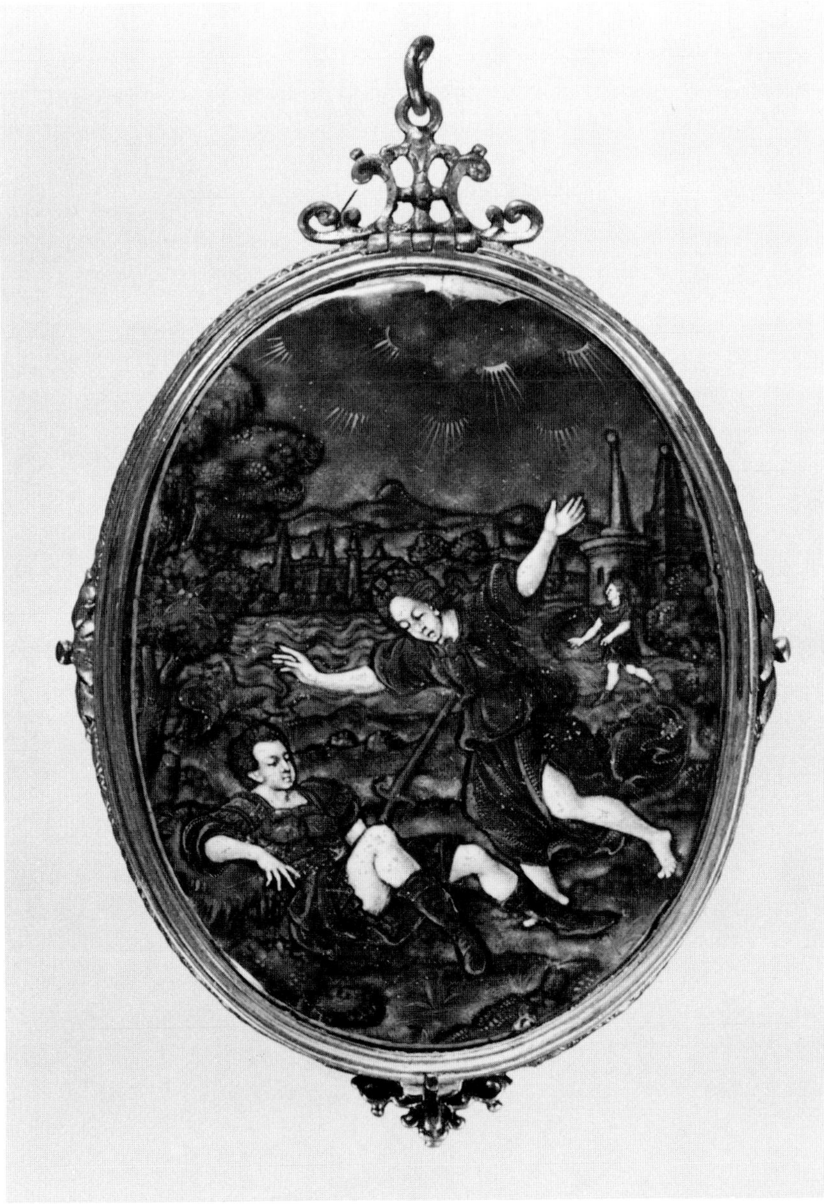

20. Attributed to Jean de Court (The Master I.C.)
The Queen of Sheba before Solomon

An oval enamelled plaque and a bevelled mirror in a gilt-metal frame

$4\frac{11}{16} \times 3\frac{5}{8}$ in. (11·9 × 9·2 cm); with frame and floriated loop, $6\frac{3}{8} \times 4\frac{5}{16}$ in. (16·2 × 10·9 cm)

Unsigned

The king's throne is ochre outlined in black, beneath a rather light blue-green curtain and a blue pelmet painted on foils. The sky is light blue, the turf blue-green and blue, the floor and the buildings mauve. The clothes are enamelled on foils. The queen, kneeling at the foot of the throne, wears a golden-yellow gown and train, a blue tunic, red sleeves and scarf and a turquoise belt. One of her attendants, who stands behind her, wears a blue dress with turquoise sleeves and turquoise jewels in her hair. The king is dressed in a short blue under-tunic beneath a turquoise tunic with brown fringes, a red scarf and boots, and a violet crown. The other figures are dressed in brown, turquoise, red and blue. The flesh is salmon-pink on a darker ground. Very delicate red lines indicate the lips. The curly beards are grey. The women's hair is ochre, covered with a fine gold sheen. The details are drawn in black. The dog is white, being modelled on a grey base.

Here again we find faces with small round chins and cheeks highlighted with touches of red, which are very close to the signed works of I.C., and this attribution is confirmed by the absence of foils under the greens, the attention paid to architectural detail, the tiled floor and the pelmet.

On a plate in the Palais des Arts at Dijon the subject is explicitly indicated by the inscription *Salomon* at the bottom. The iconography is the same, but there are differences in the number and positions of the figures. The same composition appears on *29* below.

Miss Alice de Rothschild — W1/77/8aa

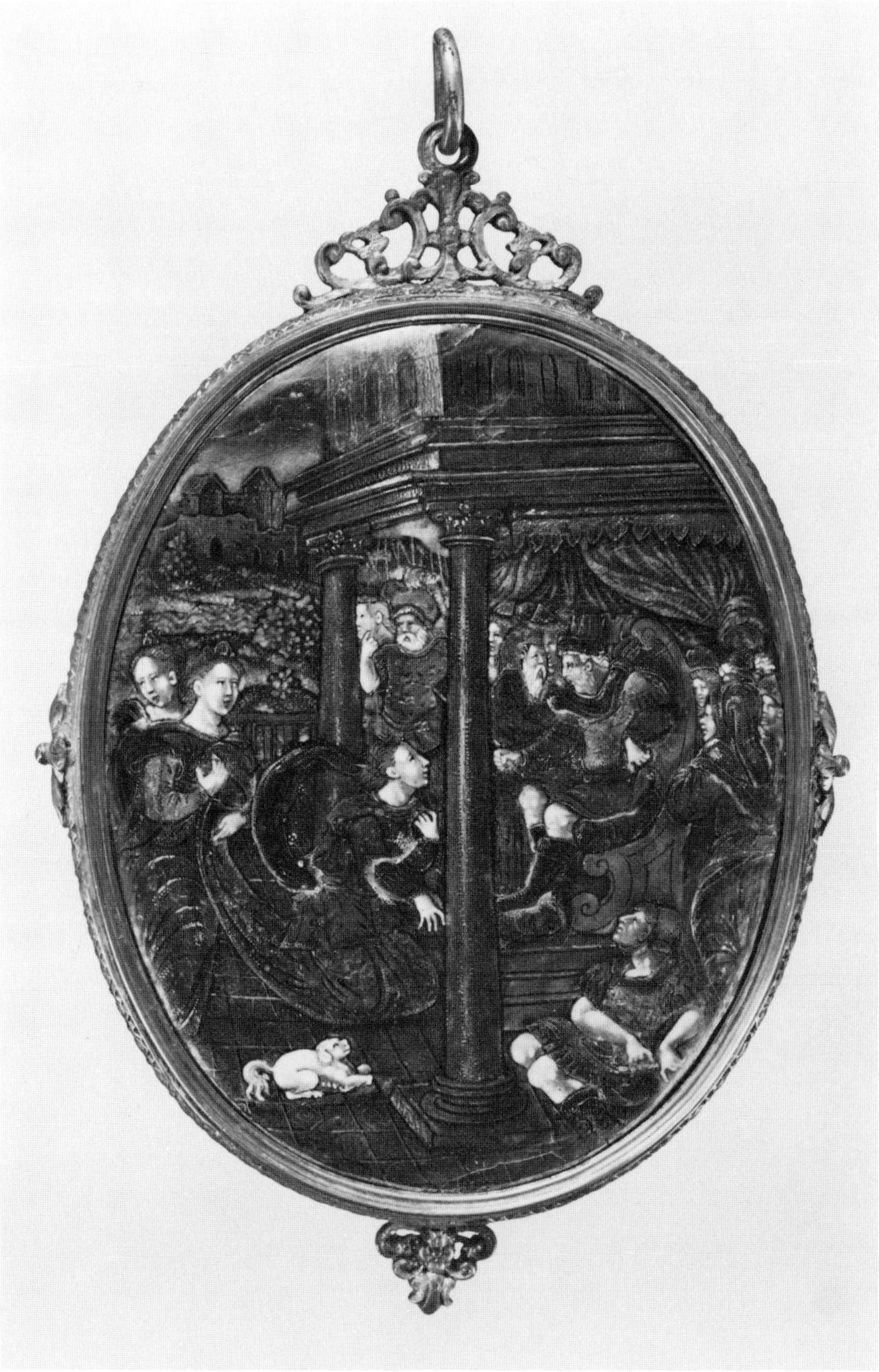

21. Attributed to Jean de Court (The Master I.C.) A female figure with a warrior in armour

An oval enamelled plaque and a mirror in a hinged gilt-metal frame; the glass has lost part of its tin

$3\frac{9}{16} \times 2\frac{11}{16}$ in. ($9 \times 6{\cdot}8$ cm); with frame and floriated loop, $4\frac{9}{16} \times 3\frac{1}{16}$ in. ($11{\cdot}5 \times 7{\cdot}7$ cm)

Unsigned

The subject of this enamel is difficult to identify as neither of the figures has any attributes. They appear to be meeting outside a city; above is a cloud, veiling the sun. The woman wears a violet gown, a blue tunic with short sleeves and scarf of green verging on turquoise. The man is dressed in a blue tunic over brown, blue boots and a turquoise helmet. The flesh is painted in grisaille strongly touched with red. The sky is dappled dark blue. The water is blue with waves marked in black, the architecture is violet, and the rest is green. Under-drawing in black can be seen clearly where the enamel is transparent.

The counter-enamel is rather thick, translucent and irregular.

The treatment of the folds, the application of gold in small parallel marks, the length of the fingers, the pointed oval faces, the transparency of the coloured enamels, are strongly reminiscent of the Master I.C.

This piece may be compared with another mirror in the Walters Art Gallery in Baltimore, Inv.44.131 (Verdier *Catalogue*, No.179), attributed to the workshop of Jean de Court, in which a peacock between the two figures clearly identifies the woman as Juno and the other figure is said to be Paris. In this version the two figures are transposed and the male figure is seated.

Miss Alice de Rothschild — WI/77/8h

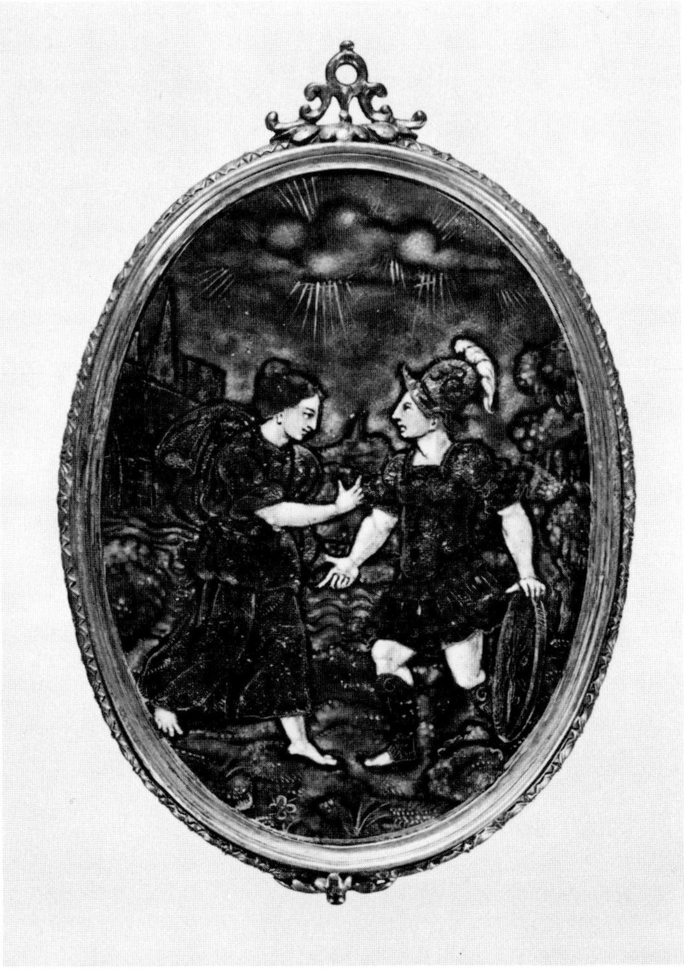

22. Attributed to Jean de Court (The Master I.C.) The death of Niobe's Children

An oval enamelled plaque and a mirror in a hinged gilt-metal frame

$2\frac{11}{16} \times 2\frac{1}{16}$ in. ($6{\cdot}8 \times 5{\cdot}2$ cm); with frame and floriated loop, $3\frac{7}{8} \times 2\frac{3}{8}$ in. ($9{\cdot}8 \times 6$ cm)

Unsigned

From the clouds Diana and Apollo shoot arrows at Niobe's children, some of whom have already been struck; the others raise their arms towards the sky trying to protect themselves from the arrows. On the right are two rearing horses. Diana, dressed in blue, and Apollo in purple, are partially set on a gold ground. The clothes are enamelled on foils, the flesh is white, lightly tinged with red.

The piece is executed on very fine copper. The counter-enamel is mainly translucent showing smooth copper. The mirror case has been crushed and the lower left section is much damaged.

A comparison of the faces with those of the Muses on mirror *9*, which is signed I.C., reveals the close similarities of their shape, the way the hair grows, and the white light in the centre of the lower – and particularly the upper – eyelids, which gives them a curious prominence beneath eyebrows which are barely indicated.

The composition is based on the engraving of the same subject in *La Métamorphose d'Ovide figurée* (cf. *9*).

Miss Alice de Rothschild — WI/77/8U

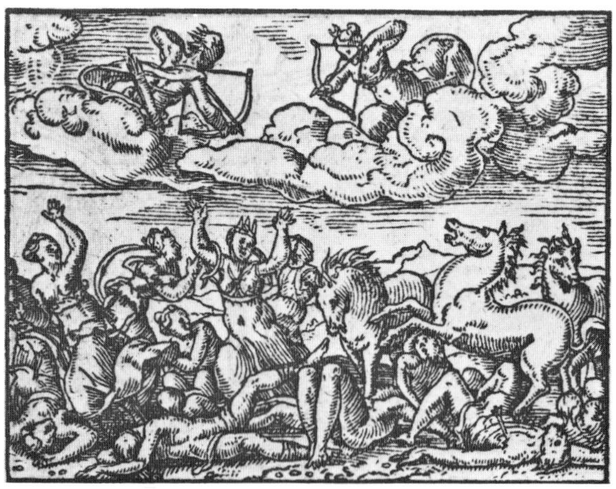

Illustration to La Métamorphose d'Ovide *by Bernard Salomon, 1557*
(Paris, Bibliothèque Nationale)

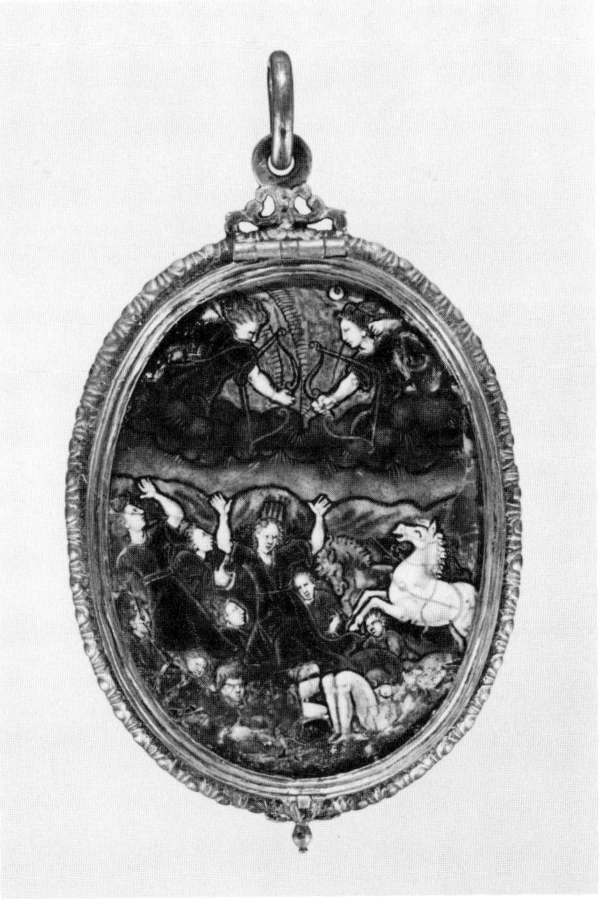

23. Attributed to Jean de Court (The Master I.C.)
Juno and the three Furies

A rectangular plaque

$4\frac{13}{16} \times 3\frac{13}{16}$ in. ($12{\cdot}2 \times 9{\cdot}6$ cm)

The letter C is visible at the centre of the top of the composition but the enamel has been damaged where the first initial or initials would have been.

Juno arrives at the gates of hell, and is met by the Furies and Cerberus. The upper part illustrates the surface of the earth.

The enamel is partly grisaille and partly polychrome. Cerberus is curiously mottled in green, blue and pink. Juno wears a brown gown, a green scarf, a short blue overskirt and a blue jacket with short violet sleeves. One of the Furies is in a green verging on turquoise, another in brown, and the third in blue. The clothes are enamelled on foils, the architecture is rendered in grisaille and the flames are dark red or gold on foils. The flesh is coloured pink, the hair in grisaille, Juno's having a gold sheen. The counter-enamel is opaque, slightly granulated, with the letter N in black.

On stylistic grounds it is safe to attribute the enamel to Jean de Court rather than Suzanne Court, and to assume that the missing letter or letters in the inscription were I or ID, rather than S.

The composition is based on the engraving of the subject in *La Métamorphose d'Ovide figurée* (cf. *9*). The engraving was also used for *27* below.

Miss Alice de Rothschild — WI/77/4

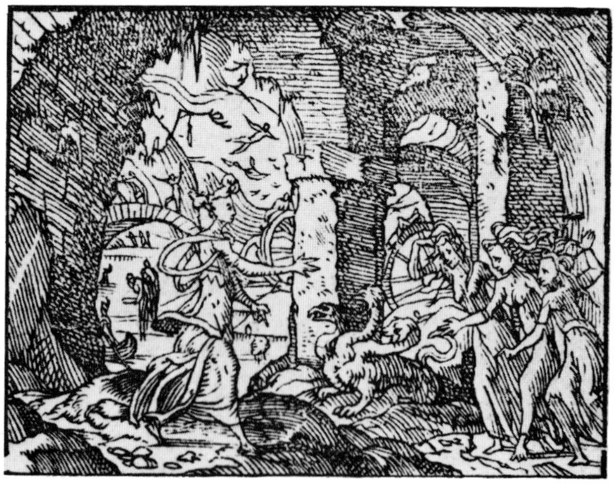

Illustration to La Métamorphose d'Ovide *by Bernard Salomon, 1557*
(Paris, Bibliothèque Nationale)

SUZANNE COURT (SUZANNE DE COURT)

Suzanne Court, who was the only female enamel worker of the Renaissance in Limoges, may have been the daughter of Jean de Court. None of her works are dated, but their style links them with the end of the sixteenth century and it is known that she was still alive in 1600, living in the Boucherie district of Limoges. Her works are however often signed, the more important with her full name SVSANNE DE COVRT, SVSANNE. DE.COVR, SVSANNE COVRT, others with her monogram S.C., and some with both.

She made enamelled ewers, plates, oval dishes, caskets, large plaques of religious subjects and specialised in small plaques and the backs of mirrors ornamented with mythological scenes and figures.

Her figures resemble those of the Master I.C. (Jean de Court) but are less thin and angular. The faces are different, with heavier jaws, and the flesh is generally milky-white. With the exception of the backs of certain plates and dishes which she decorated in grisaille like those of Jean de Court, her compositions are in colour, with a moderate use of foil, in rich harmonies of blues and greens with much delicate but not over-meticulous detail, and a judicious use of gold to lighten the rather dark general effect of the work.

25. Suzanne Court: The Annunciation

A rectangular plaque

$10 \times 7\frac{15}{16}$ in. ($25{\cdot}4 \times 20$ cm)

Signed lower left, SVSANNE COVRT. F.

Companion to *26*

The angel, who bows before the Virgin and offers her a lily, is wearing a blue gown, purple tunic and green scarf. He points towards God the Father who, clothed in blue and purple, hovers over the canopied bed in a dappled cloud which emits rays of light. The dove of the Holy Spirit appears in a gold radiance. In the background a door opens on to a highly-coloured countryside. Generally speaking the tonality is sombre though the dense white flesh tones are offset by the translucent enamels and the numerous gilt highlights. The counter-enamel is translucent but slightly worn.

The same pattern was used for an oval plaque also attributed to Suzanne Court, now in the Walters Art Gallery in Baltimore (Inv.44.191, Verdier *Catalogue*, no. 181) and for an oval plaque dating from 1605, attributed to Jean I Limosin in the same museum (No.44.346, Verdier *Catalogue*, No.188).

The plaque was part of a series of which four other pieces can probably be identified: *The Nativity* (*26* below); perhaps *The Adoration of the Magi* signed SVSANNE COVRT.F. (acquired by Allain at the Ducatel sale, Paris, 21–26 April, 1890 (25 × 19 cm)), an *Ecce Homo* signed S.C. (from the Ducloux sale in 1891, lot 20, then in the Jules S. Bache Collection and now in the Metropolitan Museum, New York, inv. 45.60.7), and a plaque representing *Pentecost* (sold from the Sir Francis Scott Collection at Sotheby's on 27 April, 1965 (lot 8)).

Miss Alice de Rothschild — WI/138/1

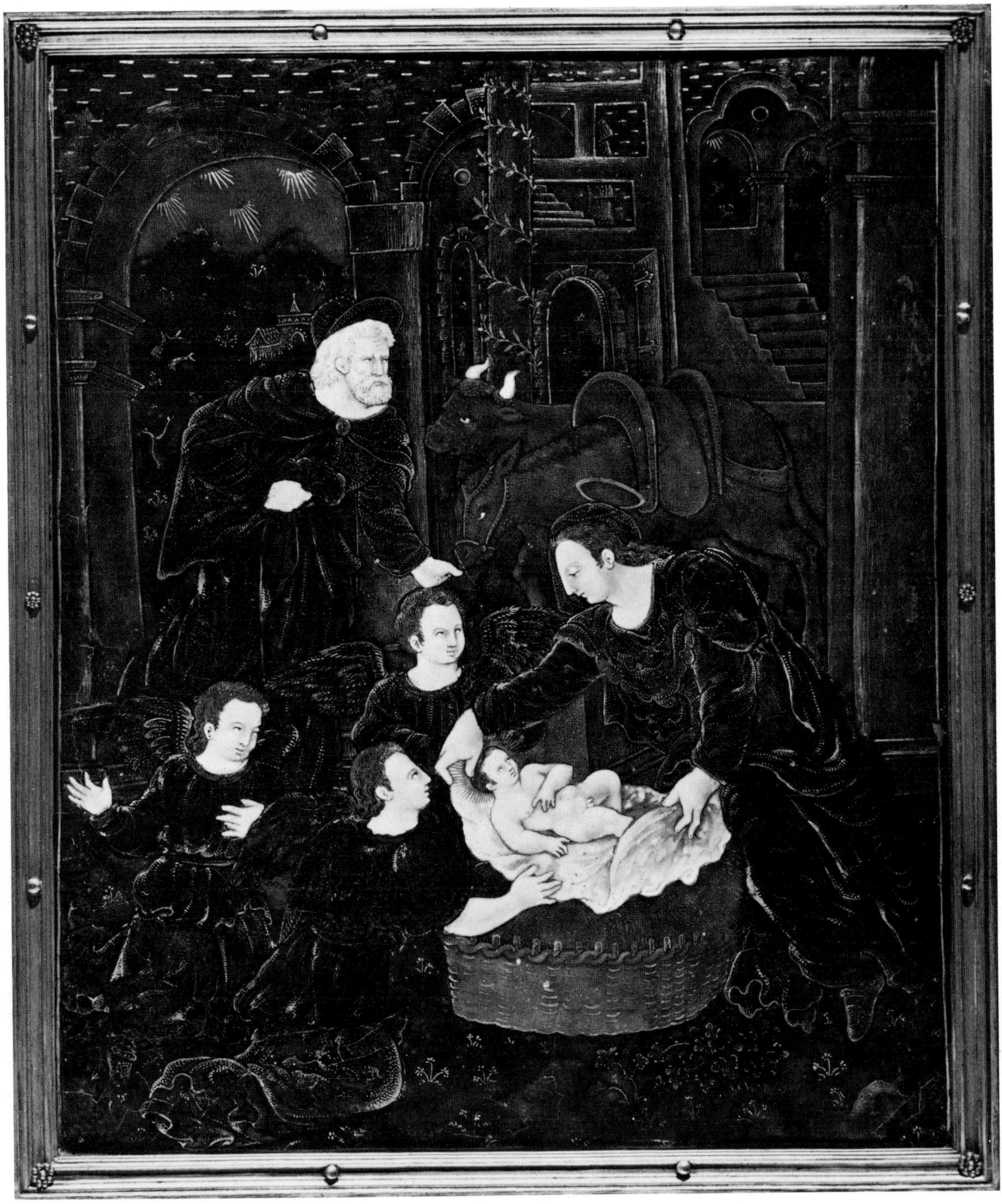

27. Suzanne Court: Juno and the Furies

An oval enamelled plaque and a bevelled mirror in a gilt-metal frame

$4\frac{7}{8} \times 3\frac{13}{16}$ in. (12·3 × 9·6 cm); with frame and floriated loop, $6\frac{5}{16} \times 4\frac{9}{16}$ in. (16 × 11·6 cm)

Signed with the monogram S.C.

The subject and the iconographical source are the same as those of plaque *23* above. The title IVNON. LES FVRIES. S.C. is inscribed at the top of the composition, slightly hidden under the frame.

Cerberus is coloured green speckled with russet. The clothing is indicated by translucent enamels, much defaced, painted on foils. The architecture is coloured blue-green, the flames bright red on foils, and the hair is painted in gold.

The water is yellowish and the waves are indicated in black. The light blue sky is transparent enough for the underlying white to show through in parts. Rather clumsy red streaks are used to highlight the flesh.

Miss Alice de Rothschild — WI/77/8bb

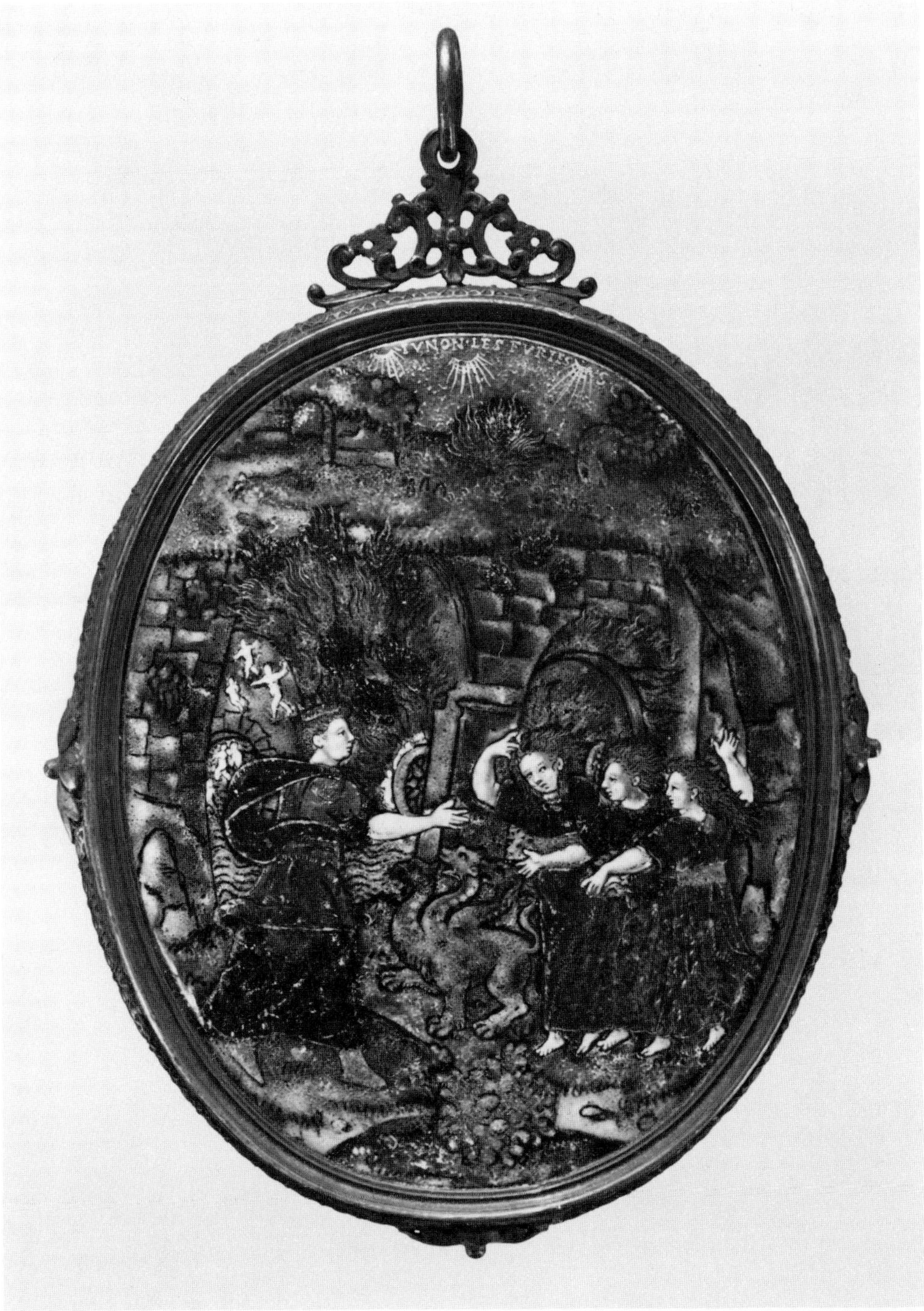
IVNON·LES FVRIES

28. Suzanne Court

Minerva visits the Muses on Mount Helicon

An oval enamelled plaque and a mirror in a hinged gilt-metal frame

$4\frac{1}{8} \times 3\frac{3}{16}$ in. (10·4×8·1 cm); with frame, $4\frac{5}{16} \times 3\frac{7}{16}$ in. (11×8·8 cm)

Signed with the monogram S.C.

Taken from the illustration by Bernard Salomon for *La Métamorphose d'Ovide* used for *9* above (*q.v.*) and *30* below. The colours are stronger, the flesh more salmon-pink than in the work of the Master I.C. The heads have the broad foreheads, with the line of the brow continued into the nose, which are characteristic of Suzanne Court.

The mirror is in a hinged frame, which allows one to see the rather light copper showing through the translucent counter enamel. The inside is lined with old-rose silk.

Another mirror, illustrating the same subject and signed with the monogram S.C., is in the Lehman Collection in the Metropolitan Museum (inv. E.15).

Provenance: This mirror may be the one recorded in the Bernal sale, Christie's, 5–29 March, 1855, lot 1505: 'Apollo and the Muses on Mount Parnassus . . . signed S.C.' of roughly the same dimensions. It was acquired by Andrew Fountaine and appeared in the Fountaine sale, 16–19 June, 1884, lot 290 (bought Davis).

Miss Alice de Rothschild WI/77/8z

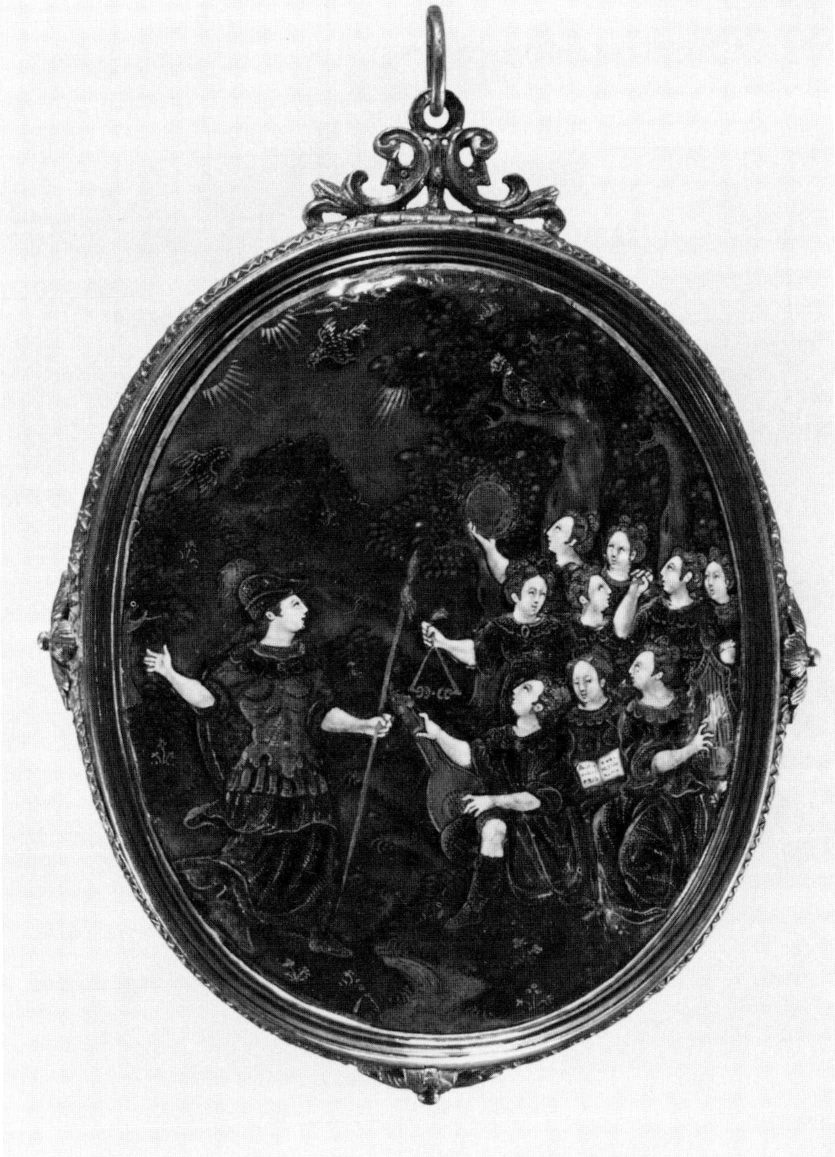

29. Suzanne Court
The Queen of Sheba before Solomon

An oval enamelled plaque with a bevelled mirror in a gilt-metal frame

4×3 in. ($10{\cdot}2 \times 7{\cdot}6$ cm); with frame and floriated loop, $5\frac{1}{8} \times 3\frac{3}{8}$ in. ($13 \times 8{\cdot}6$ cm)

Signed with the monogram S.C. at the top on the left of the canopy.

The king is seated on a throne beneath an architectural canopy. The placing of the figures is almost the same as in *20* above.

The floor is covered with grey tiles decorated in gold. The base of the throne and the columns are grey-blue and the balustrade ochre. The sky is dark blue above a green landscape background. The architecture is painted mauve. The clothing is indicated by translucent enamels on foils. The queen wears a red gown with sleeves, blue tunic, green scarf, yellow-gold crown and belt, and her attendant has a pink gown with blue tunic. Solomon wears a blue robe with green sleeves, red collar, blue crown and pink boots. The other men are dressed in pink. The colours on foils are very vivid. The flesh is white with red spots on the cheeks. The hair is covered with very fine gold lines and the beards are yellow ochre.

Miss Alice de Rothschild — W1/77/8m

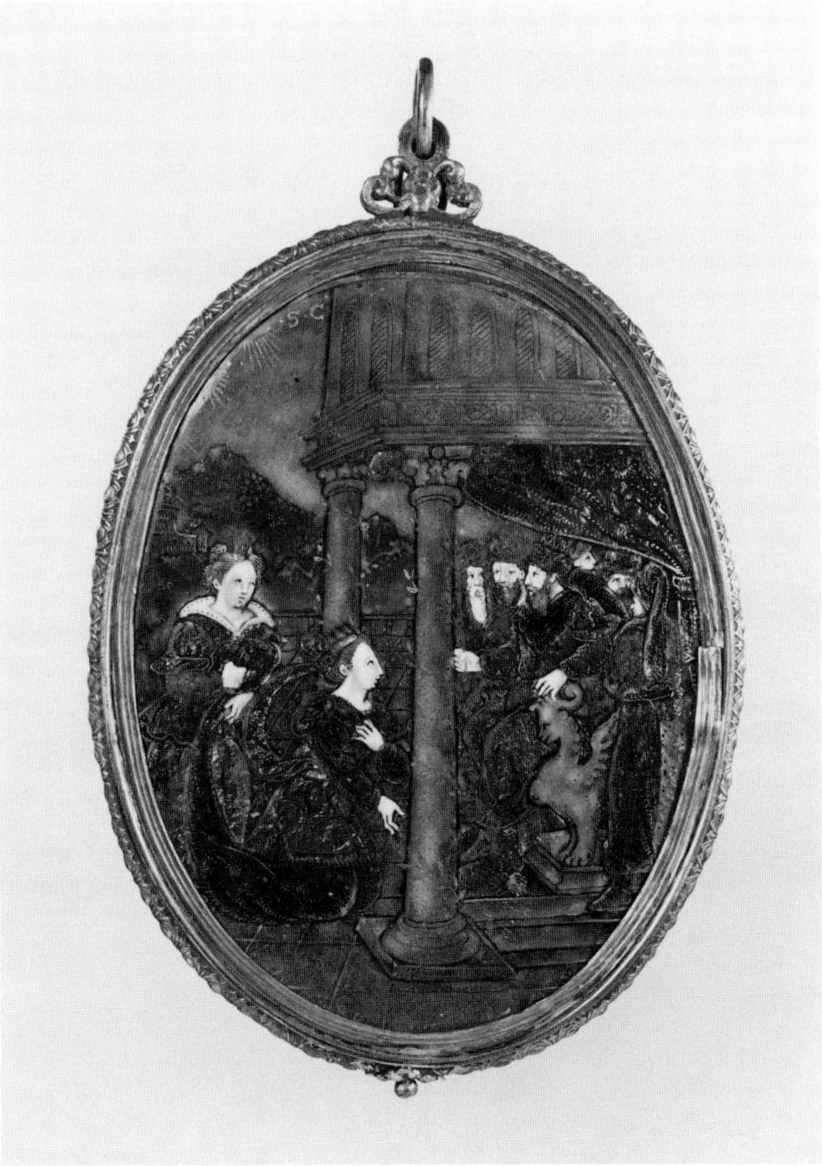

30. Attributed to Suzanne Court
Orpheus charming the animals

An oval enamelled plaque and a mirror in a hinged gilt-metal frame

$3\frac{5}{16} \times 2\frac{11}{16}$ in. (8·5 × 6·7 cm); with frame and floriated loop, $4\frac{7}{16} \times 3\frac{3}{8}$ in. (11·2 × 8·5 cm)

Unsigned

At the feet of Orpheus are a unicorn, a kneeling stag and a fox (?)

The dark background is rather indistinct. The green boots, pink robe and blue tunic are enamelled on foils. The birds perched on the branches behind Orpheus are also enamelled on foils. There is no cross hatching and little variety in tone. Blue and green predominate while the flesh is slightly salmon-pink. Certain features point to the authorship of Suzanne Court; the heavy face of Orpheus, with high broad forehead, on which the hair looks like a wig and forms little claw-shapes in front of the ears; the care in depicting ornament even on such a relatively slight enamel (for instance the buckles on Orpheus' boots).

The subject is freely interpreted from 'Orphée harpeur excellent' in *La Métamorphose d'Ovide figurée* (cf. *9*). The same subject appears, differently treated, on *31* below.

Provenance: It is possible that this was the mirror in the Charvet sale, Paris, 11–12 May, 1883, lot 2038, attributed to Suzanne Court, which had exactly the same dimensions (bought by Allain for 1,100 francs).

Miss Alice de Rothschild — W1/77/8g

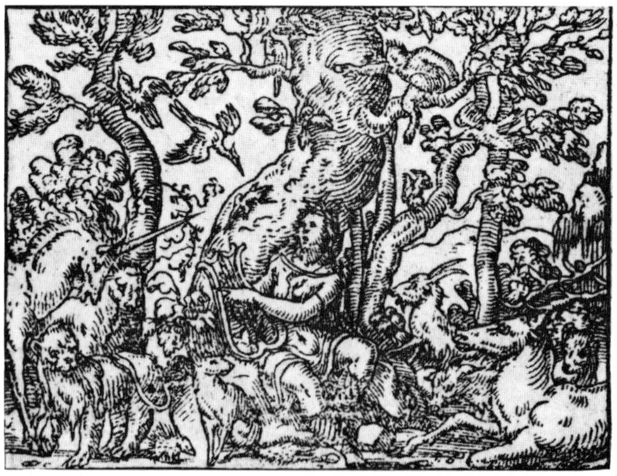

Illustration to La Métamorphose d'Ovide *by Bernard Salomon, 1557*
(Paris, Bibliothèque Nationale)

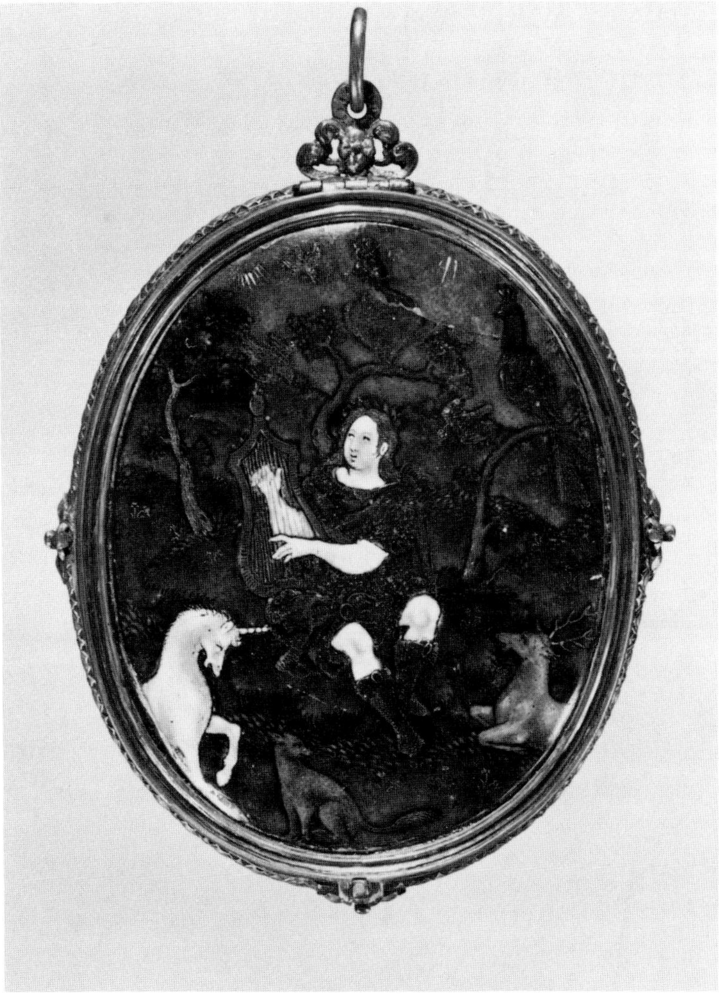

31. Studio of Suzanne Court?
Orpheus charming an animal

An oval enamel plaque with a mirror in a hinged gilt-metal frame

$4\frac{1}{4} \times 3\frac{1}{4}$ in. (10·8 × 8·2 cm); with frame and floriated loop, $5\frac{5}{16} \times 3\frac{7}{8}$ in. (13·5 × 9·8 cm)

Unsigned

The subject is painted in an oval outlined in gold, surrounded by a black border decorated with two birds and flowers on foils, and small gold motifs.

In this enamel Orpheus plays a cithara instead of the lyre which appears in *30* above.

The birds and the clothes are enamelled on foils. Orpheus wears brown boots with blue cuffs and a blue cloak lined with purplish-blue. His flesh is white slightly tinted with red. At his feet is a lion, some details of which are delicately picked out in gold. The waves are drawn in black on the blue water. Beyond is a schematically drawn mauve landscape.

The counter-enamel is translucent, the copper red.

The milky whiteness of the flesh, the hair in the shape of a little claw in front of the ear, and the delicate drawing of the lion, suggest that this enamel is from the studio of Suzanne Court.

Miss Alice de Rothschild W1/77/8t

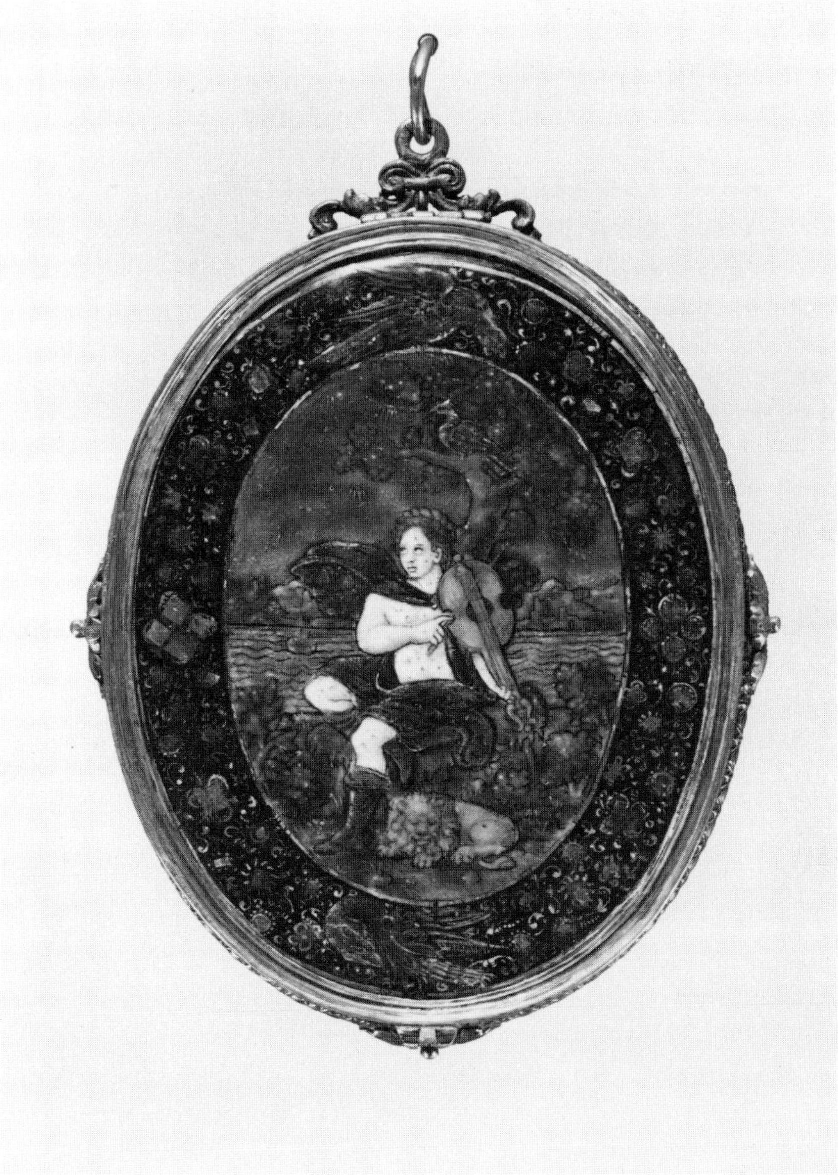

JEAN I LIMOSIN

The date of his birth is unknown but he seems to have died before 1610. His relationship with Léonard Limosin is uncertain, but his work is close to that of Suzanne Court, who seems to have exercised a strong influence on him, though his works are drier and more lifelike than hers. They are stiff and dark and there is less inventiveness and attention to detail.

Jean I Limosin's ornaments are easily recognizable but lacking in variety: black grounds loaded with almost vermiculated gold scrollwork broken here and there by small flowers, square or lozenge-shaped motifs, or birds, in cold colours on foils. In the skies there are often dappled clouds emitting gold rays.

He signed his large oval dishes with his full name IEHAN LIMOSIN, and his rare ewers, small plaques, backs of mirrors and salt cellars, with the monogram I.L. The fleur-de-lis which is sometimes placed between the two letters of the monogram may indicate that the monogram belongs to Jean II, his son or nephew.

32. Jean I Limosin

Baluster-shaped ewer with S-shaped handle

Overall height, 11 $\frac{5}{16}$ in. (28·7 cm)

Signed I.L. inside the neck where the handle springs

The upper part is filled with a panel representing a satyr blowing a horn; next comes the chariot of Cybele in front of which is Proserpine with a basket of flowers on her head. Below are episodes from the story of Abraham: the expulsion of Hagar and Ishmael, the angel appearing to Hagar, and the sacrifice of Isaac. Between the last two is a scene with a shepherd – probably one of the patriarchs – guarding his sheep.

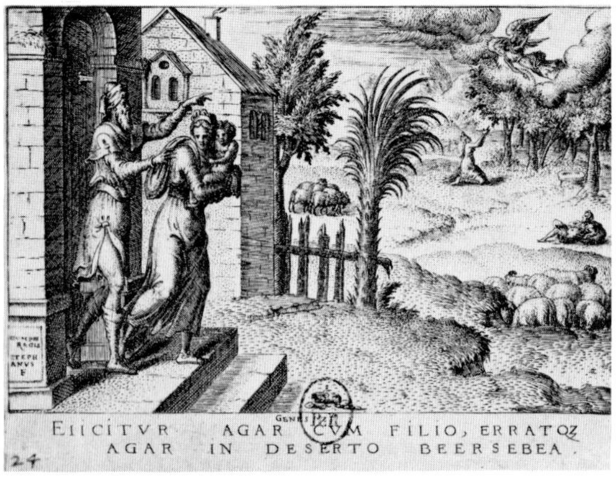

Engraving by Etienne Delaune for Histoire d'Abraham (*Paris, Bibliothèque Nationale*)

32

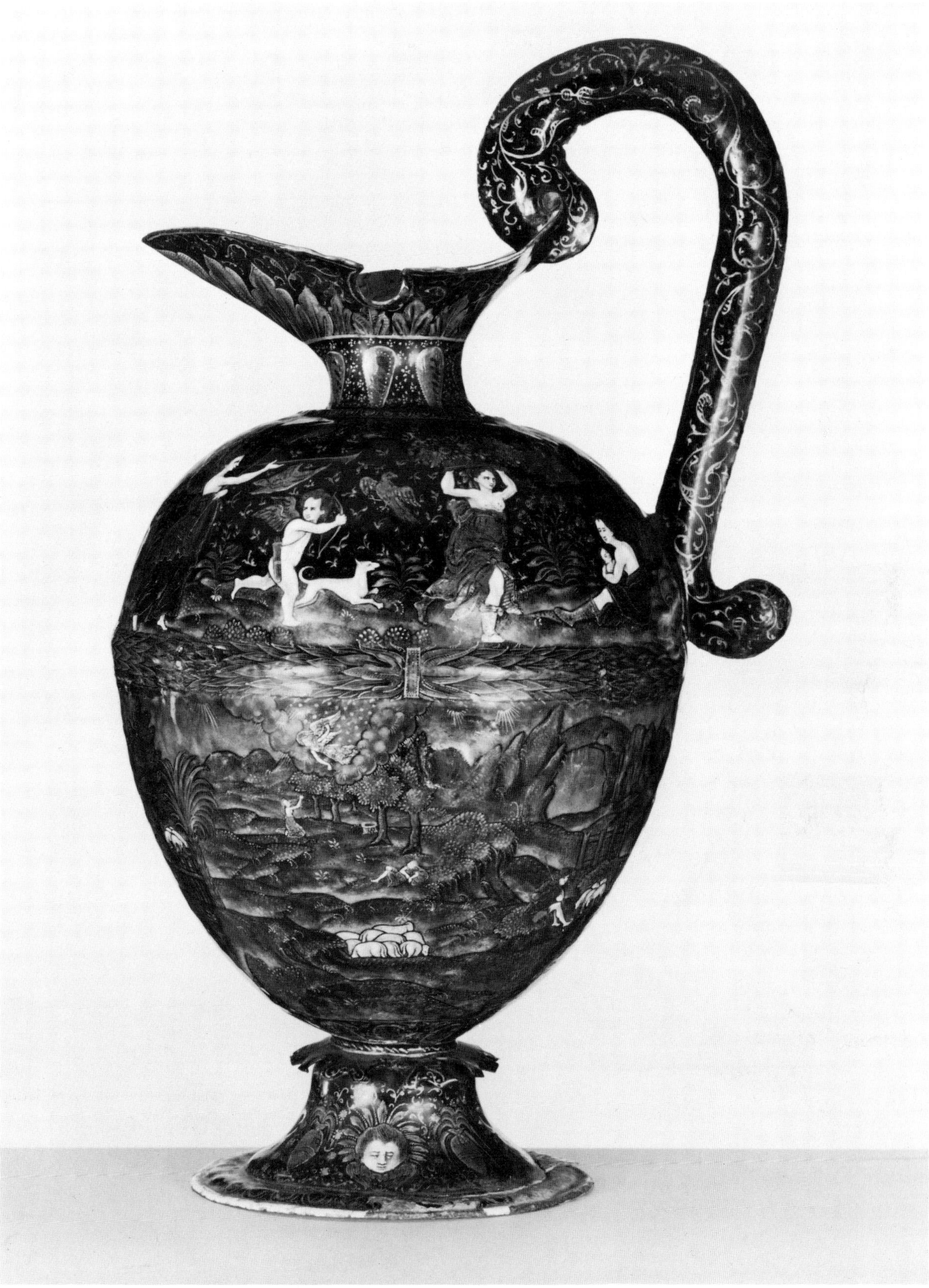

At the bottom is a band of oval motifs, and the base is decorated with masks between vases on a black ground strewn with small gold leaves. The foliage is continued on a larger scale on the handle. The neck and mouth are decorated with oval motifs and acanthus leaves, also on a black ground, while the inside is enamelled in white with gold foliage. The coloured enamels are painted on a black ground. Only the clothes and the decorative motifs such as flowers and birds are enamelled on foils. The counter-enamel is black, strewn with small flowers and fleurs-de-lis in gold.

The panels illustrating the story of Abraham are taken from engravings by Etienne Delaune (Robert-Dumesnil, 45ff.) (one reproduced on p.400).

Exhibited: Musée Rétrospectif, Paris, 1865 (No.2500) and *Exposition Universelle*, Paris, 1867 (No.3078).

Baroness James de Rothschild (Collection No.H.65); Baron Edmond de Rothschild P/1701

33. Jean I Limosin: Apollo and the Muses

An oval enamelled plaque and a mirror in a gilt-metal frame.

$3\frac{15}{16} \times 2\frac{13}{16}$ in. (10 × 7·2 cm); with frame, $4\frac{1}{4} \times 3\frac{1}{8}$ in. (10·8 × 7·9 cm)

Signed with the monogram I.L.

At the top are Apollo and Pegasus; below, the Hippocrene spring is symbolized by a bearded man instead of a water nymph. The nine Muses are at the foot of the mountain.

There are small gold motifs on the green ground, the sky is dark blue, the water turquoise. The Muses' gowns are in dark colours enamelled on foils.

The nudes are white, slightly tinted with grey by impasto. The details are rendered in extremely fine stippling.

There is a comparable plaque in the Victoria and Albert Museum (8418–1863) illustrating the same subject and probably by the same artist, but the composition is there in reverse.

Literature: Catalogue of the Spitzer Collection Paris, 1891, II, No.76.

Provenance: Spitzer sale, Paris, 17 April–16 June, 1893, lot 492.

Miss Alice de Rothschild W1/77/8a

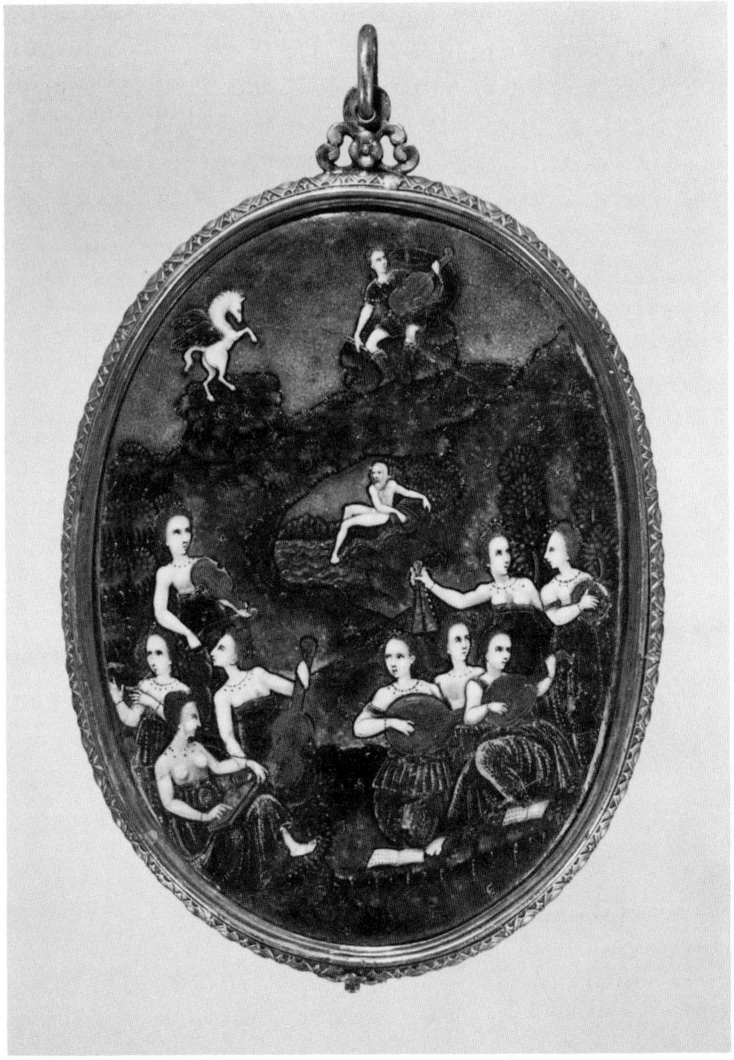

34. Jean I Limosin: Minerva

An oval enamelled plaque and a mirror in a gilt-metal frame

$3\frac{5}{8} \times 2\frac{11}{16}$ in. (9·2 × 6·8 cm); with frame and floriated loop, $4\frac{9}{16} \times 3\frac{3}{16}$ in. (11·6 × 8·1 cm)

Signed with the monogram I.L.

The goddess is standing on a pedestal supported by a console composed of a decorative floral motif. An owl is perched on each of the pieces of drapery which hang from either side of the pelmetted canopy.

The enamel is painted in delicate grisaille and colours on a black ground. The purple gown, mauve tunic and blue breastplate, dark blue cloak and collar are enamelled on foils. At the edges are gold foliated scrolls.

Miss Alice de Rothschild — W1/77/8p

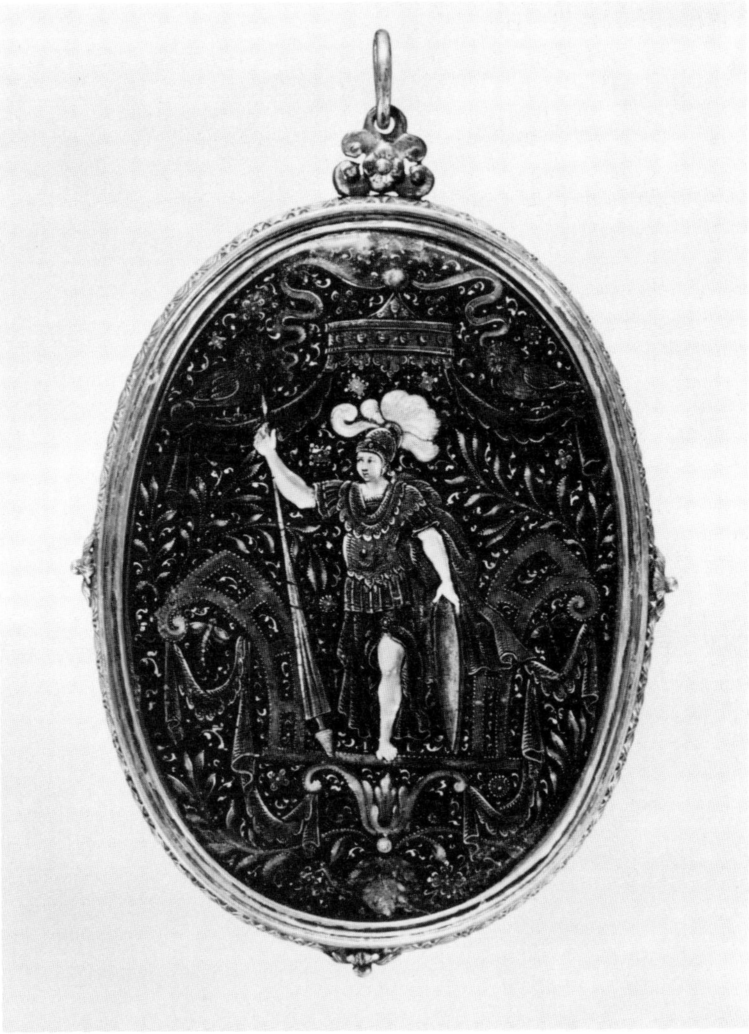

35. Jean I Limosin

Britomartis pursued by Minos throws herself into the sea

An oval enamelled plaque and a bevelled mirror in a gilt-metal frame

$3\frac{7}{8} \times 2\frac{13}{16}$ in. (9·8 × 7·1 cm); with frame and floriated loop, $4\frac{13}{16} \times 3\frac{3}{8}$ in. (12·2 × 8·6 cm)

Unsigned

For the subject see *17*.

The waves are marked in black and gold on the blue sea which verges on turquoise. The turf is bright green, and the undulations of the ground are indicated in black. The rocks are blue or purplish-blue, and the ships gold. Minos wears a short blue robe and cap, a pink cloak and pink boots with blue cuffs. Britomartis is in blue and wears a purplish-pink scarf. The wings of the little Cupid are enamelled in green, blue and pink. The flesh is painted in grisaille, heightened with red. The hair is golden, the eyebrows overshadowing the eyelids.

The model for the composition of this enamel is the same as for *17* above and *42* below.

Certain features common to this enamel and to signed works by Jean Limosin justify an attribution to him: the exaggerated forms of the muscles, the folds and borders underscored with small oblique marks, the stippling round the breasts, the subtlety of the grisaille, and the slightly purple tinge of the pink scarf.

Miss Alice de Rothschild   WI/77/8i

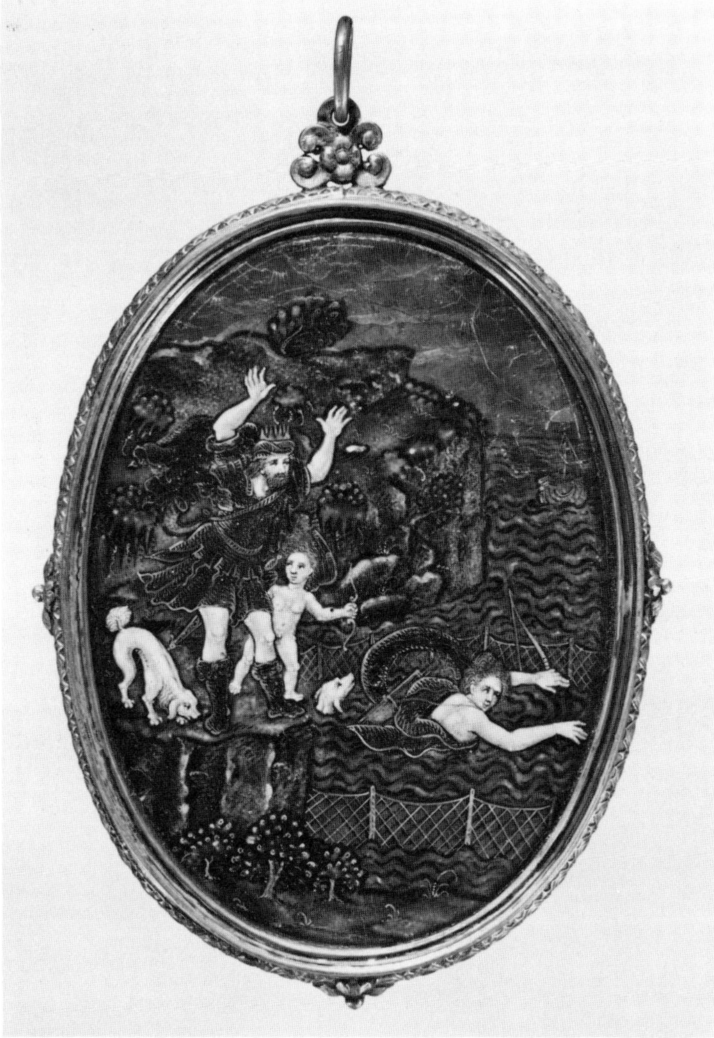

36. Jean I (or II) Limosin
A man and a woman hunting

An octagonal mirror. The glass has been broken and is missing

$3\frac{3}{16} \times 2\frac{7}{16}$ in. ($8 \times 6{\cdot}2$ cm); with frame and fleurette, $4\frac{3}{16} \times 2\frac{13}{16}$ in. ($10{\cdot}5 \times 7$ cm)

Signed with the monogram I.L. at the bottom

The scene is in an oval surrounded by a white band in a border. There is no means of deciding definitely whether it represents Venus and Adonis, Meleager and Atalanta or Dido and Aeneas, but the first is the least likely as the man is bearded, which would be very unusual for Adonis.

The clothing is indicated by enamels on foils. The woman wears a dark blue dress with a yellow belt and a turquoise scarf; the man is in mauve, pink, and turquoise, with brown boots. The ground is bluish-green with very fine clumps of grass in gold. The sky is dark and the clouds emit shafts of gold light. The flesh tints and one of the hounds are salmon pink, the other hound is white. The black ground of the border is decorated with flowers and birds on foils and gold foliated scrolls.

Miss Alice de Rothschild — W1/77/8y

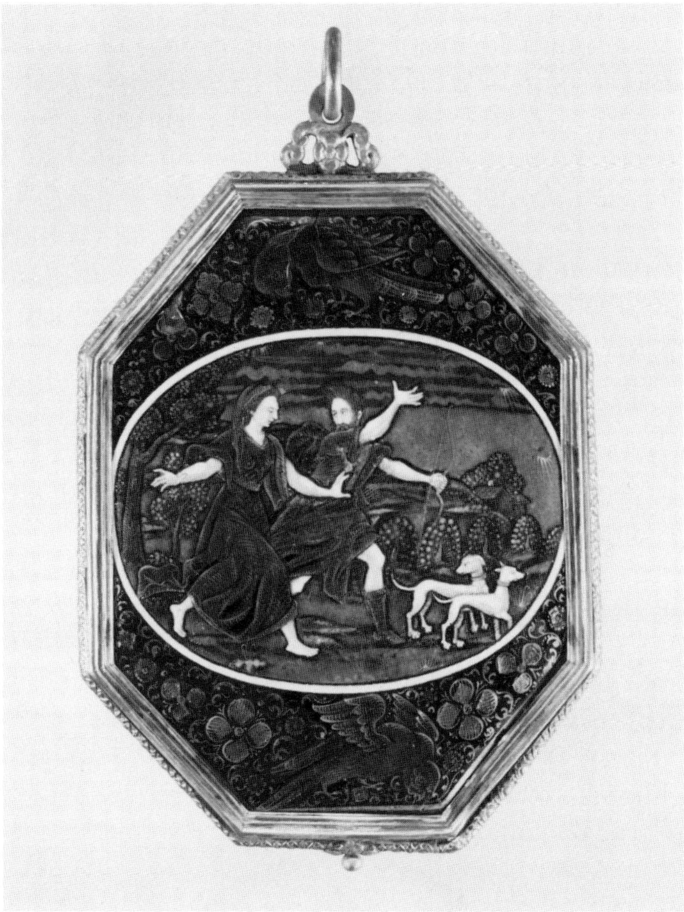

37. Jean I (or II) Limosin
Meleager and Atalanta

An octagonal enamelled plaque with a mirror in an octagonal hinged gilt-metal frame

$3\frac{3}{8} \times 2\frac{9}{16}$ in. (8·5 × 6·5 cm); with frame and floriated loop, $4\frac{1}{4} \times 2\frac{7}{8}$ in. (10·8 × 7·4 cm)

Signed with the monogram I.L. with a fleur-de-lis between the letters, at the bottom of the plaque

The principal scene is oval framed by a white band, surrounded by a black border decorated with small gold motifs and birds and flowers enamelled on foils.

The enamel shows Meleager presenting the head of the boar to Atalanta (Ovid, *Metamorphoses*, VIII, 425 ff.). It is not clear why she wears in her hair the crescent moon of Diana, who in this story was her enemy and had sent the boar to ravage her land.

The man is dressed in a blue tunic, pink boots and cloak, and a blue helmet. Atalanta wears a blue gown and a purplish-pink tunic, both enamelled on foils. The flesh is tinted salmon-pink, the sky is dark blue, the tree trunks and the head of the boar are mauve, and the rest is blue or green. The highlights are stippled in gold. The piece is executed with great delicacy. The counter-enamel is translucent showing the copper which is streaked with brown and green.

The subject is taken from the oval print by Etienne Delaune (Robert-Dumesnil 108), inscribed PREAMIA PRIMA REFERT PRIMIQVE VVLNERIS AVCTOR (referring to the fact that Atalanta had inflicted the first wound on the boar), but only as far as the figures are concerned. The same print was used for *38* below.

Miss Alice de Rothschild — W1/77/8x

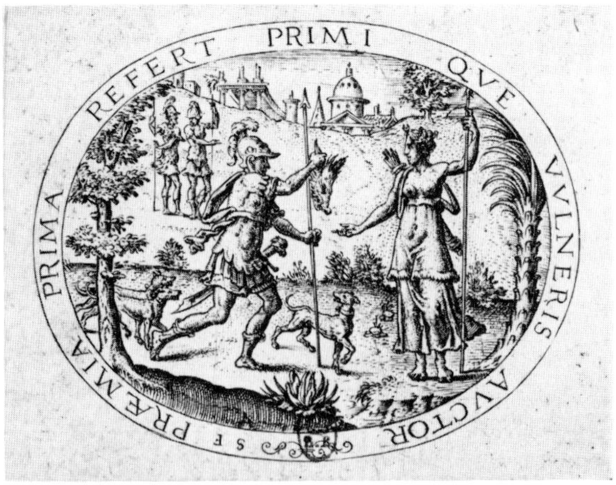

Engraving by Etienne Delaune of Meleager and Atalanta (Paris, Bibliothèque Nationale)

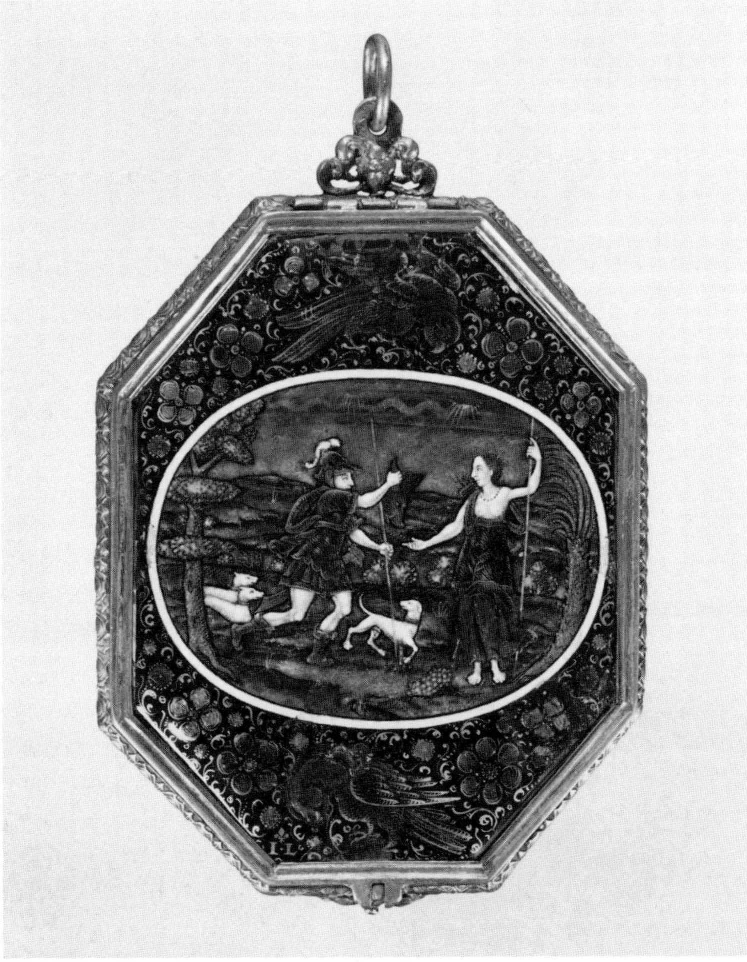

38. Attributable to Jean I (or II) Limosin?
Meleager and Atalanta

An octagonal enamelled plaque and a bevelled mirror in an octagonal gilt-metal frame

$3\frac{1}{16} \times 2\frac{9}{16}$ in. ($7{\cdot}7 \times 6{\cdot}4$ cm); with frame and fleurette, $3\frac{15}{16} \times 2\frac{7}{8}$ in. ($10 \times 7{\cdot}2$ cm)

Unsigned

With the exception of the landscape, which in this case includes a town in the distance, the scene is the same as that of *37* and is in an oval surrounded by a gold band.

On the black ground of the border are small gold motifs, and birds and lozenges enamelled in colour on foils. The clothes are enamelled on foils. Meleager wears a brown under-tunic edged in blue, a turquoise over-tunic and helmet, a blue cloak and pink boots. Atalanta is dressed in a blue gown and scarf and purple tunic. The grisaille is delicate and well modelled. There are salmon-pink highlights on the faces and the man's legs. The boar's head is brownish-purple.

The composition is based on the same engraving as *37*.

This piece has features which recall the work of one of the two Jean Limosins: the colour of the clothes, the spherical shrubs, the use of lines underscored with oblique marks to depict the rise and fall of the terrain, the exaggeration of the muscles, the mis-shapen feet with prominent ankles, the same fidelity to the detail of the original engraving, as in the stippling round Atalanta's breasts. The black border contains ornaments dear to the two Jean Limosins: brilliantly coloured birds, four-petalled flowers, and small rectangles of colour on foils, and little foliated scrolls. However, the birds are different and the main scene is more highly finished than is usual in their work.

Miss Alice de Rothschild — WI/77/8V

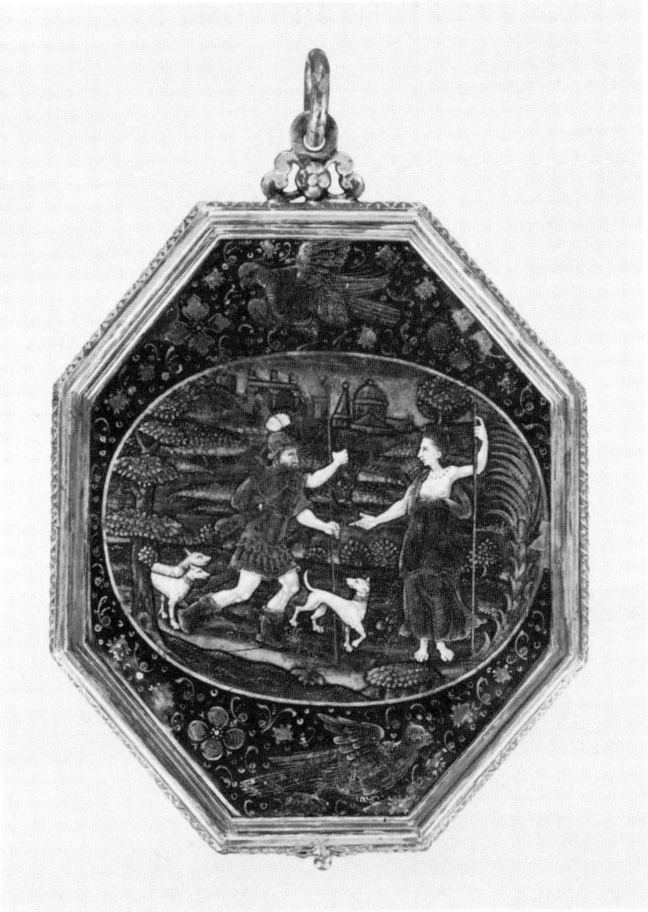

39. Attributed to Jean I (or II) Limosin, or François Limosin
Jason and Hypsipyle

An octagonal enamelled plaque in an octagonal gilt-metal mirror frame; the glass is now missing

$3\frac{1}{8} \times 2\frac{3}{8}$ in. (7·9 × 6·1 cm); with frame and floriated loop, $4\frac{1}{16} \times 2\frac{13}{16}$ in. (10·4 × 7·1 cm)

On the back is an old label inscribed 'François Limosin'

The scene shows a bearded man embracing a woman wearing a helmet and leaning her hand on a shield. In the background is a harbour with a ship at anchor. The persons represented may possibly be Jason and Hypsipyle (Hyginus, Fable XV).

The main scene is oval, surrounded by a white band and framed in a border with flowers and birds enamelled on foils and small gold motifs on a black ground. The sky is dark blue with little gold shafts of light; the dark blue sea is marked with waves; the town and the tree trunk are violet. Hypsipyle wears a purple cloak, Jason dark green, light green and brown, all enamelled on foils. The modelling of the figure is heavy, with salmon-pink highlights. Behind Jason is a banner in translucent enamel on foils.

The ornamental part of this mirror is wholly in the manner of Jean Limosin, and the clouds, which seem to unfold like streamers, are to be found in *36* and *37*, but the treatment of the flesh is quite different and it may be correct to follow the traditional attribution to François Limosin who may be either the son or the grandson of Léonard, both of whom bore this Christian name.

The mirror can be compared with the reverse of one in the Louvre which is signed I.L. and described by Darcel as 'Un vieillard assis auprès de Minerve' (Darcel, *Musée du Louvre, Notice des émaux et de l'orfèvrerie*, D.383). Marquet de Vasselot (*Musée du Louvre, Catalogue . . . de l'émaillerie*, 1914, No.744) described it as 'Pallas résiste à son père'. The plaque is appreciably smaller: 4·7 × 3·6 cm.

Provenance: Debruge-Duménil sale, 23 January to 9 February, 1850, lot 776. Heckscher sale, Christie's, 4–6 May, 1898, lot 293.

Miss Alice de Rothschild — W1/77/8S

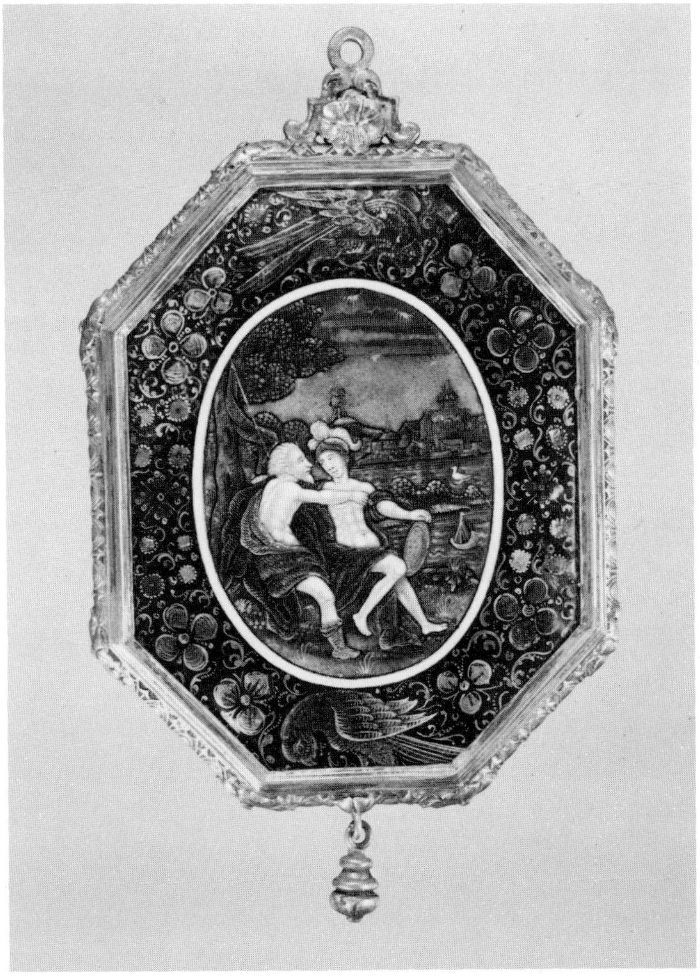

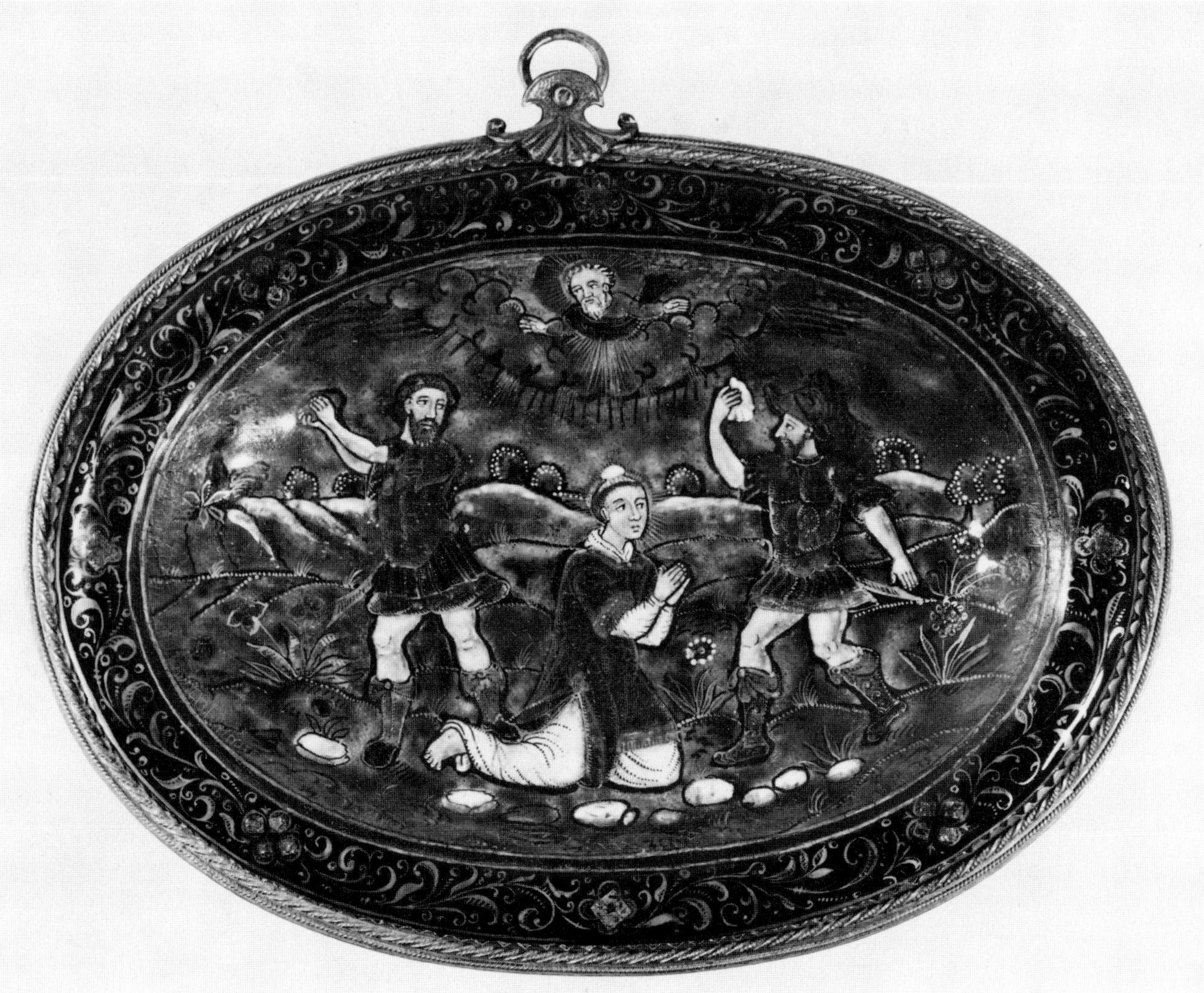

40. Limoges, end of the 16th century
The Martyrdom of St Stephen

An oval dish, decorated on both sides, in a gilt-metal frame with a loop

$5\frac{3}{16} \times 6\frac{3}{4}$ in. (13·1 × 17·1 cm); with frame, $5\frac{7}{16} \times 7\frac{1}{16}$ in. (13·7 × 17·9 cm)

Unsigned

418

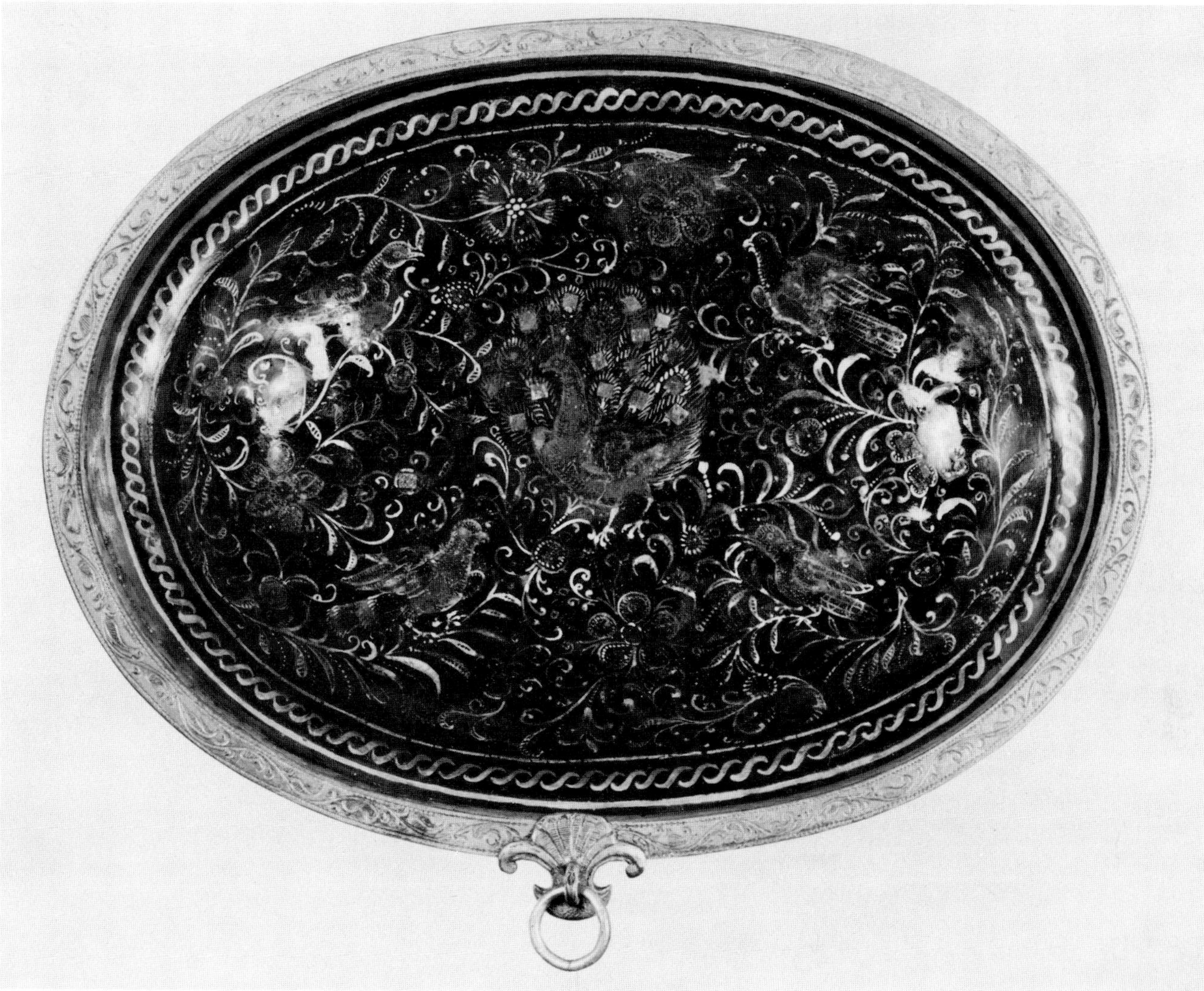

Round the principal scene on the obverse is a wide black border decorated with stylized foliated scrollwork interrupted symmetrically by four-petalled flowers and by small rectangles of colour on foils.

The figures are outlined in thick black lines, but the tones of the background are soft. The clothes are enriched by the use of foils, and the flesh is salmon-pink with reddish touches.

In the centre of the reverse is a peacock surrounded by four birds perched on slender stylized gold foliage decorated with small rectangles on foils and four or five-petalled flowers which stand out from the background. A twisted cable motif encircles the border.

Miss Alice de Rothschild WI/77/6

41. Limoges, end of the 16th century
The Annunciation

A plaque with a curved top, widening towards the base, in a gilt-metal frame with a ring

$4\frac{1}{16} \times 3\frac{1}{8}$ in. (10·3 × 7·9 cm)

Unsigned

The Virgin, who is kneeling, wears a purple gown with yellow sleeves and a blue veil. The angel who kneels on one knee before her, wears a purple gown and has pale blue wings which stand out from the background of the sky seen through the open window. A ribbed oval vase holding three white flowers stands on the floor. The clothes are enamelled on foils; there is a great variety of blues. Dark outlines accentuate the contours and gold is used for the folds and details. The translucent counter-enamel is matt.

Miss Alice de Rothschild WI/77/5

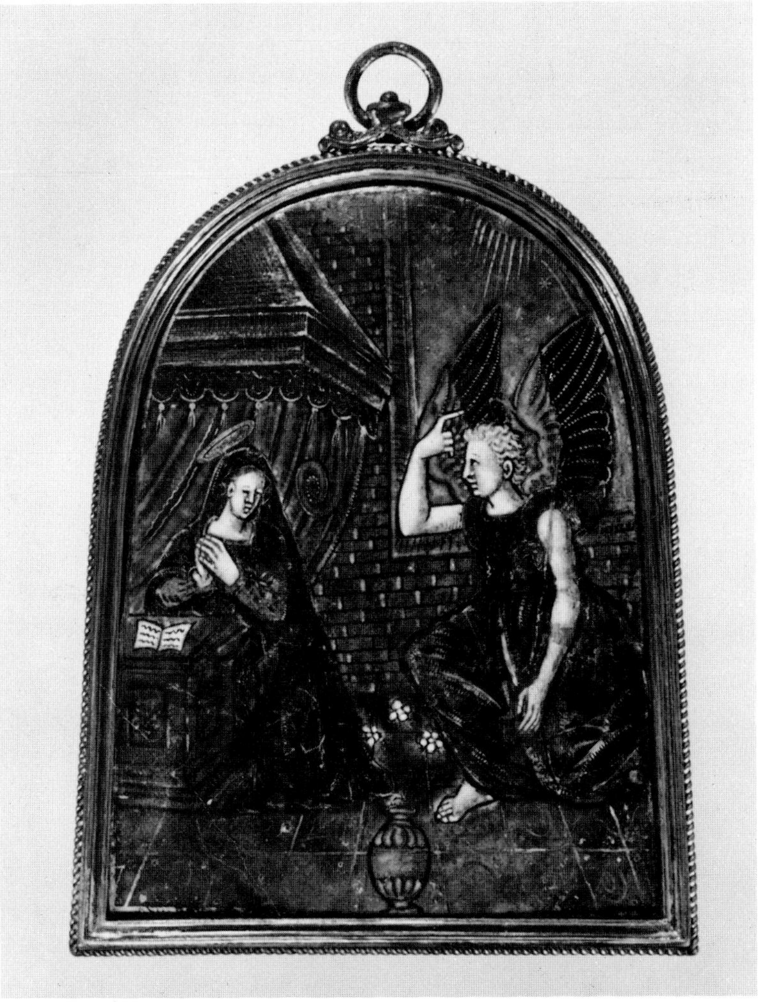

LEONARD II LIMOSIN

He may have been the son of Martin Limosin, and therefore the nephew of the great Léonard. He lived at the end of the sixteenth century and the beginning of the seventeenth. His signature consisted of two Ls., sometimes divided by a fleur-de-lis, which he applied to small oval or rectangular plaques, characterised by a dark velvety effect, an excessive use of foil, and rather light flesh colours with red details.

42. Léonard II Limosin

Britomartis, pursued by Minos, throws herself into the sea

An oval enamel plaque and a mirror in a gilt-metal frame

$3\frac{5}{8} \times 2\frac{9}{16}$ in. (9·2 × 6·5 cm); with frame and floriated loop, $4\frac{9}{16} \times 2\frac{15}{16}$ in. (11·5 × 7·4 cm)

The monogram L.L. is on the rock beneath the foot of Minos. For the subject see *17*, which is close to *42* in composition

The general effect is velvety. Minos is dressed in a blue over-tunic with a brown fringe, a purple under-tunic and coat with short green sleeves. Britomartis is in a dark purple gown and dark brownish-gold scarf, all enamelled on foils. The waves are marked in black on a dark blue sea which is lighter in places. The cliff is violet, the grass green and turquoise.

The light-coloured flesh is thickly treated – as is the spine of the dog – with some red spots on the cheeks and the knees. There are no highlights but the detailing on the clouds, the ship, the net and Cupid's bow is in dull gold.

The model for the composition of this enamel was also used for *17* and *35* above.

Provenance: Probably the mirror in the Bernal sale, Christie's, 5–29 March, 1855, lot 1504, which was exhibited in 1862 in the South Kensington Museum (No. 1868); the entry in the catalogue specifies that there is 'on a rock, the signature L.L.'. The dimensions of the Bernal mirror are given as 11·3 × 7·5 cm with the frame in the Bernal catalogue and $3\frac{1}{2} \times 2\frac{1}{2}$ in. or 8·9 × 6·3 cm without the frame in the exhibition catalogue which correspond fairly closely to those of the Waddesdon mirror.

Miss Alice de Rothschild — W1/77/8r

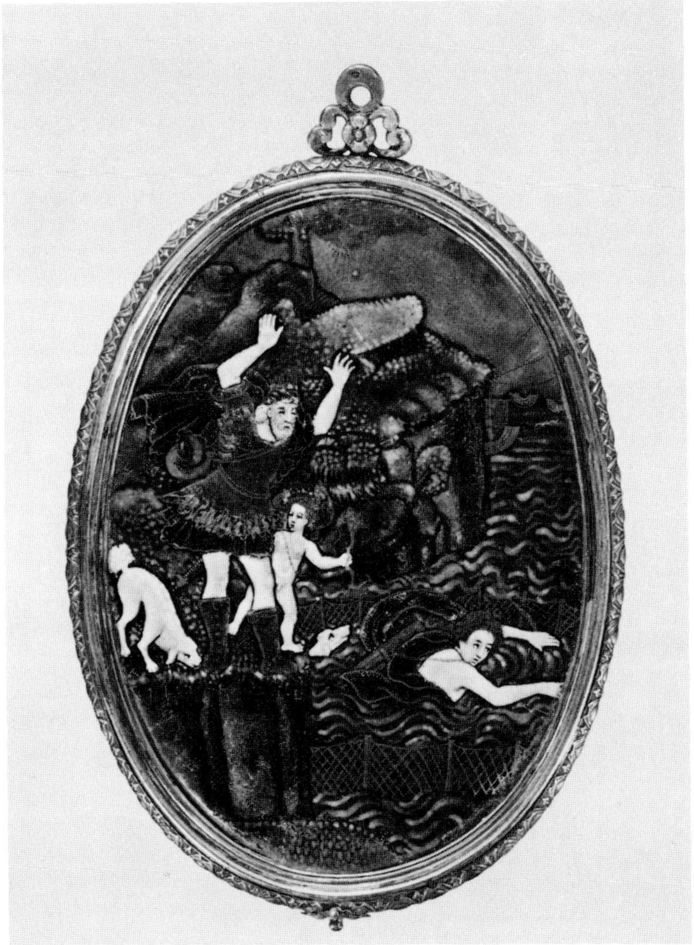

43. Léonard II Limosin
Orion pierced by Apollo's arrows

An oval enamelled plaque with a mirror in a hinged gilt-metal frame

$3\frac{3}{4} \times 2\frac{5}{8}$ in. (9·5 × 6·7 cm); with frame and floriated loop, $4\frac{11}{16} \times 3\frac{1}{4}$ in. (11·9 × 8·3 cm)

Signed L.L. in gold on a blue stone at the base of the composition

The scene is taken from the lower part of the composition of a print by Etienne Delaune (cf. *18* above). The upper section, showing Diana and Apollo as well as Orion, does not appear here. A standing nymph leans against a tree in the foreground; at her feet is a hound which lifts its muzzle towards three other seated nymphs.

The flesh tones are white, modelled in impasto and slightly tinted with pink. The hair is finely drawn in gold with a brush. The black outlines are lightly drawn. There is little or no variety of tone. Some of the clothes are enamelled on foils.

This piece may be compared with the reverse of a mirror (unframed) in the Walters Art Gallery in Baltimore, Inv.44.138 (Verdier *Catalogue*, No.193) also signed with the monogram L.L. The details of the landscape differ slightly and the Baltimore plaque is more subtle, but the figures and the hound are identical.

Miss Alice de Rothschild — W1/77/8q

Introduction

The enamels which form the subject of this section of the Catalogue were, with three exceptions, included in Mr James A. de Rothschild's bequest to the National Trust. They include a group of watches in seventeenth-century enamelled cases, five of which were previously in the collection of Baron Anselm de Rothschild of Vienna, father of Baron Ferdinand and Miss Alice de Rothschild. The set of South Staffordshire buttons (*58*) are part of the large collection of eighteenth- and nineteenth-century buttons formed by Baroness Edmond de Rothschild and bequeathed by her to Mrs James A. de Rothschild, who has given them to the National Trust.*

The art of enamelling at Limoges suffered a disastrous decline toward the end of the sixteenth century. The famous potter Bernard Palissy records in his *Moyen de Devenir Riche* that the Limoges enamellers were reduced to selling, for three sols a dozen, plaques 'so well worked and their enamels so well fired on copper that there is no painting more pleasing'.[1] Although enamel-painting lingered on there into the eighteenth century, further developments in the art were reserved to other regions of France, and to other spheres of craftsmanship.

The earliest account of these developments is given by André Félibien des Avaux in his *Des Principes de l'Architecture* . . ., published in Paris in 1676. Félibien writes of enamelling: 'Before 1630 work of this type was unknown, for it was not until two years later that Jean Toutin, a goldsmith of Châteaudun, who enamelled to perfection with the ordinary transparent enamels, and who had an assistant named Gribelin, having set himself to seek out the means of using enamels which should produce matt pigments in a variety of colours, fuse in the fire, and preserve both their evenness and their brilliance, finally found the secret and communicated it to other craftsmen, who each in turn contributed progressively to its perfection'.[2] This account was accepted by later writers, and has become the basis of the received ideas about the beginnings of jewellers' painted enamels of the seventeenth century.

Jean Toutin was born at Châteaudun in 1578, son of Estienne Toutin, a goldsmith of the same place, and was baptized into the Reformed faith, as were many French enamellers in this period.[3] By 1604 he was established as a goldsmith with his brother Joseph, in the town of

*An account of the collection of buttons by Mrs M. Ginsburg will be found in *Apollo*, June, 1977.

[1] B. Palissy, *Le Moyen de devenir riche* . . ., Paris, 1636, pp.365–6: (*Practique* discourses to *Théorique*)' . . . As tu pas veu aussi les esmailleurs de Limoges, lesquels par faute d'avoir tenu leur invention secrerte [*sic*], leur art devenu si vil qu'il leur est difficile de gaigner leur vie au prix qu'ils donnent leurs œuvres. Ie m'asseure avoir veu donner pour trois sols la douzaine de figures d'enseignes que l'on portoit aux bonnets, lesquelles enseignes estoyent si bien labourées et leurs esmaux si bien parfondus sur le cuivre, qu'il n'y avoit nulle peinture si plaisante . . .'

[2] 'Avant l'an 1630 ces sortes d'Ouvrages estoient encore inconnus, car ce ne fut que deux ans aprés, que Jean Toutin, orfévre de Chasteaudun qui émailloit parfaitement bien avec les Émaux ordinaires et transparens, et qui avoit pour Disciple un nommé *Gribelin*, s'estant mis à rechercher le moyen d'employer des Émaux qui fissent des Couleurs mates pour faire diverses teintes, se *parfondre* au feu, et conserver une mesme égalité et un mesme lustre, en trouva enfin le secret qu'il communiqua à d'autres Ouvriers, qui tous contribuerent ensuite à le perfectionner de plus en plus' (Félibien, op. cit., 2nd ed., Paris, 1690, pp.429–30).

[3] These and subsequent facts concerning Toutin's life are derived from Clouzot, *La Miniature*, pp.6, 8–14, 181–4; and id., *Dictionnaire*, pp.196–7, *s.v.* See also Verdier, p.2.

Blois, then the centre of watch-making in France, and a city in which at least two goldsmiths, one named Gribelin (possibly Jean Toutin's apprentice) and the other named Morlière, practised enamelling as early as 1632.[4]

The watch was to the seventeenth century what the snuff-box was to become for the eighteenth century, the appropriate gift for a special occasion between royalty, members of the nobility, and the *corps diplomatique*. Marie de Médicis, receiving the Venetian Ambassador in 1621, in order to take leave of him, detached from her bodice two little watches and gave them to him, saying that she 'made a present to him of the wonderful fruits which have their origin in Blois at the hands of the excellent craftsmen of the district'.[5] The essence of the art of enamelling of the 'School of Blois' was that it was executed on gold instead of copper and consisted of painting on a ground of white enamel laid on the gold. Blois enamelling concentrated with a few exceptions[6] on the embellishment of the watch and its covers, offering to the painter's brush as many as four circular panels, and a long narrow field formed by the edge of the watch-case, to say nothing of the dial itself, which was also not infrequently embellished in this way (cf. *46*, *48*, *49*).

Jean Toutin probably remained in the provinces until about 1634 or 1635, then moved to Paris. He died at Charenton in 1644, called 'Jean Toutin, master goldsmith, aged 66, native of Châteaudun . . .', and leaving two daughters and three sons, two of whom were to succeed him as enamellers. Unfortunately, no identifiable enamel by Jean Toutin is known, although he was an accomplished draughtsman and etcher, as a few surviving prints of 1618–1619 demonstrate[7]: and tradition has it that he became jeweller to the King.

Jean Toutin I was succeeded by his sons Jean II (1619–1660) and Henry (1614–after 1683). Jean went to Sweden about 1645 to become Court enamel-painter to Queen Christina, the first of several French enamellers to make their way to the northern kingdoms in the course of the seventeenth century. One or two signed watch-cases by Jean Toutin II are known[8], but his elder brother Henry, who seems never to have left France, fortunately signed a number of both watch-cases and backs for miniatures. He was established in Paris by 1636, in which year he painted and signed a portrait of Charles I of England, the earliest surviving dated miniature in painted enamel. In this field Henry Toutin was outstripped for subtlety of portraiture and richness of colour by his contemporary, the Genevan enameller Jean Petitot; but in the sphere of subject-painting, and particularly in the rendering of flowers, his work was unsurpassed.[9]

Unfortunately, little is known of the work of his contemporaries and immediate predecessors. Christofle Morlière (1604–1644) must have been an important figure amongst the goldsmiths and enamellers of Blois. In 1637 he took to Gaston d'Orléans, younger brother of Louis

[4] Ibid., p.16. Seven gold watch-cases 'enamelled with figures' (émaillées à figure) are quoted as a royal commission two years earlier (Schneeberger, p.91).

[5] Clouzot, *La Miniature*, pp.15–16 ('qu'elle lui fait don de ces fruits merveilleux qui naissent à Blois dans les mains des excellents ouvriers du pays').

[6] A number of backs and cases for miniatures were painted in enamels, as well as an occasional pomander (see, e.g., Sale *Catalogue* Christie's, Geneva, 30 April, 1974, Lot 251), and other goldsmiths' trinkets.

[7] Clouzot, *La Miniature*, pp.12–14.

[8] Clouzot, *La Miniature*, pp.73–4; id., *Dictionnaire*, p.197, *s.v.* One of Jean II Toutin's watches is illustrated by Clouzot, *La Miniature*, Pl. IV.

[9] Details of Henry Toutin's life and work are given by Clouzot, *La Miniature*, pp.23–27, 69–73, 185 (see also *n*.3); id., *Dictionnaire*, p.196, *s.v.* See also Schneeberger, pp.87, 90, 92, 94, Figs.8, 10, 12; Verdier, pp.2–7 (on which, however, see Lightbown, 'Jean Petitot and Jacques Bordier', p.89).

XIII, in Tours, twelve hundred *livres*' worth of jewels; and in 1643 he was chosen by the city fathers of Blois to offer a gift to the King's brother to mark the occasion of his marriage. He was commissioned to produce 'a watch-case of enamelled gold, with figures and people'.[10] We know from a law-suit of 1639 that he also painted landscapes on watch-cases; and from another law-suit in 1632 that he had decorated a watch-case with flowers but that the work did not come within the meaning of the term *émail peint* as it was generally understood and as Morlière commonly executed it.[11] We have here all the constituents of painted enamel as it was applied to the gold-watch. This was normally painted with figural scenes, usually derived from mythology, on the outside, and with a landscape on the inside as well as miniature landscapes in the cartouches round the edge (cf. *45–50*). Occasionally one of the inner surfaces was painted with flowers, as might be the face of the watch itself; occasionally a wreath of flowers framed one of the figural subjects on an outer cover.[12] Unfortunately, no signed work by Morlière appears to exist, only two pieces being tentatively attributed to him[13]; our knowledge of his activity is due solely to the fact that being extremely litigious he is mentioned in a number of law-suits.

In one of these – that concerning the flower-painted watch – the outside assessor called in to give expert evidence was Isaac Gribelin, perhaps the same who was Jean Toutin's apprentice (see p.428 above). Tradition had it that he was a specialist in the painting of miniatures, and watches of this period do in fact sometimes have portraits on the inner surfaces of their cases.[14] No signed work by Gribelin exists.

Further enamel-painters of the School of Blois are known by name, if not by identifiable works. One of these, however, Robert Vauquer (1625–1670), member of a well-known Blois family of watch-makers and engravers, did sign a number of pieces between 1660 and 1670. These were all portraits and religious subjects, however, and it is extremely unlikely that we should find in Vauquer the painter of the Waddesdon casket (*44*), the mainspring of this discussion.[15] The same may be said of Henry Toutin's brother, Jean II. His two signed watch-cases show no stylistic affinity with the Waddesdon casket (see p.428 above). More shadowy still is the figure of Pierre Chartier (1618–after 1683), the elder son of a Blois goldsmith. By 1651 he had moved to Paris, where he was established as a 'marchand orfèvre'. Félibien says of him that 'he set himself to doing flowers, at which he succeeded perfectly, and one soon saw a number of people in Paris applying themselves to this way of painting, in which a number of medallions and small works were produced.'[16] Even less is known of a certain Dubié, mentioned by Félibien as one of the first craftsmen to be a beneficiary of Jean Toutin's discoveries, and alleged to have come to Paris to perfect his craft, taking up quarters in the galleries of the Louvre, normally the prerogative of specially favoured artists. The name Dubié is not other-

[10] See Clouzot, *La Miniature*, pp.19–22, 187–8, 190–199; id., *Dictionnaire*, p.142.
[11] Clouzot, *La Miniature*, pp.192–4 and 18–19.
[12] See e.g., Clouzot, *Dictionnaire*, Pl. IV, c; Schneeberger, Figs. 17–18.
[13] Clouzot, *La Miniature*, p.22.
[14] See, e.g., H. P. Mitchell, 'Two French Enamelled Watches', *The Burlington Magazine*, XXIV–XXV (1908), pp.245–6, portraits of Louis XIII and Cardinal Richelieu. For details of the life and work of Isaac Gribelin, see Clouzot, *La Miniature*, pp.14, 16, 18–19, 42, 185, 189, 194–9; id., *Dictionnaire*, p.90, *s.v.*
[15] For details of the life and work of Robert Vauqu(i)er, see Clouzot, *La Miniature*, pp.22, 61–3, 187–8; Id., *Dictionnaire*, p.201, *s.v.*
[16] For Pierre Chartier, see Clouzot, *La Miniature*, pp.22, 64–6, 184; id., *Dictionnaire*, pp.47–8, *s.v.*

wise traceable, and if it is not merely a misprint, Félibien may have confused it with that of Dublé, a family of Parisian goldsmiths and jewellers.[17]

Perhaps the significance of these men lies in the fact that the centre of their activity was not Blois, but Paris, whither Henry Toutin, as has been seen, had already gone by 1636, finally to establish himself there by 1643. Apart from Toutin, the capital was blessed with the presence of the greatest of seventeenth-century enamel miniaturists, Jean Petitot. Petitot had gone to England probably in 1636, and left in the troublous days of the Civil War, probably in 1644.[18] Although we find him in Blois for his marriage to Marguerite Cuper in 1651, and his first child was born there in 1653, by 1650 at latest he was established in Paris, in the Rue de l'Université. It is most likely that he was in Paris also from the date of his arrival from England in 1644 until the Summer of 1651, when the troubles connected with the Fronde finally caused the Court to leave Paris.[19] Although in his early years Petitot had not been above painting merely decorative enamels (in London he had apparently copied in enamel a landscape by Oliver after Breughel)[20] after his English sojourn he almost certainly restricted himself to portraiture, imparting to his miniatures a finesse and a richness of colours which could not but have had an influence on his contemporaries. The exact nature and degree of the collaboration with Petitot of his life-long friend and relation by marriage, Jacques Bordier, is not established. Bordier, in his early days, had painted religious and mythological figure-subjects, but no authenticated work by him survives.[21] He appears to have executed miniature-portraits himself, but also to have collaborated with Petitot, possibly executing some of the subordinate details of the work.[22]

Where in this complicated but nebulous story does the Waddesdon casket fit? The first thing to be observed is that it is not a piece of goldsmith's work. The mounts are of gilt base metal, and although the plaques have never been detached from the casket for inspection, it is very likely that they are painted on copper and not on gold. Although, therefore, the painting of the plaques bears some resemblance to that on gold watch-cases which might loosely be associated with a 'School of Henry Toutin', the casket must have been made outside the ambit of the goldsmith's craft – perhaps in Paris by one of those who imitated the flower-painting of Pierre Chartier, referred to by Félibien. The style of flower-painting is indeed looser and less exquisite than that of Henry Toutin[23] or that attributed to Christofle Morlière[24], and begins to approximate to the more workaday decorative style current among enamel-painters, not only in

[17] See Clouzot, *La Miniature*, pp.22–3.

[18] For the most recent discussion of Petitot see R. W. Lightbown, 'Jean Petitot and Jacques Bordier . . .'; id., 'Les Origines de la Peinture en Email sur Or . . .'; id., 'Jean Petitot; Etude pour une Biographie . . .', *passim*, and previous works cited there.

[19] Clouzot, *La Miniature*, p.41.

[20] Lightbown, 'Jean Petitot . . .', pp.99–100.

[21] Lightbown, 'Jean Petitot and Jacques Bordier . . .', p.83: the attributions by Professor P. Verdier in 'An Unknown Masterpiece by Henri Toutin', pp.5–6 and Figs.2–4 are not substantiated (see also Lightbown, loc. cit. p.91, *n*.37.)

[22] See, e.g., Lightbown, 'Jean Petitot . . .', p.82 (citing Dezallier d'Argenville, *Supplément* to the *Abrégé de la vie des plus fameux peintres*, published in Paris, 1752): see also Clouzot, *Dictionnaire*, pp.34 and 218, *s.v.*

[23] See, e.g., Schneeberger, Fig.8.

[24] See, e.g., a watch-case in the Metropolitan Museum of Art, New York (G. C. Williamson, *Catalogue of the Collection of Watches the Property of J. Pierpont Morgan*, London, 1912, pp.82–85, Nos.77–78).

France, in the latter years of the seventeenth century.[25] There is no guarantee that the figure-subjects on the casket were executed by the same hand, or even in the same workshop, although this is likely, but there can be no doubt that they are French. The derivation of some of their subjects from the paintings of Simon Vouet, and perhaps in one instance of Laurent de la Hyre, seems to justify the conclusion, no less than the apparent dependence of the auricular-style mount on the engravings of Daniel Rabel (*c.*1578–1637). The delicate *pontilliste* rendering of these figure-subjects quite clearly derives from the work of the 'School of Blois', and is executed by an artist of no mean ability. By far the most likely attribution of this casket, therefore, is to a Paris workshop established in the quarter-century after 1650.

The remaining seventeenth-century enamels at Waddesdon are all in the form of embellishments of watches or watch-cases, and all stem from the school of enamelling which grew up in Geneva in the course of the seventeenth century. In the sixteenth century Geneva had established itself as an independent Protestant city-state, and in the seventeenth century it became a bastion of the Reformed Faith and a haven of refuge for those escaping the religious persecutions inspired in Catholic countries by the Counter-Reformation. The tiny state encouraged manufacturing industries, and was a centre for weavers, pewterers, silversmiths and, especially, watch-makers.[26] The same factors that had made Blois a centre of enamelling in France militated in favour of a parallel development in Geneva. The craft of jeweller's enamelling had been practised there in the sixteenth century, almost exclusively by craftsmen of French origin,[27] and in the seventeenth century too citizens of Geneva trained in the first place as goldsmiths – Jean Petitot and Paul Prieur, for example[28] – were to make their mark among the finest enamel-painters of the age: paradoxically, however, it seems most likely that they learned the finer points of this craft in France, on their way to fame in the countries of northern Europe – Petitot in England, and Prieur in Denmark. Their experience, however, suggests that the soil was prepared, and the interchange of craftsmen with France, particularly with Châtellerault (between Tours and Poitiers), provided a cross-fertilization. Thus, in 1629 a Genevan watch-maker sent his son as apprentice to another Genevan established in Châtellerault, a watch-maker named Marc Grangier: a watch by this maker is known which is enamelled in white with flowers in relief on a blue ground. Marc Grangier's son Louis in his turn migrated to Geneva in 1656.[29] In the same way Pierre Huaud, born in 1612 as the son of a Protestant goldsmith in Châtellerault, moved in 1630 to Geneva to take up his apprenticeship with Laurent Légaré, another goldsmith. He settled, married in 1643, and had three sons – Pierre II (b.1647), Jean-Pierre (b.1655) and Amy (b.1657).[30] No enamelled work can confidently be ascribed to him, but in 1661 he contracted to teach enamel-painting to Jean André (1646–1714), who was destined in his turn to become the head of a family of enamel-painters.[31]

25 See, e.g., Schneeberger, Figs.9, 13; or a plaque representing the *Baptism of Christ*, with lobed floral border, in the Victoria and Albert Museum (C.52–1923), all attributed to Nuremberg.
26 Schneeberger, p.100.
27 Ibid., p.99.
28 For Prieur, see Clouzot, *La Miniature*, pp.78–81; id., *Dictionnaire*, pp.168–9, *s.v.*; Schneeberger, pp.136–140, etc.
29 Schneeberger, pp.99–100.
30 Details of the lives and work of the Huaud family may be found in Clouzot, *La Miniature*, pp.103–114, 211–219; id., *Dictionnaire*, pp.100–107; Schneeberger, pp.101–113.
31 For the André family, see Clouzot, *La Miniature*, p.114; id., *Dictionnaire*, pp.3–4, *s.v.*; Schneeberger, pp.101, 103, 114–116.

From this, and from the fact that by far the greater part of his sons' work was done on watches and watch-cases, we may assume that Pierre I Huaud also enamelled gold watches.

Four of the six enamelled watches at Waddesdon are signed by the Huaud brothers (*46–49*), and they were unquestionably the most prolific identifiable enamellers of the late seventeenth century. The eldest, Pierre II, went in 1685 to the Court of the 'Great Elector' of Brandenburg (Frederick William), returning to Geneva in the following year. In 1689 he went again to Berlin, and in 1691 was appointed enamel-painter to the new Elector (later King Frederick I). He died in Berlin before 1698. A watch at Waddesdon signed simply, and exceptionally, 'Huaut p.' may be by him (*46*). The second brother, Jean-Pierre, who for a while signed himself 'Huaud le puisné' (*47*) to differentiate him from his brother Pierre 'l'aisné', in 1682 entered into association with his younger brother Amy for a term of six years. In 1686 the two brothers were appointed as enamel-painters to the Great Elector, no doubt in place of Pierre, and there they stayed until August, 1700. One of the Waddesdon watches (*48*) is signed 'les deux frère [*sic*] Huaut pintre [*sic*] de son A(ltesse) E(lectorale) de B(randebourg)' and must mark this phase of their activity.

The work of the Huaud brothers is marked out from that of the earlier French enamels already discussed by the greater warmth and richness of its palette. By this date the rose-purple enamel based on gold has come into its own, to supplant the drier iron-red of the School of Blois; the blue and yellow are deeper and richer, providing a harmony to match the lush painting of mythological and other figural scenes which characterize the Geneva School. To the pieces signed by the Huaud brothers must be added others (e.g. *50* here) perhaps attributable to pupils and affiliates of the Huauds, such as members of the André and Mussard families,[32] who share with the Huauds a variety of recurring themes of decoration.

Of an entirely different technical character is the egg-shaped box *51*, perhaps a *bonbonnière*. It is decorated with reliefs made by shaping thin gold sheet over a die, then firing it – presumably by means of some enamel-like composition which fills the hollow of the moulded leaf – to the body of the box. The gold is itself further embellished by tiny patches of green, and in one instance pink, translucent enamel. This technique is best known in Germany, where it was certainly used in the Berlin workshop of Pierre and (after 1738) Alexander Fromery; and almost certainly also at Dresden.[33] Even closer technically are the gold-reliefs produced by the gilder and 'arcanist' Christopher Conrad Hunger, who was at Meissen about 1715 and went subsequently to Vienna, Venice, Stockholm, Copenhagen and finally Russia, in the hope of founding porcelain-factories in each.[34] His knowledge of the secrets of porcelain-manufacture may have been defective, but he was unquestionably an adept in the technique of decorating both porcelain and enamel by means of gold-reliefs. A porcelain bowl in the Österreichisches Museum für Angewandte Kunst, Vienna, is signed 'Hunger F', and is decorated with gold reliefs heightened by the use of green and crimson translucent enamels, exactly as on the Waddesdon box[35]; and an enamel portrait, partly in relief-gilding on a turquoise ground, was

[32] For the Mussard family, see Clouzot, *La Miniature*, p.115; id., *Dictionnaire*, pp.145–6; Schneeberger, pp.141–3, etc. Pierre Huaud married a Françoise Mussard. See also *n*.30.

[33] W. Holzhausen, 'Email mit Goldauflage in Berlin und Meissen nach 1700', *Der Kunstwanderer*, 12. Jahrgang, 1930, pp.4–12, 78–9.

[34] W. B. Honey, *European Ceramic Art, Dictionary of Factories . . .*, London, 1953, *s.v.*

[35] G. E. Pazaurek, *Deutsche Fayence- und Porzellan-Hausmaler*, Leipzig, 1925, Pl.10.

signed by Hunger at Rörstrand (Stockholm) some time between 1729 and 1733.[36]

There is no comparable documentary evidence for the manufacture of this type of enamel in France, but several circumstances speak for it. First, there are a number of enamels of this type which, like the Waddesdon box, bear Parisian goldsmiths' marks.[37] It has been argued that this may mean no more than that German enamels were imported and merely mounted in France.[38] To this, however, there are counter-arguments other than the fact that the boxes are mostly not very like the known German examples. There is in the James Ward Usher collection (No.53) in the Usher Art Gallery in Lincoln a watch-case decorated with gold-reliefs in incipient *rococo* style, and signed 'Guillaume Bouvier F', the watch itself being by D. Cochin.[39] Bouvier may have come from a family of Genevan enamellers, and little is known about him: but it seems likely that this watch-case was made in France.[40] Secondly, the same style of decoration, including the use of green and pink translucent enamels, occurs on rare examples of Saint-Cloud porcelain.[41] Thirdly, although Hunger was Thuringian by birth, he apparently arrived in Meissen from France.[42] It should not be forgotten either that the Fromery family was also French; while another firm in Berlin doing the same kind of work was Jordan and Lautier,[43] the second partner of which at least must have been of French origin. Despite the fact that there are no examples which can be cited with confidence as filling the gap between the early years of the eighteenth century[44] and the 1730s and 1740s, when most of the hall-marked French examples occur, it seems by no means impossible that this technique was as much at home in France as it was in Germany; and there seems no reason for supposing that the Waddesdon box is anything other than French.

36 C. Hernmarck, 'Schwedisches Porzellan im 18. Jahrhundert', *Berichte der Deutschen Keramischen Gesellschaft*, Heft 9, 36, 1959, p.281, Fig.1.

37 See, e.g., Solange de Plas, *Tabatières*, Paris, *n.d.* about 1975, Fig. on p.37; Christie's sale, 16 December, 1975, Lots 151–4.

38 Clare Le Corbeiller, *European and American Snuff Boxes, 1730–1830*, London, 1966, Figs.574, 579–80. A box in typical Fromery style in the Victoria and Albert Museum (Mus.No.1955–1898) bears a French mark used on imported plate, indicating that it was imported already mounted.

39 According to G. H. Baillie, *Watchmakers and Clockmakers of the World*, London, 1929, p.74, and second edition, 1947, p.62, Daniel Cochin was in Geneva in 1732, but later perhaps in Paris until 1770. He was apparently a fine *repoussé* worker.

40 Clouzot, *Dictionnaire*, p.38, *s.v.*, records a Pierre-Louis Bouvier who was born in Geneva in 1766 and died there in 1836. G. H. Baillie, op. cit., in *n.*39, records (1929, p.43; 1947, p.35) that Guillaume Bouvier was active in Paris about 1740 as an enamel-painter. F. J. Britten, *Old Clocks and Watches and their Makers*, London, 6th ed., 1932, p.708, states that Guillaume Bouvier was a well-known painter of watch-cases in enamel, about 1740. Both authorities also record other members of the Bouvier family in Geneva (see also C. Brun, *Schweizerisches Künstlerlexikon*, I, Frauenfeld 1905, *s.v.*). The technique of raised gilding, however, was not even restricted to Germany and France. A box in the British Museum with relief-gilding on a turquoise ground bears an inscription suggesting that it was made in Milan.

41 W. B. Honey, *French Porcelain of the 18th century*, London, 2nd ed., 1972, pp.11–12 and Pl.17, D. It is also worth bearing in mind that later in the century the Sèvres porcelain factory took up essentially the same technique in its 'jewelled' decoration (ibid., pp.12, 39).

42 J. F. Hayward, *Vienna Porcelain of the Du Paquier Period*, London, 1952, Appendix III, p.186.

43 Holzhausen, loc. cit., p.11.

44 It is worth bearing in mind, however, that the mounts on some of the 'jewelled' pieces of Saint-Cloud porcelain bear marks giving dates as early as 1723 (Honey, *French Porcelain*, p.12). It is perhaps also worth citing a mother-of-pearl box with raised gold and enamel decoration which bears Paris marks for 1722–27 (A. Kenneth Snowman, *Gold Boxes of Europe*, Fig.125).

The study of English painted enamels was long bedevilled by the notion that most mid-eighteenth century specimens were likely to have been made at the Battersea factory, in London, about which a certain body of tradition had survived from the eighteenth century. Horace Walpole had listed in his *Description of the Villa . . . at Strawberry Hill*, 1774: 'George II and Frederick, Prince of Wales, in Battersea enamel', and 'a king-fisher and ducks of the Battersea enamel: it was a manufacture stamped with a copper-plate, supported by Alderman Jansen, but failed'. The involvement of Sir Stephen Theodore Janssen in the Battersea concern is securely documented,[45] but in fact the known plaques transfer-printed with the kingfisher and ducks theme are evidently of a different character from the plaques decorated with certainly Battersea prints, such as the *George II* and *Frederick Prince of Wales*[46] recorded by Walpole in his collection; and were undoubtedly made elsewhere.[47] The 'manufacture stamped with a copper-plate' characterizes succinctly the peculiarity of Battersea – that its products were transfer-printed (a process by which a pull on paper taken from a copper-plate is 'transferred' to the surface of pottery, porcelain or enamel, the impression made permanent by the use of a metallic pigment subsequently fired in a kiln). In a letter written in 1755 to his friend Richard Bentley,[48] Walpole again puts his finger on this vital point: 'I shall send you a trifling snuff-box, only for a sample of the new manufacture at Battersea, which is done from copper-plates'.

Before 1924, when Bernard Rackham published a new edition of the *Catalogue* of the enamels in the Schreiber Collection at the Victoria and Albert Museum,[49] almost all English enamels were attributed to Battersea. Rackham, however, realizing that on stylistic grounds the vast majority must have been made after 1756, in which year the Battersea concern was known to have gone bankrupt, re-attributed these post-1756 enamels to the South Staffordshire towns of Bilston and Wednesbury, documentary evidence existing to prove that the manufacture of enamels was established there in the second half of the eighteenth century. The process of re-allocation was hastened by the discovery that the factory cannot have got going at York House, Battersea, until 1753; and thus could only have had at most three years of activity there.[50] For some thirty-five years after the date of this discovery, the main efforts of research in the field of English enamels were directed towards establishing the nature of the objects made at the Battersea factory.[51] In 1964, however, it was established that John Brooks, an Irish mezzotint engraver who was known to have been Janssen's partner in the Battersea concern had, as early

45 See e.g., Mew, *Battersea Enamels*, pp.2 ff.

46 Ibid., Fig.3.

47 See Cook, *Hancock*, Item 58.

48 ed. W. S. Lewis and others, *Horace Walpole's Correspondence*, 35, London/New Haven, 1973, p.252, letter dated 18 September, 1755.

49 Victoria and Albert Museum, *Catalogue of English Porcelain Earthenware Enamels and Glass collected by Charles Schreiber, Esq. and the Lady Charlotte Elizabeth Schreiber . . .*, III, *Enamels and Glass*, by Bernard Rackham, London, 1924.

50 Honey, 'New Light', pp.84–5.

51 See, e.g., Mew, *Battersea Enamels*; and the articles by Cyril Cook, H. W. Hughes, B. Rackham and A. J. Toppin cited in the Bibliography.

ENGLISH

Bernard Rackham, 'The Engraver Hancock and Battersea Enamels', *Burlington Magazine*, Vol. XXVI, 1915, pp. 155–6.

London, Victoria and Albert Museum, *Catalogue of English Porcelain Earthenware Enamels and Glass collected by Charles Schreiber Esq MP . . . and the Lady Charlotte Elizabeth Schreiber . . .*, Vol. III, *Enamels and Glass* (by Bernard Rackham), London, 1924. ('Rackham, *Schreiber Catalogue*').

Wolverhampton Municipal Art Gallery and Museum, *Old Wolverhampton and District Exhibition*. Wolverhampton, 1925 – Classes A, B and C *Bilston Enamels*, by G. P. Mander.

Egan Mew, *Battersea Enamels*, London, 1926. ('Mew, *Battersea Enamels*').

Bernard Rackham, 'Battersea Enamels', *Connoisseur*, March, 1927, pp. 133–43.

Egan Mew, 'Battersea Enamels and the Anti-Gallican Society', *Apollo*, May, 1928, pp. 216–21.

W. B. Honey, 'New Light on Battersea Enamels', *Connoisseur*, February–March, 1932, pp. 82–9 ('Honey, "New Light"').

Bernard Rackham, 'Porcelain as a Sidelight on Battersea Enamels' *Transactions of the English Porcelain Circle*, Vol. IV, 1932, pp. 45–58.

Bernard Rackham, 'Supplementary Notes on the Battersea Factory', *Transactions of the English Porcelain Circle*, Vol. IV, 1932, pp. 69–74.

Herbert Read, 'Cross-currents in English Porcelain Glass and Enamels', *Transactions of the English Porcelain Circle*, Vol. IV, 1932, pp. 7–17.

A. J. Toppin, 'Notes on Janssen and the Artists of the Battersea Factory', *Transactions of the English Porcelain Circle*, Vol. IV, 1932, pp. 58–68.

H. W. Hughes, 'Authorship of some Designs on Porcelain and Enamel and Robert Hancock's Connection with Battersea and Bow', *Transactions of the English Ceramic Circle*, Vol. I, no.3, 1935, pp.85–96.

Cyril Cook, *The Life and Work of Robert Hancock*, London, 1948. ('Cook, *Hancock*').

G. B. and T. Hughes, *English Painted Enamels*, London, 1951.

Cyril Cook, 'James Gwynn and his Designs on Battersea Enamels', *Apollo*, March, 1952, pp. 66–9.

Cyril Cook, 'John Brooks and his Designs on Battersea Enamels', *Apollo*, May, 1952, pp. 147–9, 153.

Cyril Cook, 'Louis F. Boitard and his Designs on Battersea Enamels', *Apollo*, March, 1953, pp. 72–5.

Cyril Cook, *Supplement* to *The Life and Work of Robert Hancock*, Tonbridge, 1955.

Cyril Cook, 'Simon François Ravenet and his Engravings on Battersea Enamels' *Transactions of the English Ceramic Circle*, Vol.3, part 4, 1955, pp. 185–191.

Manchester City Art Gallery, *The Raby Collection of Battersea and South Staffordshire Enamels*, Manchester, *n.d.*, 1958.

Cyril Cook, 'The Art and Artists of the Battersea Wine Label', *Connoisseur*, March 1959, p. 101.

Bernard Watney and R. J. Charleston, 'Petitions for Patents concerning Porcelain Glass and Enamels with special reference to Birmingham . . .', *Transactions of the English Ceramic Circle*, Vol.6, part 2, 1966, pp. 57–123. ('Watney and Charleston, "Petitions . . ."').

Bernard Watney, 'English Enamels in the 18th century', *Antiques International*, London, 1966, pp. 287–96.

R. J. Charleston, 'Battersea, Bilston – or Birmingham?', *Victoria and Albert Museum Bulletin*, January 1967, pp. 1–12. ('Charleston, "Battersea, Bilston – or Birmingham?"').

Eric Benton, 'John Brooks in Birmingham', and 'The Bilston Enamellers' *Transactions of the English Ceramic Circle*, Vol. 7, part 3, 1970, pp. 162–190.

Eric Benton, 'The London Enamellers' *Transactions of the English Ceramic Circle*, Vol. 8, part 2, 1972, pp. 137–163. ('Benton, "London Enamellers"').

Bernard Watney, 'Battersea Wine Labels – the present Position', *Wine Label Circle Journal*, Vol.5, No.6, December, 1973, pp. 133–140.

Wolverhampton Art Gallery, *Catalogue of English Painted Enamels; 18th and 19th century*, by Mary S. Morris, 1973.

CATALOGUE OF PAINTED ENAMELS OTHER THAN LIMOGES

44. Writing-casket, painted enamel on copper, mounted in gilt metal

French (probably Paris); third quarter of the 17th century

Length, $5\frac{1}{2}$ in. (14 cm); height, $3\frac{1}{2}$ in. (9 cm); width, $3\frac{3}{4}$ in. (9·5 cm)

The casket is formed of oval and spandrel-shaped panels held together by fillets of gilt metal. On the top of the cover is an oval panel painted in colours with the figure of a woman in a blue cloak tasting a cherry, while at her feet a monkey samples a basket of fruit, symbolic of *Taste*. This panel is enclosed by a frame of gilt metal decorated in low relief with lions' masks and claws combined with scrollwork in 'Ohrmuschel' style. Between this and the laurel frieze which forms the outer frame of the box are four spandrel-shaped enamel panels, painted with flowers in polychrome. The sides of the cover are formed by four narrow oval panels in wreath-like frames of gilt metal, each between two fitting enamel panels painted with flowers in polychrome: the oval panels are painted with mythological figures emblematic of *The Seasons*, as follows:— on the front, a seated youth in sandals and a red robe shaded with yellow, wearing a chaplet of flowers and having beside him a basket of flowers (*Spring*); on the right side, a youth in a red garment reclining on sheaves of corn and wearing ears of corn in his hair, a sickle at

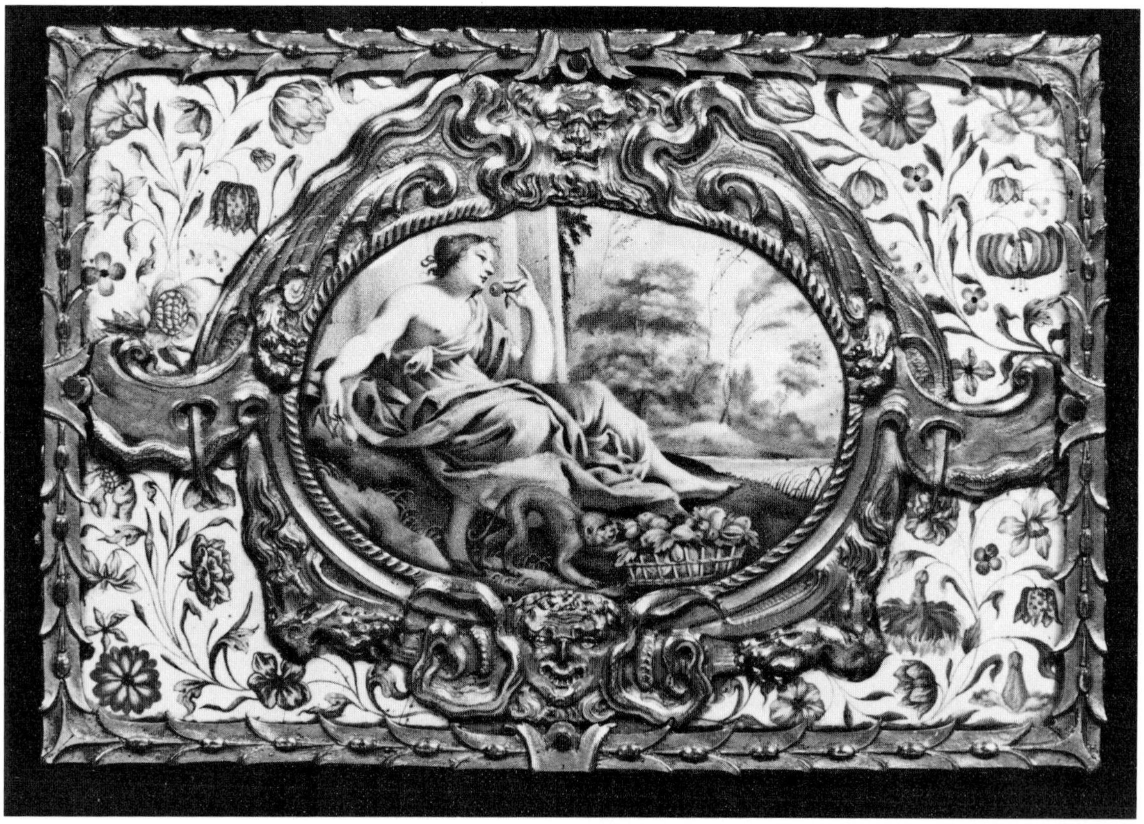

The top of the writing-casket

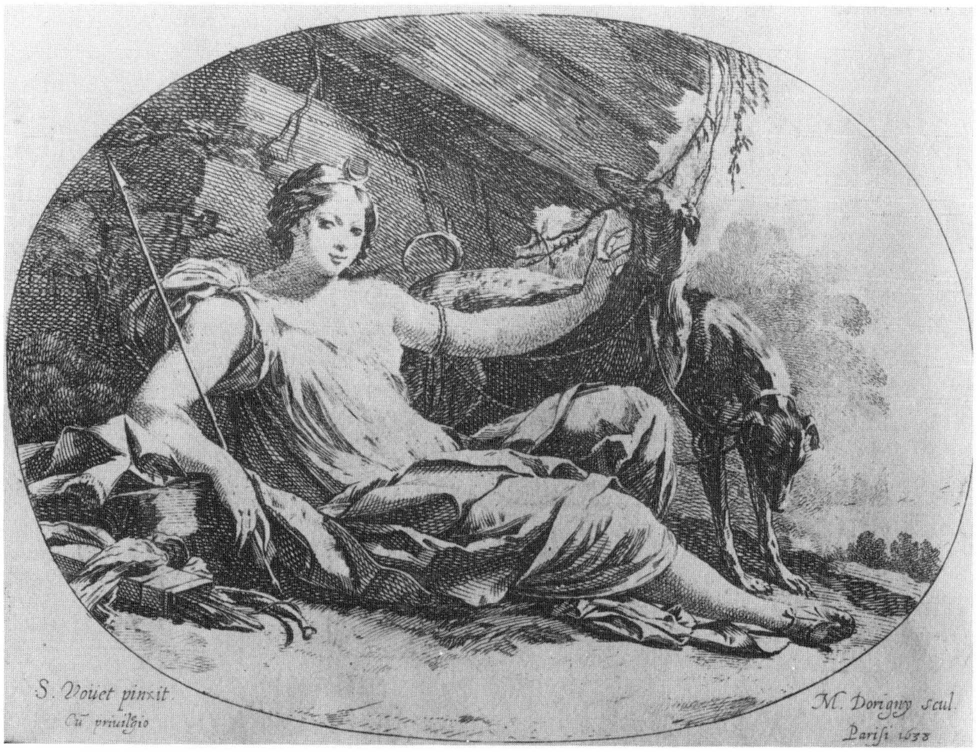

Touch : *engraving by Michel Dorigny, 1638, after a painting by Simon Vouet, dated 1637*

'S. Vouet pinxit cũ privilegio M. Dorigny scul. Parisi 1638') represents Diana reclining with an arrow in her right hand, and with two hounds in the background. These attributes and Diana's halfmoon emblem have been omitted from the enamel, to adapt the subject to represent *Touch*. The remaining figures painted on the box are probably also derived from originals by or after Simon Vouet. The figure of a woman representing *Sight*, however, may be after a painting by Laurent de la Hyre, who was in any case Vouet's pupil.

The design of the ormolu cartouche on the cover closely resembles those of cartouches engraved by D. Rabel (1578–1673).[2]

Baron Ferdinand de Rothschild — WI/23/2

[2] *Cartouches de Differantes Inventions Utiles a plusieurs Sortes de Persones*. A Paris chez F(rançois) L(anglois) D(e) Chartres . . . (n.d.) in the Victoria and Albert Museum – see also Peter Ward-Jackson, 'Some main streams and tributaries in European Ornament 1500–1750: Part 3', *Victoria and Albert Museum Bulletin*, III, No.4, October, 1967, pp.131–2.

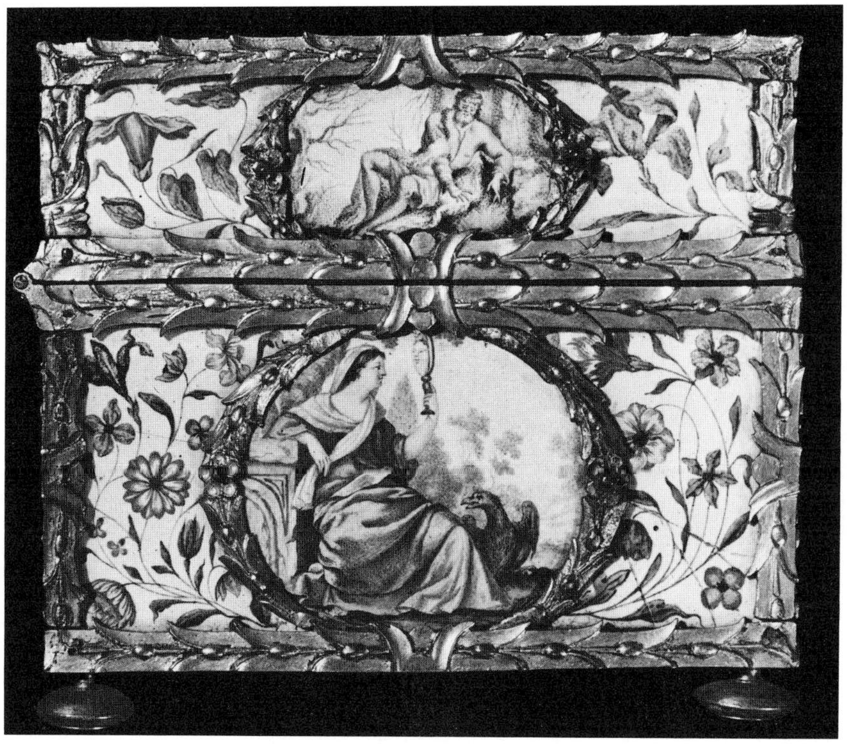

The left side of the writing-casket, 44

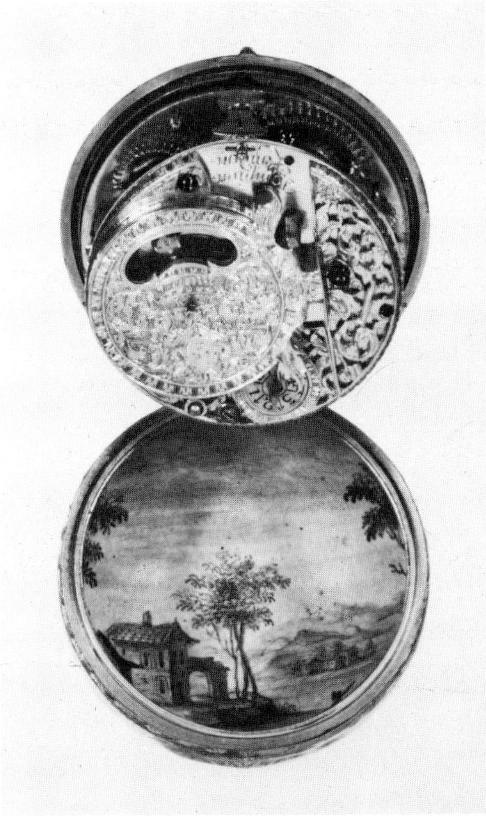

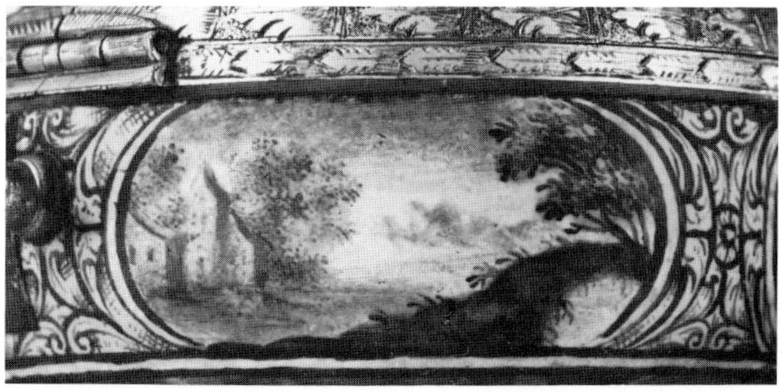

Detail of the side, enlarged to twice size

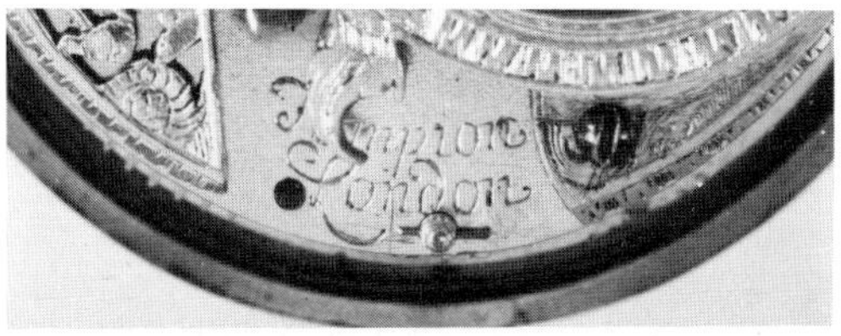

The signature on the backplate, enlarged to twice size

47. Watch, gold, copper and painted enamel

The watch probably English, the case Swiss (Geneva); about 1675–82

Height, $2\frac{3}{16}$ in. (5·5 cm); diameter, $1\frac{1}{2}$ in. (3·8 cm)

A circular watch, with a case enamelled *en plein* with the following scenes: on the front, the three Graces with Mercury standing behind; round the sides, four oval landscapes against a blue background and a cartouche with the inscription *Huaud le puisné fecit*. Inside the case is a landscape with a ruined building.

The dial is enamelled in white with a circle of Roman numerals indicating the hours, which encloses two half-length figures. The two hands are of gold, the glass convex. The movement is anonymous; it has plain pillars, a fusee and chain, verge escapement, balance-wheel and spring and Tompion-type regulator. The balance-cock has a single foot.

The inscription on the side of the case provides a precise attribution to Jean-Pierre Huaud, 1655–1723, the second of three brothers, who signed himself *le puisné* (to distinguish him from his brother Pierre, 'l'aîné') until 1682, when he entered into partnership with his brother Amy, after which the two brothers normally signed as 'les deux frères Huaut'.

The principal scene on the case reproduces an engraving dated 1642, by Michel Dorigny after a lost painting, *Mercury and the Three Graces* by Simon Vouet (William R. Crelly, *The Painting of Simon Vouet*, New Haven and London, 1962, p.252, No.228, Fig.187: A. P. F. Robert-Dumesnil, *Le Peintre-Graveur français*, IV, Paris, 1839, p.285, No.89). The two figures on the face may also be after a composition by Vouet, but no precise model has been traced.

In the workshop of Jean-Pierre Huaud at his death in 1723 was 'Un grand rond pour (f)ond de boete qui est Les Graces avec Mercure' (Clouzot, *Miniature*, p.217).

For other watches signed and enamelled by the Huauds, see *46*, *48* and *49*.

See colour illustration on page 449

Baron Anselm de Rothschild (Schestag No.168)
Baron Ferdinand de Rothschild — WI/12/10

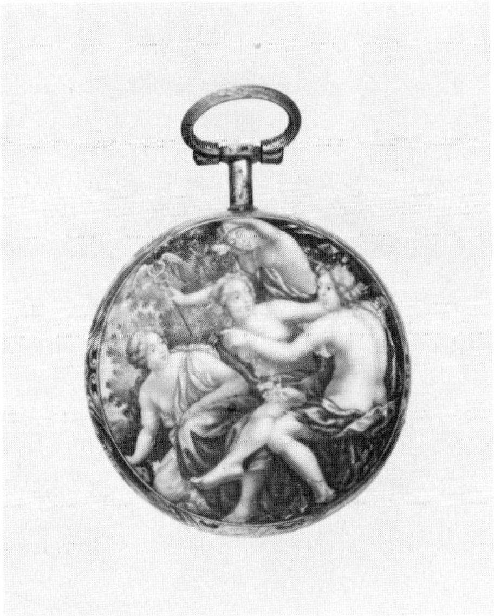

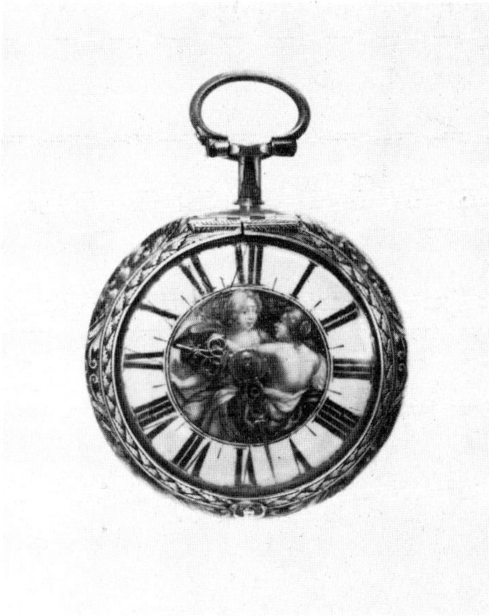

The inscription on the side of the case, enlarged to twice size

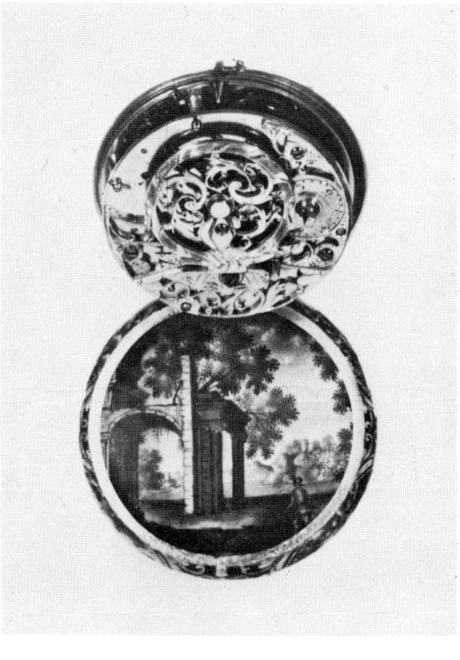

On the right: an engraving dated 1642 by Michel Dorigny after a lost painting, Mercury and the Three Graces, *by Simon Vouet*

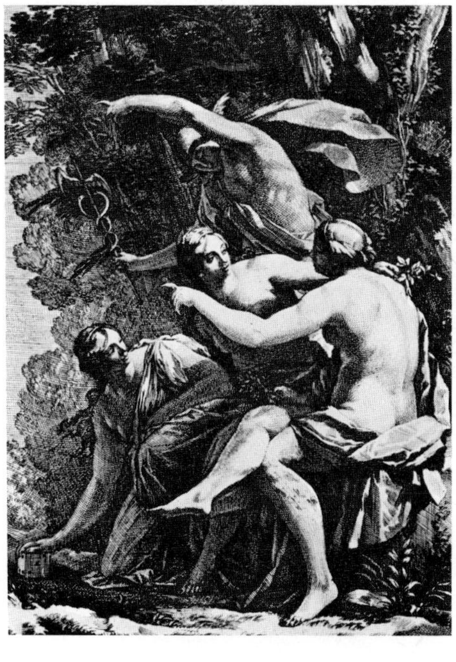

48. Watch, gold, copper and painted enamel

The watch Dutch (Amsterdam), middle of the 18th century; the case German (Berlin), between 1686 and 1700

Height, $2\frac{7}{16}$ in. (6·2 cm); diameter, $1\frac{11}{16}$ in. (4·2 cm)

Circular in shape, with a gold case enamelled *en plein* with scenes painted in colours. On the front is the *Judgement of Solomon*; on the sides four landscapes with ruins on a blue background, and a yellow cartouche with the inscription 'les deux frere Huaut pintre de son A.E de B.' (*son Altesse Electeur de Brandebourg*). The inside of the case is decorated with another landscape with a ruin.

The dial is enamelled in white with a ring of black Roman numerals for the hours enclosing a scene of two women and a crowned and turbaned figure (? *Solomon and the Queen of Sheba*). The single hand is of gilt copper, as are the movement and the cock, which is decorated with pierced-work scrolls. The backplate of the movement is signed 'Amsterdam Johannes Oosterwyck' and the plates are separated by broad Egyptian pillars. The movement is of a conventional Continental type, with verge escapement, steel balance-spring and Tompion-type regulator, and the balance-cock is of the bridge type.

This watch-case dates from the period when the two brothers Jean-Pierre and Amy Huaud were in Berlin, between 1686, when they were appointed painters in enamel to the Elector of Brandenburg, and 1700, when they returned to their native town of Geneva.

G. H. Baillie states that the period of greatest activity of the watchmaker Johannes Oosterwijk was the middle of the eighteenth century (*Watchmakers and Clockmakers of the World*, London, 2nd ed., 1947, p.238). Thus the movement would have been made some decades after the case.

The subject of the *Judgment of Solomon* is taken from an engraving – probably that by Boethius Adams Bolswert (F. W. H. Hollstein, *Dutch and Flemish Etchings Engravings and Woodcuts*, III, Amsterdam, *n.d.*, ?1951, p.61) – after Rubens' picture in the Royal Gallery in Copenhagen (A. Rosenberg, *P. P. Rubens*, Berlin and Leipzig, *n.d.*, (?1921), p.128). The landscapes recall the type much used by Pierre Patel the Elder (*c.*1620–*c.*1676) and his son Pierre (1648–1707), who was active when the watch was enamelled.

For other watches by the Huauds, see *46*, *47* and *49*.

See colour illustration on page 449

Baron Anselm de Rothschild (Schestag No.166)
Baron Ferdinand de Rothschild   W1/13/2

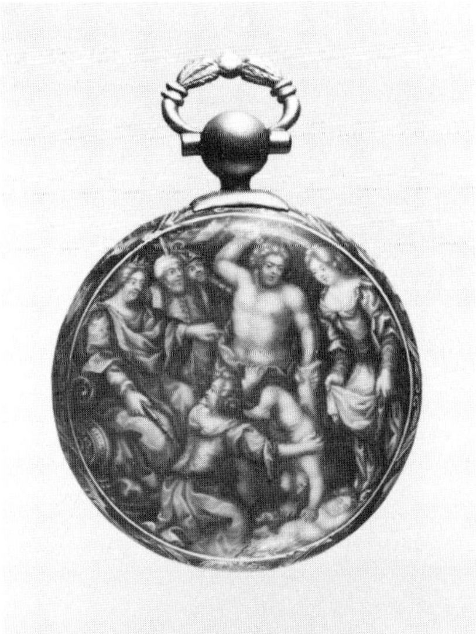

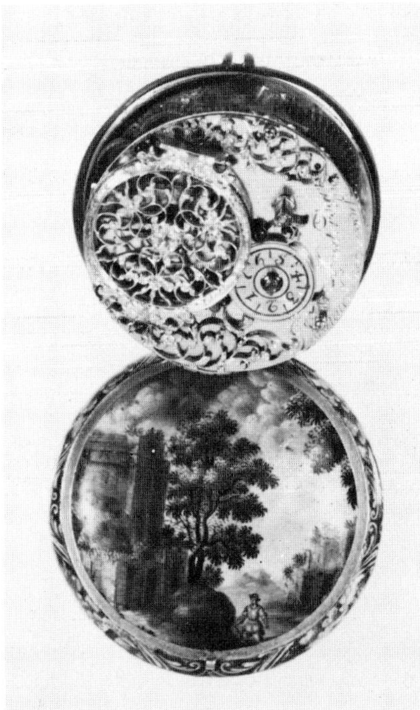

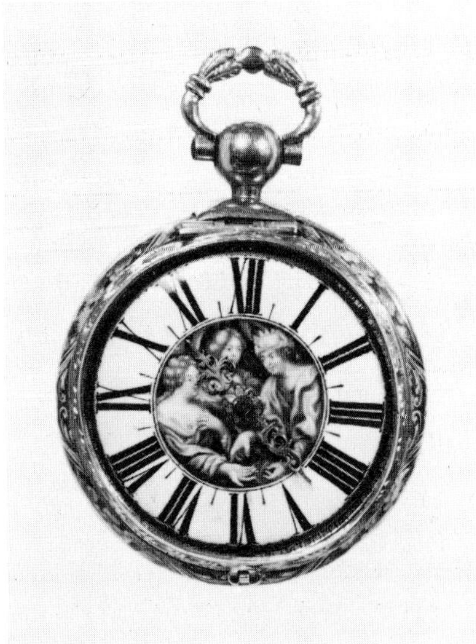

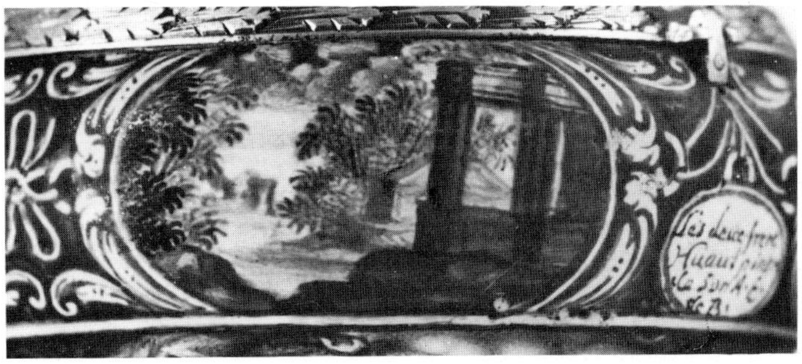

Detail of the side, enlarged to twice size

On the left: inscription of the Huaud brothers, enlarged to twice size. On the right: inscription of the watchmaker enlarged to twice size

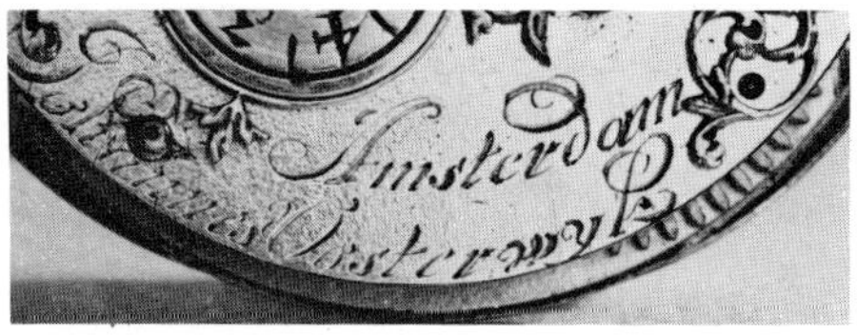

49. Watch, gold, copper and painted enamel

The watch Dutch (Amsterdam), about 1700; the case Swiss (Geneva), about 1700–23

Height, $2\frac{1}{8}$ in. (5·4 cm); diameter, $1\frac{9}{16}$ in. (4 cm)

Circular in shape, with a gold case enamelled *en plein* with scenes painted in colours. On the front is represented the theme of *Roman Charity*; on the sides, four landscapes with cottages against a blue background, and a cartouche bearing the inscription 'fretres [*sic*] Huaut pinxerunt'. There is another landscape with a cottage on the inside of the case.

The dial is enamelled in white with a circle of black Roman numerals for the hours enclosing two half-lengths of women, one of them holding a serpent (? *Cleopatra*). The single hand is of blued steel, probably a renewal. The glass is convex. The backplate of the movement is signed 'Josephus Norris, Amsterdam', a maker who was at work about 1700[1]. The plates are separated by tulip pillars and the movement has a fusee and chain, verge escapement, steel balance-spring and Tompion-type regulator. The balance-cock has a single foot.

The theme of *Roman Charity* depicted on the face of the watch is taken from a story recorded by Valerius Maximus in his chapter in filial piety (V, 4, 1) concerning Pero, who kept alive her father, Cimon, unjustly imprisoned, by giving him her breast when she visited him. This story seems to have been a great favourite with the Huaud brothers and many watches are known decorated by them with representations of it. These watches contain movements supplied by contemporary watchmakers of different nationalities. Thus, apart from Joseph Norris of Amsterdam, who signed the Waddesdon watch, the names of the following watchmakers occur on watches decorated with this subject: Louys Gautier de Blois (Jubinal de Saint-Albin Sale, Paris, 1 April 1938, Lot 64, pl.XI); Romilly, of Paris (Metropolitan Museum, New York, Pierpont Morgan gift); Christin, of Berne (Jubinal de Saint-Albin sale, Lot 66, Pl. XI); J. G. Ulrich, of Hamburg (Shandon Collection, Robert Napier sale, Christies', 11 April–12 June 1877, Lot 2087); Charles Helot, of Amsterdam (Toulouse, Saint-Raymond Museum, Gélis gift); John Ellicott, of London (Metropolitan Museum, New York, Pierpont Morgan gift); Massy, also of London (Edward Tuck gift, Petit Palais, Paris); Bacenot, of Geneva (Victoria and Albert Museum, M3–1957) and there are many others.[2] Mention may be made of two gold watches by Jean V. Mussard of Geneva, on which the subject of *Roman Charity* was painted on enamel, one in the British Museum and the other in the Delle Piane Collection in Milan. The

[1] Josephus Norris (active about 1700) was one of a dynasty of English clockmakers but worked in Amsterdam at the end of the 17th century. His movements are mostly found in watches enamelled by the Huaud brothers. (G. H. Baillie, *Watchmakers and Clockmakers of the World*, 2nd ed., London, 1947, p.234; F. J. Britten, *Old Clocks and Watches and their Makers*, 8th ed. (edited by Cecil Clutton *et al.*), London, 1973, p.460).

[2] Examples with contemporary but unmarked movements are in the Victoria and Albert Museum (M208–1919; 4858–1901); the British Museum (Ilbert Collection) signed by Pierre Huaud and with a probably Dutch movement; the Cabrier Collection, London, No.278 (Christie's Sale, 10 March 1964, Lot 117, signed 'Huaud le puisné'); the François Mercier Collection (Christie's Sale, 5 July 1971, signed 'Les frères Huaud').

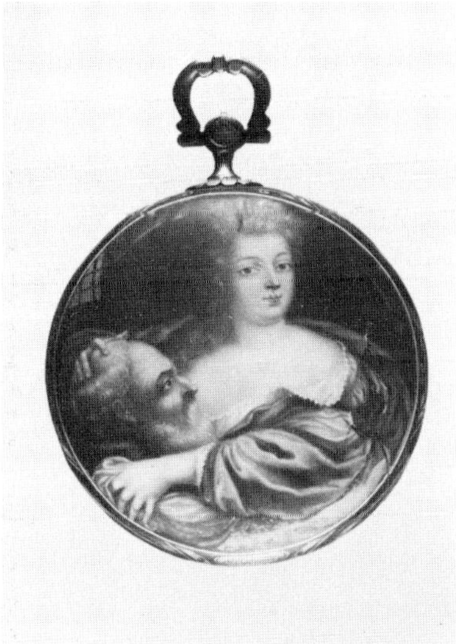

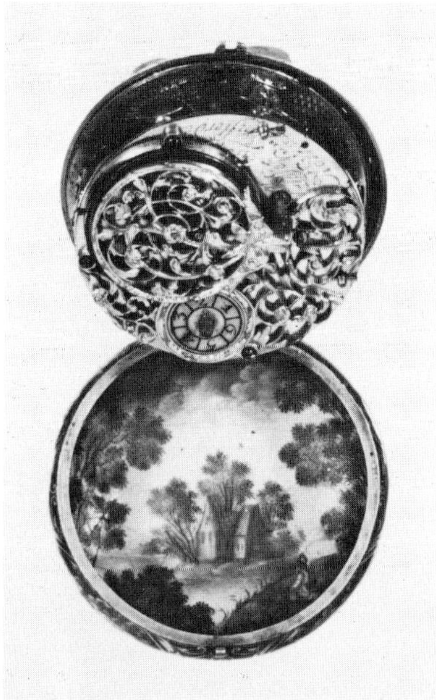

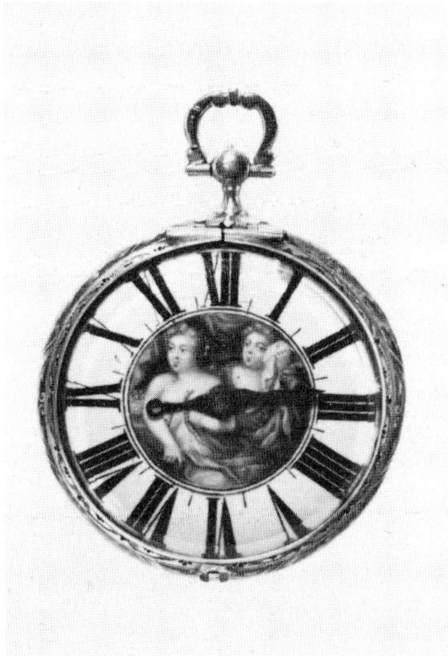

Below, right: detail of the side, enlarged to twice size

Below, left: inscription of the Huaud brothers

Below, centre: inscription of the watchmaker. Both enlarged to twice size

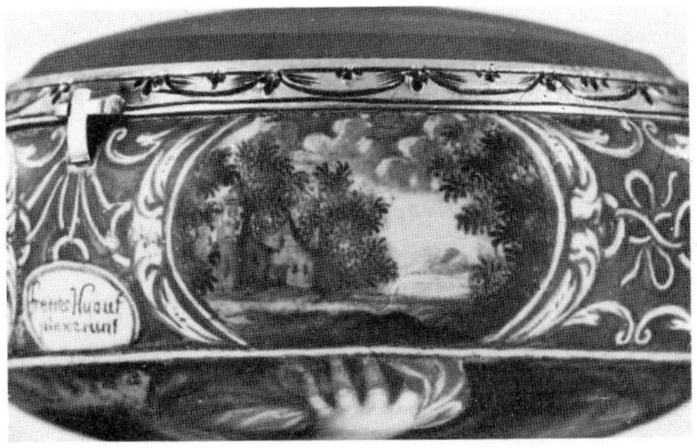

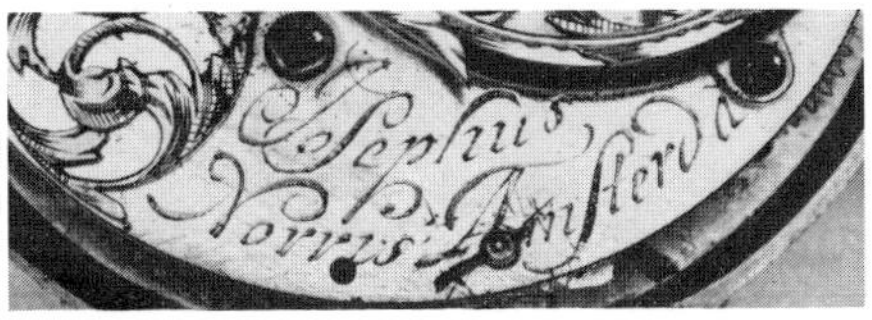

subject is taken from an engraving by Claude Mellan (1598–1688)[3] after a painting by Simon Vouet in the Museé des Beaux-Arts, Nantes.[4]

In the workshop of Jean-Pierre Huaud at his death in 1723 was 'Une Charité romaine avec sa garniture de Vermeil', appraised at 150 florins (Clouzot, *La Miniature sur Email en France*, p.216). The scene on the face of the watch probably represents *The Death of Cleopatra*. A replica of the Waddesdon watch, bearing the signature of the Dutch watchmaker Joseph Norris, and made by the two younger Huaud brothers, appeared in the Maurice M. Sternberger Sale at Christie's, 27 November 1937, Lot 72.

For other watches enamelled by the Huaud brothers, see *46–8*.

See colour illustration on page 449

Baron Ferdinand de Rothschild W1/13/3

[3] A. de Montaiglon, *Catalogue raisonné de l'œuvre de Claude Mellan*, Abbeville, 1856, p.134, No.125 'La Charité romaine', inscribed 'Si. Voüet Paris pinx – Cl.Mellan Gal.Sculp.Romae'.

[4] William R. Crelly, *The Painting of Simon Vouet*, Newhaven, 1962, pp.182–3, No.76 'Caritas Romana'.

50. Watch, gold, copper and painted enamel

The watch probably French, the case perhaps Swiss (Geneva); second half of the 17th century

Height, $1\frac{3}{4}$ in. (4·4 cm); diameter, $1\frac{7}{16}$ in. (3·6 cm)

No marks or signatures

A circular watch with a gold case enamelled *en plein* with compositions painted in colour. On the front is a pair of lovers seated beneath a tree beside a hound held by a winged cupid; on the sides are four landscapes in yellow with blue skies and green trees, separated by four black and white rosettes. Inside the case is a landscape with a classical temple.

The dial is enamelled in white (? later) with a double circle of numerals, Roman for the hours (I to XII) and Arabic for the minutes (5 to 60). The glass is convex. The movement is unsigned but probably French. It has plain pillars, a fusee, verge escapement, balance-wheel with spring and Tompion-type regulator marked R and A with a pierced bridge cock, this last being a Continental feature. Winding is through a hole in the dial.

The two figures on the front of the case represent *Venus and Adonis* and are probably taken (reversed) from an engraving by Antonio Tempesta with the addition of Cupid. The style of the enamel suggests that it comes from the general circle of the Huaud brothers (see *46–9*). A watch of enamelled gold with a variant of the scene with the two lovers and bearing the signature of Henry Arlaud (of Geneva and London) was once in the Olivier Collection, which belongs now to the Louvre and is at present deposited at the Musée des Arts Décoratifs (Inv. A.8444).

This watch is very close in style to those of the Huaud brothers, but differs from them in the predominance of a yellow-green and a yellow, while the scrollwork between the small panels on the periphery is painted in black on white instead of in the mainly blue and yellow of the Huaud pieces. It may possibly be by D. André, a signed watch by whom is in the John Gersholm Parkington Memorial Collection (No.63), Bury St Edmunds (S. Benson Beevers, 'The John Gersholm Parkington Memorial Collection of Time Measurement Instruments", *Connoisseur Year Book*, 1958, pp.115–126, with movement by François Mercier, about 1690–1700). This shows very similar scrollwork between the peripheral panels and is otherwise painted in a manner very close to that of the Huaud brothers.

Baron Anselm de Rothschild (Schestag No.169)
Baron Ferdinand de Rothschild — W1/13/4

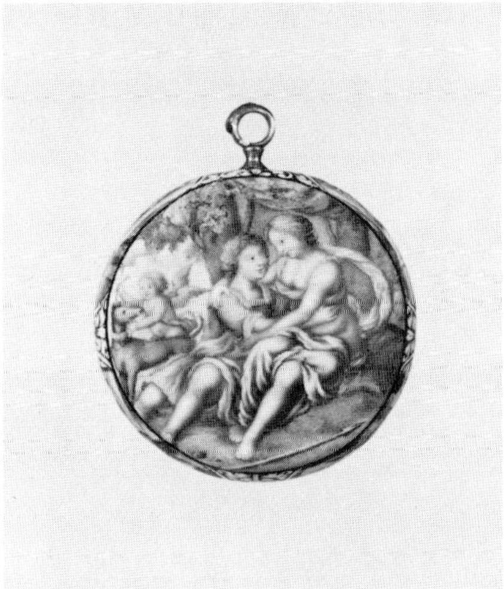

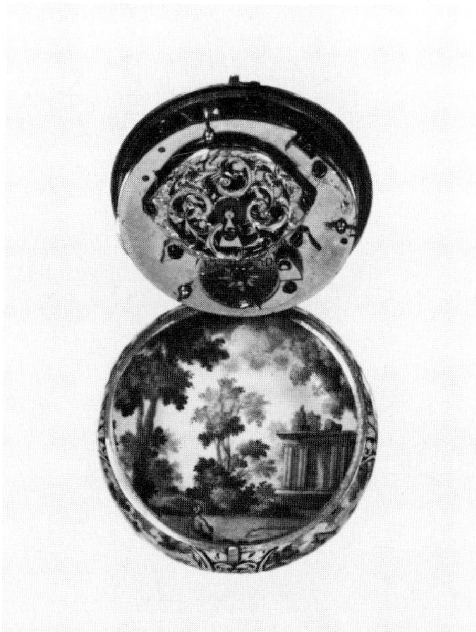

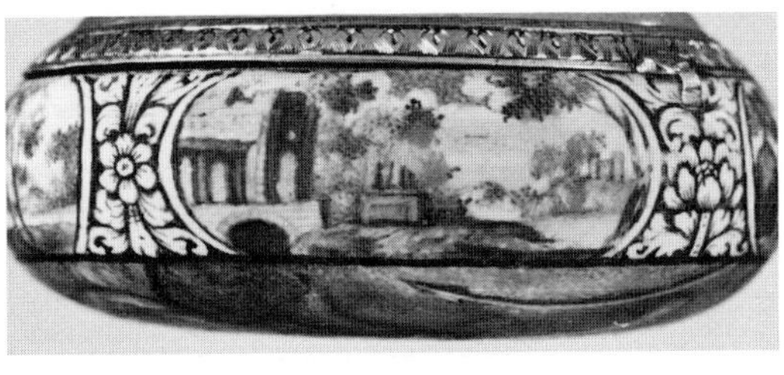

Detail of the side, enlarged to twice size

51. Box in the form of an egg, white enamel on copper, decorated in relief gold and green enamel, with gold mounts

French; about 1740

Height, $2\frac{1}{2}$ in. (7 cm); diameter, $1\frac{3}{4}$ in. (4·5 cm)

Mark: Discharge mark (a fox's head) of the *sous-fermier* Louis Robin, 1738–44 (*reproduced, enlarged four times, below*) (Henry Nocq, *Le Poinçon de Paris*, IV, Paris, 1931, p.233: see also A. Kenneth Snowman, *18th Century Gold Boxes of Europe*, London, 1966, Fig.113 and p.87).

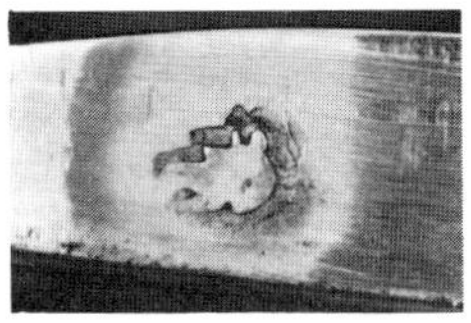

Entirely covered in white enamel inside and outside. The decoration in low relief is in stamped gold-leaf partly enamelled in translucent green (with one spot of pink), depicting poodles begging in front of winged cupids playing stringed instruments and trumpets beneath garlands of flowers. The lid is hinged, the mount of gold.

For a discussion of this type of enamels, see Introduction, p.433.

Eggs made by goldsmiths and jewellers of gold, silver, enamelled or lacquered metal, porcelain, rock crystal, agate, or jasper, were supplied to *marchands-merciers* for resale to a rich clientele. Lazare Duvaux supplied 'deux oeufs de cristal de roche garnis d'or' to Monsieur Duflot (probably Augustin Duflos, jeweller to the King) for 264 *livres* on 18 June 1750, and on 14 May the following year 'deux œufs, l'un en jaspe, l'autre en agathe, garnis & couverts d'or ciselé' for 336 *livres* to Madame de Pompadour (Courajod, Vol.II, 1873, pp.53 and 84, Nos.538 and 806). The purpose of these objects is doubtful. Some may have been destined to contain gold thimbles, others sweetmeats, or even perfume as has been suggested more than once.

A further comparison may be made with a gold egg enamelled in white with red and green bands which was in the Edward Joseph Collection in New Bond Street, London (Christie's Sale, 6 May, 1890, Lot 129).

A Meissen porcelain egg-shaped box of mid-eighteenth century date in the Wrightsman Collection, New York, has been fitted out, perhaps at a later date than the box itself, as a *nécessaire* (F. J. B. Watson and Carl Dauterman, *The Wrightsman Collection*, III, *Furniture, Snuffboxes, Silver*, New York, 1970, p.219, No.47: this entry also draws attention to an egg-shaped Meissen *confiture* box decorated with a coaching-scene, once in the C. H. Fischer Collection, Dresden). Boxes of this shape were later (about 1760) manufactured in England in enamel on copper (see, e.g., Schreiber Collection *Catalogue*, III, Nos.90, 180, 211–13, called

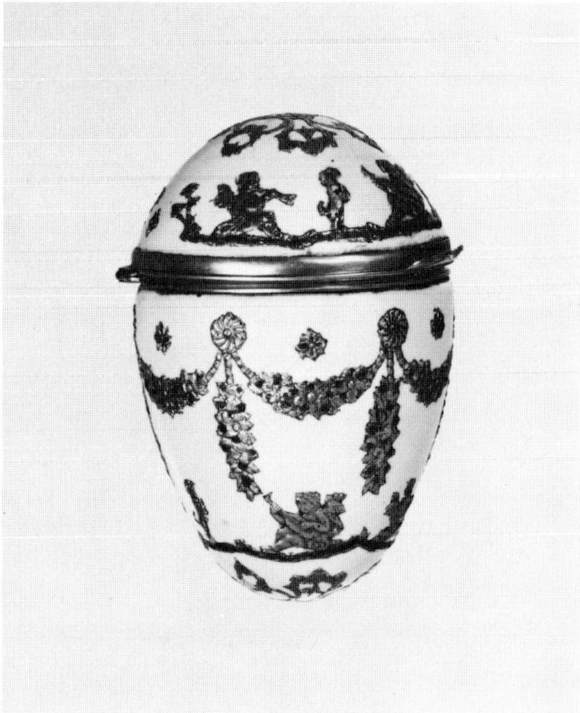

bonbonnières), in *piqué* and blue glass, and in semi-precious stones mounted in gold[1]. Smaller in format were enamel eggs housing nutmeg-graters (ibid., Nos.188–9, 279, 289).

The figure of a begging dog on this box resembles that in a drawing attributed to Pillement, and perhaps intended as a box-design, in the Victoria and Albert Museum (E.66–1938 – see A. K. Snowman, op. cit., Fig.50).

Baron Ferdinand de Rothschild　　　　　　　　　　　　　　　　　　　　　　w3/8/4

[1] See, e.g., two boxes in the Waddesdon Collection (Serge Grandjean et al., *Gold Boxes and Miniatures of the 18th century*, Fribourg, 1975, pp.74–5).

52. Box in the form of an egg, painted enamel on copper with gold-leaf embellishments

Probably French (Paris); second half of 19th century

Painted enamel on copper, with applied gold-leaf and 'jewels' of opalescent white glass, mounted in gilt metal

Height, $2\frac{1}{8}$ in. (5·5 cm)

Egg-shaped box of turquoise enamel on copper, the upper end painted with flowers and leaves in colours within a star-shaped panel edged with 'jewels' of opalescent white glass, and reserved in white on the turquoise ground, which is further enlivened by scrolls in gold-leaf connecting patches of opaque-white and opaque-black enamel, the latter having a red dot in the centre.
For a discussion of this shape, see *51*.

Miss Alice de Rothschild

W3/6/7

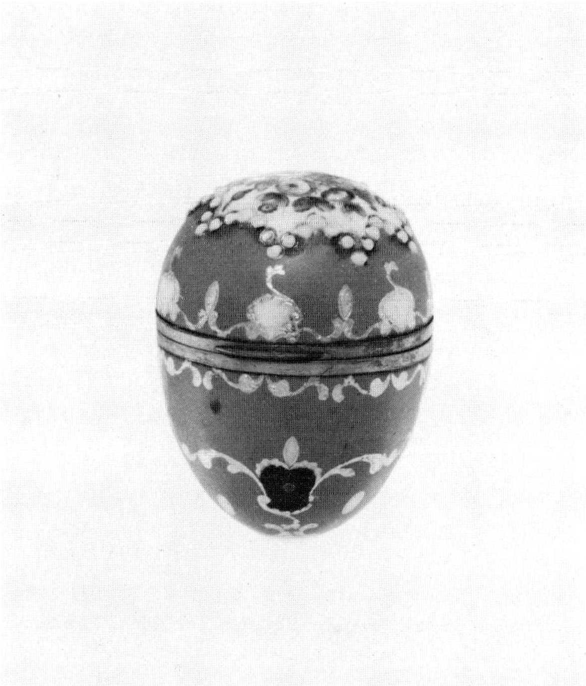

53. Toilet-casket, painted enamel on copper, mounted in gilt metal

English (probably London); about 1755–60

Length, $5\frac{1}{8}$ in. (13 cm); height, 5 in. (12·7 cm); width, $3\frac{3}{4}$ in. (9·5 cm)

The casket is formed of five enamel panels painted with sprays of flowers in colours on an apple-green ground, held together by gilt metal strips elaborately moulded and chased with rococo scrollwork and flowers. The cover is held in a deep gilt-metal border of concave profile and is secured by a lock and key, the escutcheon of the former being integral with the top front frame of the body, and similarly decorated. The casket stands on four openwork bracket feet of gilt metal pierced with a design of rococo scrollwork.

The sprays of flowers decorating the top and the front panels of the casket are built up with layers of enamel to form a considerable relief. The inside of the cover is painted with a spray of natural flowers on a white ground.

The contents of the casket, which is lined with cherry-coloured silk and a gold-thread gimp, comprise: three larger and two smaller bottles of colourless cut glass for perfumes or toilet-waters, and a silver-gilt funnel with which to fill them; the bottles are square in section, mould-blown and then ground and polished on the wheel, with facetted shoulders and facet-cut stoppers fitting into gilt-metal collars.

The funnel bears the following mostly indistinct marks: (1) maker's mark 'CR' (unidentified); (2) sovereign's head facing right; (3) lion passant guardant; (4) old English letter, either A, K or R; (5) leopard's head, apparently uncrowned. The most likely interpretation of this mark is that it was struck in 1836 at the Goldsmiths' Company, long after the date of the casket and its contents, although the funnel itself may be a later addition.

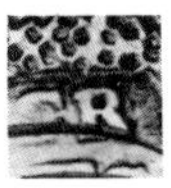
1
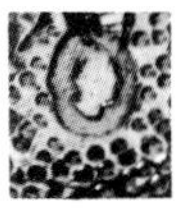
2
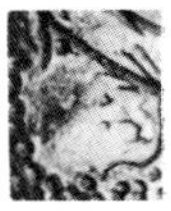
3
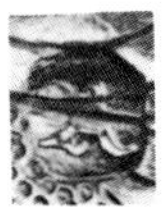
4
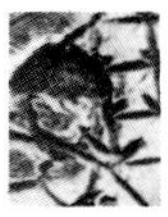
5

The marks are reproduced four times actual size

The style of the painting of this casket suggests a London origin – see E. Benton, 'London Enamellers', pp. 137ff, and particularly Pl. 97b.

Miss Alice de Rothschild — E2/12/3b

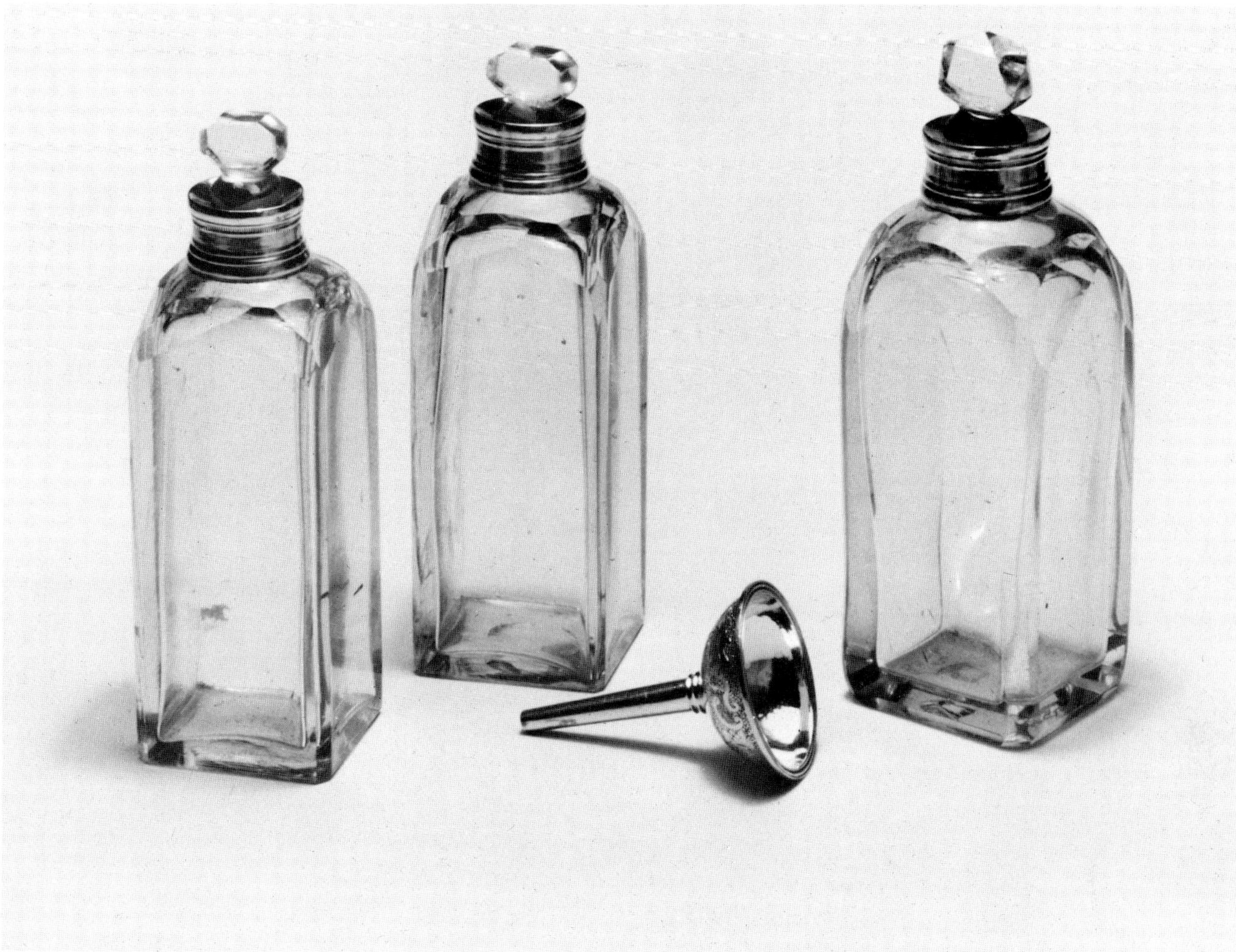

The bottles and funnel contained in the casket

54A. Set of four candlesticks, printed and painted enamel on copper

English (South Staffordshire); about 1765

Height, $11\frac{1}{4}$ in. (28·5 cm)

Table-candlesticks formed of three moulded sections: candle-holder, tall inverted baluster stem, and domed foot, the first two elements being held together by a gilt metal fillet, and the stem being originally probably joined to the foot by a lap-joint under the enamel. The edge of the foot is encircled by a gilt metal band with waved edge. All three sections have moulded decoration in relief, usually picked out with raised gilding. The candle-holder has a gadrooned lower part, above which raised vertical ribs rise to the rim, each embellished with a vertical gilt spray of flowers and leaves. The upper part of the stem has four relief panels, each of which is converted by means of raised gilt rococo scrollwork into a cartouche for painting: on the lower part of the stem four vertical raised fields terminate below in offset swelling panels, each embellished with a raised gilt scallop-shell. The foot is decorated with five relief-panels, each outlined with raised gilt rococo scrollwork to form a cartouche for printing and painting.

The cartouches of the upper part of the stem are filled with sprays of flowers and leaves painted in polychrome, while those of the foot enclose subjects as follows, painted in polychrome over underlying prints in soft purple – a gentleman with a cane conversing with a girl holding a basket; a peacock on a balustrade with two stone balls, a house and trees in the background; a boy and a girl seated beside a pedestal topped with a garden urn, in an avenue of trees terminating in a garden-temple; a gentleman and a lady seated in a park landscape, with water in the foreground and garden urns in the background; a lady and a gentleman seated in a park landscape with water in the foreground and a circular garden-temple with a statue behind.

The grease-pans have a waved edge of six lobes lined with a gilt relief cordon and fit into the candle-holders by means of a tube integral with the pan. The upper surface is decorated with sprays of flowers and leaves painted in colours, alternating with sprigs in raised gilding.

Very similar candlesticks were in the Hon. Mrs Ionides' Collection (G. B. Hughes, *English Painted Enamels*, Fig. 19), and others were sold at Christie's on 6 February 1962, Lot 91, and at Sotheby's on 23 November 1970, Lot 19. The theme of a peacock on a balustrade with two stone balls, identically rendered, is found on a double-sided box once in the Hon. Mrs Ionides' Collection (G. B. Hughes, *English Painted Enamels*, Fig. 13a); and that of a couple sitting in an avenue of trees, on a box in the Metropolitan Museum of Art, New York (C. Le Corbeiller, *European and American Snuff-boxes*, Fig. 613).

Damaged and repaired.

Baron Ferdinand de Rothschild WI/30/5

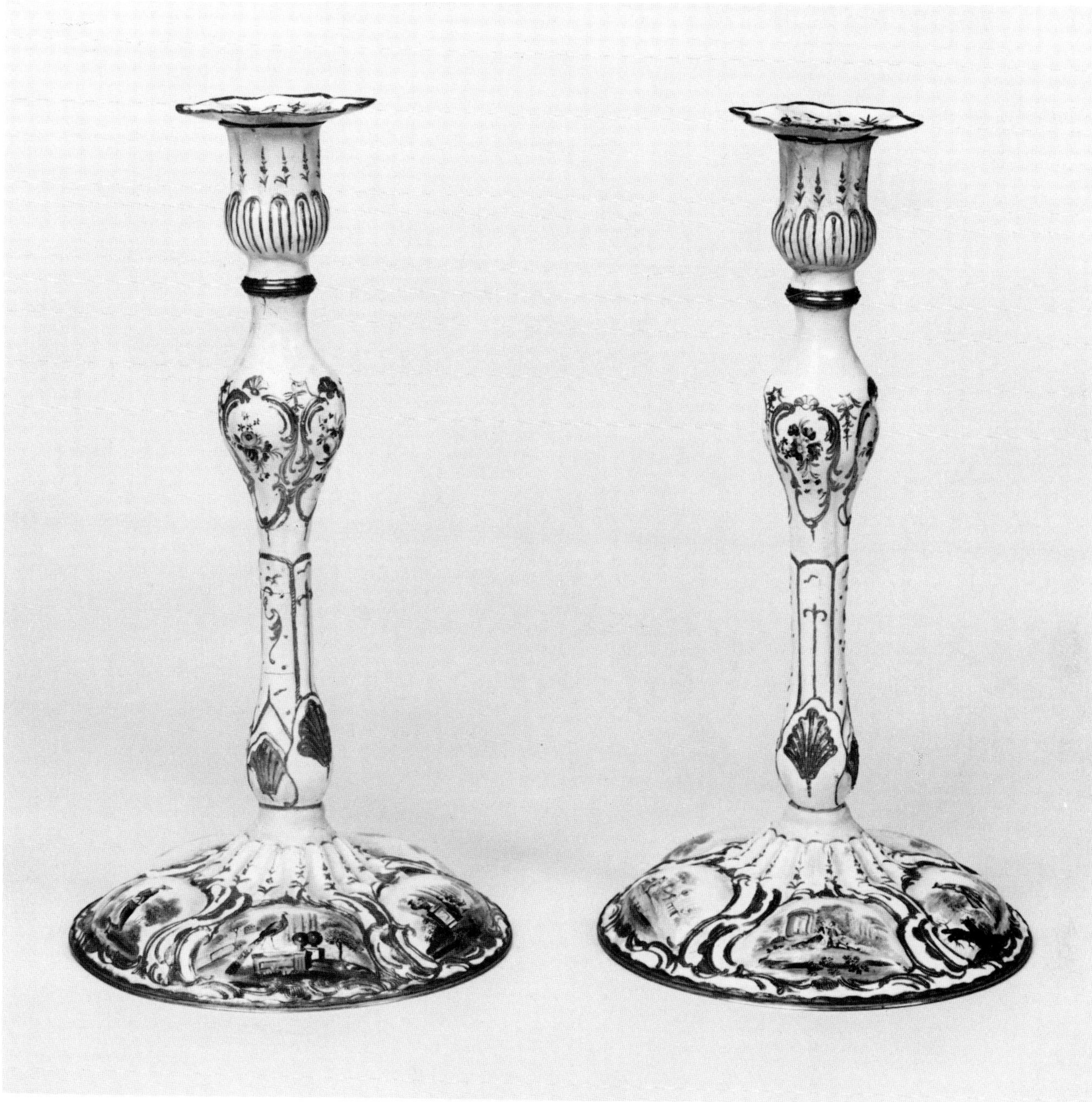

Two of the candlesticks, 54a

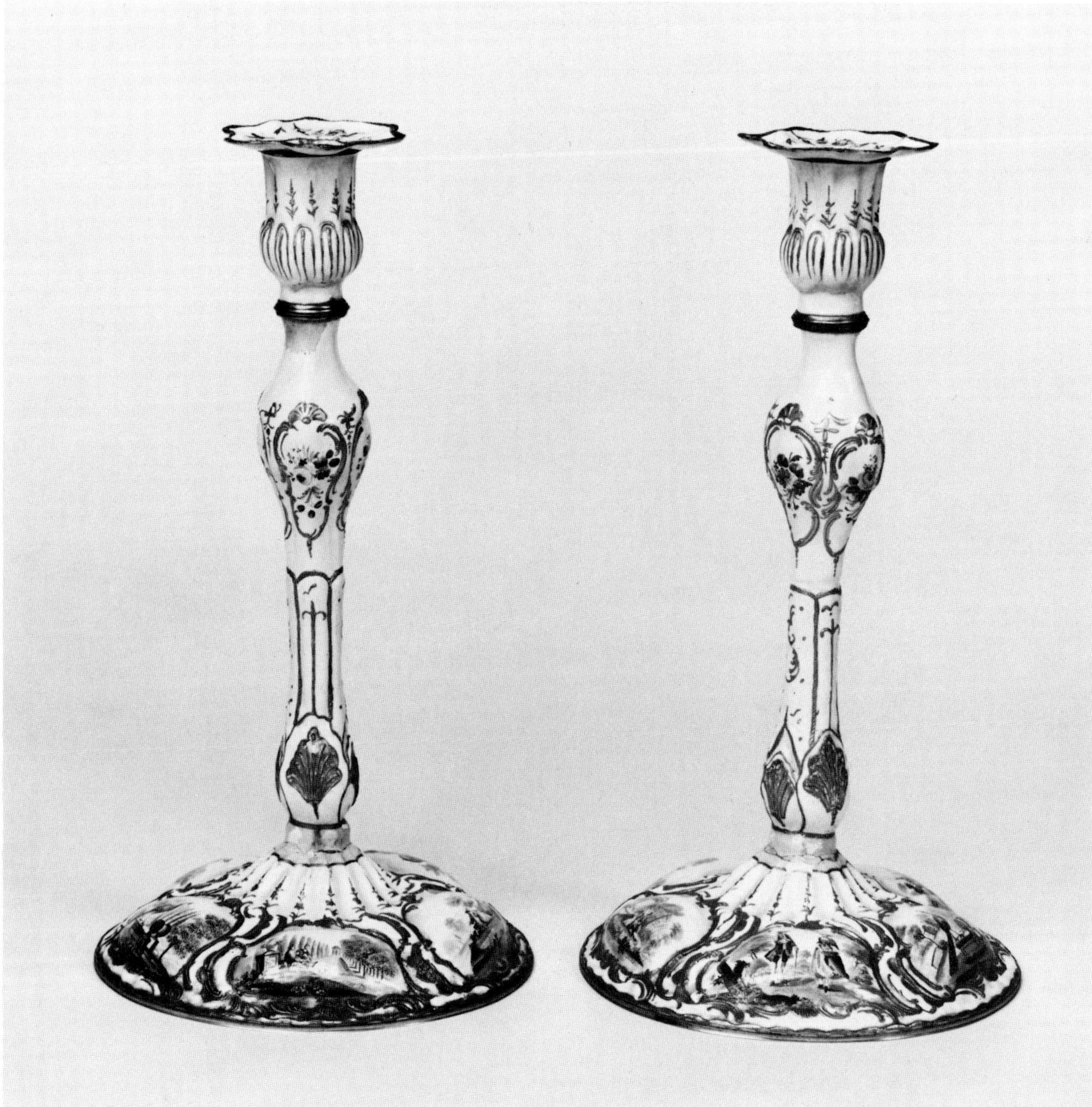

Two more of the set of candlesticks, 54a

Opposite: details of the feet of the candlesticks, 54a

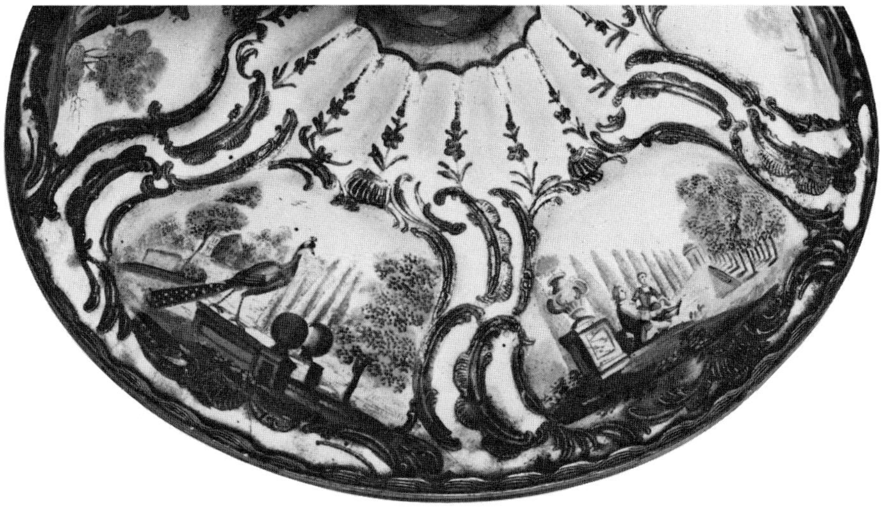

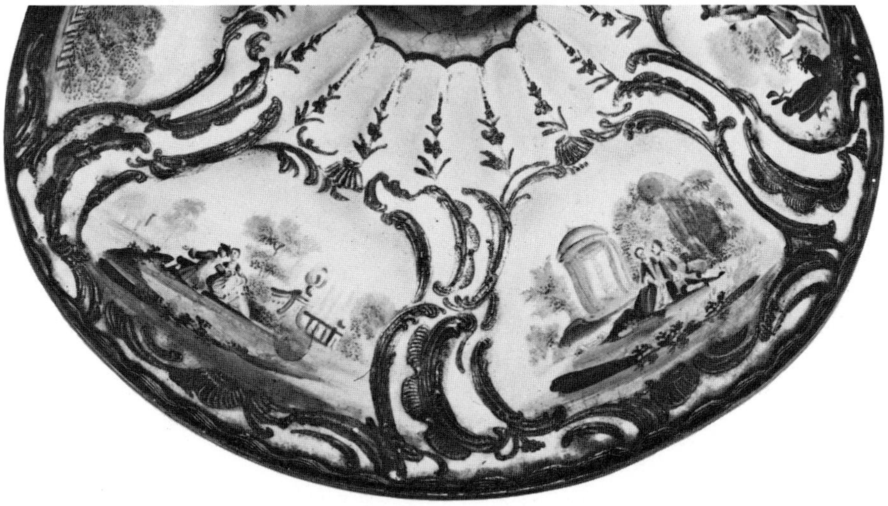

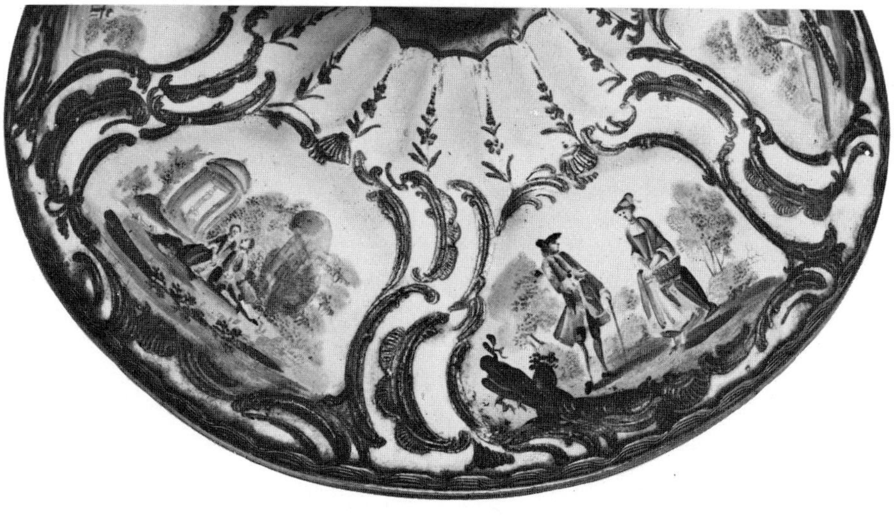

54B. Pair of candlesticks, printed and painted enamel on copper

English (South Staffordshire); about 1765

Height, $11\frac{1}{4}$ in. (28·5 cm)

These candlesticks match those of *54A* except for the designs in the cartouches on the foot. These run as follows: (on the left-hand candlestick) (1) a lady and gentleman seated in a landscape, facing to right, with a dog to their right, trees in the background; (2) a girl holding out her skirts and dancing, and on the right a man seated on a rock and playing the violin, trees in the background; (3) a lady and a gentleman seated together on the ground, apparently on an island, a sleeping dog beside them, a pair of goats to the right and trees in the background; (4) a fishing party composed of a lady and two gentlemen with fishing rods on a river bank surrounded by reeds, the lady and one gentleman seated on a fallen tree-trunk to the right, the other gentleman standing on the left; (5) a lady and a gentleman walking to the left, she wearing a black shawl and he carrying a cane, in a landscape with water, and trees in the background: (on the right-hand candlestick) (6) a man with a gun under his arm, and a pointer moving to the left before a landscape with trees; (7) as (5) above; (8) as (3) above; (9) as (4) above; (10) as (2) above.

A number of these subjects are found on other contemporary South Staffordshire enamels, but usually reversed. Thus, the theme of a girl dancing (2 above, probably adapted from Plate 32 of *The Ladies' Amusement* of about 1760) is found on an *étui* once in the Ionides Collection (G. B. Hughes, *English Painted Enamels*, Fig.33, *right*) and on two snuff-boxes in the Schreiber Collection (Rackham, *Schreiber Catalogue*, Nos.148 and 357): it also occurs on Liverpool delftware printed tiles made by the firm of John Sadler (A. Ray, *English Delftware Tiles*, London, 1973, pp.248–9, No.653). The 'Fishing Party' (4 above) is probably also adapted, reversed, from the same Plate of *The Ladies' Amusement*: it is found the right way round on two *étuis* and a scent-flask once in the Ionides Collection (Hughes, op. cit., Fig.33, *left*, Plate IV, *centre*, Fig.65, *top centre*) and four snuff-boxes in the Schreiber Collection (Rackham, op. cit., Nos.148, 149, 152 and 155), where the subject is combined with others from the same source. One of these is the lady and gentleman walking, he with a long cane (No.5 above). This occurs on a scent-flask once in the Ionides Collection (Hughes, op. cit., Fig.65, *top right*) and on a *bonbonnière* and a medallion in the Schreiber Collection (Rackham, op. cit., Nos.185 and 336). On a double box in the Untermyer Collection, now in the Metropolitan Museum, New York, it occurs on the cover, while the peacock-on-balustrade theme of the four candlesticks *54A* decorates the base (Y. Hackenbroch, *Chelsea and other English Porcelain, Pottery and Enamels in the Irwin Untermyer Collection*, London, 1957, p.255, Fig.325; see also Watney and Charleston, 'Petitions . . .', p.116).

The theme of a man with a gun (No.6 above), also derived from the same print, is also found, reversed, on an oval medallion in the Schreiber Collection (Rackham, op. cit., No.336, Pl.45), where it is mounted with a medallion of the subject just discussed (No.5 above).

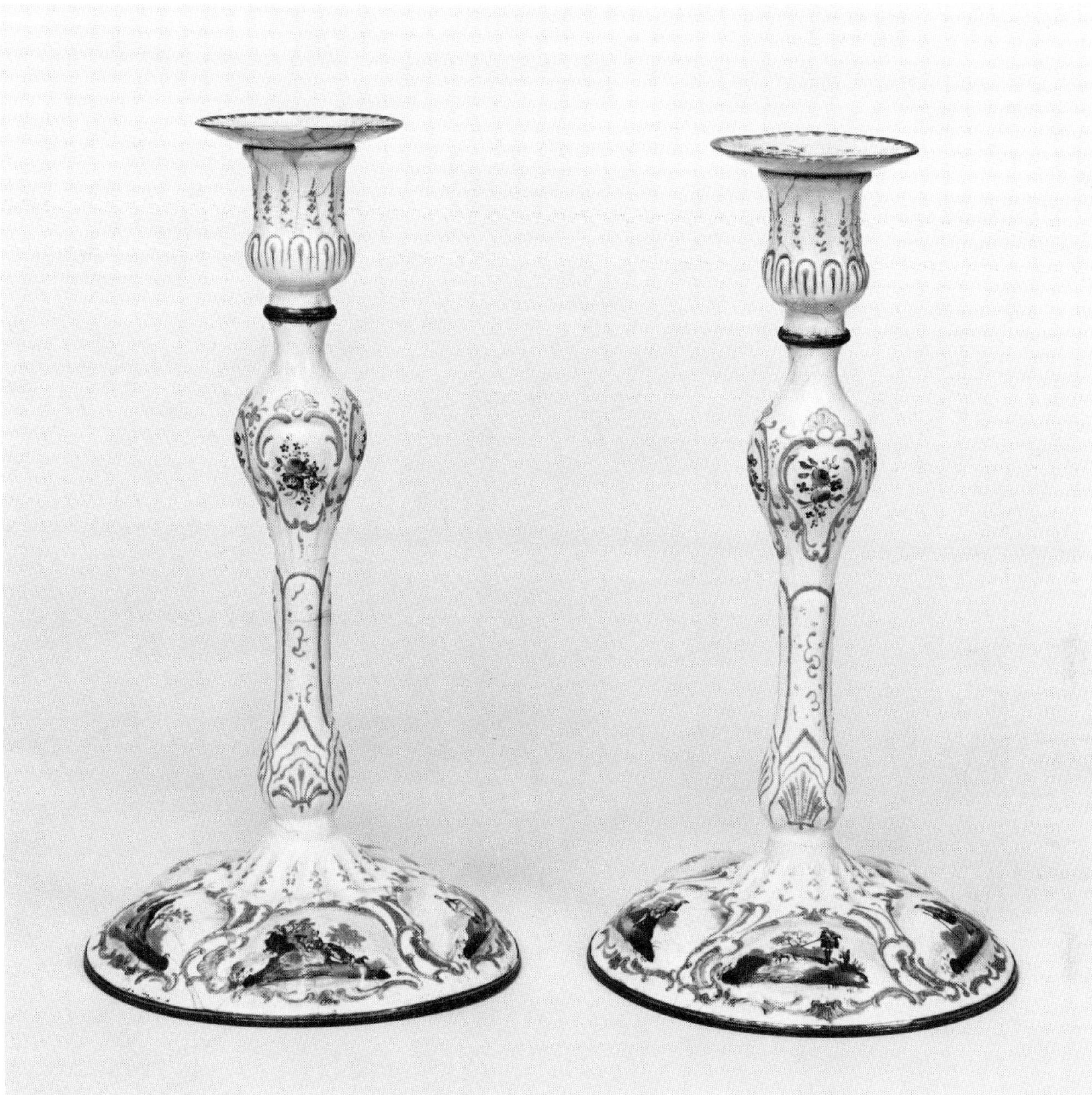

The theme of a pair of seated figures with a dog (No.1 above) is also derived from the same plate of *The Ladies' Amusement*. These subjects, reversed from the sense in which they usually occur, clearly form a close-knit group within the repertory of the firm which made these candlesticks.

Damaged and repaired.

Baron Ferdinand de Rothschild WI/30/5

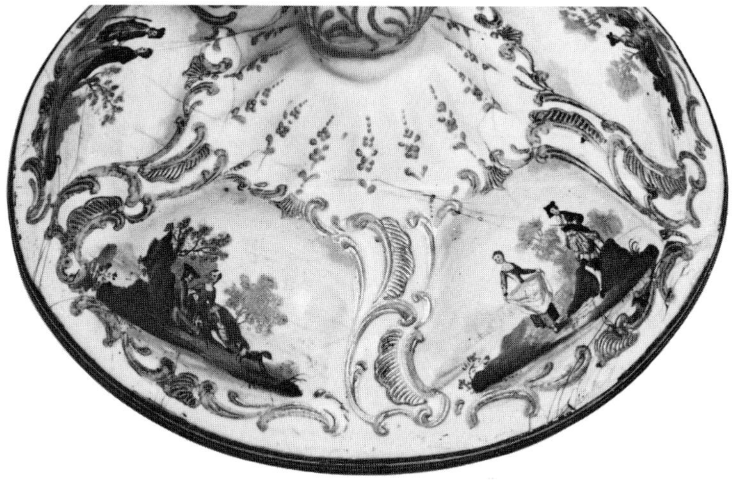

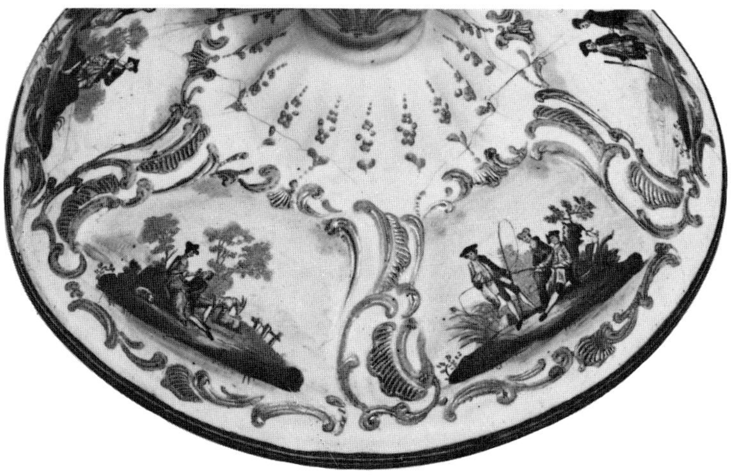

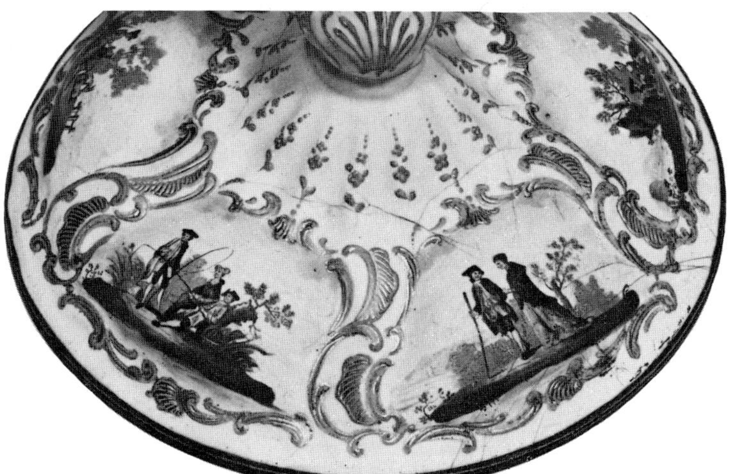

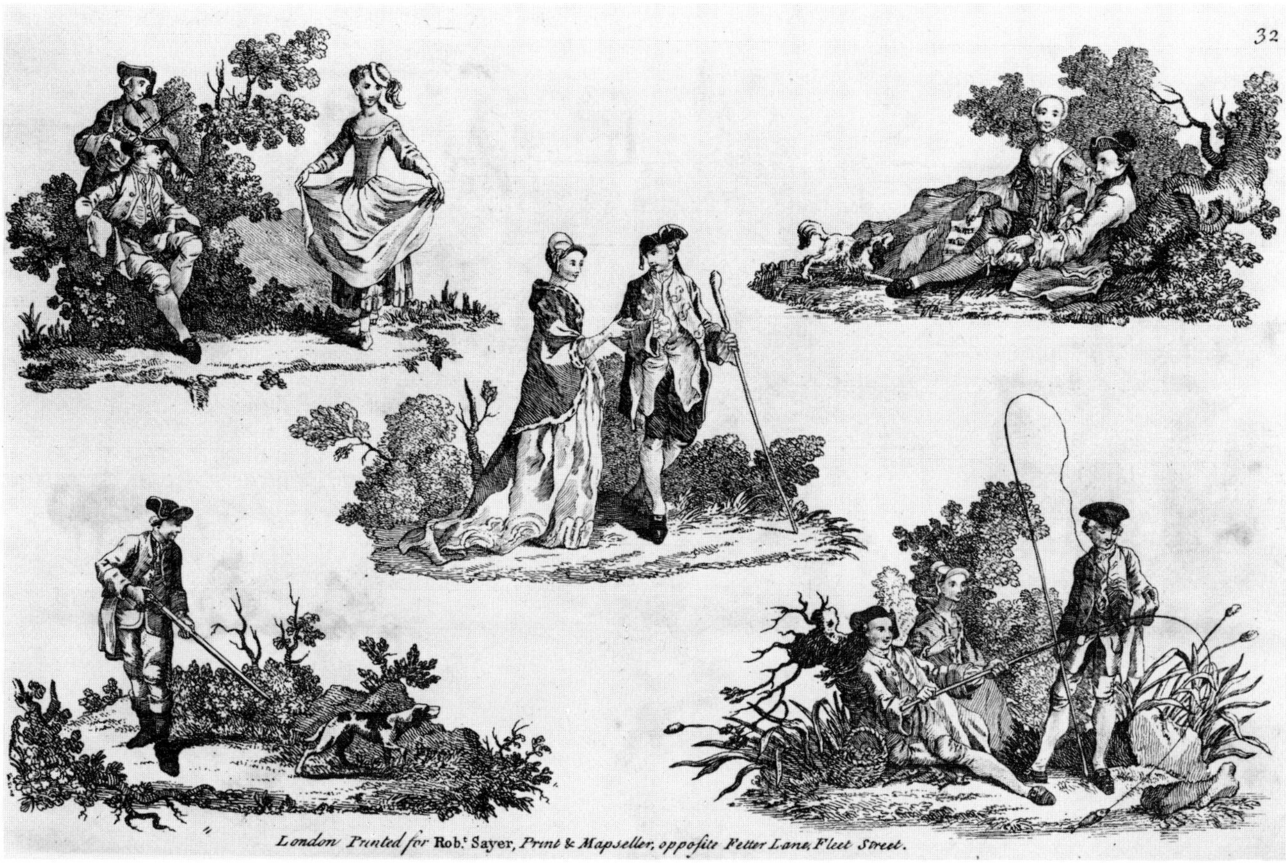

Engraving from The Ladies' Amusement, *c.1760* (*Victoria and Albert Museum*)

Opposite: details of the feet of the pair of candlesticks, 54b

55. Casket, containing three tea-caddies, painted enamel on copper mounted in gilt metal

English (South Staffordshire); about 1765

Length of casket, $8\frac{1}{4}$ in. (21 cm); width, $4\frac{3}{4}$ in. (11·5 cm); height, $5\frac{1}{2}$ in. (14 cm). Length of caddies, $3\frac{1}{2}$ in. (8·8 cm); width, $2\frac{1}{4}$ in. (5·7 cm); height of central caddy, $3\frac{3}{4}$ in. (9·5 cm); height of side caddies, $4\frac{1}{8}$ in. (10·5 cm)

The rectangular casket is formed of five convex panels held together at the edges by fillets of gilt metal decorated with chased and punched rococo scrollwork. The cover is secured by a lock and key, the escutcheon decorated in the same way as the fillets; on top is a swivel-handle formed of gilt metal similarly embellished. The case is lined with green velvet.

The cover-panel is painted with the figures of a standing gentleman conversing with a seated lady in a landscape with tall feathery trees: behind them is a man seated fishing in a lake, and in the background a pink building in front of which stand two shepherds with their sheep. The front of the casket is decorated with a coastal landscape dominated on the right by headlands surmounted by fortified buildings, while on the left a group of Italianate buildings is separated from the main scene by tall trees with feathery foliage. On the back is another coastal scene with two male figures conversing on a road in the foreground, while in the middle distance there are men fishing in meadows with cattle, a drover with his donkey, etc., these groups separated by trees with feathery foliage. At the sides are similar pastoral scenes. All these panels are framed with rococo scrollwork in raised gilding interspersed with diaper designs in raised white enamel.

The central caddy inside is hinged on the short axis and the whole top lifts as a cover. The side caddies have a central circular hole over which fits a cap with a milled gilt metal collar. All the available surfaces of these caddies are decorated with pastoral scenes painted in colours within irregular panels framed by rococo scrollwork in raised gilding, outside which are areas of diaper designs in raised white enamel.

Closely similar caddy-sets were in Queen Mary's Collection, and in the Hon. Mrs N. Ionides' Collection (G. B. Hughes, *English Painted Enamels*, p.94, Fig.53 (sold at Sotheby's 9 October 1967, Lot 209), and p.138, Fig.73, both with pink ground).

Baron Ferdinand de Rothschild — WI/7/5

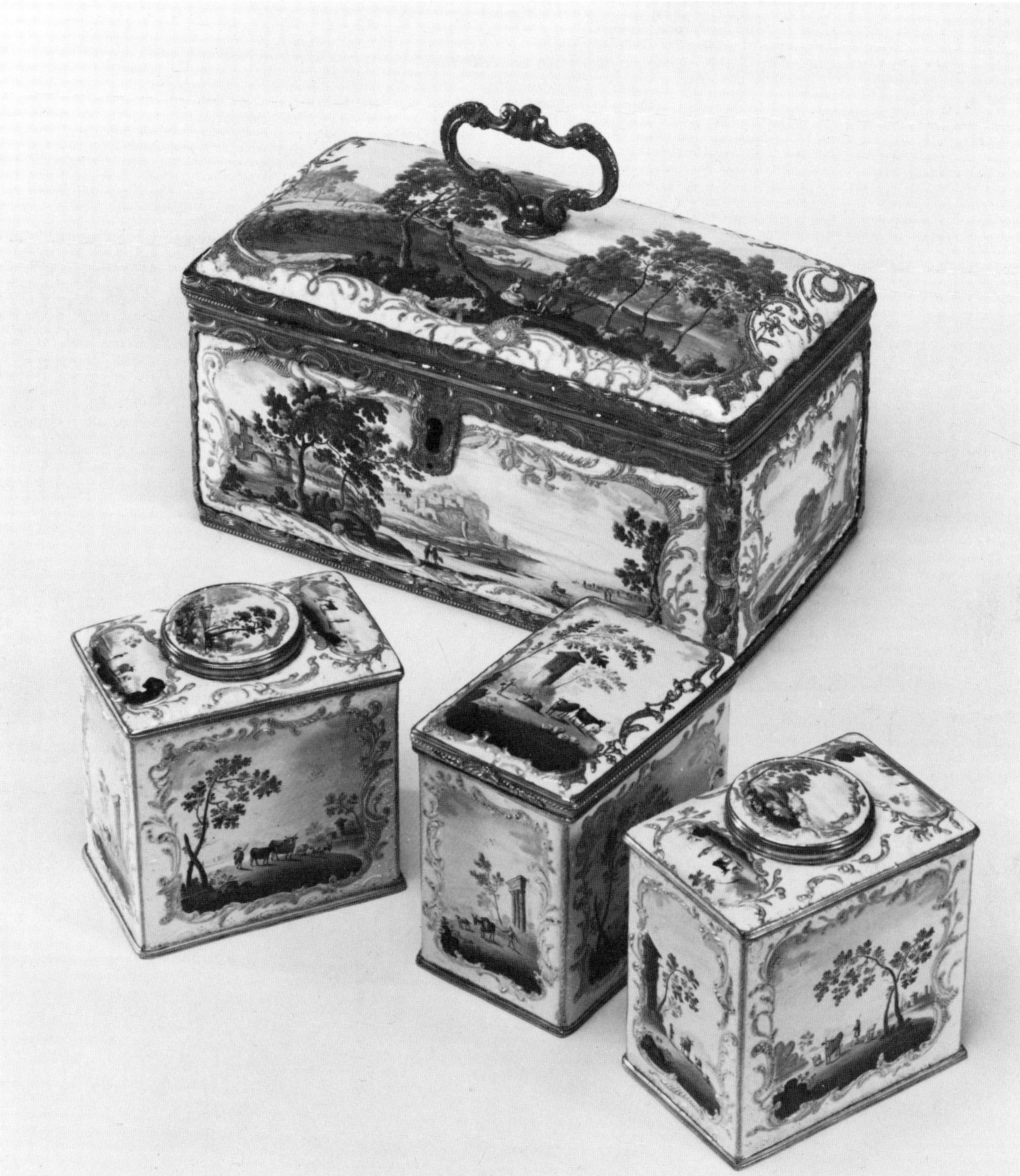

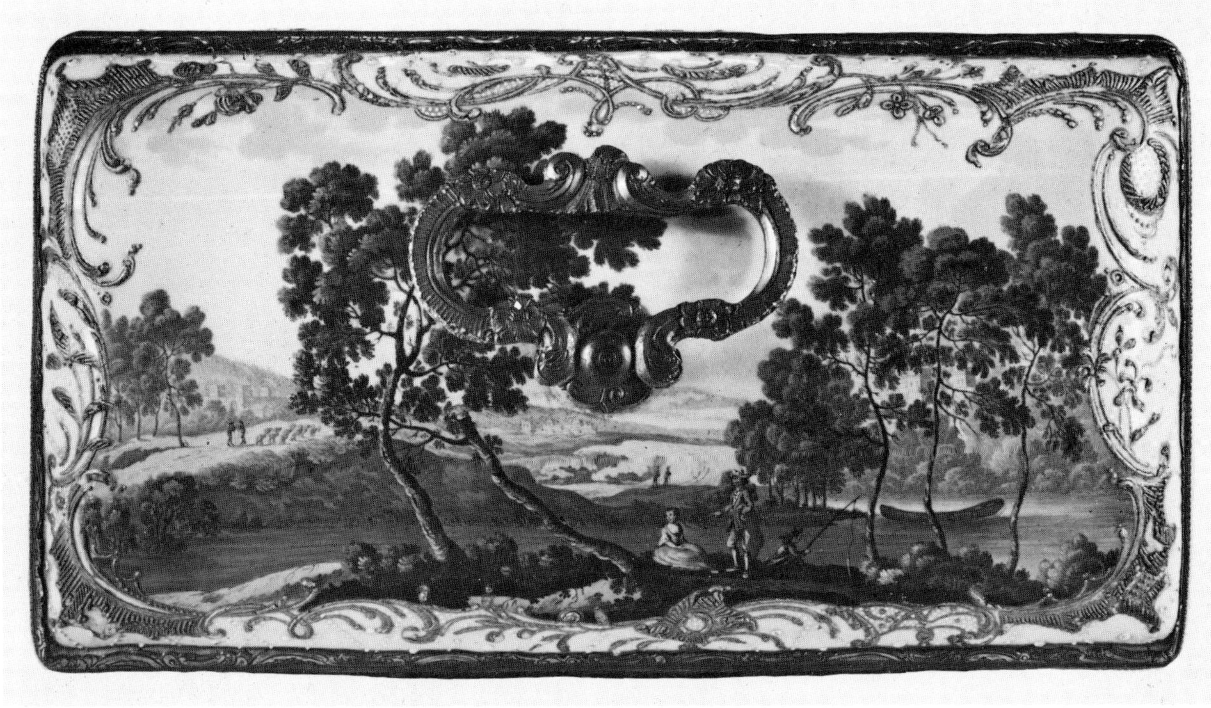

Above: lid of the casket. Below: back of the casket. 55

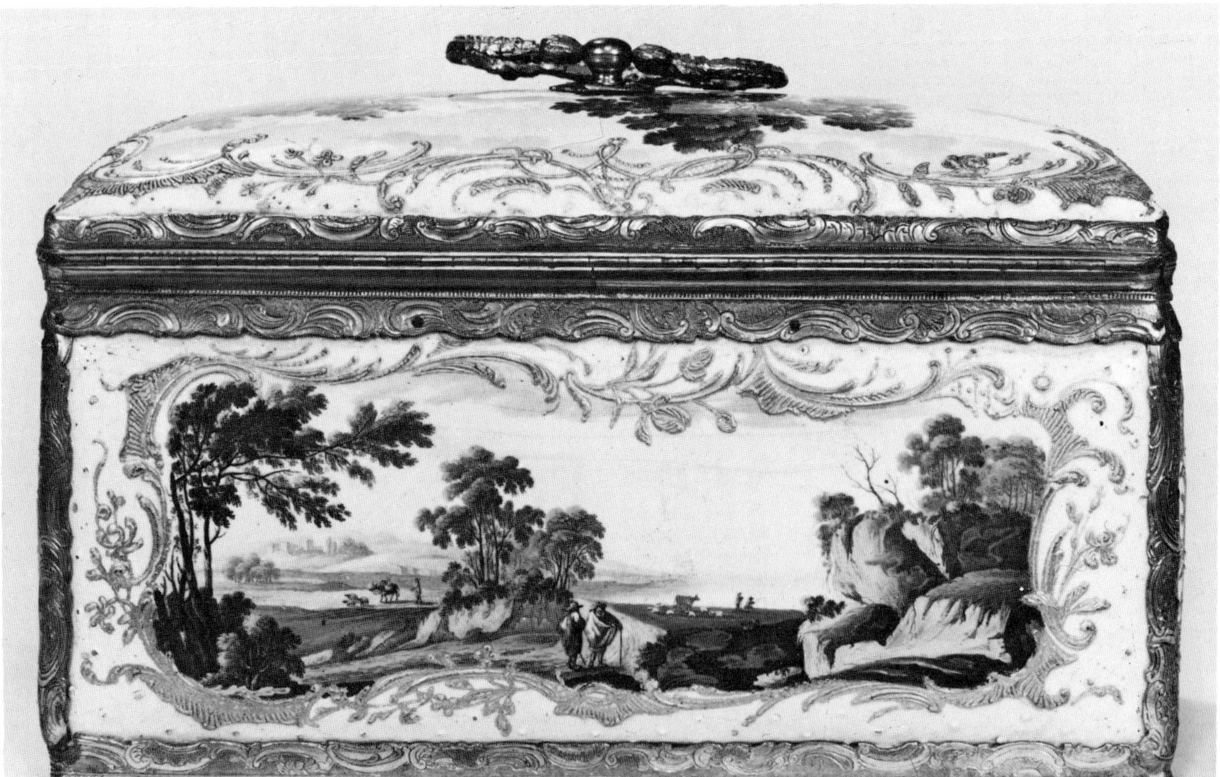

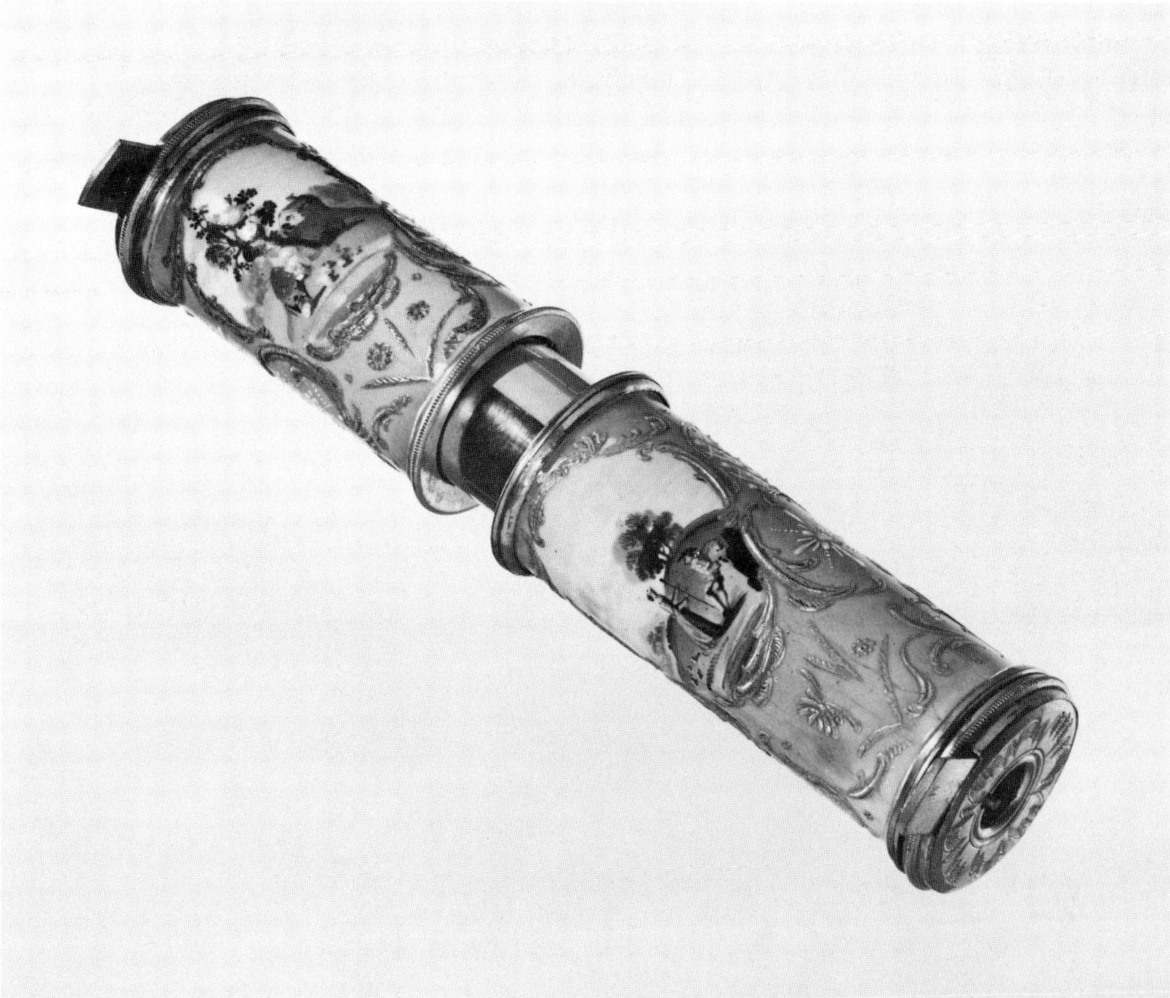

56. Spy-glass, painted enamel on copper, mounted in gilt metal

English (South Staffordshire); about 1765

Length when closed, $5\frac{1}{8}$ in. (13 cm)

The spy-glass is formed of a shorter and a longer cylinder painted in colours with pastoral scenes within irregular panels outlined by rococo scrollwork in raised gilding, and reserved on a turquoise-green ground enlivened by sun-burst motifs and diaper patterns in raised white enamel. These cylinders are joined by a gilt metal tube fixed to the longer section and on which the shorter section slides to give the right focal length. The cylinders are mounted at both ends with gilt metal fillets, and the lenses are protected by discs of gilt metal embellished with 'rose engine turning'. These have central apertures covered by slides which can be drawn to one side when it is desired to use the spy-glass.

Similar portable spy-glasses are also found in precious metals (e.g., the gold and enamel telescope sold at Christie's on 26 November 1968, Lot 94 (English), or that sold by Sotheby Parke Bernet, Monaco, on 29 November 1975, Lot 155 (Swiss). English examples in enamel on copper are sometimes found combined with *étuis* – e.g. a piece in the late Queen Mary's Collection, illustrated G.B. Hughes, *English Painted Enamels*, p.71, Fig.35c; also Rackham, *Schreiber Catalogue*, Nos.265 and 266: No.249 in the same collection has identical ornamentation, but on a pink ground.

An Italian enamel spy-glass working on a slightly different principle, facilitating a sideways look into a neighbouring box, is in the Victoria and Albert Museum (No.C.451–1927). It is inscribed: 'Il Poleidoscopio ossia il nobile esploratore'.

Baron Ferdinand de Rothschild — W3/7/3

57. Combined bodkin-case, scent-flask and thimble, painted enamel on copper mounted in gilt metal

English (South Staffordshire); about 1765

Length, 5⅛ in. (13 cm)

The bodkin-case is formed of a shorter and a longer cylinder of enamel on copper, painted in polychrome with pastoral scenes in irregular panels framed with rococo scrollwork in raised gilding, and reserved on a pink ground enlivened by a diaper design of groups of dots in raised white enamel. The two parts of the bodkin-case are edged with fillets of gilt metal and are held together by a gilt tube affixed to the longer section on which the shorter section slides. At the top of the shorter section is contrived a cavity for perfume, stopped by a stopper of gilt metal with a scroll finial: over this screws a cap in the form of an enamel thimble decorated *en suite* with the bodkin-case and furnished with a gilt metal tip.

A closely similar bodkin-case, but with the thimble missing, is in the Schreiber Collection at the Victoria and Albert Museum (Rackham, *Schreiber Catalogue*, No.251, cf. ibid., Nos.252–3, 291, 361–3).

Baron Ferdinand de Rothschild — W3/7/10

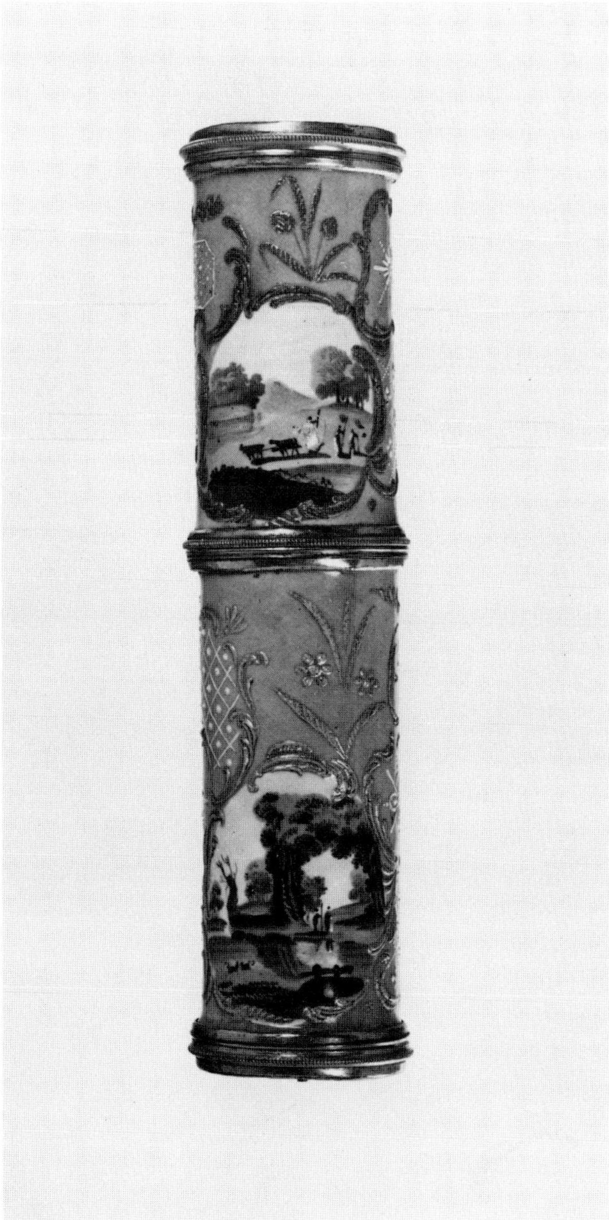

Spy-glass, 56

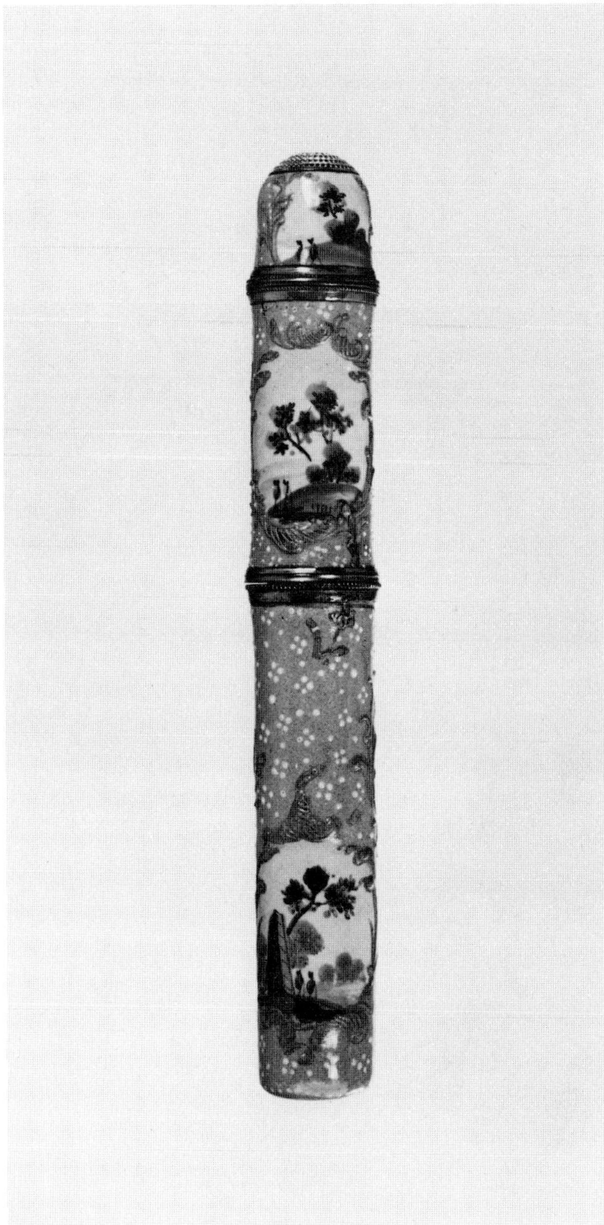

Bodkin-case, scent-flask and thimble, 57

58. Set of four coat-buttons, printed enamel on copper

English (South Staffordshire, perhaps Birmingham); about 1765–75

Diameter, $1\frac{5}{8}$ in. (3·3 cm)

Circular, slightly convex, buttons consisting of an enamelled disc held in a copper frame, the shanks formed from two crossed copper wires which pass through holes in the mount and are secured behind it.

The buttons are decorated with a series of equestrian prints in sepia on the white ground, as follows: (1) a lady riding side-saddle to the left; (2) a stallion in his stable, turning his head back to bite his offside back leg; (3) a mounted huntsman with horn, whipping off his pack; (4) a man wearing a tricorn hat and riding-boots and holding a whip in his right hand, standing to the left of an unsaddled horse.

A number of circular medallions of similar diameter and with printed decoration exist, some of which might originally have served for buttons (e.g., Rackham, *Schreiber Catalogue*, Nos.380–83, 386, 395). A trade-catalogue of about 1780 in the Victoria and Albert Museum shows similar plaques being used indifferently for 'cloak-pins' (coat-pegs), 'chimney-piece knobs' and 'Line Knobs' (presumably to secure blindcords). Several of them have hunting and equestrian subjects analogous to those on the Waddesdon buttons, and others show ladies' heads, classical urns, etc., such as appear on the plaques in the Schreiber Collection (G. B. Hughes, *English Painted Enamels*, Fig.83). A pair of buttons (of $1\frac{1}{4}$ in. diameter) in the Schreiber Collection are decorated with red prints washed over in colours and showing horses being tended by one and two grooms respectively (Rackham, *Schreiber Catalogue*, no.364).

Buttons of all sorts were a speciality of the Midlands and especially of Birmingham, where the great manufacturer John Taylor (1711–75) alone was said to have made £800-worth of buttons per week. It was said of Birmingham (1758): 'that it seems there as if God had created man only for making buttons' (see Watney and Charleston, 'Petitions . . .', p.64). A number of firms in Birmingham are recorded specifically as makers of enamelled buttons – notably Edward Carr of Carr's Lane (1767 at latest – 1770 or later); Richard Jefcoat of Great Charles Street (1777 at latest – 1800 or later); Basil Palmer of Weaman Street (1767 at latest – date unknown); Samuel Porter of Mount Pleasant and later of 38, Temple Street (1767 at latest – 1770 or later). The firm of Gibbons and Gibbs, of 56–57, Snow Hill were 'Enamel Manufacturers and Metal Button Makers' from at latest 1767 until probably the 1780's, and it is difficult to imagine that the two branches did not overlap; the same is probably true of Thomas Humstone, 'enameller and Button-maker' at Lower Priory from at latest 1767 until 1770, or later (see Watney and Charleston, 'Petitions . . .' pp.81, 117, 119–123).

The nineteenth century cardboard box which contained these buttons has a label with the Nazi inventory number R4232/2 (see p.7).

Baroness Edmond de Rothschild. No.59 of the button collection

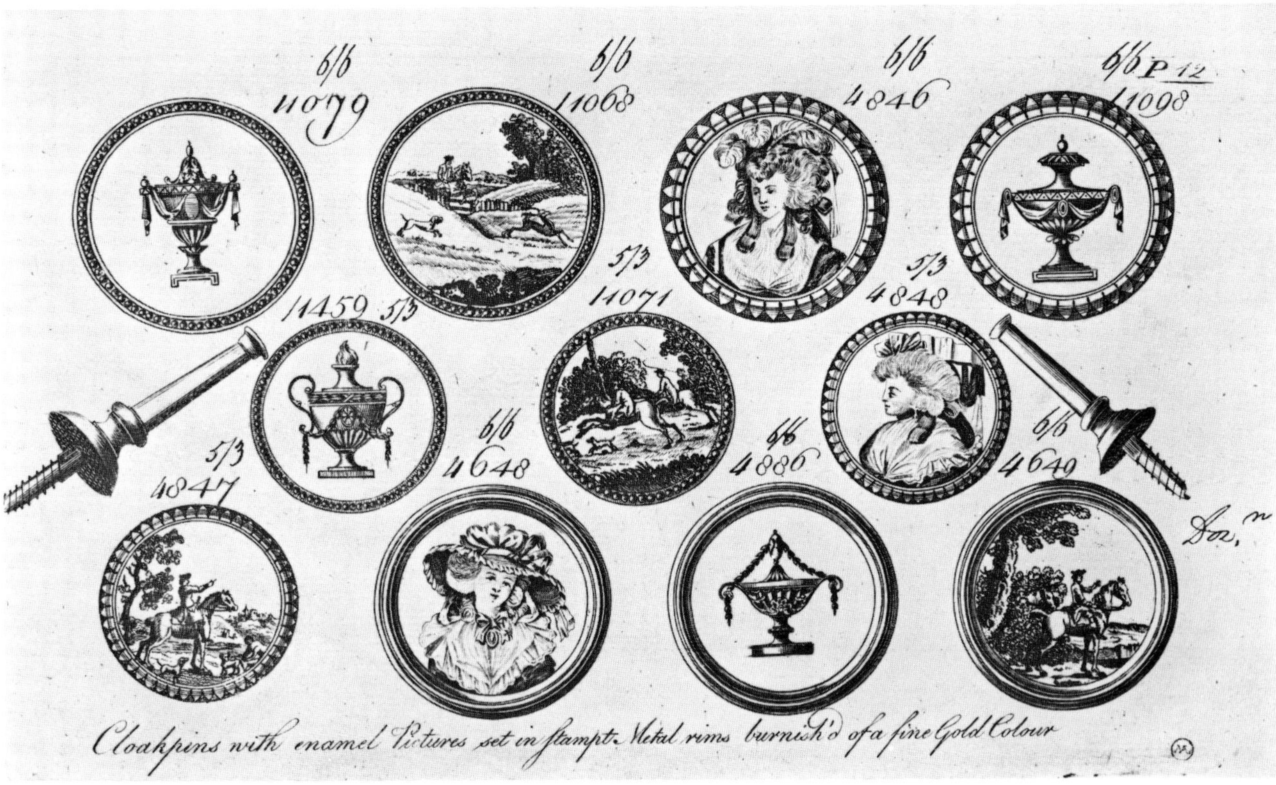

Above: page from a metalworker's pattern book, c.1780
(Victoria and Albert Museum)

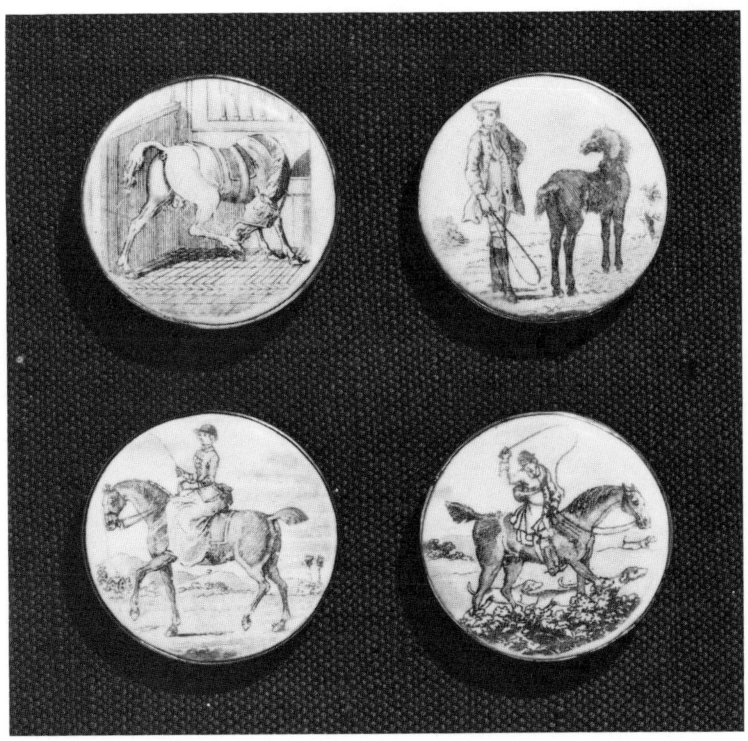

59. Portrait of Admiral Lord Nelson, painted enamel on copper, in a carved gilt wooden frame

Probably French (Samson of Paris); second half of the 19th century

Enamel, height, $6\frac{7}{8}$ in. (17·5 cm); width, $5\frac{3}{4}$ in. (14·5 cm)
Frame, height, $11\frac{1}{4}$ in. (28·5 cm); width, $8\frac{5}{8}$ in. (22 cm)

Portrait of Lord Nelson facing three-quarters right, wearing the uniform of an Admiral of the Fleet, various Orders on his breast, the whole on a brown ground.

The gilt wooden frame is carved with anchors at the corners, between which run borders of egg-and-dart and acanthus ornament, the whole surmounted by an elaborate bow with ribbons.

The *contremail* is white with a few greenish patches.

A closely comparable portrait of the American naval commander, Paul Jones, in the Captain Woollett Collection, and previously in the J. H. Barnes Collection, bore on the reverse the interlaced double-S mark used on enamels by the firm of Samson of Paris.

Baron Ferdinand de Rothschild kept this portrait on his yacht, the *Rona*. After his death it was taken to Eythrope.

Baron Ferdinand de Rothschild — E1/22/6

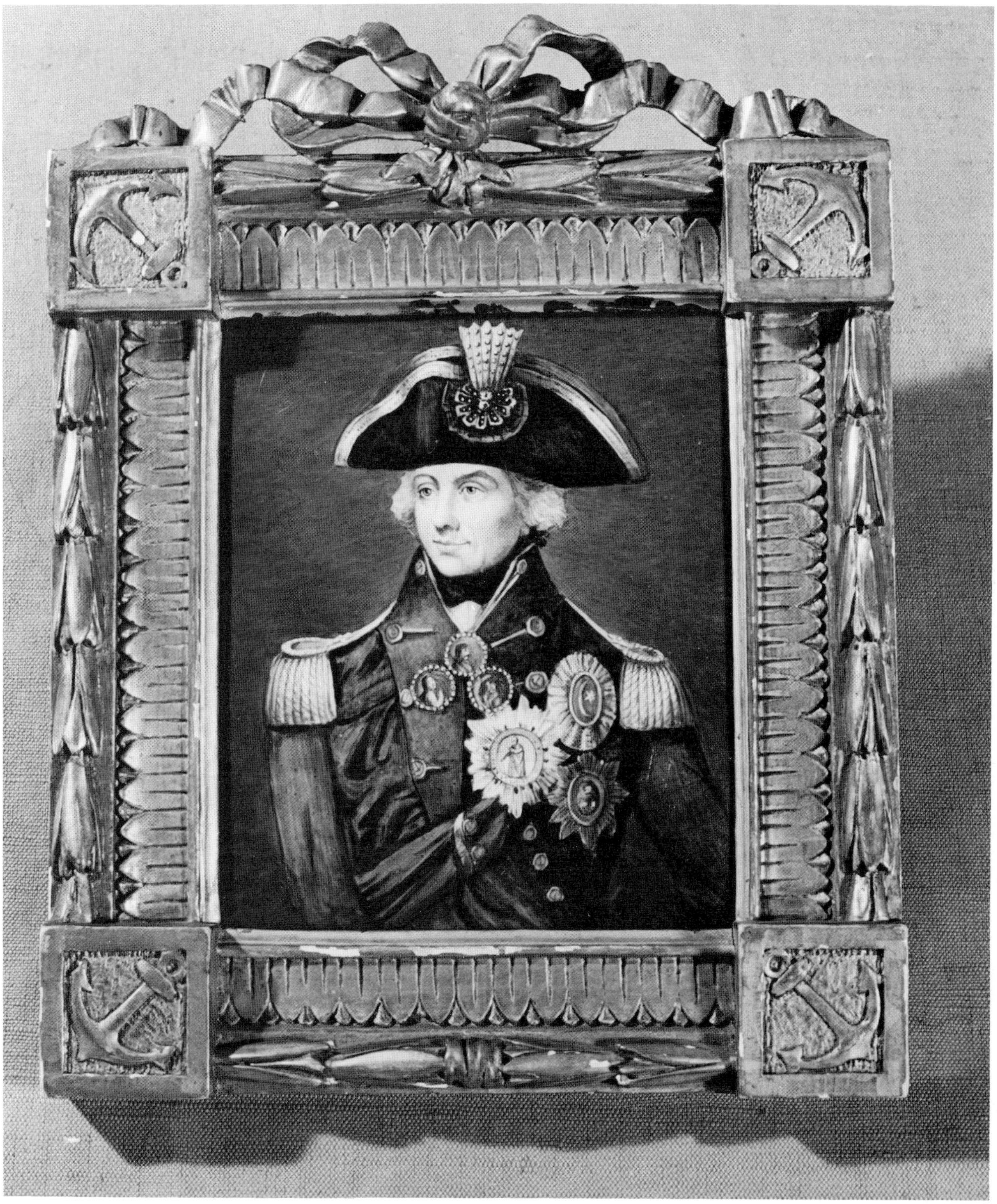

Index

References in the Index are to pages, not entries